BRADY

ifsta ®
INTERNATIONAL FIRE SERVICE
TRAINING ASSOCIATION

W9-ABH-571

FIRE SERVICE
FIRST RESPONDER

Daniel Limmer

Michael Grill

Validated by: The International Fire Service Training Association

IFSTA Senior Editor: Michael A. Wieder

Medical Editor
Edward T. Dickinson, MD, NREMT-P, FACEP

Brady/Prentice Hall Health
Upper Saddle River, NJ 07458

Library of Congress Cataloging-in-Publication Data

Limmer, Daniel.
    Fire service first responder / Daniel Limmer, Michael Grill.
        p. cm.
    Includes bibliographical references and index.
    ISBN 0-8359-5314-9 (pbk.)
    1. Medical emergencies.   2. First aid in illness and injury.
3. Fire fighters.   I. Grill, Michael.   II. Title.
    RC86.7.L563   1999
    616.02′5—dc21   99-35632
                                                                          CIP

PUBLISHER: Julie Alexander
ACQUISITIONS EDITOR: Laura Edwards
MANAGING DEVELOPMENT EDITOR: Lois Berlowitz
DEVELOPMENT EDITOR: Josephine Cepeda
DIRECTOR OF PRODUCTION AND MANUFACTURING: Bruce Johnson
MANAGING PRODUCTION EDITOR: Patrick Walsh
SENIOR PRODUCTION MANAGER: Ilene Sanford
PRODUCTION EDITOR: Julie Boddorf
CREATIVE DIRECTOR: Marianne Frasco
INTERIOR DESIGNER: Sue Walrath
COVER DESIGNER: Bill Smith Studio
COVER PHOTOGRAPHER: Craig Jackson
MANAGING PHOTOGRAPHY EDITOR: Michal Heron
ASSISTANT PHOTOGRAPHY EDITOR: Mary Jo Robertiello
PHOTOGRAPHERS: Michael Gallitelli, Michal Heron, Richard
    Logan
MARKETING MANAGER: Tiffany Price
MARKETING COORDINATOR: Cindy Frederick
COMPOSITION: The Clarinda Company
PRINTING AND BINDING: The Banta Company

ISBN 0-8359-5314-9

PRENTICE-HALL INTERNATIONAL (UK) LIMITED, *London*
PRENTICE-HALL OF AUSTRALIA PTY. LIMITED, *Sydney*
PRENTICE-HALL CANADA, INC., *Toronto*
PRENTICE-HALL HISPANOAMERICANA, S.A., *Mexico*
PRENTICE-HALL OF INDIA PRIVATE LIMITED, *New Delhi*
PRENTICE-HALL OF JAPAN, INC., *Tokyo*
PRENTICE-HALL ASIA PTE. LTD., *Singapore*
EDITORA PRENTICE-HALL DO BRASIL, LTDA., *Rio de Janeiro*

Chapter opening photos were supplied as follows: Chapter 4—Sarasota Fire Department/Don Hall Productions; Chapters 8 and 26—Craig Jackson/In the Dark Photography; Chapter 14—Mark C. Ide; Chapter 15—Shout Picture Library; Chapter 22—Harvey Eisner; Chapter 25—Howard M. Paul/Emergency! Stock.

NOTICE ON CARE PROCEDURES

It is the intent of the author and publisher that this textbook be used as part of a formal First Responder education program taught by qualified instructors and supervised by a licensed physician. The procedures described in this textbook are based upon consultation with EMT and medical authorities. The author and publisher have taken care to make certain that these procedures reflect currently accepted clinical practice; however, they cannot be considered absolute recommendations.

The material in this textbook contains the most current information available at the time of publication. However, federal, state, and local guidelines concerning clinical practices, including without limitation, those governing infection control and universal precautions, change rapidly. The reader should note, therefore, that the new regulations may require changes in some procedures.

It is the responsibility of the reader to familiarize himself or herself with the policies and procedures set by federal, state, and local agencies as well as the institution or agency where the reader is employed. The authors and the publisher of this textbook and the supplements written to accompany it disclaim any liability, loss, or risk resulting directly or indirectly from the suggested procedures and theory, from any undetected errors, or from the reader's misunderstanding of the text. It is the reader's responsibility to stay informed of any new changes or recommendations made by any federal, state, and local agency as well as by his or her employing institution or agency.

NOTICE ON GENDER USAGE

The English language has historically given preference to the male gender. Among many words, the pronouns "he" and "his" are commonly used to describe both genders. Society evolves faster than language, and the male pronouns still predominate in our speech. Thus, in many instances, male pronouns may be used in this book to describe both males and females solely for the purpose of brevity. This is not intended to offend any readers of the female gender.

NOTICE RE "ON SCENE"

The names used and situations depicted in "On Scene" scenarios throughout the text are fictitious.

To Sarah Katherine, my Little Miss Magic—DL

Dennis Eaton, thanks for being a mentor, guide, and confidant.

Jeff Dyar, thanks for believing in me and giving me a chance years ago. My wife Sam and kids Moe and Lacey Ella, thanks for being patient and loving me. Behind every successful man is a supportive wife and a very surprised mother-in-law!—MG

Brief Contents

Preface

This textbook is unique. *Fire Service First Responder* is the product of a joint venture by Brady Publishing and the International Fire Service Training Association (IFSTA), two leading producers of fire service training manuals. It is the first textbook created specifically for you—a member of the fire service who is studying to be a First Responder. Your role will be a special one. While the emergency medical services (EMS) system is comprised of a talented team of many individuals at different certification levels, as a First Responder, you are the one who will arrive on the scene of an emergency first. This is a special responsibility for which you are to receive special training in emergency medical care, as well as personal safety.

FEATURES OF THE TEXTBOOK

To help you in your role as a First Responder in the fire service, this textbook includes features intended to make teaching and studying easier:

◆ *Objectives.* Objectives are listed at the beginning of each chapter. They cover—verbatim—all of the objectives in the DOT's "First Responder: National Standard Curriculum," plus objectives for enrichment materials. You should be able to master the cognitive, affective, and enrichment objectives by reading the text; psychomotor objectives also require hands-on practice with an instructor. The page references correlate the objectives with text.

◆ *On Scene.* Each chapter opens with a scenario in which fire service First Responders provide emergency care. These case studies describe scene safety, patient assessment, and patient care in a way that helps tie the textbook discussion of a topic to its application in the real world.

◆ *Fire Drill.* This feature appears whenever a firefighter's personal knowledge and experience can be drawn upon to help make an EMS concept clear.

◆ *Company Officer's Note.* This feature appears from time to time throughout the text to highlight safety and patient-care considerations for personnel in charge at emergency scenes.

◆ *Chapter Review.* Each chapter ends with a selection of review materials:

- *Firefighter Focus* boils the chapter down to the ideas and insights that will be most important to you in the field.
- *Fire Company Review* offers review questions for each section of a chapter and includes references to textbook pages on which you can find answers.
- *Resources to Learn More* offers titles of books you can find to help you dig deeper into subjects in which you may be interested.
- *At a Glance* offers you flow charts and other visual summaries of key information from the chapter.

NATIONAL STANDARD CURRICULUM AND ENRICHMENT MATERIALS

Fire Service First Responder meets the guidelines established for training of First Responders set by the U.S. Department of Transportation (DOT) in its most recent "First Responder: National Standard Curriculum." Some of the information required by the DOT has been slightly rearranged and some has been elaborated on to take into account the distinct needs and requirements of members of the fire service. In addition, the DOT acknowledges that many instructors will add enrichment materials to their lessons. That is, materials not covered in the National Standard Curriculum may be covered in your class. This is due to state or regional mandates or to special situations in your area. So, there are many places where we have chosen to add extra, or "enrichment," materials to this textbook. They include:

◆ *A full chapter on Fireground Rehabilitation.* Because firefighters are often exposed to the possibility of heat- and stress-related emergencies, this chapter covers the basic set up and procedures of a rehabilitation section/group area.

◆ *A full chapter on automated external defibrillation.* This life-saving skill is being taught to many First Responders. The ability of First

Responders to deliver an electrical current to the chest of a patient whose heart has stopped beating is the most significant advance in the emergency medical services in many years.

♦ *Expanded coverage of patient assessment.* If in your EMS system you are required to take a patient's vital signs, including blood pressure, you will find the topic covered in the "Patient Assessment" chapter.

♦ *Oxygen administration.* From the basics of handling an oxygen cylinder to the basics of applying oxygen to a patient—all are included in the "Airway" chapter.

♦ *Expanded coverage of common medical emergencies.* Cardiac and respiratory complaints are two of the most frequent reasons you will be called as a First Responder. Though they are not covered in the National Standard Curriculum, you will find them in this textbook, as well as coverage of diabetic emergencies, alcohol and drug emergencies, stroke, poisoning, seizures, and abdominal pain.

♦ *Splinting and spinal immobilization.* Find basic how-to information for these topics in the chapters covering musculoskeletal injuries.

We believe that you will be pleased with *Fire Service First Responder*. We always welcome your comments, which may be mailed to:

Brady Marketing Department
c/o Tiffany Price
Brady/Prentice Hall Health
One Lake Street
Upper Saddle River, NJ 07458

Acknowledgments

PROGRAM DEVELOPMENT

We would like to thank IFSTA, our Medical Editor, and the many members of the Brady team for the skill and dedication they have brought to developing *Fire Service First Responder*. The Brady team works together with a strong sense of professionalism, one that we greatly admire.

IFSTA Validating Committee The process of creating this book involved an extensive review and validation process. The IFSTA validation team met three times. (Please see a list of their names on page x.) During our meetings, we shared ideas and laughs. It was a great process that involved some of the finest people the fire service has to offer. Thank you. Thanks also to Mike Wieder of IFSTA for his support and organization during the creation of this textbook.

Medical Editor Edward T. Dickinson, MD, NREMT-P FACEP, is Medical Editor for *Fire Service First Responder*. We are always grateful for the keen eye that Ed brings to his review of material. His advice and observations have helped us stay current in a time of rapidly changing medicine and new curricula. Ed's knowledge, street experience, and high standards (combined with the fact that he was an English major in college) have helped us tremendously. Ed's contributions to this text have been invaluable. Our special thanks to Ed for contributing the chapter on Fireground Rehabilitation, an increasingly relevant and essential topic in today's firefighter training programs.

Brady/Prentice Hall Health *Fire Service First Responder* is the result of many talented people working together. We gratefully acknowledge the tireless effort and attention to detail Jo Cepeda, our editor, adds to this and every project she works on. Editors make authors look good, and we are glad to have Jo working with us. At Brady, Lois Berlowitz is an advocate, a taskmaster, and a true professional in every sense of the word. A day cannot go by without appreciating Jo and Lois and what they have done for this project.

Photos are an important part of every textbook. Photos provide instruction, feeling, and examples. A book full of great words still is not complete without great photographs. Michal Heron, Managing Photography Editor, works tirelessly to ensure accurate, quality photographs and a clean, refreshing look to the text. Thank you. Anyone who has ever worked a photo shoot with Michal can attest to her quest for perfection. We must also thank all who worked as models and technical advisors. Working on the photos in this text involved long, hot days in the sun (and sometimes in fully encapsulated suits). As authors, we appreciate the efforts of everyone who made the book look great.

The production crew for this text, including Pat Walsh, Julie Boddorf, and Ilene Sanford are usually acknowledged, but perhaps not enough. Writing a book is only one part of the process that brings this book to you. Assuring the book is printed, that the photos are in the right places, and that it looks as good as we envision rests largely on these people. And they do so well.

During the production of this textbook there were changes in publishers. We worked with Susan Katz, our friend and publisher, for many years. It was Susan's vision that brought this book and others about. Susan's way of "doing business" brought a style to publishing that is unique and certainly appreciated over the many years we have had the pleasure of working with her. Susan made the business of publishing an art form. She will be truly missed. We wish Susan happiness and success in the future. Carol Sobel, Susan's assistant, was a constant voice behind the scenes. We talked to and relied on Carol for many things over the years. Her helpfulness and kindness will never be forgotten.

Julie Alexander has the responsibility of taking over a job held by Susan. These were "big shoes to fill," and Julie has done well. We have enjoyed working with Julie as she begins what we are sure will be a long and productive stay at Brady. Julie's energy and ideas are strong, even contagious. We look forward to a long and positive relationship with our new publisher.

Our marketing and sales force must also be commended. Tiffany Price and Cindy Frederick are driven, creative, and appreciated. Judy Streger has worked both marketing and editorial functions during the creation of this book. Judy's input and friendship is always appreciated. Judy Stamm and the sales reps work to get our book out there, and we thank all of you.

Last, but certainly not least, we would like to acknowledge our friends and colleagues Brent Hafen and Keith Karren. Brent and Keith began in the book business long ago. Their efforts and vision in bringing First Responder books from their infancy to where they are today must be noted. We gratefully acknowledge the contributions made to this text and to the field of EMS.

IFSTA VALIDATING COMMITTEE

Acknowledgment and special thanks are extended to the members of the IFSTA validating committee who contributed their time, wisdom, and knowledge to a thorough review of the manuscript for this book.

IFSTA/Fire Protection Publications Staff Liaison
Michael A. Wieder
Fire Protection Publications
Oklahoma State University
Stillwater, OK

Mike Buscher
Omaha Fire Department
Omaha, NE

Stephen W. Carrier, Sr.
Riverbend Career & Technical Center
Bradford, VT

Gary Davis
Oklahoma City Fire Department
Oklahoma City, OK

Beverly Deister
Reno County EMS
Hutchinson, KS

Jerome Harvey
City of Lead Fire Department
Lead, SD

William J. Mackreth
Wasilla, AK

Mark Monroe
Reichhold Chemicals, Inc.
Valley Park, MO

Bill Roth
Hemet Fire Department
Hemet, CA

REVIEWERS

Our thanks to the many people involved in manuscript review for their feedback and suggestions.

Gary Ferrucci, EMT-CC PC
Nassau County Police Department
Mineola, NY

Capt. Krista Wyatt
Lebanon Fire Division
Lebanon, OH

Capt. William Seward, III
Director of Training
Department of Fire Service
City of New Haven, CT

Lonnie D. Inzer
Lieutenant—Colorado Springs Fire Department
Fire Science Coordinator—Pikes Peak
    Community College, CO

Michael Zanotti, CEM, NREMT-P
Baker Heights Fire Department, WV

Stanley C. Vinson
Fire Service Captain
Mobile Fire-Rescue Department, AL

Al Lewin, EMT-P/Instructor
Auburn Emergency Squad, IL

Deputy Chief Bill Madison
Lincoln Fire Department
Lincoln, NE

Bradley Golden, RN, EMT-P, AAEMT
Jackson Fire Rescue
Jackson, MO

Sean Wilson AEMT-I-CIC
EMS Program
Fulton Montgomery Community College
New York State Academy of Fire Science
    Adjunct-CIC, NY

David W. Akers, EMT-B
Coordinator/Instructor
Sierra Fire Academy
Tollhouse, CA

Stephen Bardwell
Mississippi State Fire Academy
Jackson, MS

Lynn Lybrook, EMT-P, EMS Field Training
    Coordinator
Alabama Fire College
Tuscaloosa, AL

Dennis Matty, Lieutenant
Miramar Fire Rescue, FL

Robert C. Hecker, Fire Captain, Paramedic
    Instructor
St. Tammany Parish Fire District #4, LA

Capt. Tony C. Watson, EMT-P
Instructor/Coordinator
Pigeon Forge Fire Department, TN

James B. Miller, EMT-P
EMS Coordinator
Fire and Emergency Services
Fort Sam Houston, TX

Capt. Steve Moffitt
EMS Division
Alabama Fire College
Tuscaloosa, AL

Mark Stewart
Louisiana State University Fire and Emergency
    Training Institute, LA

Chip Boehm, RN, EMT-P/FF
EMS Education/QI Officer
Portland Fire Department
Portland, ME

PHOTO ACKNOWLEDGMENTS

All photographs not credited adjacent to the photograph were photographed on assignment for Brady/Prentice Hall Health.

Organizations: We wish to thank the following organizations for their valuable assistance in creating the photo program for this edition:

American Medical Response, Hemet Valley
    Ambulance Service: Laurie Hunter, Director
    of Government Affairs; Jack Hansen,
    Operations Manager; Art Durbin, EMT-P,
    RN, BS, Clinical Manager

Fire Protection Publications, IFSTA: Michael A.
    Wieder

City of Hemet Fire Department, Hemet, CA:
    David A. Van Verst, Battalion Chief

Plano Fire and Rescue, Plano, Texas: Chief Bill
    Peterson; EMS Coordinator Ken Klein;
    Monique Cardwell, Public Education

Reichhold Chemicals, Inc., Newark, NJ:
    Ron Kurtz, Jack Connolly, EMS Technician
    TACTRON Incident Control Products,
    Sherwood, Oregon

Technical Advisors: Our thanks to the following people for providing extraordinary assistance and valuable technical support during the photo schoots:

Art Durbin, EMT-P, RN, BS, Clinical Manager,
    Hemet Valley Ambulance Service,
    American Medical Response, Hemet, CA

Jack Connolly, EMS Technician,
    Reichhold Chemicals, Inc., Newark, NJ

Ken Klein, RN, EMT-P Coordinator, Kenneth
    C. Larsen EMTP-TACT, Plano Fire and
    Rescue, Plano, TX

Daniel Pohan, Savox Lifeline Communications,
    Ridgefield, NJ

Brian Rathbone, Hazmat/Rescue Training
    Consultant, The Mechanical Advantage,
    Hackettstown, NJ

Captain William C. Roth, and Jim Snodgrass,
    Battalion Chief, City of Hemet Fire
    Department, Hemet, CA

Happy Snodgrass, Officer, City of Hemet
    Police Department, Hemet, CA

CORRELATION

U.S. DOT "First Responder: National Standard Curriculum" with
FIRE SERVICE FIRST RESPONDER

DOT National Standard Curriculum	FIRE SERVICE FIRST RESPONDER
1-1 Introduction to EMS Systems 1-2 The Well-Being of the First Responder 1-3 Legal and Ethical Issues 1-4 The Human Body 1-5 Lifting and Moving Patients 1-6 Evaluation: Preparatory	1 Introduction to the EMS System 2 Scene Safety and the Well-Being of the First Responder 3 Legal and Ethical Issues 4 The Human Body 21 Lifting and Moving Patients
2-1 Airway 2-2 Practical Lab: Airway 2-3 Evaluation: Airway	5 Airway
3-1 Patient Assessment 3-2 Practical Lab: Patient Assessment 3-3 Evaluation: Patient Assessment	8 Scene Size-up 9 Patient Assessment
4-1 Circulation 4-2 Practical Lab: Circulation 4-3 Evaluation: Circulation	6 Circulation 7 Automated External Defibrillation
5-1 Medical Emergencies	10 Cardiac and Respiratory Emergencies 11 Other Common Medical Complaints 12 Environmental Emergencies 13 Psychological Emergencies and Crisis Intervention
5-2 Bleeding and Soft Tissue Injuries	14 Bleeding and Shock 15 Traumatic Injuries 16 Burn Emergencies
5-3 Injuries to Muscles and Bones 5-4 Practical Lab: Illness and Injury 5-5 Evaluation: Illness and Injury	17 Musculoskeletal Emergencies 18 Injuries to the Head, Neck, and Spine
6-1 Childbirth 6-2 Infants and Children 6-3 Practical Lab: Childbirth and Children 6-4 Evaluation: Infants and Children	19 Childbirth 20 Infants and Children
7-1 EMS Operations 7-2 Evaluation: EMS Operations	22 Multiple-Casualty Incidents and Incident Management 23 EMS Operations 24 Hazardous Materials 25 Fireground Rehabilitation 26 EMS Rescue Operations

Introduction to the EMS System

*I*NTRODUCTION   *The North American fire service is the single largest provider of prehospital care. Emergency medical services have been provided by fire departments since the 1930s, when municipalities such as New York City began providing first aid. As new life-saving techniques were developed, fire departments in Baltimore, Seattle, Los Angeles, and Milwaukee—to name just a few— were among the first to implement them in the field. In fact, the Miami Fire Department was the first EMS provider trained in the use of defibrillation outside a hospital setting. According to a recent*

survey (Phoenix Fire Department, 1993), more than 80% of all fire departments in the United States now perform some level of emergency medical service (EMS).

This textbook will help fire department personnel acquire the knowledge, skills, and attitudes needed to provide quality patient care. To begin, your instructor will describe what you can expect in the course. He or she will inform you of required immunizations and physical exams, and outline your state and local certification requirements. Your instructor also may explain local policies concerning the Americans with Disabilities Act (ADA) and the implications of harassment in the classroom environment.

OBJECTIVES

Cognitive, affective, and psychomotor objectives are from the U.S. DOT's "First Responder: National Standard Curriculum." Enrichment objectives, if any, identify supplemental material.

Cognitive

1-1.1 Define the components of Emergency Medical Services (EMS) systems. (p. 5)

1-1.2 Differentiate the roles and responsibilities of the First Responder from other out-of-hospital care providers. (pp. 6, 8–10)

1-1.3 Define medical oversight and discuss the First Responder's role in the process. (p. 10)

1-1.4 Discuss the types of medical oversight that may affect the medical care of a First Responder. (p. 10)

1-1.5 State the specific statutes and regulations in your state regarding the EMS system.

Affective

1-1.6 Accept and uphold the responsibilities of a First Responder in accordance with the standards of an EMS professional. (pp. 9–10)

1-1.7 Explain the rationale for maintaining a professional appearance when on duty or when responding to calls. (p. 9)

1-1.8 Describe why it is inappropriate to judge a patient based on a cultural, gender, age, or socioeconomic model, and to vary the standard of care rendered as a result of that judgment. (p. 10)

Psychomotor

No objectives are identified.

Enrichment

* Describe the various methods by which the public can access EMS. (p. 5)

ON SCENE

DISPATCH

Our engine company was dispatched to an injured person at 59 James Place. I was very excited, as this was my first call since I had joined our community's volunteer fire department. I felt very nervous, especially with the lights and siren on. Looking over at the other three members of our engine company, however, I felt comfortable knowing I was with "veterans."

SCENE SIZE-UP

When we approached the scene, our driver-engineer turned off the lights and siren. We grabbed protective gloves and eye protection and left our engine. We noticed that the police were already on scene. Even so, we kept alert for signs of danger as we got closer to the house. When I was about to enter, I noticed bright red spots of blood on the front porch. Our captain commented, "Well, that's not a good sign."

INITIAL ASSESSMENT

Inside the house, we saw our patient sitting on a chair. Our general impression was that he was conscious and speaking with the police officers. I noticed he held a wash cloth to his forehead. Almost immediately we manually stabilized his head and neck and began to assess his ABCs—airway, breathing, and circulation.

Our captain asked what happened. The patient answered that he had been hit in the head with a baseball bat. Another member of our engine company asked if he could see the wound. When the patient removed the wash cloth from his forehead, I saw an open wound that was oozing blood. It looked as if that part of his forehead was pushed in. I also noted his right eye was swollen shut. Our captain gave me a look, which let me know that he believed the patient could be in serious condition.

I radioed the dispatcher to update the incoming unit. Another member of our engine company administered oxygen to the patient while keeping him calm and explaining exactly what it was he was doing and why.

> *You will encounter a wide range of calls as a First Responder. Some may be trauma (injury) calls such as this one. Others may involve medical conditions. You will learn how these fire service First Responders handled their patient's emergency at the end of this chapter.*

1 EMERGENCY MEDICAL SERVICES

An ill or injured patient may need immediate medical care to prevent permanent disability or death. Too often those who arrive first at the scene of the emergency are not trained to give proper care. As a result, a patient's condition can deteriorate and he or she can die.

The "Golden Hour" is a term trauma experts often use when they speak of severely injured patients who require surgery. It refers to the belief that such patients have the highest survival rates when they are on the operating table within 60 minutes of injury. It is within this "Golden Hour" that lives can be saved by proper emergency care.

For cardiac arrest, research has shown that for every minute the heart does not beat, the patient loses another 10% chance of surviving. As you can see, your role as the first medically trained rescuer on scene is very important.

In general, the **emergency medical services (EMS) system** is a network of resources that provides emergency medical care to victims of sudden illness or injury. This network may include any specific arrangement of emergency medical personnel, equipment, and supplies designed to function in a coordinated fashion.

For example, in an emergency, anyone can call for help by dialing 9-1-1 or another emergency number. The EMS dispatcher may then give that person some basic instructions for patient care. When fire service First Responders arrive on scene, they assess the situation, provide emergency care, and request additional resources if needed. Usually, fire service EMS personnel with higher levels of training arrive next. They continue care and transport the patient to the hospital. There the patient is transferred to emergency department personnel and, finally, to the in-hospital care system (Figure 1-1).

Figure 1-1a A member of the public calls 9-1-1 for help.

Figure 1-1b The EMS dispatcher alerts EMS and fire units and gives the caller basic instructions for patient care.

Figure 1-1c Fire service First Responders assess the situation, provide emergency care, and request additional resources if needed.

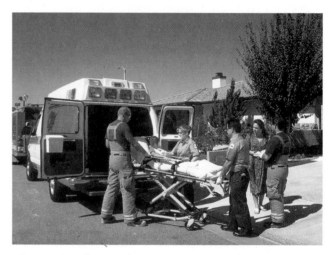

Figure 1-1d EMS rescuers at a higher level of training continue emergency care and transport the patient to the hospital.

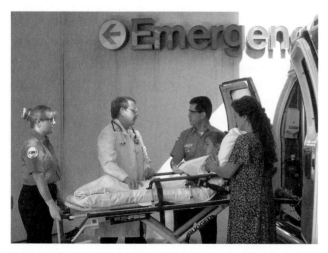

Figure 1-1e The patient is then transferred to the hospital emergency department staff.

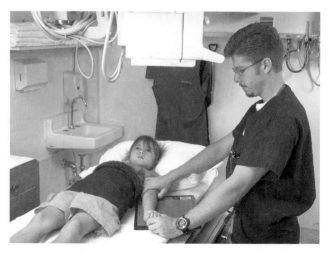

Figure 1-1f Finally, the patient is transferred to the in-hospital care system.

Classic Components of EMS

Each state in the U.S. has control of its own EMS systems. However, standards are set by the National Highway Traffic Safety Administration, U.S. Department of Transportation (DOT). Those standards include the following classic components of any EMS system:

◆ *Regulation and policy.* Each state must have laws, regulations, policies, and procedures that govern its EMS system. It also is required to provide leadership and guidance to local jurisdictions.

◆ *Resources management.* Each state must have central control of EMS resources so all patients have equal access to acceptable emergency care.

◆ *Human resources and training.* Ambulances must be staffed by personnel trained to at least the EMT-Basic (EMT-B) level.

◆ *Transportation.* Patients must be safely and reliably transported by ground or air ambulance.

◆ *Facilities.* Every seriously ill or injured patient must be delivered in a timely manner to an appropriate medical facility.

◆ *Communications.* A system for public access to the EMS system must be in place. Communication among dispatcher, fire service personnel, other EMS providers, and medical facilities must also be possible.

◆ *Public information and education.* EMS personnel should participate in programs designed to educate the public. The programs are to focus on the prevention of injuries and how to properly access the EMS system.

◆ *Medical oversight.* Each EMS system must have a physician as a medical director accountable for the activities of EMS personnel within that system.

◆ *Trauma systems.* Each state must develop a system of specialized care for trauma patients, including one or more trauma centers and rehabilitation programs. It must also develop systems for assigning and transporting patients to those facilities.

◆ *Evaluation.* Each state must have a quality improvement system in place for continuing evaluation and upgrading of its EMS system.

Access to EMS

There are two general types of telephone systems by which the public can access EMS: *9-1-1* and *non-9-1-1*. Currently, about 25% of the U.S. geography—and almost 80% of the U.S. population—is covered by a universal number: 9-1-1. It is used to access police, fire, rescue, and ambulance. Generally, calls are received at a **public safety answering point (PSAP).** There a dispatcher decides which agency is to be activated and alerts that agency (Figure 1-2).

There are two main benefits of a universal telephone number. First, the PSAP should be staffed by trained technicians. They may offer medical advice over the phone while the patient waits for rescuers to arrive. This is referred to as "emergency medical dispatching." The second benefit of a universal number is that it minimizes delay. Callers do not have to look up a number and it is easy enough to remember by even the youngest caller.

About 79% of the most populous U.S. cities utilize **enhanced 9-1-1.** With enhanced 9-1-1, or E-9-1-1, the EMS dispatcher is able to see the street address and phone number of the caller on a computer screen. This is valuable when a patient becomes unconscious before giving an address.

In areas not served by 9-1-1, callers either call a dispatch center or the specific agency they need (police, fire, etc.). Probably the most serious drawback of a non-9-1-1 system is the potential delay in reaching the appropriate services.

Figure 1-2 The EMS dispatcher receives calls at the public safety answering point (PSAP).

Levels of Training

Across North America, more than 40 different levels of emergency medical technician (EMT) certification exist at the state or provincial level. However, the "National Emergency Medical Services Education and Practice Blueprint" recognizes four levels of emergency medical services training: **First Responder, EMT-Basic, EMT-Intermediate,** and **EMT-Paramedic.** Note that responsibilities for each level may vary from state to state. However, minimum certification guidelines are published by the U.S. Department of Transportation:

◆ *First Responder.* The First Responder is the first person on the scene with emergency medical training. He or she may be a police officer or firefighter, a truck driver or school teacher, an industrial health officer, or a community volunteer (Figures 1-3 and 1-4). Training includes:

  ◆ Airway care and suctioning.

  ◆ Patient assessment.

  ◆ Cardiopulmonary resuscitation (CPR).

  ◆ Bleeding control.

  ◆ Stabilization of injuries to the spine and extremities.

  ◆ Care for medical and trauma emergencies.

  ◆ Use of a limited amount of equipment.

  ◆ Assisting other EMS providers.

  ◆ Other skills and procedures, such as use of an automated external defibrillator (AED) as permitted by local or state regulations.

Figure 1-3 First Responder patch from the National Registry of Emergency Medical Technicians. *(Source: National Registry of Emergency Medical Technicians)*

COMPANY OFFICER'S NOTE

The term "first responder" has an additional meaning. Besides denoting a level of certification, it may also refer to a fire department response when an engine company or rescue vehicle is the "first" unit on the scene of an emergency. In many cases, this "first responder" holds a level of certification at the EMT-Basic, Intermediate, or Paramedic level. In fact, the concept of paramedic-engine companies as the first rescuers on scene of emergencies has increased in popularity in the last 10 years.

◆ *EMT-Basic.* The EMT-B can do all that the First Responder does. In addition to having more extensive training in patient assessment, he or she also may restrain patients, staff and drive ambulances, use an automated external defibrillator (AED), and assist patients in taking certain medications (Figure 1-5).

◆ *EMT-Intermediate.* The EMT-I can do all that the two previous levels do. He or she also can perform a limited number of advanced techniques and administer a few medications. In some states, the EMT-I may also be trained as a cardiac technician (Figure 1-6).

◆ *EMT-Paramedic.* The paramedic has the most advanced EMS training. He or she can do all that the three previous levels do, plus administer more medications and perform more advanced techniques such as cardiac monitoring (Figure 1-7).

The National Registry of Emergency Medical Technicians was formed in 1970. It offers examinations for certification of First Responders and EMTs. If your state does not recognize or require national registration, it may still be helpful, especially when you move to another state. It also may be considered desirable by private employers. Ask your instructor about getting national certification or contact:

National Registry of Emergency Medical Technicians

6610 Busch Boulevard

Columbus, OH 43229

614-888-4484

Figure 1-4 First Responders may be firefighters, police officers, educators, truck drivers, industrial workers, or community volunteers.

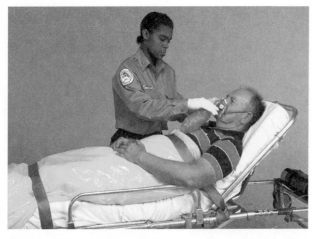

Figure 1-5 An EMT-Basic.

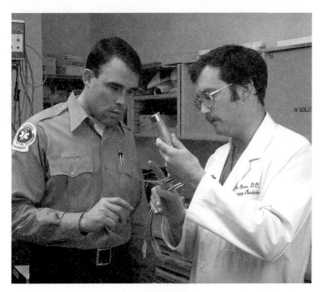

Figure 1-6 An EMT-Intermediate.

In-Hospital Care System

First Responders and EMTs provide **prehospital care,** or emergency medical treatment, before transport to a medical facility. In some areas, the term **out-of-hospital care** is preferred. It reflects a trend toward providing care on the scene with or without transport to a hospital. (Your instructor will provide information on how these terms apply to your EMS system.)

The most familiar destination of EMS patients is the local hospital emergency department. There a staff of physicians, nurses, and other allied health professionals stabilize the patient and prepare him or her for further care elsewhere in the hospital.

Specialized facilities to which some patients may be taken include:

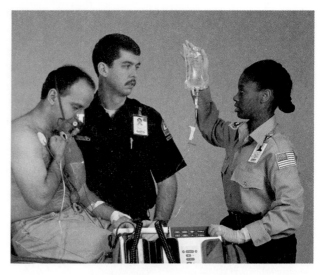

Figure 1-7 An EMT-Paramedic.

◆ *Trauma center*—for injury treatment that may exceed what a general hospital can do.

◆ *Burn center*—for treatment of burns, often including long-term care and rehabilitation.

◆ *Pediatric center*—for treatment of infants and children.

◆ *Perinatal center*—for high-risk pregnant patients.

Traditionally, information and advice on how to treat poisoning victims may be obtained by phone from the local poison control center.

SECTION

2

THE FIRST RESPONDER

As a fire service First Responder, you may be called to emergencies where you are the only trained rescuer on scene. At other times, specialized rescue teams, fire personnel, and law enforcement may all be involved.

Your Role

Generally, your role as a First Responder includes the following:

◆ *Protect your safety and the safety of your crew, the patient, and bystanders.* This is your first and most important priority. Remember that you cannot help the patient if you are injured.

You also do not want to endanger other rescuers by forcing them to rescue you. Avoid becoming part of the problem. Once scene safety is assured, the patient's needs become your primary concern.

◆ *Gain access to the patient.* In some emergencies, you may need to move one patient in order to gain access to a more critically injured one.

◆ *Assess the patient to identify life-threatening problems.* Always perform an initial assessment to help you identify any threats to a patient's life, such as a blocked airway, cardiac arrest, or severe bleeding.

◆ *Alert additional EMS resources.* If you determine that a patient requires advanced care or rescue personnel for extrication, summon those resources early in your care.

◆ *Provide care that is based on assessment findings.* While you are waiting for EMS resources to arrive, you must provide patient care based on the needs you identified during patient assessment.

◆ *Assist other EMS personnel.* When requested, assist other EMS personnel with patient care as needed.

◆ *Participate in record keeping and data collection.* You may be required by state law or your local EMS system to document your calls, especially if a patient refuses care.

◆ *Act as liaison with other public safety workers.* They may include local, state, or federal law enforcement, fire department personnel, and other EMS providers.

COMPANY OFFICER'S NOTE

To help you remember priorities on scene, you might want to keep the acronym *L.A.S.T.* in mind. It stands for: locate, access, stabilize, treat.

Your Responsibilities

First Responder responsibilities vary from state to state. However, they always include assuring scene safety and maintaining a professional attitude and appearance, compassion for your patients, and up-to-date skills. Specifically, you should:

◆ *Guard your personal health and safety.* Drive safely at all times. Use a seat belt whenever you drive or ride. Remove yourself from hazards such as gas leaks, chemical spills, and so on, and follow the directions of specialized rescuers at those scenes. Never enter a crime scene or an angry crowd until it has been controlled by the police. Locate or create a safe area in which you can care for patients. Stay away from high-traffic areas. Always wear the proper personal protective equipment (PPE). (See Chapter 2 for details.)

◆ *Maintain a caring and compassionate attitude.* Often you will arrive at an emergency scene to find the patient, family, and bystanders in a state of panic or chaos. These are normal reactions. Reassure and comfort them. Identify yourself, and let them know that more help is on the way.

◆ *Maintain your own composure.* After all, you did not create the emergency. You are there to stop it from getting worse. Many calls will be routine, and patient care will be simple. However, some calls will involve life-threatening or emotionally charged problems. In those cases, it is critical that you stay calm so you can get an accurate picture of the scene and properly establish your priorities.

◆ *Keep a neat, clean, professional appearance.* Excellent personal grooming and a crisp, clean appearance help instill confidence in patients. Being clean also helps to protect your patients from contamination from dirty hands or soiled clothing. Respond to every call in uniform and other appropriate dress. Portray the positive image you want to project. Remember that you are on a medical team. Your appearance can send the message that you are competent and trustworthy.

◆ *Maintain up-to-date knowledge and skills.* New research often shows us better ways of doing things. So take every opportunity to continue your education, including taking refresher courses offered through your local EMS system.

◆ *Maintain current knowledge of local, state, and national issues affecting the EMS system.*

Attend conferences and read professional journals dedicated to EMS issues.

◆ *Maintain patient confidentiality.* Hold in confidence all information obtained in the course of your work, unless required by law to share it.

You will be expected to accept and uphold the responsibilities of a First Responder in accordance with the standards of an EMS professional. Note that as a First Responder you will come in contact with people of both sexes, as well as different ages, cultures, and socioeconomic backgrounds. It is your responsibility to meet the standard of care for all your patients. (Standard of care is discussed in Chapter 3.)

Figure 1-8 The medical director may give instructions to an EMS rescuer on scene.

3 MEDICAL OVERSIGHT

Medical Director

A formal relationship exists between a community's EMS providers and the physician who is responsible for out-of-hospital emergency medical care. This physician is often referred to as the **medical director.** He or she is legally responsible for the clinical and patient-care aspects of an EMS system.

Every EMS system must have a medical director. He or she must provide guidance to all emergency care and rescue personnel. The medical director is also responsible for reviewing and improving the quality of care in an EMS system.

Medical Control

Two basic types of medical oversight are **direct medical control** and **indirect medical control.** Direct medical control occurs when the medical director or another physician directs an EMS rescuer at the scene of an emergency (Figure 1-8). This may be done via telephone, radio, or in person. This usually occurs when an EMS rescuer asks for help with a patient. Note that in different areas, direct medical control may also be called "on-line," "base station," "immediate," or "concurrent."

Indirect medical control may also be called "off-line," "retrospective," or "prospective." It includes such things as system design and quality management. Through indirect medical control, standing orders and protocols spell out the accepted practice for fire service First Responders

in your area. They tell you things like whether or not you can give oxygen to a patient or how to respond to a family who refuses your help. They also tell you how to document each call, participate in reviews, gather feedback, and maintain your skills.

The First Responder

In general, fire service First Responders are the designated agents of the medical director. If this is true in your area, the care you render by law may be considered an extension of the medical director's authority. Your instructor will tell you what the law is in your area.

> **COMPANY OFFICER'S NOTE**
>
> It can be challenging for fire departments to start up EMS delivery. One of the greatest challenges is acceptance by the fire chief that authority for medical issues now rests with an "outsider," which in this case is a physician. The tradition for centuries has been for the fire chief to have ultimate authority over personnel, answering only to a governing body. The information in Table 1-1 can ease the transition. It describes the general areas of authority for a medical director.

Table 1-1

AUTHORITY OF MEDICAL DIRECTOR AND FIRE CHIEF

	Medical Director	Fire Chief
Medical Issues	Has overall authority.	Makes recommendations.
EMS Operations	Makes recommendations.	Has overall authority.

ON SCENE FOLLOW-UP

At the beginning of this chapter, you read that fire service First Responders were caring for a patient who had been injured by a blow to his forehead. To see how chapter material applies to this emergency, read the following. It describes how the call was completed.

PATIENT HISTORY

The man told us he was Ron York, 41 years old. He said he had an argument with a neighbor across the street. He stated his neighbor became angry and struck him twice—once on the forehead and once across the back of his head—with a wood baseball bat. He stated he did not lose consciousness and was able to walk back to his home and call 9-1-1. He told us his head hurt, but that he thought he would be okay. The patient stated he was on no medications, and was not allergic to anything he knew of.

I noticed our patient seemed to slur his words when speaking. I also noted a sweet odor whenever I was close to his mouth. I mentioned it to my captain. He asked the patient if he was a diabetic. The patient said no. Josh, another member of our engine company, asked the patient if he had any alcohol to drink. The patient said he had consumed about a pint of vodka in the last two hours. He also stated he had eaten some pizza while drinking the vodka.

PHYSICAL EXAMINATION

The patient denied any neck or back pain. We noted blood continuing to ooze from the wound on his forehead, as well as blood dripping from his nose. Checking his pupils, we noted that

both were equal and reactive to light, although we had to manually assist opening the right eye due to the swelling. On the back of his head we noted swelling, which looked like a "goose egg," but no bleeding. The patient denied any other pain, and he kept saying he was "okay." We took his vital signs and found that his pulse was 130, strong and regular. His respirations were 20 and unlabored. Blood pressure was 140/102. Listening to his chest we heard adequate air entering on both sides. We applied oxygen and made sure the EMTs were on the way.

ONGOING ASSESSMENT

We remained concerned because the patient was slurring his speech and his face seemed to be more swollen than it was before. We verified that he was still alert. We asked him what day it was, and he gave us the correct date. Oxygen continued to flow through a nonrebreather mask. We made sure the patient was as comfortable as he could be and tried to reassure him as best we could. We finished another set of vitals just as the ambulance pulled up.

PATIENT HAND-OFF

Since my captain had responsibility for patient care during this call, he gave the EMTs the hand-off report:

"We have a 41-year-old male patient, Mr. York. He was struck in the forehead and again in the back of the head with a wood baseball bat during an altercation. He denies loss of consciousness and states he walked from where the confrontation occurred to his residence across the street, where we found him sitting in a chair with a washcloth applied to his forehead. The wound was bleeding. There is also a trickle of blood coming from his nose. We noticed a possible small indentation in the front part of the forehead where he was struck with the bat. He has a "goose egg" on the back of his head, but there was no bleeding from that injury. He denies any neck or back pain. His right eye is swollen shut, and the swelling appears to be getting more noticeable. His vital signs are 140/102, pulse is 130 strong and regular, respirations are 20 and unlabored. He has no medical history and denies any allergies. We noted a fruity, sweet odor on his breath. He states he is not a diabetic, but he does tell us he consumed about a pint of vodka along with some pizza approximately two hours ago. We've loosely applied a sterile dressing to his forehead and an ice pack to his face. The bleeding has been controlled. He is also receiving oxygen via nonrebreather mask at 15 liters per minute."

After we made sure the EMTs didn't need us any longer, we radioed dispatch to say we were available and headed back to our station.

> *It takes all of the resources of an EMS system working together to help a patient survive an illness or injury. As a fire service First Responder, not only will you probably be the first EMS personnel on scene, but your actions also will help determine whether or not a patient lives or dies. Because of this, you are one of the most valuable resources in the EMS system.*

Chapter Review

As a fire service First Responder, you play a vital role in the emergency medical care of patients experiencing an illness or injury. Perhaps the most important reason your role is so crucial is that you are responsible for the first few minutes with the patient. The EMS system depends on your actions during this time to set the foundation for the remainder of the call.

It is during the first few minutes that correcting a breathing problem or stopping bleeding will actually save a life. You also will help patients who are not in critical condition when you prevent further injury, perform the proper assessments, gather a medical history, and prepare for the arrival of the EMTs or paramedics who may transport the patient.

FIRE COMPANY REVIEW

Page references where answers may be found or supported are provided at the end of each question.

Section 1

1. What is the purpose of the EMS system? (p. 3)

2. What is the typical sequence of events that occurs from the time an accident occurs and EMS is activated? (p. 3)

3. What are five of the 10 classic components of an EMS system? (p. 5)

4. What are two basic ways to access the EMS system? (p. 5)

5. How many levels of training are offered in the EMS system? Briefly describe each one. (p. 6)

Section 2

6. What is the First Responder's role at the scene of an emergency? (pp. 8–9)

Section 3

8. What are the two types of medical oversight? Describe each one. (p. 10)

RESOURCES TO LEARN MORE

Paris, P.M., R.N. Roth, and V.P. Verdile, eds. *Prehospital Medicine: The Art of On-line Medical Command.* St. Louis: Mosby, 1996.

Steele, S. *Emergency Dispatching: A Medical Communicator's Guide.* Upper Saddle River, NJ: Brady/Prentice-Hall, 1992.

TYPICAL ACCESS TO AN EMS SYSTEM

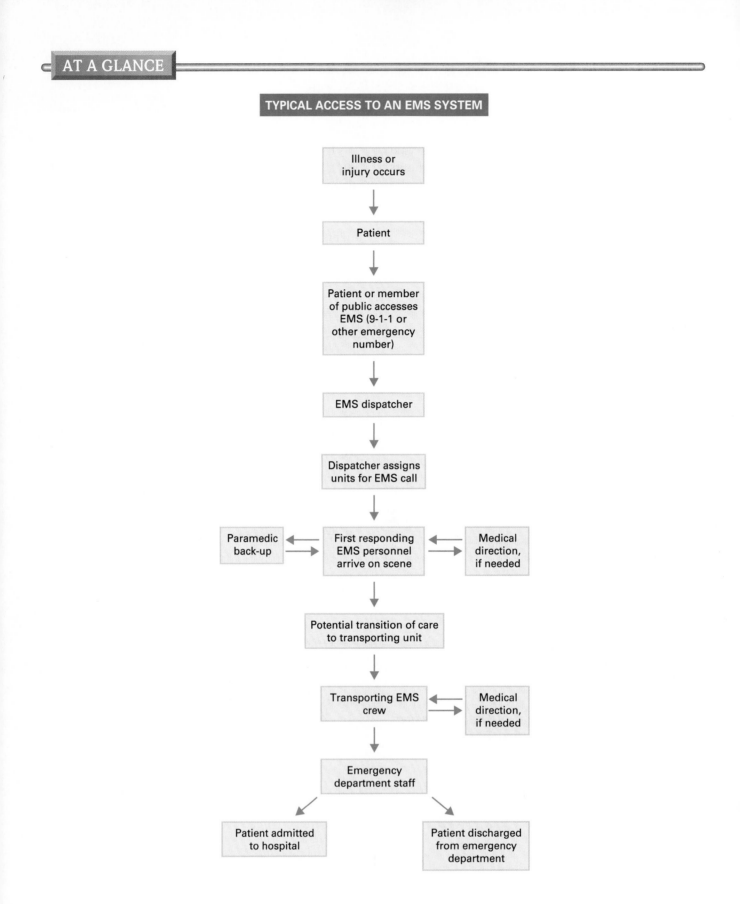

Illness or injury occurs

Patient

Patient or member of public accesses EMS (9-1-1 or other emergency number)

EMS dispatcher

Dispatcher assigns units for EMS call

Paramedic back-up → First responding EMS personnel arrive on scene ← Medical direction, if needed

Potential transition of care to transporting unit

Transporting EMS crew ← Medical direction, if needed

Emergency department staff

Patient admitted to hospital

Patient discharged from emergency department

Scene Safety and the Well-Being of the First Responder

*I*NTRODUCTION *As a fire service First Responder, your safety always comes first. It comes before the patient and before any bystander at the scene. The reason is simple. If you are injured, you lose the ability to help those who need you and, instead of providing emergency care, you end up needing it.*

Many elements make up rescuer safety. Most basic training programs teach how to react safely to a variety of environmental threats, such as fire and flood. But the most common threat to a rescuer is his or her attitude towards safety. This chapter outlines the basic steps you should take to maintain your well-being. It includes an introduction to scene safety, how to anticipate and handle the emotional aspects of emergencies, and how to protect yourself against infection.

OBJECTIVES

Cognitive, affective, and psychomotor objectives are from the U.S. DOT's "First Responder: National Standard Curriculum." Enrichment objectives, if any, identify supplemental material.

Cognitive

1-2.1 List possible emotional reactions that the First Responder may experience when faced with trauma, illness, death, and dying. (pp. 17–18)

1-2.2 Discuss the possible reactions that a family member may exhibit when confronted with death and dying. (pp. 19, 20–22)

1-2.3 State the steps in the First Responder's approach to the family confronted with death and dying. (pp. 20–22)

1-2.4 State the possible reactions that the family of the First Responder may exhibit. (p. 19)

1-2.5 Recognize the signs and symptoms of critical incident stress. (pp. 17–20)

1-2.6 State possible steps that the First Responder may take to help reduce/alleviate stress. (pp. 18–19)

1-2.7 Explain the need to determine scene safety. (pp. 27–30)

1-2.8 Discuss the importance of body substance isolation (BSI). (pp. 22–27)

1-2.9 Describe the steps the First Responder should take for personal protection from airborne and bloodborne pathogens. (pp. 22–27)

1-2.10 List the personal protective equipment necessary for each of the following situations: (pp. 22–30)
— Hazardous materials
— Rescue operations
— Violent scenes
— Crime scenes
— Electricity

— Water and ice
— Exposure to bloodborne pathogens
— Exposure to airborne pathogens

Affective

1-2.11 Explain the importance of serving as an advocate for the use of appropriate protective equipment. (p. 26)

1-2.12 Explain the importance of understanding the response to death and dying and communicating effectively with the patient's family. (pp. 21–22)

1-2.13 Demonstrate a caring attitude towards any patient with illness or injury who requests emergency medical services. (p. 24)

1-2.14 Show compassion when caring for the physical and mental needs of patients. (pp. 21–22)

1-2.15 Participate willingly in the care of all patients. (p. 24)

1-2.16 Communicate with empathy to patients being cared for, as well as with family members, and friends of the patient. (pp. 20–22)

Psychomotor

1-2.17 Given a scenario with potential infectious exposure, the First Responder will use appropriate personal protective equipment. At the completion of the scenario, the First Responder will properly remove and discard the protective garments. (pp. 25–26)

1-2.18 Given the above scenario, the First Responder will complete disinfection/cleaning and all reporting documentation. (pp. 25–27)

DISPATCH

My partner and I are First Responders with our volunteer fire department. We were finishing up our monthly training one Saturday and were just about to leave the station, when dispatch called. There was an assault with one person bleeding at 99 Snyder Drive.

We got in the rescue truck. It was my partner's turn to drive. I got out the town map and located the residence. On the way, I mentally went over the procedures for body substance isolation and bleeding control.

SCENE SIZE-UP

En route, we asked dispatch if the police were responding. She stated they were. About one block from the patient's house, we decided to stage our vehicle until the police could determine if the scene was safe to enter. After a few minutes, dispatch informed us that the police were on scene and were requesting our response.

Arriving at the address, we parked our vehicle behind the police squad cars. Exiting our vehicle, I immediately heard men yelling and glass breaking. Loud thumps and scuffling made it obvious that an altercation was occurring. It became clear to me and my partner that, although the police had requested our response, things were not quite under control yet.

> *Caution will help you to recognize potential dangers at an emergency scene. But is violence the only kind of danger you may face? Consider your answer as you read this chapter.*

EMOTIONAL ASPECTS OF EMERGENCY CARE

Stress is any change in the body's internal balance. It occurs when outside demands are greater than the body's resources. High-stress situations include multiple-casualty incidents, injury to an infant or child, death of a patient, an amputation, violence, abuse, and injury or death of a coworker.

A First Responder's initial response to stress may include weakness, nausea, vomiting, or fainting. You can help avoid these reactions by using the following techniques:

◆ Close your eyes and take several long, deep breaths. Focus on counting each breath. When you feel more in control, return to giving emergency care.

◆ Change your thought patterns. Silently hum a soothing tune to yourself, or visualize a happy outcome for the call.

◆ Eat properly to maintain your blood sugar. Low blood sugar can add to a fainting problem.

Stress Management

As a fire service First Responder, you will be exposed to a great deal of stress when meeting the needs of your patients. Some First Responders tend to feel completely responsible for everything that happens at the scene, even things clearly out of their control. A few become so involved that their self-image is actually based on job performance.

Chronic stress at work plus an emotionally charged environment can lead to a state of exhaustion and irritability. Beware. That state can markedly decrease your effectiveness. Even some of the very best fire service personnel have had to leave EMS because of it.

Recognize Warning Signs

One of the best ways to manage chronic stress and prevent burnout is to be aware of the warning signs. The quicker they are spotted, the easier they are to remedy. The warning signs include (Figure 2-1):

◆ Irritability with coworkers, family, and friends.

◆ Inability to concentrate.

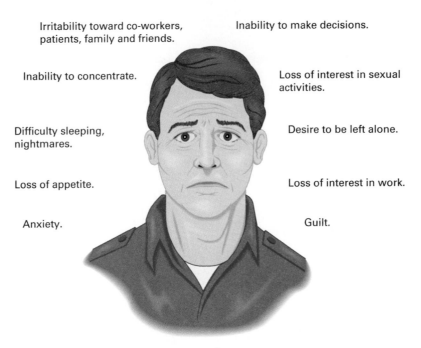

Irritability toward co-workers, patients, family and friends.

Inability to concentrate.

Difficulty sleeping, nightmares.

Loss of appetite.

Anxiety.

Inability to make decisions.

Loss of interest in sexual activities.

Desire to be left alone.

Loss of interest in work.

Guilt.

Figure 2-1   The warning signs of stress.

◆ Difficulty sleeping, nightmares, excessive sleeping.

◆ Anxiety.

◆ Inability to make decisions.

◆ Guilt.

◆ Loss of appetite.

◆ Loss of sexual desire.

◆ Isolation.

◆ Loss of interest in work.

In addition, the following general signs and symptoms have been identified with stress:

◆ *Cognitive*—confusion, inability to make judgments or decisions, loss of motivation, memory problems, loss of objectivity.

◆ *Psychological*—depression, excessive anger, negativism, hostility, defensiveness, mood swings, feelings of worthlessness.

◆ *Physical*—constant exhaustion, headaches, stomach problems, dizziness, pounding heart, high blood pressure, chest pain.

◆ *Behavioral*—overeating, increased use of drugs or alcohol, grinding teeth, hyperactivity, lack of energy.

◆ *Social*—frequent arguments, decreased ability to relate to patients.

Make Lifestyle Changes

Certain lifestyle changes can help you deal with chronic stress. Diet is one of them. Certain foods, such as sugar, fatty foods, caffeine, and alcohol increase the body's response to stress. So while at work, try to eat low-fat carbohydrates. Also, eat often but in small amounts.

Exercise more often (Figure 2-2). It offers all kinds of benefits, including physical release for pent-up emotions. It also breaks down the natural chemicals our bodies dump into our systems any-

Figure 2-2   As a First Responder, you must safeguard your own health.

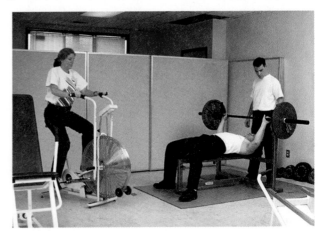

time we are under stress. Without exercise, these chemicals tend to build up, causing potentially harmful effects.

Finally, learn to relax. Meditation and visual imagery are helpful techniques. You also may want to try to cut loose a little bit. Watch a funny movie, read a good book, go dancing, or go to a concert.

Keep Balance in Your Life

Balance is an important principle of the world. We have all heard the terms "balance of nature," "balance of trade," and "balance of power." Yet, it is easy for First Responders to become unbalanced in their personal lives. One way to balance work, recreation, family, and health is by recognizing each role you play on a daily basis. For example, you not only have a role as a firefighter, you may also have a role as a parent, spouse, son or daughter, aunt or uncle, grandson or granddaughter, teacher or student, sibling, little league coach, or Girl Scout leader. It is critical for you to assess each and every one. Try this: Take a few minutes to list all your roles on paper. Most of us have at least five or six. After listing them, ask yourself this question: "Am I investing time and energy in one role at the expense of another?"

If you find you are out of balance, try the following:

◆ Each week write one or more goals to accomplish. Do it for each of your roles. This is a type of time management. Yet, most time-management tools do not consider all of our roles; they consider only our "work" role.

◆ Synergize roles, or combine actions, whenever possible. For example, if one of your goals is to exercise three times a week, try inviting your spouse or child. Swimming laps in a pool can be fun and aerobic when it becomes a family event. Or, when teaching a CPR class, have your spouse come in and be the "victim."

◆ It is important to recognize that your role as a fire service First Responder is not all there is to life! Has anyone ever heard of a firefighter whose last words were "I wish I had spent more time at the firehouse"? When you are on duty (at the station), give 100% of yourself. When you are off duty (away from the station), give 100% to the people you are with

and the activities you are involved in. In other words, wherever you are, be there.

◆ Finally, try to stay away from the firehouse. When you are off duty, stay off duty unless absolutely necessary.

It is important to remember that the support of your family and friends is essential to how well you manage stress. Keep in mind, though, that they suffer from stress related to your job, too. Their stress factors include the following:

◆ *Lack of understanding.* Families typically have little if any knowledge about prehospital emergency care.

◆ *Fear of separation or of being ignored.* Long hours can take their toll and increase your family's distress over your absences. You may hear, "the fire department is more important to you than your family!"

◆ *Worry about on-call situations.* Stress at home may increase because your family may focus on the danger you face when you respond to emergency calls.

◆ *Frustrated desire to share.* It may be too difficult for you to talk about what happened on certain calls. Even though your family and friends understand that, they may still feel frustrated in their desire to help and support you.

If at all possible, you can help to keep balance in your life by changing your work environment. Request work shifts that allow for more time to relax with family and friends. Ask for a rotation of duty to an assignment that is less stressful. Take periodic breaks to exercise and to support and encourage coworkers. If you are a volunteer, take some time off from responding to calls.

Seek Professional Help

Mental health professionals, social workers, and clergy can help you realize that your reactions—and your family's—are normal. They also can mobilize your best coping strategies and suggest more effective ways to deal with stress.

Critical Incident Stress Management (CISM)

A **critical incident** is any event that causes unusually strong emotions that interfere with your ability to function either during the incident or later. This type of stress requires aggressive and immediate management, including:

◆ Pre-incident stress education.

◆ On-scene peer support.

◆ One-on-one support.

◆ Disaster support services.

◆ Follow-up services.

◆ Spouse and family support.

◆ Community outreach programs.

◆ Other general health and welfare programs, such as wellness programs.

There are two basic critical incident stress management (CISM) techniques: debriefing and defusing.

Debriefing

The cornerstone of most CISM programs is called a **critical incident stress debriefing (CISD).** CISD combines a team of peer counselors with mental health professionals (Figure 2-3). It is successful because it helps rescuers vent their feelings quickly. Its nonthreatening environment also encourages rescuers to feel free to air their concerns and reactions.

CISD includes anyone involved in an incident—police, firefighters, EMS personnel, dispatchers, doctors, and so on. In some cases, it may also include their families.

Ideally, the debriefing is held within 24 to 72 hours of a critical incident. It is not an investi-

Figure 2-3 A critical incident stress debriefing (CISD).

gation or an interrogation. Everything that is said at a debriefing is confidential. Rescuers are urged to explore any physical, mental, or emotional symptoms they are having. CISD counselors and mental health professionals then evaluate the information and offer suggestions on how to cope with the stress resulting from the incident.

After multiple-casualty incidents (such as earthquakes or explosions), a number of CISD meetings may be needed.

Defusing

Much shorter and less formal than a debriefing, a defusing is usually held within hours of the critical incident. It is attended only by those most directly involved and lasts only 30 to 45 minutes. A defusing gives rescuers a chance to vent their feelings and get information they may need before the larger group meets. It may either eliminate the need for a formal debriefing, or it may enhance a later debriefing.

Accessing CISD

Generally, your agency or organization will organize a CISD. Attend if you have been involved in:

◆ Serious injury or death of a rescuer in the line of duty.

◆ Multiple-casualty incident.

◆ Suicide of an emergency worker.

◆ An event that attracts media attention.

◆ Injury or death of someone you know.

◆ Any disaster.

◆ Injury or death of an infant or child.

Also access CISD after any event that has unusual impact on you. That may include an incident in which injury or death of a civilian was caused by a rescuer (an ambulance colliding with a car, for example). A death of a patient, child abuse or neglect, an event that threatens your life, or one that has distressing sights, sounds, or smells—all may be cause for accessing CISD.

Ask your instructor about the CISD programs available through your EMS system.

Death and Dying

Death and dying are inherent parts of emergency care. If you think about it, the fear of death is one of the most useless fears we have. Why? Death is one of the few things in life all of us will sooner or later experience. If we are afraid of death, we might as well be afraid of gravity or breathing or "The Brady Bunch" reruns.

Be that as it may, it is most important to care for a dying patient's emotional needs. If the patient dies suddenly, his or her family becomes your patient. Be prepared to provide for their emotional needs and those of bystanders as well.

The Grieving Process

There are five general stages that dying patients—and those who are close to them—experience. The five stages are called the **grieving process.** Each person progresses through the stages at his or her own rate and in his or her own way.

Patients with nonfatal emergencies also may go through a grieving process. For example, a patient who loses both legs in a factory accident will grieve the loss of his limbs.

As a fire service First Responder, you will not witness all five stages during emergency care. For example, a critically injured patient who is aware that death is imminent may just begin the process. A terminally ill patient who is more prepared may be at the final "acceptance" stage. The key is to accept all emotions as real and necessary. Respond accordingly.

The stages of the grieving process occur as follows:

◆ *Denial ("Not me!").* At first, the patient may refuse to accept the idea that death is near. This refusal creates a buffer between the shock of approaching death and the need to deal with the illness or injury. Families of dying patients often are at the denial stage.

◆ *Anger ("Why me?").* Watch out. You may be the target of this anger. But remember that it is a normal part of the grieving process. Do not take it personally. Be tolerant. Try to understand, and use your best listening and communication skills.

◆ *Bargaining ("Okay, but first let me . . .").* In the patient's mind a bargain, or agreement of sorts, will postpone death. For example, a patient may mentally determine that if he is allowed to live, he will patch up a long-standing break with his parents.

◆ *Depression ("Okay, but I haven't . . .").* As reality settles in, the patient may become silent, distant, withdrawn, and sad. He usually is thinking about those he leaves behind and all the things left undone.

◆ *Acceptance ("Okay, I am not afraid.").* Finally, the patient may appear to accept the fact that he is dying, though he is not happy about it. At this stage, the family usually needs more support than the patient does.

Dealing with the Dying Patient

It is one of your jobs to help a patient and his or her family through the grieving process. Keep in mind that they may progress through the stages of grief at different rates. Whatever stage they are in, their needs include dignity, respect, compassion, sharing, communication, privacy, and control. To help reduce their emotional burden, consider the following:

◆ *Do everything possible to maintain the patient's dignity.* Avoid negative statements about the patient's condition. Even an unresponsive patient may hear what you say and feel the fear in your words. Talk to the patient as if he or she is fully alert. Explain the care you are providing.

◆ *Show the greatest possible respect for the patient.* Do this especially when death is imminent. Families will be extra sensitive at this time. Even attitudes and unspoken messages are perceived. So explain what you are doing, and assure family members that you are making every possible effort to help the patient. It is important for them to know with certainty that you never simply "gave up."

◆ *Communicate.* Help the patient become oriented to the surroundings. If necessary, explain several times what happened and where. Explain who you are and what you and others are planning. Assure the patient that you are doing everything possible and that you will see that he or she gets to a hospital as quickly as possible. Without interrupting care, communicate the same message to the family. Explain any procedure you need to carry out. Answer their questions. Do not guess. Report only what you know to be true.

◆ *Allow family members to express rage, anger, and despair.* They should be able to scream, cry, or vent grief but in a way that is not dangerous to you or others. Be tolerant. If they vent their anger at you, do not get angry or hostile.

◆ *Listen with empathy.* Many dying people want messages delivered to survivors. Take notes. Assure the patient that you will do whatever you can to honor his or her requests. Then follow through on your promise. If necessary, stay with the family to listen to their concerns and answer their questions.

◆ *Do not give false assurances, but allow for some hope.* Be honest but tactful. If the patient asks if he is dying, do not confirm it. Patients who do the most poorly are often the ones who feel hopeless. Instead, say something like "We are doing everything we can. We need you to help us by not giving up." If the patient insists that death is imminent, say "That might be possible, but we can still try the best we can, can't we?"

◆ *Use a gentle tone of voice.* Be kind to both the patient and the family. Explain the scope of the injury and your medical care. When necessary, do so as gently and kindly as you can in terms they will understand.

◆ *Use a reassuring touch,* if appropriate.

◆ *Do what you can to comfort the family.* Arrange for them to briefly see or talk to the patient. However, do not interrupt your emergency care or delay transport.

◆ *If the patient is deceased* and no further medical interventions are indicated, and the family asks you to pray with them, do so. Call clergy or a chaplain, if the family requests it. Allow family members to touch or hold the body. The exception is at a crime scene. Never clean the patient or remove any blood from the patient or the crime scene itself. Finally, stay with the family until law enforcement, the medical examiner, coroner, or funeral home arrives.

SECTION 2 PREVENTING DISEASE TRANSMISSION

As a fire service First Responder, you will come in contact with patients who are sick. The following section will explain how diseases are transmitted and discuss the ones of most concern to the fire service First Responder. It also will describe ways for you to protect yourself.

How Diseases Are Transmitted

Diseases are caused by **pathogens,** microorganisms such as bacteria and viruses. An **infectious disease** is one that spreads from one person to another (Figure 2-4). It can spread *directly* through blood-to-blood contact (bloodborne), contact with open wounds or exposed tissues, and contact with the mucous membranes of the eyes and mouth. An infectious disease also can spread *indirectly* by way of a contaminated object, such as a needle, or by way of infected droplets inhaled into the respiratory tract (airborne).

Diseases also can be "vector-borne," or spread through contact with another organism. Examples include mosquito bites, which can spread malaria, or ticks, which can transmit Lyme disease.

Some pathogens are transmitted easily, such as the viruses that cause the common cold. Others need specific routes of contact. The tuberculosis bacteria, for example, is transmitted by droplets from a cough or sneeze of an infected patient. Poor nutrition, poor hygiene, crowded or unsanitary living conditions, and stress all make diseases easier to acquire.

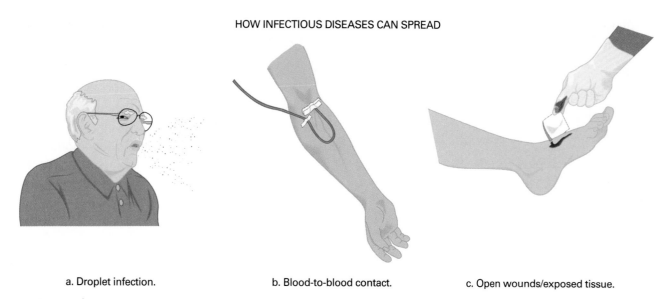

a. Droplet infection.   b. Blood-to-blood contact.   c. Open wounds/exposed tissue.

Figure 2-4   Infectious diseases can spread from one person to another.

In order to protect yourself, you must *always* make use of the appropriate body substance isolation (BSI) equipment, including barrier devices such as a pocket face mask, *every time* you ventilate a patient (Figure 2-5). Make sure barrier devices used for ventilation have a one-way valve that prevents fluid from backing up.

Diseases of Concern

As a fire service First Responder, every patient can potentially expose you to an infectious disease. Three diseases of most concern are described below.

Hepatitis B

Hepatitis B (HBV) is a potentially fatal virus directly affecting the liver. A serious disease that

Figure 2-5   A pocket face mask with one-way valve and carrying case.

can last for months, HBV is contracted through blood and body fluids. A major source of the virus is the "chronic carrier," a person who carries the virus for years, often unaware that he or she is infected. When signs and symptoms do appear, they may include:

◆ Fatigue.

◆ Nausea.

◆ Loss of appetite.

◆ Abdominal pain.

◆ Headache.

◆ Fever.

◆ Yellowish color of the skin and whites of eyes.

The best defense against HBV infection is BSI equipment. The second best defense is the HBV vaccination, which most agencies and employers offer free of charge. If you suspect you have been exposed to HBV, report the incident to your infectious disease officer immediately or contact a physician or your local public health agency for care.

Tuberculosis

Tuberculosis (TB) is back with a vengeance. In fact, researchers are worried because of the development of new drug-resistant strains. The pathogen that causes TB is found in the lungs and other tissues of the infected patient. You can be infected from droplets in a patient's cough or from infected sputum. The signs and symptoms of TB include:

◆ Fever.
◆ Cough.
◆ Night sweats.
◆ Weight loss.

OSHA has adopted standards for rescuer protection against TB which include the use of special masks. One type of mask is called the **N-95 respirator** (Figure 2-6). Another type is called the **HEPA respirator,** or high efficiency particulate air respirator. Regardless of which mask your fire department chooses to use, it is critical that all First Responders wear one whenever a patient is suspected of having TB, or whenever a patient is coughing for an unknown reason.

Acquired Immune Deficiency Syndrome (AIDS)

Fortunately, acquired immune deficiency syndrome (AIDS) cannot be spread by touching the skin, coughing, sneezing, sharing eating utensils, or other indirect ways. *Transmission requires inti-mate contact with the body fluids of infected persons.* Infection may occur with:

◆ Sexual contact involving the exchange of semen, saliva, blood, urine, or feces.
◆ Infected needles.
◆ Infected blood or blood products, especially when it comes in contact with another person's eyes, mucous membranes, or broken skin.
◆ Mother to child during pregnancy, birth, or breast-feeding.

The human immunodeficiency virus (HIV) destroys the body's ability to fight infection. HIV can lead to AIDS. AIDS victims get infections caused by viruses, bacteria, parasites, and fungi—the same organisms that usually cause no harm in people with healthy immune systems. The infections involve many organs of the body, causing a countless array of signs and symptoms.

Not everyone infected by HIV develops AIDS. However, people who carry HIV are still able to spread the infection to others. Therefore, every patient you come in contact with could be infected with HIV or other diseases. To protect yourself, follow all the precautions described below at all times and with all patients.

Figure 2-6 Wear an N-95 respirator when you care for a patient with TB.

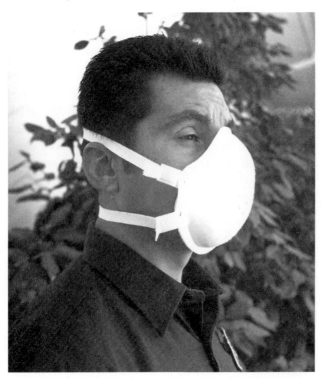

Body Substance Isolation (BSI)

For many years, OSHA guidelines required all EMS personnel to take steps, or "universal precautions," to protect themselves against diseases transmitted through contact with blood. In the late 1980s, the Centers for Disease Control (CDC) published new guidelines. They make the assumption that *all blood and body fluids are infectious,* and therefore require all EMS personnel to practice a strict form of infection control with every patient. That strict form of infection control is called **body substance isolation (BSI).**

With BSI precautions, it is possible to take care of all patients safely, even those with infectious diseases. However, you must remember that it is not possible to tell if a patient is infectious by the way he looks, speaks, or acts. Treat all patients the same way. Always wear the proper BSI equipment with every patient.

BSI precautions include handwashing; proper cleaning, disinfection, or sterilization of equipment; and the use of BSI equipment.

Handwashing

Handwashing is the single most important thing you can do to prevent the spread of infection. According to the U.S. Public Health Service, most contaminants can be removed from the skin with 10 to 15 seconds of vigorous lathering and scrubbing with plain soap.

Always wash your hands after caring for a patient, even if you were wearing gloves. *For maximum protection, begin by removing all jewelry from your hands and arms.* Then lather up and rub together all surfaces of your hands. Pay attention to creases, crevices, and the areas between your fingers. Use a brush to scrub under and around your fingernails (Figure 2-7). (Note that it is a good idea for you to keep your nails short and unpolished.) If your hands are visibly soiled, spend more time washing them. Wash your wrists and forearms, too. Rinse thoroughly under a stream of water and dry well. Use a disposable towel if possible.

If you do not have access to soap and running water, you can use a foam or liquid washing agent. As soon as you can, wash your hands again using the procedure described above.

Cleaning Equipment

Cleaning, disinfecting, and sterilizing are related terms. **Cleaning** is simply the process of washing a soiled object with soap and water. **Disinfecting**

Figure 2-8 Discard contaminated items in the proper receptacle.

is cleaning plus using a chemical like alcohol or bleach to kill many of the microorganisms on an object. **Sterilizing** is a process in which a chemical or other substance (like superheated steam) kills all of the microorganisms on an object.

Generally, disinfection is used on items that make contact with intact skin, such as stethoscopes and blood pressure cuffs. Items that come in contact with open wounds or mucous membranes should be sterilized.

Whenever possible, use disposable equipment. Never reuse disposable items. Instead, place them in a plastic bag that is clearly labeled "infectious waste" (Figure 2-8). Then seal the bag. Disposable items used with patients who are known to have HBV or HIV should be double-bagged.

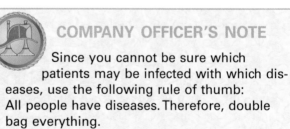

COMPANY OFFICER'S NOTE

Since you cannot be sure which patients may be infected with which diseases, use the following rule of thumb: All people have diseases. Therefore, double bag everything.

After each use, clean nondisposable equipment. Wash off all blood, mucus, tissue, and other residue. Be sure to wear a good pair of utility

Figure 2-7 Washing your hands is the most important action you can take against the spread of infection.

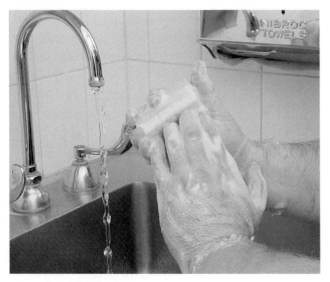

gloves while doing so. Then disinfect or sterilize it as per local protocols.

Wash items that do not normally touch the patient. Rinse them with clear water and dry thoroughly. Clean walls or window coverings in an ambulance or rescue vehicle when they get soiled. Then use a hospital-grade disinfectant or a solution of household bleach and water to clean up any blood or body fluids.

If your clothes get soiled with body fluids, remove, bag, and label them. Wash them in hot soapy water for at least 25 minutes. Do not mix contaminated and non-contaminated clothing in the washer. Take a hot shower yourself and rinse thoroughly.

BSI Equipment

Always use personal protective equipment (PPE) as a barrier against infection. Such items will keep you from coming into contact with a patient's blood and body fluids. They include eye protection, gloves, gowns, masks, and protection for the head and feet.

- *Eye protection.* Use eye shields to protect yourself from blood and body fluids splashing into your eyes. Several types are available. Clear plastic shields cover the eyes or the whole face. Safety glasses have side shields. If you wear prescription glasses, attach removable side shields. Form-fitting goggles are also available.

- *Gloves* (Figure 2-9). Wear high-quality vinyl or latex gloves whenever you care for a patient. Never reuse them. Put on a new pair for each patient to avoid exposing one patient to another's infection. Soiled gloves must be changed as soon as it is practical to do so. If a glove accidentally tears, remove it as soon as you can do so safely. Then wash your hands and replace the torn glove with a new one.

- *Gowns.* Wear a gown when there might be significant splashing of blood or body fluids. Generally, you will need a gown during childbirth or major injury. Whenever possible, use a disposable one. It also is recommended that you change your clothes if the gown gets soiled.

- *Masks.* Wear a disposable surgical-type face mask to protect against possible splatter of

Figure 2-9 Wear protective gloves whenever you care for a patient.

blood or body fluids. An N-95 or HEPA respirator is recommended for use with suspected tuberculosis patients. An alternative is to place a properly fitted N-95 or HEPA respirator on your patient as well as on yourself.

- *Head and foot protection.* Under certain circumstances, head protection and shoe covers may be required to protect you from excessive splashing or contamination from fluids. Hospital head and footwear usually worn in the hospital operating room may be one type of protective gear used for this purpose. Structural firefighting gear, such as impervious boots and helmets with face shields, also may be appropriate. Always check local protocols.

Use BSI equipment yourself and be sure to remind other EMS personnel at the scene to wear it, too. Make gloves and other PPE available to others who arrive to help.

Immunizations

Before you begin providing First Responder care, have a physician make sure you are adequately protected against common infectious diseases.

Health-care workers wearing latex gloves on a daily basis have a risk of developing latex allergy that is greater than the general population. Unfortunately, it is not possible to predict when and if any individual will develop a latex allergy. The most common symptom is a poison ivy-like rash on the hands. However, serious allergic reactions—including death—have occurred in a number of cases.

Fire service First Responders can avoid this problem by using non-latex gloves, such as those made of nitrile. However, if latex gloves are used as the appropriate protection, they should be reduced-protein, powder-free latex gloves. Note that powder used as a lubricant in some gloves can increase exposure to the allergy.

If you use latex gloves, always ask your patients if they are allergic to latex prior to touching them, if possible. Several groups are at high risk for latex allergy, including children with serious congenital defects, such as spina bifida, and individuals who have already been sensitized.

The following immunizations are recommended for active-duty fire service First Responders:

◆ Tetanus prophylaxis (every 10 years).

◆ Hepatitis B vaccine.

◆ Influenza vaccine (every year).

◆ Polio.

◆ Rubella (German measles).

◆ Measles.

◆ Mumps.

Because some immunizations offer only partial protection, have your physician verify your immune status against rubella, measles, and mumps. Remember always practice BSI precautions, even after being vaccinated.

Have a tuberculin skin (tine) test at least once every year you are on duty. It will tell you if you have been exposed to TB. If you have been exposed, you will require more frequent testing. Your agency's medical director or your physician can advise you.

Reporting Exposures

Immediately report any suspected exposure to blood or body fluids to your infectious disease officer, to medical control, and to your immediate supervisor, especially if the patient is HIV-positive, has hepatitis B, or is in a high-risk category for infection. Include in your report the date and time of the exposure, the type of body fluid involved, the amount, and details of the incident. State laws vary, so be sure to follow all local protocols.

SECTION 3

SCENE SAFETY

The number-one priority at each and every scene is to protect yourself. Be part of the solution, not part of the problem.

The only valid assumption regarding scene safety is to assume the worst. We all know there is absolutely no place on the fireground for the "hero mentality." The same is true on any EMS scene. No scene should ever be considered "secure" as long as there are people involved. That means every scene! Remember, it's people who shoot guns, wield knives, drive vehicles, and swing bats. As an instructor once said, "I never thought much of the courage of a lion-tamer. Inside the cage he is at least safe from people."

As you approach a scene, turn off your lights and siren to avoid broadcasting your arrival, if it is safe to do so. Take a good look at the neighborhood. If possible, do not park directly in front of the call address. This is so you can size up the scene unnoticed and save a place for an ambulance to park.

Then decide whether or not it is safe to approach the patient. If any of the following exists on scene, you may need to call for help:

◆ Motor-vehicle or airplane crashes.

◆ Presence of toxic substances or low levels of oxygen.

- Crime scenes.
- Presence of a weapon of any kind.
- Possible drug or alcohol use.
- Arguing, threats, violent behavior, broken glass, overturned furniture.
- Unstable surfaces, such as water and ice.
- Any type of entrapment.

While each emergency scene is unique, a general rule applies to all. *If the scene is unsafe, make it safe before you enter.* Otherwise, wait for help to arrive. Specialized personnel will have the training, equipment, and protective gear needed to enter an unstable scene safely.

Once a scene is secure, take measures to protect the patient from hazards. That includes fire, structural instability, gasoline leaks, chemical spills, oncoming traffic, and extremes in temperature. Bystanders should also be protected from illness or injury.

Figure 2-10 A self-contained breathing apparatus (SCBA).

In general, specialized team members should wear protective clothing, such as a self-contained breathing apparatus (Figure 2-10) and a "hazmat" suit (Figure 2-11). Although it is important for you to learn how to access your local hazmat team, you should at least be trained to the hazmat "awareness" level. (Hazardous materials are discussed in more detail in Chapter 24.)

Figure 2-11 Typical hazardous materials protective suit.

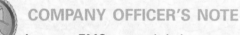

COMPANY OFFICER'S NOTE

At every EMS scene, it is important for you to remember to perform a size up that is similar to the one performed when arriving at a burning building. Just like a size up for a reported structure fire, this size up can begin at the time of dispatch. Is it day or night? Is it during the week or on the weekend or a holiday, when alcohol or drugs may be involved or a large gathering of people? Does the dispatcher report the call as an "unknown problem?" Could this be a potential crime scene? Is there the possibility of violence? If so, is law enforcement en route? If this is a reported car crash, is it on a 65-mph (100 km/h) interstate, which requires additional resources "just in case," or is it on a town street with a posted speed limit of 25 mph (40 km/h)? Remember, the scene size-up is important in accomplishing your number one objective: your safety!

Hazardous Materials

Do not enter a hazardous materials scene. Call a specialized team of rescuers to secure the scene first. Provide emergency care only after the scene is safe and the patient has been decontaminated.

Motor-Vehicle Collisions

Some car crashes lead to situations that threaten the life of both patients and rescuers. For example:

◆ Downed power lines or other potential sources of electrocution.

◆ Fire or the potential for fire such as leaking gasoline.

◆ Explosion or the potential for explosion.

◆ Hazardous materials.

◆ Oncoming traffic.

◆ Unstable vehicles.

When there is life-threatening danger on scene, call for specially trained personnel *before you enter.* Also call for special teams when a complex or extensive rescue is needed. Once a scene is safe, make sure you are wearing the proper personal protective equipment before you enter, including turnout gear, puncture-resistant gloves with latex or non-latex gloves underneath, helmet, and eye protection (Figure 2-12). Follow local protocols.

Figure 2-13 Body armor, or bullet-proof vest.

Figure 2-12 Turnout gear, plus helmet, eyewear, puncture-resistant gloves, and boots.

Initial positioning of rescue vehicles can help in protecting fire department personnel from oncoming traffic. Check local protocols or standard operating procedures for information on proper apparatus positioning.

Violence

You may face violence without warning, from a patient, bystander, or perpetrator of a crime. If you suspect potential violence, call law enforcement before you enter the scene. Never enter the scene to give patient care until it has been adequately controlled by the police.

Always call for law enforcement in cases of domestic disputes, street or gang fights, bar fights, potential suicides, crime scenes, and incidents involving angry family or bystanders. Remember, violence does not only occur in dark alleys. Case in point: violence is the leading cause of trauma in the workplace.

No matter where you work, consider using **body armor** (Figure 2-13). It is made of Kevlar™

or other synthetic materials that resist penetration by bullets. The amount of protection it offers depends on the tightness of the weave and the number of layers. While it will protect you, body armor will not make you invincible. You are still vulnerable in the areas not covered, and you can still be killed by the blunt force of a bullet, even if it does not penetrate the armor. Never take chances you would normally avoid just because you are wearing armor.

If you need to treat patients at a crime scene, you must preserve the chain of evidence needed for investigation and prosecution. A general rule is to avoid disturbing the scene unless absolutely necessary for medical care. Basic guidelines include:

◆ Never wipe away blood. It can be used as evidence.

◆ Touch only what you need to touch to provide patient care.

◆ Move only what you need to move to protect the patient and to provide proper care.

◆ Do not use the telephone unless the police give you permission to do so.

◆ Observe and document anything unusual at the scene.

◆ If possible, do not cut through holes in the patient's clothing. The holes could have been caused by bullets or other penetrating weapons.

◆ Do not cut through any knot in a rope or tie. Knots are used as evidence. Cut the rope or tie somewhere away from the knot.

◆ If the crime is rape, attempt to discourage the patient from washing, changing clothes, using the bathroom, or taking anything by mouth. Any of these actions could damage valuable evidence.

ON SCENE FOLLOW-UP

At the beginning of this chapter, you read that fire service First Responders were at a potentially dangerous scene. To see how chapter skills apply to this emergency, read the following. It describes how the call was completed.

SCENE SIZE-UP (CONTINUED)

My partner and I looked at each other. We decided to move our vehicle away from the front of the building and radio dispatch to double check that the police had given us the "OK" to enter the building.

We moved our vehicle out of sight, and waited. Finally, dispatch informed us that the police had secured the scene. Meanwhile, my partner tapped me on the shoulder and pointed to the house. I saw an officer standing outside the front door waving at us. We approached the scene, but with caution.

The officer told us that two brothers had been in a fist fight and that both had consumed

"too much nitwit." My partner smiled and told me that "nitwit" was street jargon for alcohol. One of the brothers said his hand went through a window. Both of the brothers appeared to be intoxicated.

INITIAL ASSESSMENT

After we put on eye protection and latex-free gloves, we introduced ourselves to Hubert, the brother who needed medical help. He seemed calm. Hubert had moderate bleeding from his left forearm. His airway and breathing were good. He denied any other injuries or falls.

PHYSICAL EXAMINATION

We decided to do a physical exam anyway. Sometimes people get into fights and get so excited that they don't know they're hurt (especially when they've had too much "nitwit"). Since there were two of us, my partner controlled the bleeding while I checked Hubert's head, neck, chest, abdomen, and extremities.

While we were waiting for an ambulance, Hubert's brother became loud and abusive to the police. We had the patient walk outside with us to get away from the potential danger. There we checked his pulse and respirations.

PATIENT HISTORY

Hubert admitted to drinking six or eight cans of beer before the fight started. He told us that he had asthma, but wasn't feeling any respiratory distress or problems. He took an inhaler for his asthma, but didn't have it with him. He continued to deny any injuries other than the cut to his forearm.

ONGOING ASSESSMENT

We reassessed Hubert's airway and breathing. They were okay. Hubert didn't have any changes in mental status. The bleeding was controlled and the bandages were secure. We didn't get to recheck the vitals before the ambulance arrived.

PATIENT HAND-OFF

We advised the EMTs:

"The patient's name is Hubert. He's a 22-year-old male who has sustained a laceration to his left forearm from going through a window. The wound has been dressed and bandaged. The bleeding was moderate and was easily controlled. Hubert has been drinking but has been alert and oriented throughout the call. His airway and breathing are good. He denies any other injuries. He has a history of asthma and uses an inhaler, which he doesn't have with him. The physical exam was negative for injuries anywhere other than his arm. His pulse is 82, strong, and regular. His respirations are 20 and adequate. We were about to recheck his vitals as you pulled up."

Life safety—first yours and then your patient's —are your top priorities. Without them, you cannot be an effective First Responder. Throughout this textbook, you will find reminders about taking BSI and other safety precautions. Make note of them.

Chapter Review

Unfortunately for all rescue workers, television and movies have led the public to believe that it is okay for fire and EMS personnel to rush into calls without regard for their safety, fostering a "hero mentality." Nothing could be further from the truth. As a fire service First Responder, you are there to keep the problem from getting worse. If you become injured or ill, you become part of that problem, which helps no one.

One of the most important factors to be considered in the first 10 minutes of a call is your own personal safety. Practice body substance isolation (BSI) precautions on every call. And always remain alert for violence, hazardous materials, and other unsafe conditions.

You owe it to yourself to stay safe when responding to every call, regardless if it is a structure fire or an EMS call. You also owe it to your family, your department, and to the community you serve. Although unforeseen events do occur on every emergency scene, it is your personal responsibility—not your company officer's, not your fire chief's, and not your partner's—to take the necessary precautions to protect yourself. It is a foolhardy person who declares it "my duty" to help a citizen without first protecting himself and the other rescuers with him.

Do not underestimate the importance of the information in this chapter. There will be reminders in every chapter in this book.

FIRE COMPANY REVIEW

Page references where answers may be found or supported are provided at the end of each question.

Section 1

1. What are three techniques you can use to help you avoid responses like nausea or fainting in an emergency situation? (p. 17)

2. What are the five stages of the grieving process? (p. 21)

3. What can you do to help a dying patient, in addition to providing medical care? (pp. 21–22)

4. What are five signs of chronic stress and burnout? (pp. 17–18)

5. What are some of the negative feelings a First Responder's family may have about the job? (p. 19)

6. What are three examples of situations that may cause critical incident stress? (p. 20)

7. What is a critical incident stress debriefing (CISD)? (p. 20)

Section 2

8. How does an infectious disease spread from person to person? (pp. 22–23)

9. What equipment is needed to take BSI precautions? (p. 26)

Section 3

10. What general rule applies to all unsafe emergency scenes? (p. 27)

EMS Safety: Techniques and Applications. Federal Emergency Management Agency/U.S. Fire Administration, FA-144. Washington, D.C.: April 1994.

Kübler-Ross, E. *On Death and Dying.* New York: Collier Books/Macmillan, 1969.

West, K.H. *Infectious Disease Handbook for Emergency Care Personnel.* Philadelphia: J.B. Lippincott, 1987.

Wieder, M.A., ed. *Fire Department Pumping Apparatus Driver/Operator,* Eighth Edition. Stillwater, OK: International Fire Service Training Association, 1999.

Wieder, M.A., ed. *Hazardous Materials for First Responders,* Second Edition. Stillwater, OK: International Fire Service Training Association, 1994.

Wieder, M.A., ed. *Principles of Extrication.* Stillwater, OK: International Fire Service Training Association, 1990.

Zimmerman, L., M. Neuman, and D. Jurewicz. *Infection Control for Prehospital Providers,* Second Edition. Grand Rapids, MI: Mercy Ambulance, 1993.

THE CISM PROCESS

CRITICAL INCIDENT OCCURS

Produces strong emotional response in emergency workers

↓

NEED FOR CISM RECOGNIZED

Usually the company officer arranges for CISM, but any emergency worker involved in incident can request one

↓

CISD SCHEDULED

Usually held within 24 to 72 hours of incident

↓

THE CISD

Participants include those involved in incident, trained peer counselors, mental health professionals; process involves the following seven phases:

↓

PHASE 1: INTRODUCTION

Sets goals for the CISD. Assures confidentiality.

↓

PHASE 2: FACTS

Sets out details of what occurred at the incident

↓

PHASE 3: FEELINGS

Encourages participants to explore feelings the incident raised in them

↓

PHASE 4: SYMPTOMS

Encourages participants to note any physical reactions the incident may have caused in them

↓

PHASE 5: TEACHING

Allows professionals to help participants sort through feelings; provides opportunity to reinforce that extreme reactions are normal in such situations

↓

PHASE 6: RE-ENTRY

Offers suggestions for coping after the CISD; may include an action plan with goals and activities to reduce stress

↓

PHASE 7: FOLLOW-UP

Explores how participants are coping months or weeks later

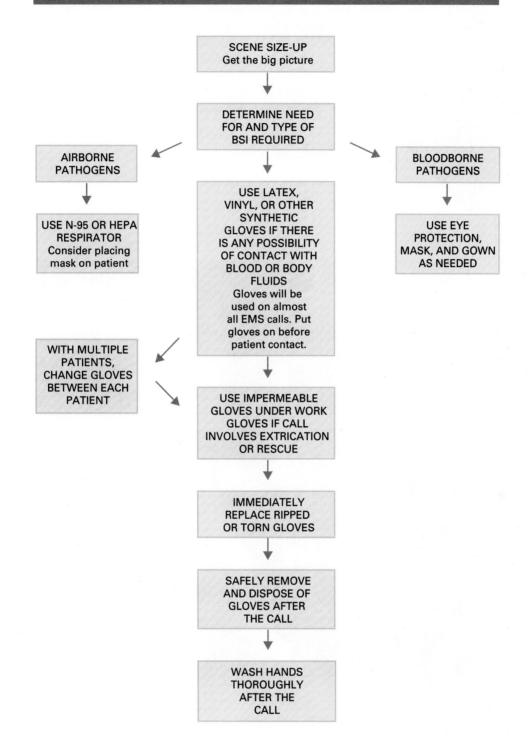

Legal and Ethical Issues

*I*NTRODUCTION   *Legal and ethical issues—both personal and professional—are a critical part of a fire service First Responder's life. You may already have some questions. For example, should you stop to treat an accident victim when you are off duty? If you are dispatched to a car crash and pass a house on fire, should you stop? Can patient information be released to an attorney over the phone? May a child be treated without a parent's consent?*

This chapter will help to answer your questions. It will describe your scope of practice and what it means to have a duty to act. It will define patient consent and explain advance directives. Finally, it will provide you with an overview of several legal issues that may affect you as a First Responder.

OBJECTIVES

Cognitive, affective, and psychomotor objectives are from the U.S. DOT's "First Responder: National Standard Curriculum." Enrichment objectives, if any, identify supplemental material.

Cognitive

1-3.1   Define the First Responder scope of care. (pp. 38–39)

1-3.2   Discuss the importance of Do Not Resuscitate [DNR] (advance directives) and local or state provisions regarding EMS application. (pp. 40–41)

1-3.3   Define consent and discuss the methods of obtaining consent. (pp. 39–40)

1-3.4   Differentiate between expressed and implied consent. (p. 39)

1-3.5   Explain the role of consent of minors in providing care. (pp. 39–40)

1-3.6   Discuss the implications for the First Responder in patient refusal of transport. (pp. 41, 43)

1-3.7   Discuss the issues of abandonment, negligence, and battery and their implications to the First Responder. (pp. 43–44)

1-3.8   State the conditions necessary for the First Responder to have a duty to act. (pp. 43–44)

1-3.9   Explain the importance, necessity and legality of patient confidentiality. (p. 44)

1-3.10 List the actions that a First Responder should take to assist in the preservation of a crime scene. (pp. 44–45)

1-3.11 State the conditions that require a First Responder to notify local law enforcement officials. (pp. 39–40, 44, 45)

1-3.12 Discuss issues concerning the fundamental components of documentation. (pp. 43, 44, 45)

Affective

1-3.13 Explain the rationale for the needs, benefits and usage of advance directives. (pp. 40–41)

1-3.14 Explain the rationale for the concept of varying degrees of DNR. (p. 41)

Psychomotor

No objectives are identified.

Enrichment

◆ Explain the "Good Samaritan" laws and how they affect First Responders. (p. 44)

◆ Describe a First Responder's role in the process of organ retrieval. (pp. 45–46)

 ON SCENE

DISPATCH

Our engine company was responding to a report of an injured child. I remember thinking how calls involving kids often upset me. I briefly pictured my two children and how much I loved them. I hoped we could help this child.

SCENE SIZE-UP

We noticed the patient sitting up against a tree and crying. It appeared he had fallen off his bike.

After taking BSI precautions, we hopped off the rig to find a police officer kneeling next to the boy. The officer motioned for us to approach. He said, "Josh here was riding his bike when his dog Musket ran out in front of him. He fell off his bike, using his arms to break his fall. He did not hit his head." He went on to tell us that he witnessed the incident, and that the boy had been wearing a helmet.

We introduced ourselves to Josh. He said he was nine years old. While the rest of my engine company performed an initial assessment, I spoke to the police officer to see if contact had been made with the boy's parents.

Why is the First Responder asking about the child's parents? Can't they just treat the child's injuries? These questions revolve around the issue of consent—one of many legal and ethical issues you will face as a First Responder. As you read Chapter 3, consider how you might answer them.

1 SCOPE OF CARE

Emergency care has changed a lot since its early days. One improvement has been in the quality of EMS training. People expect to receive the best possible care from competent and compassionate First Responders.

Legal Duties

Every state and province determines a First Responder's scope of practice, or **scope of care** (actions that are legally allowed). All First Responders are responsible for providing care to the best of their ability. For example, providing CPR when needed is within your scope of care. But opening a patient's chest with a pen knife to provide direct cardiac massage is probably not. Provide the emergency care that is defined by your scope of care. Doing otherwise is illegal. It also may be dangerous to your patients.

The law is enhanced by your local medical director. In fact, your legal right to act as a First Responder depends on medical oversight. In the United States, law is further enhanced by the "First Responder: National Standard Curriculum," which is established by the U.S. Department of Transportation.

When providing emergency care, you should:

◆ Follow standing orders and protocols as approved by local medical direction. (Do not

follow what your buddy in another fire department says he is allowed to do.)

◆ Consult medical direction. Use a phone or radio any time there is a question about the scope of care.

◆ Communicate clearly and completely with medical direction.

◆ Follow orders the medical director gives.

Ethical Responsibilities

A code of ethics is a list of the rules for ideal conduct. Basically, if you make the welfare of a patient a priority, you rarely will do anything unethical. Your ethical responsibilities are:

◆ First, do no more harm.

◆ Make the physical and emotional needs of the patient a priority. Serve those needs with respect for human dignity and with no regard to nationality, race, sex, creed, or status.

◆ Practice your skills to the point of mastery. Show respect for the competence of other medical workers.

◆ Continue your education and take refresher courses. Stay on top of changes in EMS. Help define and uphold professional standards.

◆ Critically review your performance. Seek ways to improve response time, patient outcome, and communication.

◆ Report with honesty. Hold in confidence all information obtained in the course of your work, unless required by law to share it.

- Work in harmony with other First Responders, EMTs, and other members of the health care team.

COMPANY OFFICER'S NOTE

There is no right way to do a wrong thing. As a guide to whether or not an action is ethical, ask these questions.

- Is what I am about to do legal?
- Is what I am about to do balanced? In other words, will it create a win-win situation or will it unnecessarily create a win-lose situation?
- How will it make me feel about myself? How will I feel if my decision were published in the local newspaper? Basketball coach John Wooden said it best: "There is no pillow as soft as a clear conscience."

SECTION 2 PATIENT CONSENT AND REFUSAL

Patient Competence

A **competent** adult is one who is of proper mind and able to make an informed decision about medical care. He understands your questions. He understands how he will benefit if he allows you to treat him. He also understands what bad things could happen if he decides against emergency care. A patient must be competent in order to refuse treatment. So determine competence in every adult you intend to treat.

Generally, consider an adult incompetent if he or she:

- Is under the influence of alcohol or drugs.
- Has an altered mental status.
- Has a serious illness or injury that could affect judgment.
- Is mentally ill or mentally retarded.

Patient Consent

By law, a patient must first **consent,** or give permission, before you can provide emergency care. In order for that consent to be valid, the patient must be competent and the consent must be informed. It is your responsibility, therefore, to fully explain the care you plan to give, as well as the related risks.

There are two general types of consent: **expressed consent** and **implied consent.**

Expressed Consent

Expressed consent may consist of verbal consent, a nod, or an affirming gesture from a competent adult. To get it, you must explain your plan for emergency care in terms the patient can understand. Be sure to include the risks, too. In other words, the patient needs a clear idea of all the factors that would affect a reasonable person's decision to either accept or refuse treatment.

You must get a responsive, competent adult's expressed consent before you render treatment. To do so, first tell the patient who you are. Identify your level of training. Then carefully explain your plan for emergency care. Make sure you identify both the benefits and the risks. To make sure the patient understands, question him or her briefly. If the replies are appropriate, you may proceed with treatment. If not, the consent you may have to use is implied consent.

Implied Consent

In an emergency, when an unresponsive patient is at risk of death, disability, or deterioration of health, the law assumes that he or she would agree to emergency care. This is called "implied consent." It applies when you assume that a patient who cannot consent to life-saving care, would if he or she were able to. It applies to patients who initially refuse care but then become unresponsive. It also applies to patients who are not competent to refuse care.

Children and Mentally Incompetent Adults

Most areas define a **minor** as any person under the age of 18 or 21. A parent or legal guardian must give consent before you can treat a minor. The same is true for a mentally incompetent adult. However, if a life-threatening condition exists and the parent or guardian is not available, implied consent may be used to provide emergency care.

An **emancipated minor** is one who is married, pregnant, a parent, in the armed forces, or financially independent and living away from home with permission of the courts. You do not need the consent of a parent or legal guardian to treat this patient. You only need the patient's consent.

If it should ever happen that a parent refuses emergency care for a child who needs it, call law enforcement or your jurisdictional child welfare agency.

Advance Directives

There may be a time when you are called to treat a terminally ill patient. He may ask you to let him die if his heart or lungs stop working. Legally, if he is competent, he has the right to make this request.

An **advance directive** is written in advance of an emergency. It must be signed by both the patient and a physician. Legal in many areas, it is

Figure 3-1   Example of an EMS "Do Not Resuscitate" order.

PREHOSPITAL DO NOT RESUSCITATE ORDERS

ATTENDING PHYSICIAN

In completing this prehospital DNR form, please check part A if no intervention by prehospital personnel is indicated. Please check Part A and options from Part B if specific interventions by prehospital personnel are indicated. To give a valid prehospital DNR order, this form must be completed by the patient's attending physician and must be provided to prehospital personnel.

A) _____ **Do Not Resuscitate (DNR):**
No Cardiopulmonary Resuscitation or Advanced Cardiac Life Support be performed by prehospital personnel

B) _____ **Modified Support:**
Prehospital personnel administer the following checked options:
_____ Oxygen administration
_____ Full airway support: intubation, airways, bag/valve/mask
_____ Venipuncture: IV crystalloids and/or blood draw
_____ External cardiac pacing
_____ Cardiopulmonary resuscitation
_____ Cardiac defibrillator
_____ Pneumatic anti-shock garment
_____ Ventilator
_____ ACLS meds
_____ Other interventions/medications (physician specify)
_____

Prehospital personnel are informed that (print patient name)_____
should receive no resuscitation (DNR) or should receive Modified Support as indicated. This directive is medically appropriate and is further documented by a physician's order and a progress note on the patient's permanent medical record. Informed consent from the capacitated patient or the incapacitated patient's legitimate surrogate is documented on the patient's permanent medical record. The DNR order is in full force and effect as of the date indicated below.

_____   _____

Attending Physician's Signature

                                           _____

_____   _____

Print Attending Physician's Name   Print Patient's Name and Location
                                           (Home Address or Health Care Facility)

_____

Attending Physician's Telephone

_____   _____

Date   Expiration Date (6 Mos from Signature)

commonly called a **Do Not Resuscitate (DNR) order.** It documents the wish of the chronically or terminally ill patient not to be resuscitated. In some areas it also allows the fire service First Responder to legally withhold resuscitation.

When you are given an advance directive, determine to the best of your ability if it is valid. Usually it is accompanied by a doctor's written instructions. Check to see that they are written clearly and concisely. They should be typed or written legibly on professional letterhead. Phrases like "no heroics" or "no extraordinary treatment" are not clear enough to be legal.

In many areas, a standard form is used for a DNR order (Figure 3-1). In some areas, the law also requires patients to wear a DNR insignia on their bodies where emergency personnel will be sure to find it.

By its very nature, a DNR order is best suited to a hospital or nursing home. There all personnel know the patient and his or her doctor. If the DNR order is needed, it can be found and verified quickly. However, in the field there may be problems. In many areas, for example, a second physician must verify the patient's condition. This can be difficult in an emergency, even if the DNR order is on hand. Another problem is the time it takes to verify a DNR order. Precious, life-saving moments can be lost.

There are varying degrees of DNR orders. For example, one may explain that a patient wants all medical care except long-term life support. Another might say that the patient specifically does not allow the use of a respirator.

If you are ever in doubt about the validity of an advance directive, you must begin full resuscitation immediately. However, always consult your medical director or the emergency department physician before you decide either to follow or put aside a DNR order.

Be sure to review and follow the law and local protocols on this issue.

Patient Refusal

A competent adult has the right to refuse treatment for himself or his child. Under the law, he must first be informed of the treatment, fully understand it, and completely comprehend the risks involved in refusing it. He may refuse verbally, by pulling away, shaking his head, gesturing, or pushing you away.

A competent adult has the right to withdraw from treatment after it has started. This is true for a patient who initially gave consent but then changes his mind. It is also true for the patient who at first was unresponsive, but then wakes and asks you to stop. Note that you may be required to provide some level of care while withholding the procedure that is refused.

A patient's legal refusal of treatment or transport must follow the rules of expressed consent. That is, the patient must be mentally competent and of legal age. The patient also must be informed of all risks in language he or she can fully understand. When in doubt, always err in favor of providing care.

Make every reasonable effort to persuade the patient or guardian to give consent for care. If he

Figure 3-2 Example of a patient refusal statement.

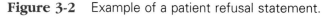

REFUSAL OF TREATMENT AND TRANSPORTATION

I, THE UNDERSIGNED HAVE BEEN ADVISED THAT MEDICAL ASSISTANCE ON MY BEHALF IS NECESSARY AND THAT REFUSAL OF SAID ASSISTANCE AND TRANSPORTATION MAY RESULT IN DEATH, OR IMPERIL MY HEALTH. NEVERTHELESS, I REFUSE TO ACCEPT TREATMENT OR TRANSPORT AND ASSUME ALL RISKS AND CONSEQUENCES OF MY DECISION AND RELEASE GOLD CROSS AMBULANCE COMPANY AND ITS EMPLOYEES FROM ANY LIABILITY ARISING FROM MY REFUSAL.

_____
SIGNATURE OF PATIENT

_____
WITNESSED BY

_____
DATE SIGNED

EMS PATIENT REFUSAL CHECKLIST

PATIENT NAME: _____   AGE: _____

LOCATION OF CALL: _____   DATE: _____

AGENCY INCIDENT #: _____   AGENCY CODE: _____

NAME OF PERSON FILLING OUT FORM: _____

I.   ASSESSMENT OF PATIENT (Circle appropriate response for each item)

    1.   Oriented to: Person?   Yes   No

                        Place?   Yes   No

                        Time?   Yes   No

                        Situation?   Yes   No

    2.   Altered level of consciousness?   Yes   No

    3.   Head injury?   Yes   No

    4.   Alcohol or drug ingestion by exam or history?   Yes   No

II. PATIENT INFORMED (Circle appropriate response for each item)

    Yes   No   Medical treatment/evaluation needed

    Yes   No   Ambulance transport needed

    Yes   No   Further harm could result without medical treatment/
                    evaluation

    Yes   No   Transport by means other than ambulance could be
                    hazardous in light of patient's illness/injury

    Yes   No   Patient provided with Refusal Information Sheet

    Yes   No   Patient accepted Refusal Information Sheet

III. DISPOSITION

    _____   Refused all EMS services

    _____   Refused field treatment, but accepted transport

    _____   Refused transport, but accepted field treatment

    _____   Refused transport to recommended facility

    _____   Patient transported by private vehicle to _____

    _____   Released in care or custody of self

    _____   Released in care or custody of relative or friend

              Name: _____   Relationship: _____

    _____   Released in custody of law enforcement agency

              Agency: _____   Officer: _____

    _____   Released in custody of other agency

              Agency: _____   Officer: _____

IV. COMMENTS: _____

_____

_____

Figure 3-3   Example of a patient refusal checklist.

still refuses, insist that additional EMS personnel evaluate the patient.

Complete and accurate records are key to protecting yourself from liability. So before you leave the scene:

◆ *Try again to persuade the patient to accept treatment or transport.* Tell him clearly why it is essential. Be especially clear when you explain what could happen if he refuses care. Write down what you tell him. Then have the patient read it aloud to see if he understands.

◆ *Be sure the patient is able to make a rational, informed decision.* Note that a patient who is seriously ill or injured may only appear to be competent. Such a patient may be emotionally, intellectually, or physically impaired and may not be able to absorb all the information you give.

◆ *Consult medical direction* as required by local protocol.

◆ *Have the patient sign a refusal or "release from liability" form* (Figures 3-2 and 3-3). It must be signed by the patient and a witness. If the patient refuses to sign, indicate that on the form and have a witness sign it. Many areas use official documents for patients to sign. Check local protocols.

◆ *Before you leave, encourage the patient to seek help if certain symptoms develop.* Be specific. Avoid using terms the patient may not understand. For example, you might tell a patient to go to the emergency department "if you get a burning pain in your stomach" or "if you start having shortness of breath." Then document the fact that you did.

◆ *Advise the patient to call EMS again immediately* if he changes his mind or if symptoms become worse.

OTHER LEGAL ASPECTS OF EMERGENCY CARE

Assault and Battery

There is no single definition of assault and battery. Traditionally, threatening physical harm is **assault.** Actual unlawful physical contact is **battery.** In the context of emergency care, you can be charged with assault and battery if you touch a patient's body or clothing without first getting consent.

Abandonment and Negligence

Simply stated, **abandonment** means you stopped providing care to a patient without making sure that the same or better care would be provided. Under the law, once you begin emergency care, you must continue until another health care professional with at least as much expertise as you takes over or until the police order you to leave the scene.

Negligence is defined as carelessness, inattention, disregard, inadvertence, or oversight that was accidental but avoidable. You may be charged with negligence if your care deviates from the accepted standard of care and results in further injury to the patient.

To establish negligence, the court must decide that all four of the following are true:

◆ *First Responder had a duty to act.* The concept known as duty to act refers to your contractual or legal obligation to provide care. That is, while you are on duty, you must care for a patient who needs it and consents to it.

◆ *There was a breach of duty.* A breach of duty exists when a First Responder either fails to act or fails to act appropriately. That is, the First Responder violated the standard of care reasonably expected of a First Responder with similar background and training.

◆ *Patient was injured physically or psychologically.*

◆ *First Responder caused the injury.* It must be proven that the First Responder's breach of duty is what caused or contributed to the patient's injury.

Note that the duty to act also means that you must render care to a patient to the best of your ability. You must follow accepted guidelines for care, and you must act as any other prudent person would in the same situation.

In some cases a duty to act refers to an implied contractual or legal obligation. For example, a patient may call for EMS. The dispatcher confirms that help will be sent. All EMS members who respond—including fire service First Responders—then have a legal obligation to provide treatment to the patient.

In most areas, you do not have a duty to act when you are off duty or driving an emergency vehicle outside your company's service area. (Check your state or provincial laws.) However, you may feel a certain moral or ethical obligation to help. In such cases, take extra steps to protect yourself. Carefully document all aspects of the call and treatment you give, including a patient's refusal of care. If you come upon an emergency in another jurisdiction, notify authorities in that area.

In general, your best defense against negligence is to have a professional attitude, to provide a consistently high standard of care, and to correctly and completely document the care you provide.

Confidentiality

A patient's history, condition, and emergency care are confidential. To release this information, you must have a written form signed by the patient or legal guardian. Never release any patient information unless you are authorized to do so in writing.

By law, you are allowed to release information without permission only if:

◆ Another health care provider needs it in order to continue medical care.

◆ You are requested by the police to provide it as part of a crime investigation. In some areas, for example, the law requires you to report rape, abuse, gunshot wounds, and certain other crimes.

◆ You are required by legal subpoena to provide it in court.

COMPANY OFFICER'S NOTE

It is fairly common for firefighters to "cuss and discuss" a call back at the firehouse. Technically, they can breach confidentiality if a patient's identity is revealed during such a discussion. Be sure to caution your personnel about using the name of a patient during a critique of a "good call."

Good Samaritan Laws

Many states and provinces have "Good Samaritan" laws. The first of these laws was enacted in 1959 in California. It was designed to protect doctors who render emergency care from civil suits. Most areas now have laws of their own, some of which cover EMS personnel. *Be sure to learn your local laws.*

Generally, these laws protect a First Responder from liability for acts performed in good faith unless those acts are grossly negligent. Under these laws, the person suing must prove that emergency care was a breach of the standard of care. **Standard of care** is defined as the care that would be expected to be provided to the same patient under the same conditions by another First Responder who had received the same training. (This is referred to as the "reasonable man" test.)

If you are sued and the case goes to court, a tort proceeding will be held. This is a civil court action, not a criminal one. It determines whether or not the natural rights of an individual have been violated. In a tort proceeding, it must be proven that you are guilty of gross negligence (conduct that deviates so greatly from acceptable standards that a reasonable First Responder would find it offensive).

A Good Samaritan law does not prevent you from being sued. But it may give you some protection against losing the lawsuit, if you have performed according to the standard of care for a First Responder. So while on and off duty, your best defense against lawsuits is prevention. Always render care to the best of your ability. Do no more or less than your scope of care allows. If you keep your patient's best interests in mind, you will seldom—if ever—go wrong.

Preservation of Evidence

Whenever fire service personnel are called to a potential crime scene, dispatch should also notify the police. In general, a **potential crime scene** is any scene that may require police support. That includes a potential or actual suicide, homicide, drug overdose, domestic dispute, abuse, hit-and-run, riot, robbery, or any scene involving gunfire or a weapon.

Your first concern should always be your own safety. If you suspect that a crime is in progress or a criminal is active at the scene, do not try to provide care to any patient. Wait until the police arrive and tell you that the scene is safe. Once the scene is safe, your priority is patient care.

When on scene, do not disturb any item that may be evidence. Basic guidelines include:

♦ Observe and document anything unusual at the scene.

♦ Touch only what you need to touch.

♦ Move only to provide emergency care and only what you need to move to protect yourself, other rescuers, and the patient. If possible, remember the position of anything you move.

♦ Do not use the telephone unless the police give you permission to do so. They may wish to find out who the last caller was.

♦ Move the patient only if he or she is in danger or must be moved in order for you to provide emergency care.

♦ If possible, do not cut through holes in the patient's clothing. They may have been caused by bullets or stabbing.

♦ Do not cut through any knot in a rope or tie. Knots are often used as evidence.

♦ If the crime is rape, attempt to discourage that patient from washing, changing clothes, using the bathroom, or taking anything by mouth. Doing any of these things can destroy valuable evidence.

♦ Communicate your actions and observations of the scene to law enforcement personnel.

Special Documentation

In general, physicians and nurses must report suspected child, elder, and spouse abuse. Some areas require others—such as teachers and First Responders—to report them as well. Related laws often grant immunity for libel, slander, or defamation of character as long as the report is made in good faith.

Fire service personnel may be required to report an injury that may be the result of a crime, such as gunshot wounds, knife wounds, and poisonings. Your state or province may also require you to report an injury that you suspect was caused by sexual assault.

In some areas fire service personnel must report all suspected infectious disease exposure. That includes TB, hepatitis B, and AIDS. Other situations to report may include use of restraints on a patient, attempted suicides, dog bites, and burn injuries. Learn your local requirements.

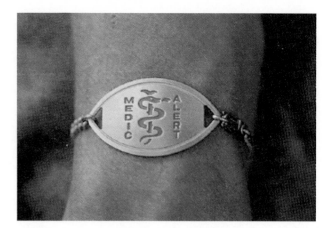

Figure 3-4   A medical identification tag.

Special Situations

Medical Identification Tags

Some patients may wear or carry a **medical identification tag** (Figure 3-4). Such tags may be found on bracelets, necklaces, or cards carried in a wallet. They identify a specific medical condition such as an allergy, epilepsy, or diabetes. Look for them whenever you examine a patient. Many list a phone number you can call for detailed information. Note that going through a patient's wallet or purse should be done in front of witnesses and only when absolutely necessary.

Donor and Organ Harvesting

In general, organs may be donated only if there is signed permission to harvest them. A signed donor card is a legal document. So is the sticker on some driver licenses (Figure 3-5).

Figure 3-5   An organ donor sticker on a driver's license. (*Source: Don Hall Productions*)

Treat a potential organ donor the same way you would treat any other patient. Remember, the person is a patient first, an organ donor last. So, in addition to providing the appropriate emergency care, you can:

◆ *Identify the patient as a potential donor.* Patients who are about to die or who have died within hours may be organ donors. In each case, the hospital staff and the patient's family must make the ultimate decision.

◆ *Communicate with medical direction.* You can begin the process by alerting the EMS personnel who take over patient care. They will in turn alert the hospital emergency staff.

◆ *Provide life-saving emergency care,* such as CPR. It will help to maintain vital organs. This is best accomplished by treating every patient equally well.

ON SCENE FOLLOW-UP

At the beginning of this chapter, you read that a nine-year-old boy fell off his bike. To see how chapter information applies to this emergency, read the following. It describes how the call was completed.

INITIAL ASSESSMENT

We confirmed that the patient's ABCs—airway, breathing, and circulation—were adequate. There was no obvious external bleeding. Our general impression was that of a nine-year-old boy who had fallen off his bike at a slow speed. His chief complaint was pain to his arm.

After the dispatcher contacted Josh's mother, she quickly arrived on scene. She gave consent for his care, and our engine company continued an assessment on Josh.

PHYSICAL EXAMINATION

The mother's presence calmed the boy. We checked Josh carefully, examining his head, neck, chest, abdomen, and extremities. The only sign of injury was to his arm. Though we had been told that Josh's head was not implicated, we checked his helmet for damage anyway. We found none. His pulse was 88, strong and regular. Respirations were 18 and adequate. Blood pressure was 108/72. My partner manually stabilized the arm to prevent further injury. I spoke to Josh's mother about his medical history.

PATIENT HISTORY

I was told that Josh was basically healthy, with no allergies. He had a heart murmur since birth, but no related problems or complications. He does not take medications. He had a full lunch consisting of macaroni and cheese.

ONGOING ASSESSMENT

Josh continued to be responsive. He responded well to his mother, an important sign in a child. He told her he was wondering if he was going to have a cast on his arm for his friends to sign. His ABCs were still fine. His pulse, respirations, and skin color and temperature were unchanged. We determined that Josh needed to be seen at the hospital. His mother said she would like an ambulance to transport him to the emergency department.

While our captain was determining the ambulance's ETA, I went back to our engine's medical cabinet and found one of the stuffed animals we carry for calls such as this. I returned to where Josh was being treated and presented the toy to him. He immediately smiled and asked, "Can I keep it?"

I replied, "Sure, but you have to promise me one thing. You have to give him a good name, OK?" Josh's smile turned into a toothy grin as he nodded his head.

PATIENT HAND-OFF

We introduced Josh and his mom to the EMTs who would be taking him to the hospital. We told the EMTs that Josh had fallen off his bike and then went on with our report:

"He complains of pain to his right arm, which we are now stabilizing. We have not taken spinal precautions because the incident was witnessed, and there was no involvement of the head. He was wearing a helmet, which we checked to find no damage. Josh never lost consciousness and has no other complaints. His pulse is 88, respirations 18, BP 108/72. He had macaroni and cheese for lunch. He has a heart murmur but no problems with it. No meds, no allergies."

The EMTs thanked us and took over care. Josh asked for us to stay because he liked to look at the fire truck, and it seemed to keep him calm. I manually stabilized Josh's arm while an EMT-Basic applied a splint. Josh's mom thanked us as she got into the ambulance with her son.

Josh did indeed need a cast for his broken arm. He (and "Cheesey," his stuffed animal friend) was able to pay us a visit at the firehouse where we could all sign his cast. While this might not have been a critical emergency, just by stabilizing Josh's arm, we prevented further injury or problems that could have been with him his whole life.

> *Consent is one of many legal and ethical issues you will face as a First Responder. Learn the laws related to you and your EMS system.*

Chapter Review

Legal issues are a large part of the first few minutes of a call. For example, it is during this time that you must obtain consent from your patient. You also may be faced with legal documents such as "DNR" orders.

Consider this scenario: You walk into a house and find a patient lying on a couch. The relative who meets you says, "I think he's dead. He's had cancer, and we got this paper from his doctor so he can die in peace." Picture the situation. You

have a patient who needs CPR, and a relative who presents you with some type of form. Every second you spend trying to figure out what it is and what to do about it is time you could be using to help the patient.

This does not have to happen to you. Learn well the regulations that affect your duties as a First Responder before you are faced with a situation like this.

FIRE COMPANY REVIEW

Page references where answers may be found or supported are provided at the end of each question.

Section 1

1. What is a First Responder's scope of care? (p. 38)

Section 2

2. What are two types of consent? Define each one. (p. 39)

3. What is a DNR order? What should a First Responder do if presented with one? (pp. 40–41)

4. How should you handle a patient's refusal of treatment? (pp. 41, 43)

Section 3

5. What must happen in order for a First Responder to be liable for abandonment or negligence? (pp. 43–44)

6. What does it mean for a First Responder to have a duty to act? (p. 43)

7. Under what conditions may a First Responder release confidential patient information? (p. 44)

8. What are some ways a First Responder can help to preserve evidence at a crime scene? (pp. 44–45)

9. What are the situations a First Responder may be required to report? (p. 45)

RESOURCES TO LEARN MORE

George, J.E. *Law and Emergency Care.* St. Louis: C.V. Mosby, 1980.

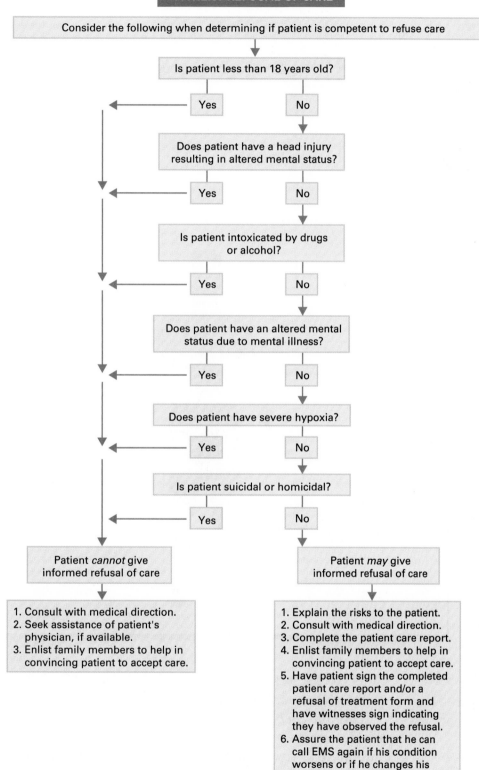

PATIENT REFUSAL OF CARE

Consider the following when determining if patient is competent to refuse care

Is patient less than 18 years old?

Yes — No

Does patient have a head injury resulting in altered mental status?

Yes — No

Is patient intoxicated by drugs or alcohol?

Yes — No

Does patient have an altered mental status due to mental illness?

Yes — No

Does patient have severe hypoxia?

Yes — No

Is patient suicidal or homicidal?

Yes — No

Patient *cannot* give informed refusal of care

1. Consult with medical direction.
2. Seek assistance of patient's physician, if available.
3. Enlist family members to help in convincing patient to accept care.

Patient *may* give informed refusal of care

1. Explain the risks to the patient.
2. Consult with medical direction.
3. Complete the patient care report.
4. Enlist family members to help in convincing patient to accept care.
5. Have patient sign the completed patient care report and/or a refusal of treatment form and have witnesses sign indicating they have observed the refusal.
6. Assure the patient that he can call EMS again if his condition worsens or if he changes his mind about treatment.

CHAPTER 4

The Human Body

I NTRODUCTION   *As a fire service First Responder, you must be able to recognize illness and injury quickly and know how to care for each. You also must be able to tell other health care providers about a patient's problem quickly and accurately. In order to do all this, you need a solid foundation of basic knowledge. In this chapter, you will study **anatomy** (the structure of the body) and **physiology** (how the body works). You also will be introduced to common anatomical terms.*

Cognitive, affective, and psychomotor objectives are from the U.S. DOT's "First Responder: National Standard Curriculum." Enrichment objectives, if any, identify supplemental material.

Cognitive

1-4.1 Describe the anatomy and function of the respiratory system. (p. 59)

1-4.2 Describe the anatomy and function of the circulatory system. (pp. 59, 61)

1-4.3 Describe the anatomy and function of the musculoskeletal system. (pp. 56, 59)

1-4.4 Describe the components and function of the nervous system. (pp. 61, 63)

Affective

No objectives are identified.

Psychomotor

No objectives are identified.

Enrichment

◆ Use anatomical terms correctly, including terms of position, direction, and location. (pp. 52–53)

◆ Describe the main body cavities. (p. 53)

◆ Describe skin and its components. (pp. 63, 66)

◆ Describe the digestive system and its components. (p. 66)

◆ Describe the urinary system and its components. (pp. 66–67)

◆ Describe the endocrine system and its components. (p. 67)

◆ Describe the reproductive system and its components. (p. 67)

ON SCENE

DISPATCH

Our engine company was dispatched to the airport for a "fall injury." A man had missed a step on an escalator, fell, and hurt his right leg.

SCENE SIZE-UP

Upon arrival, we took BSI precautions, simultaneously scanning the crowd, which was standing and moving around the patient. There were no problems. The patient was clear of the escalator. I made a mental note to check with the patient about the mechanism of injury. Did he fall just a step or two or down the whole flight?

INITIAL ASSESSMENT

The man was alert. He denied any loss of consciousness or injury to his head or spine. He had no breathing problems and no obvious bleeding. He told us that his right shin bone hurt near his ankle.

PHYSICAL EXAMINATION

We conducted a head-to-toe exam, finding no injuries other than the one to the right lower leg. It was deformed and swollen, but there was no bleeding. We stabilized the leg manually to prevent further damage. We checked his pulse and found it to be 88, strong, and regular. His respirations were 16, regular and deep. Blood pressure was 110/82. We checked to make sure there were pulses, movement, and sensation below the injury site.

SECTION 1 — ANATOMICAL TERMS

Effective communication at the scene of a fire or other major disaster relies on the proper use of terminology. This same principle is true for describing the location of a patient's pain or bleeding. Proper use of key topographical terminology is very important to documenting and communicating with other members of the health-care team.

Terms of position include the following (Figure 4-1):

◆ **Anatomical position.** In this position, a patient's body stands erect with arms down at the sides, palms facing you. "Right" and "left" always refer to the patient's right and left.

◆ **Supine position.** The patient is lying face up on his or her back. (Think of the supine position as "stomach up.")

◆ **Prone position.** The patient is lying face down on his or her stomach.

◆ **Lateral recumbent position.** In this position, the patient is lying on the left or right side. This is also known as the "recovery position."

Terms of direction and location are as follows:

◆ **Superior** means toward, or closer to, the head. **Inferior** means toward, or closer to, the feet.

◆ **Anterior** is toward the front. **Posterior** is toward the back.

◆ **Medial** means toward the midline, or center of the body. **Lateral** refers to the left or right of (away from) the midline.

Figure 4-1a   Supine position.

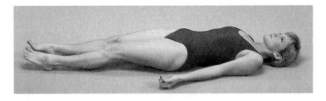

Figure 4-1b   Prone position.

Figure 4-1c   Right lateral recumbent position.

Figure 4-1d   Left lateral recumbent position.

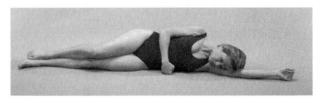

◆ **Proximal** means close, or near the point of reference. **Distal** is distant, or far away from the point of reference. The point of reference is usually the torso. For example, a wound to the forearm is proximal to the wrist because it is closer to the torso than the wrist. That same wound is distal to the elbow because it is farther away from the torso than the elbow.

◆ **Superficial** is near the surface. **Deep** is remote, or far from the surface.

◆ **Internal** means inside. **External** means outside.

Anatomical regions and topography are the internal and external landmarks of the body (Figures 4-2 and 4-3). During assessment of a patient, refer to these landmarks. They will help make the description of a patient's condition clear to others, particularly when you use a radio.

The organs of the body are located in certain body cavities (Figure 4-4). The main body cavities include:

◆ ***Thoracic cavity.*** One of the most protected cavities of the body, it is also called the chest cavity. This is where the lungs and heart are found. The diaphragm, a muscle that moves up and down during respiration, separates this cavity from the abdomen.

◆ ***Abdominal cavity.*** It contains organs of digestion and excretion, including the liver, gallbladder, spleen, pancreas, kidneys, stomach, and intestines.

Think of the abdomen of a patient who is facing you as if it were divided into four parts, or "quadrants." Health-care workers often refer to it that way. The quadrants are formed by imaginary lines. One line is drawn horizontally through the navel. The other line is drawn vertically through the midline of the body. (See Figure 4-5.)

◆ ***Pelvic cavity.*** Its bounds are the lower part of the spine, the hip bones, and the pubis. It protects the lower abdomen, including the bladder, rectum, and internal reproductive organs.

Figure 4-2   Anatomical regions.

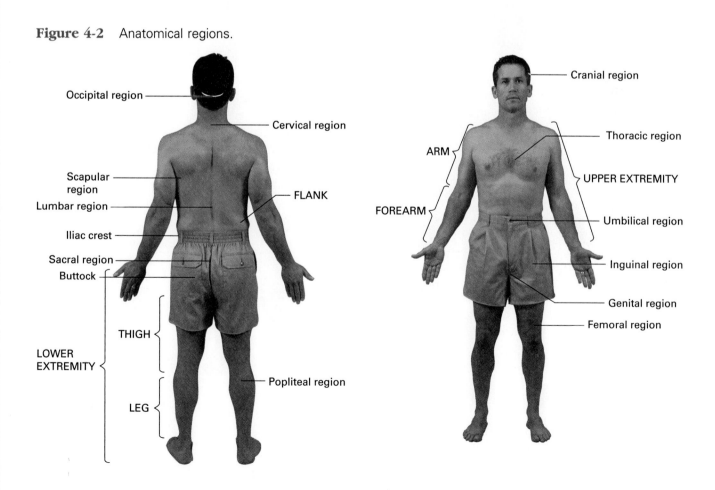

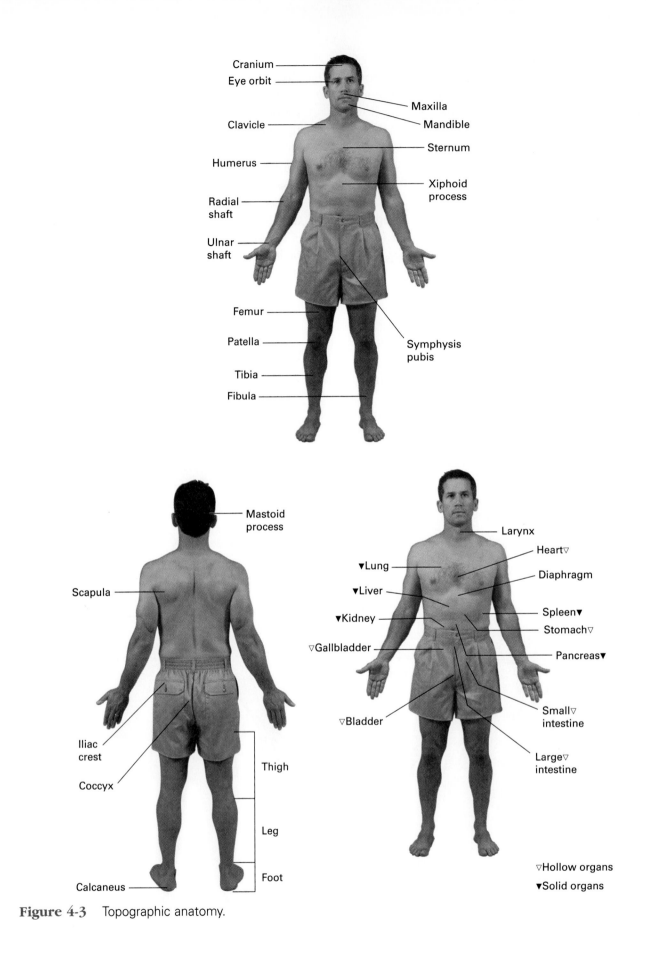

Figure 4-3 Topographic anatomy.

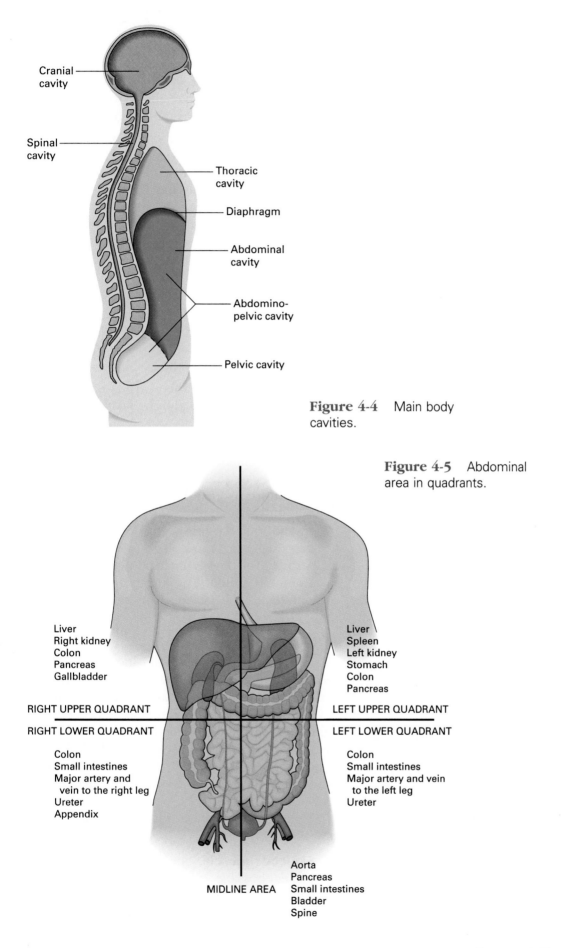

Figure 4-4 Main body cavities.

Cranial cavity

Spinal cavity

Thoracic cavity

Diaphragm

Abdominal cavity

Abdomino-pelvic cavity

Pelvic cavity

Figure 4-5 Abdominal area in quadrants.

Liver
Right kidney
Colon
Pancreas
Gallbladder

Liver
Spleen
Left kidney
Stomach
Colon
Pancreas

RIGHT UPPER QUADRANT

LEFT UPPER QUADRANT

RIGHT LOWER QUADRANT

LEFT LOWER QUADRANT

Colon
Small intestines
Major artery and
 vein to the right leg
Ureter
Appendix

Colon
Small intestines
Major artery and vein
 to the left leg
Ureter

MIDLINE AREA

Aorta
Pancreas
Small intestines
Bladder
Spine

Musculoskeletal System

The musculoskeletal system is made up of the skeleton and muscles. Each helps to give the body shape and protects internal organs. The muscles also provide for movement.

Skeleton

The human body is shaped by its bony framework (Figure 4-6). Bone is composed of living cells and nonliving matter. The nonliving matter contains calcium compounds that help make bone hard and rigid. Without bones, the body would collapse.

The adult skeleton has 206 bones, which are found in various sizes and shapes, from the eight small bones that allow the wrist to flex and turn to the single long bone found in the thigh. The skeleton must be strong to support and protect, jointed to permit motion, and flexible to withstand stress. It is held together mainly by **ligaments, tendons,** and layers of muscle. (Ligaments connect bone to bone. Tendons connect muscle to bone.) Bone ends fit into each other at joints. The three kinds of joints are: immovable like the skull, slightly movable like the spine, and freely movable like the elbow or knee.

The major areas of the skeleton include the following:

◆ *Skull.* The skull has a number of broad, flat bones that form a hollow shell. The top (including the forehead), back, and sides of the shell make up the **cranium.** It houses and protects the brain. There are 14 small bones that form the face. They give shape to the face and permit the jaw to move. The major features of the face are the nose, ears, eyes, cheeks, mouth, and jaw. Injuries to the face are often associated with injuries to the cervical spine.

◆ *Spinal column.* This houses and protects the **spinal cord.** It is the central supportive bony structure of the body. It consists of 33 bones known as **vertebrae.** The spine is divided into five sections: the **cervical spine** (the neck, formed by 7 vertebrae) the **tho-**

racic spine (the upper back, formed by 12 vertebrae), the **lumbar spine** (the lower back, formed by 5 vertebrae), the **sacrum** (the lower part of the spine, formed by 5 fused vertebrae), and the **coccyx** (the tail bone, formed by 4 fused vertebrae).

◆ *Thorax.* The thorax, or rib cage, protects the heart and lungs—vital organs of the body. They are enclosed by 12 pairs of ribs that are attached at the back to the spine. The top 10 are also attached in front to the **sternum,** or breastbone. The lowest portion of the sternum is called the **xiphoid process,** which serves as a key landmark when performing CPR.

◆ *Pelvis.* The pelvis, or hip bones, consists of the **ilium, pubis,** and **ischium.** Iliac crests form the "wings" of the pelvis. The pubis is the anterior portion of the pelvis. The ischium is in the posterior portion.

◆ *Shoulder girdle.* This consists of the **clavicle** (the collarbone) and the **scapulae** (shoulder blades).

◆ *Extremities.* The upper extremities extend from the shoulders to the fingertips. The arm (shoulder to elbow) has one bone known as the **humerus.** The bones in the forearm are the **radius** and **ulna.** The lower extremities extend from the hips to the toes. The bone in the thigh, or upper leg, is known as the **femur.** The bones in the lower leg are the **tibia** and **fibula.** The knee cap is called the **patella.**

Muscles

There are over 600 muscles in the human body. Movement of the body depends on the work performed by the muscles. Muscles have the ability to contract (become shorter and thicker) when stimulated by a nerve impulse. Each muscle is made up of long threadlike cells called "fibers," which are closely packed or bundled. Overlapping bundles are bound by connective tissue. (See Figure 4-7.)

There are three basic kinds of muscles (Figure 4-8):

◆ **Skeletal muscle,** or **voluntary muscle,** makes possible all deliberate acts such as walking and chewing. It helps shape the body and form its walls. In the trunk this type of

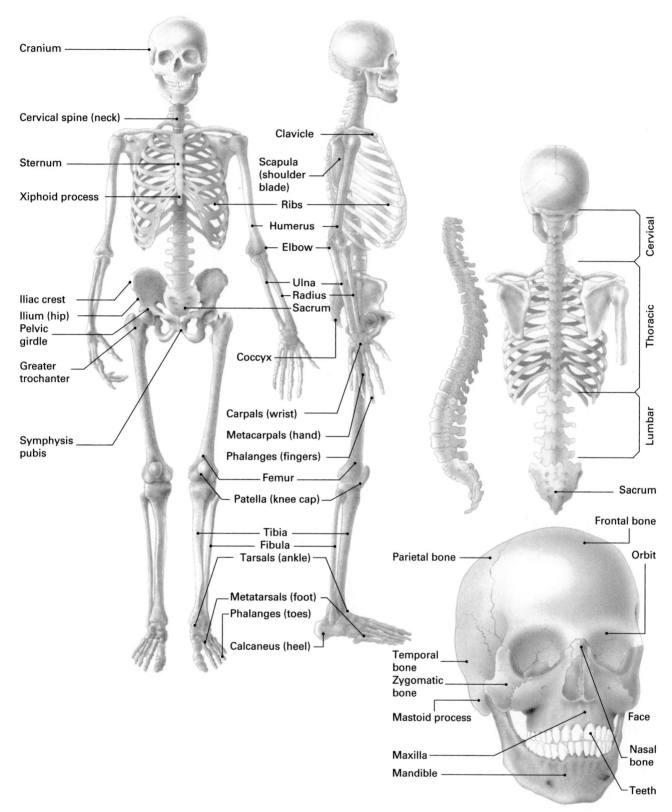

Figure 4-6 Skeletal system.

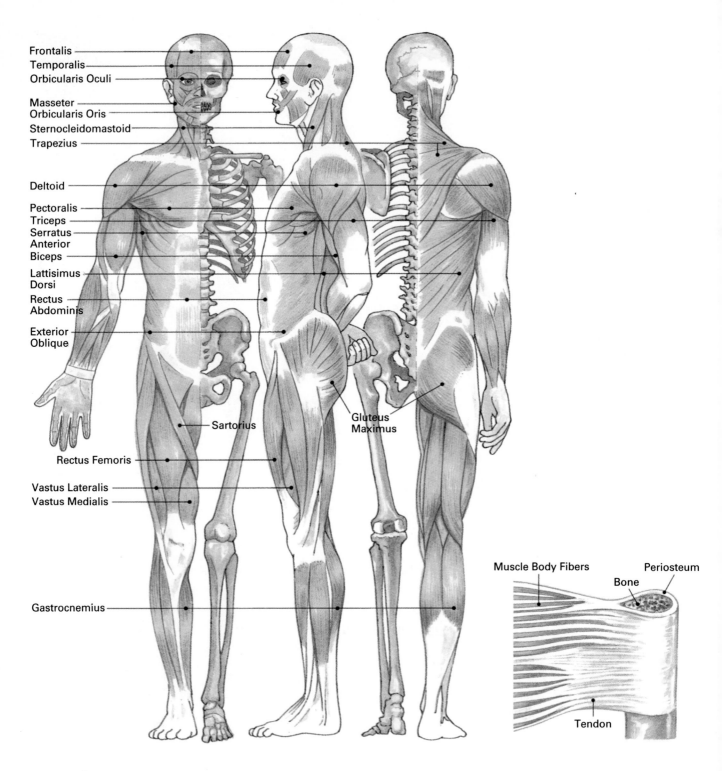

Frontalis
Temporalis
Orbicularis Oculi

Masseter
Orbicularis Oris
Sternocleidomastoid
Trapezius

Deltoid

Pectoralis
Triceps
Serratus
Anterior
Biceps

Lattisimus
Dorsi

Rectus
Abdominis

Exterior
Oblique

Sartorius

Rectus Femoris

Vastus Lateralis
Vastus Medialis

Gastrocnemius

Gluteus
Maximus

Muscle Body Fibers   Periosteum
                    Bone

Tendon

Figure 4-7 The muscular system.

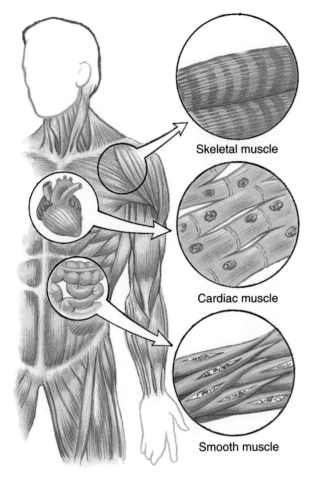

Figure 4-8 Three types of muscles.

Skeletal muscle

Cardiac muscle

Smooth muscle

muscle is broad, flat, and expanded. In the extremities, it is long, rounded, and attached to the bones by tendons.

◆ **Smooth muscle,** or **involuntary muscle,** is made of longer fibers. It is found in the walls of tubelike organs, ducts, and blood vessels. It also forms much of the intestinal wall. A person has little or no control over this type of muscle.

The diaphragm is a muscle fitting into both the voluntary and involuntary categories. It is attached to the spinal column and ribs like skeletal muscle, but it functions involuntarily with each breath. Though we can intentionally breathe faster or slower or take a deep breath, we can do this only for a short time. Ultimately, the involuntary function takes over and resumes the regular pattern of breathing in response to the body's need for oxygen.

◆ **Cardiac muscle** makes up the walls of the heart. It is able to stimulate itself into contraction, even when disconnected from the brain.

Respiratory System

The human body may get enough nutrition from food to last for several weeks. It can store water to last for several days. But it can only store oxygen for a few minutes. The body depends on a constant supply of oxygen. The respiratory system delivers oxygen to the body, as well as removes carbon dioxide from the body.

The passage of air into and out of the lungs is called **respiration.** Breathing in is called **inspiration,** or inhaling. Breathing out is called **expiration,** or exhaling.

During inspiration, the muscles of the thorax contract, moving the ribs outward and up. The diaphragm contracts and lowers. These movements expand the chest cavity and cause air to flow into the lungs. During expiration the opposite happens. The muscles of the chest relax and cause the ribs to move inward. The diaphragm relaxes and moves up.

The respiratory system consists of the organs that let us breathe (Figure 4-9). When air enters the body, it does so through the mouth and nose. The area posterior to the mouth and nose is called the **pharynx,** which is divided into the **oropharynx** and **nasopharynx.** Air then travels down through the **larynx** (voice box) and into the **trachea** (windpipe). The trachea is the air passageway to the lungs. It is made of cartilage rings and is visible in the anterior portion of the neck.

To keep food from entering the trachea each time you swallow, there is a small flap of tissue called the **epiglottis.** Food or drink reaching the back of the throat triggers the epiglottis to close, allowing safe passage into the stomach. When the epiglottis fails to close, food or drink enters the trachea, resulting in a coughing or choking reaction.

The trachea splits into two **bronchi.** These air passages gradually become smaller and smaller until they reach the **alveoli,** where carbon dioxide and oxygen are exchanged with blood.

Circulatory System

The circulatory system delivers oxygen and nutrients to the body's tissues and removes waste

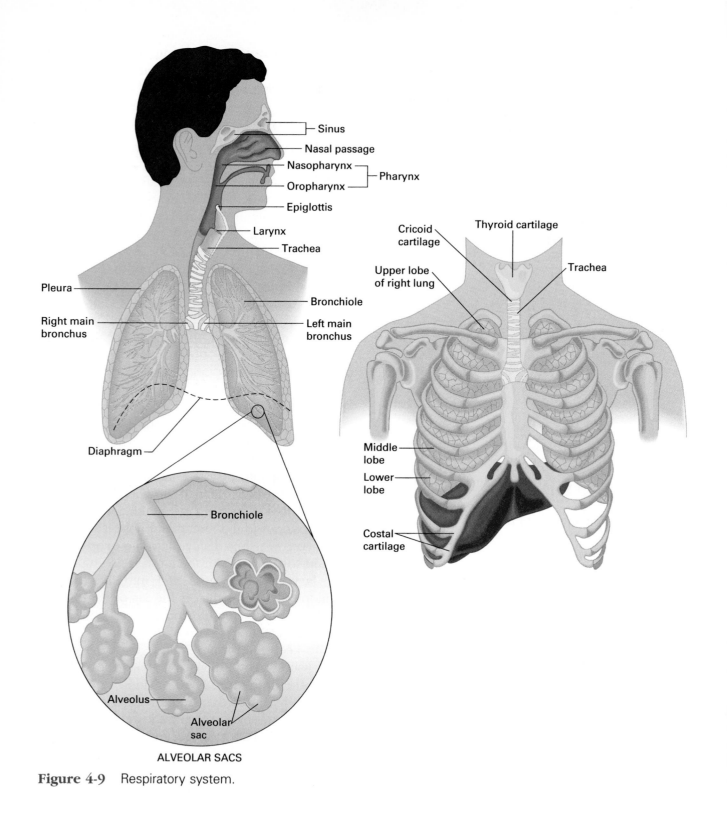

Figure 4-9 Respiratory system.

products. It consists of the heart, blood vessels, and blood. (See Figure 4-10.)

The heart is a muscular organ that is responsible for pumping blood through the body. The adult heart contracts between 60 and 80 times per minute when at rest and faster when under stress. Problems with the heart account for many of the emergencies you will encounter as a First Responder.

The heart is divided into four chambers. The upper chambers are called **atria.** The lower chambers are called **ventricles.** The heart has a left and right side, each of which has an atrium and a ventricle. Each atrium is separated from a ventricle by a valve, allowing blood to flow in one direction—from the atrium to the ventricle. The right side of the heart receives blood from the body and pumps it to the lungs. The left side of the heart receives oxygenated blood from the lungs and pumps it to the body.

Because the right side of the heart does not have to circulate blood nearly as far as the left side does, it pumps at a much lower pressure than the left side.

When the heart pumps blood from the left ventricle, blood enters the arteries. This pumping action causes a wave of pressure that can be felt as a "pulse." There are many points where a pulse can be felt in the body. The most common are:

◆ *Carotid pulse point,* felt on either side of the neck.

◆ *Brachial pulse point,* felt on the inside of the arm between the elbow and the shoulder.

◆ *Radial pulse point,* felt on the thumb side of the wrist.

◆ *Femoral pulse point,* felt in the area of the groin in the crease between the abdomen and thigh.

The **blood vessels** are a closed system of tubes through which blood flows. **Arteries** and **arterioles** take blood away from the heart. The **capillaries** are distributors. They are the smallest vessels through which the exchange of fluid, oxygen, and carbon dioxide takes place between blood and tissue cells. The **venules** and **veins** are collectors. They carry blood back to the heart from the rest of the body.

Nervous System

The nervous system is composed of the brain, the spinal cord, and nerves (Figure 4-11). It has

MAJOR ARTERIES

MAJOR VEINS

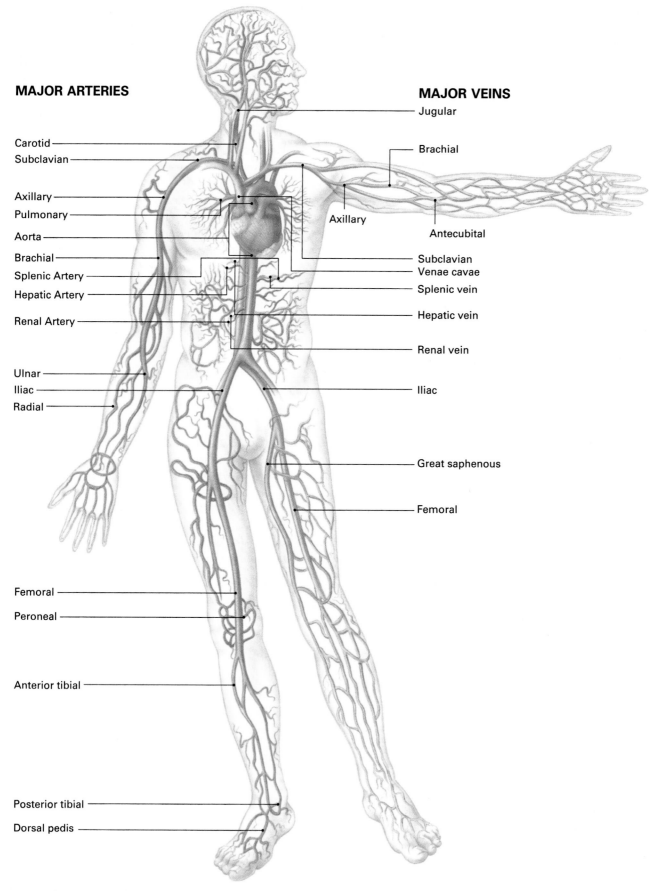

Carotid

Subclavian

Axillary

Pulmonary

Aorta

Brachial

Splenic Artery

Hepatic Artery

Renal Artery

Ulnar

Iliac

Radial

Femoral

Peroneal

Anterior tibial

Posterior tibial

Dorsal pedis

Jugular

Brachial

Axillary

Antecubital

Subclavian

Venae cavae

Splenic vein

Hepatic vein

Renal vein

Iliac

Great saphenous

Femoral

Figure 4-10   Circulatory system.

network of nerve tissue that regulates functions we normally pay no attention to, such as how quickly or slowly the heart beats.

two major functions—communication and control. It lets a person be aware of and react to the environment. It coordinates the body's responses to stimuli and keeps body systems working together.

The nervous system has two main parts—the **central nervous system** and the **peripheral nervous system.** The central nervous system consists of the brain and spinal cord. The peripheral nervous system consists of the nerves, which carry information back and forth from the body to the spinal cord and brain.

The nervous system may also be broken down by function, or voluntary and involuntary components. Voluntary components are under our control. They are responsible for movement, such as walking or throwing a ball. Involuntary components, such as breathing, are handled by the **autonomic nervous system.** This system is a

The Skin

The skin is considered to be the largest organ in the body, covering between 2500 to 3000 square inches (16 129 mm^2 to 19 355 mm^2). It separates the human body from the outside world. It protects the deep tissues from injury, drying out, and invasion by bacteria and other foreign bodies. The skin helps to regulate body temperature, and it helps to prevent dehydration. It also acts as the receptor organ for touch, pain, heat, and cold. (See Figure 4-12.)

The **epidermis** is the outermost layer of skin. It contains cells that give the skin its color. The **dermis,** or second layer, contains a vast network

THE BRAIN

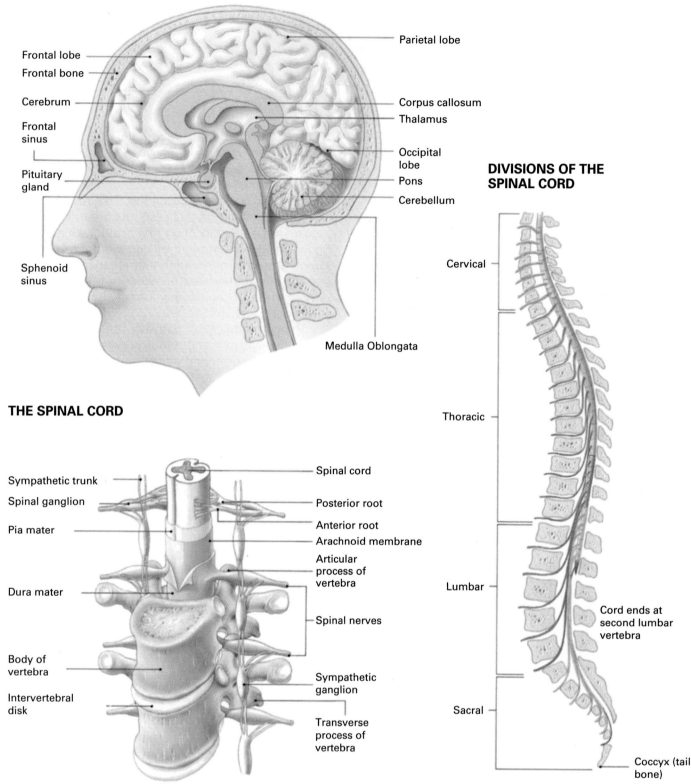

Parietal lobe

Frontal lobe

Frontal bone

Cerebrum

Frontal sinus

Pituitary gland

Sphenoid sinus

Corpus callosum

Thalamus

Occipital lobe

Pons

Cerebellum

Medulla Oblongata

DIVISIONS OF THE SPINAL CORD

Cervical

Thoracic

Lumbar

Sacral

Cord ends at second lumbar vertebra

Coccyx (tail bone)

THE SPINAL CORD

Sympathetic trunk

Spinal ganglion

Pia mater

Dura mater

Body of vertebra

Intervertebral disk

Spinal cord

Posterior root

Anterior root

Arachnoid membrane

Articular process of vertebra

Spinal nerves

Sympathetic ganglion

Transverse process of vertebra

Figure 4-11a   Nervous system.

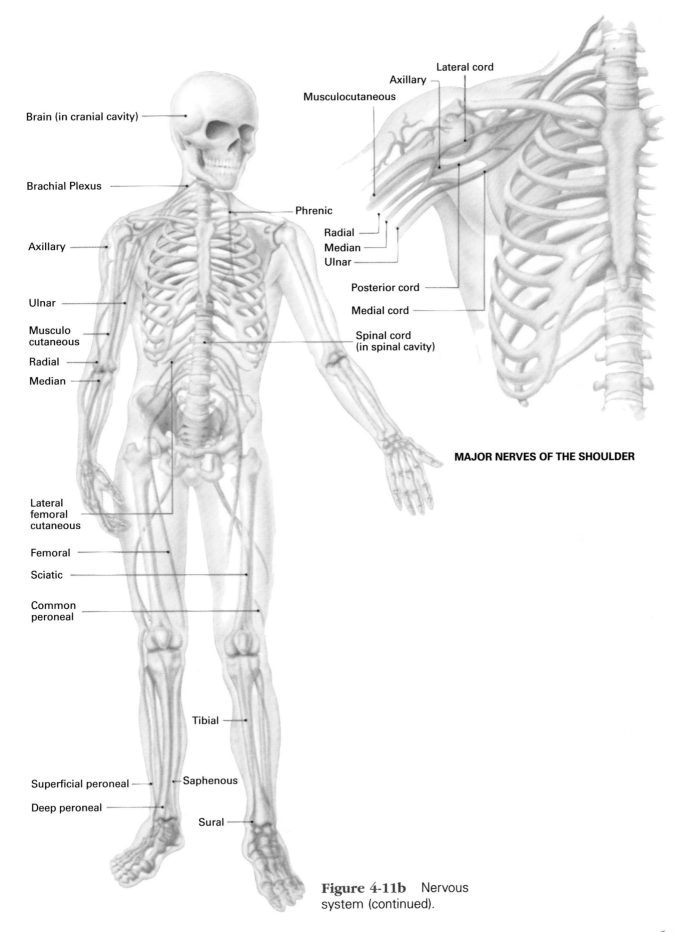

Brain (in cranial cavity)

Brachial Plexus

Axillary

Ulnar

Musculo cutaneous

Radial

Median

Lateral femoral cutaneous

Femoral

Sciatic

Common peroneal

Tibial

Superficial peroneal

Deep peroneal

Saphenous

Sural

Phrenic

Spinal cord (in spinal cavity)

Lateral cord

Axillary

Musculocutaneous

Radial

Median

Ulnar

Posterior cord

Medial cord

MAJOR NERVES OF THE SHOULDER

Figure 4-11b Nervous system (continued).

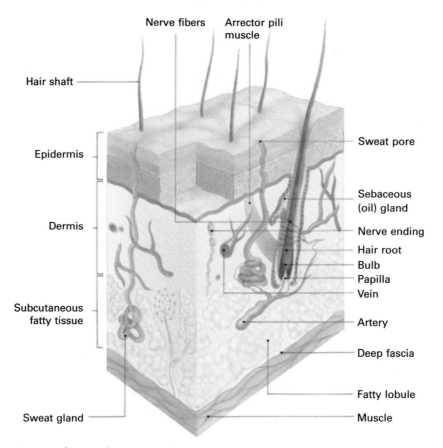

Figure 4-12   Structure of normal skin.

of blood vessels, hair follicles, sweat and oil glands, and sensory nerves. Just below the skin is a layer of fatty tissue, which varies in thickness. For example, it is extremely thin in the eyelids, but thick over the buttocks.

Digestive System

The digestive system is composed of the **alimentary tract** (food passageway) and accessory organs (Figure 4-13). Its main functions are to ingest food and get rid of waste. Digestion consists of two processes—mechanical and chemical.

The mechanical process includes chewing, swallowing, the rhythmic movement of matter through the tract, and defecation (the elimination of waste). The chemical process consists of breaking food into simple components that can be absorbed and used by the body.

Except for the mouth and **esophagus,** the organs of this system are in the abdomen. They include the stomach, pancreas, liver, gallbladder, small intestine, and large intestine.

Urinary System

The urinary system filters and excretes waste from the body. It consists of two kidneys and two ureters, one urinary bladder, and one urethra (Figure 4-14). The kidneys filter the blood and create urine. The ureters take urine from

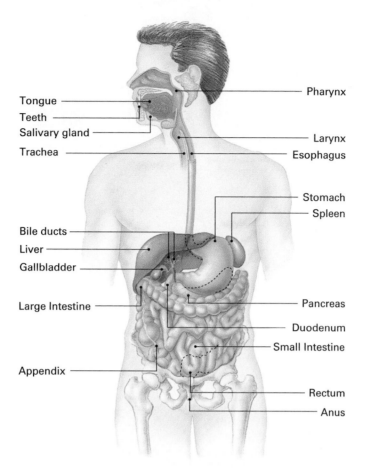

Figure 4-13 Digestive system.

the kidneys to the next part of the system, the bladder. The bladder stores urine until it is passed through the urethra and excreted from the body.

The urinary system helps the body maintain the delicate balance of water and chemicals needed for survival. During the process of urine formation in the kidneys, wastes are removed and useful products are returned to the blood.

Endocrine System

The endocrine glands regulate the body by secreting chemical hormones directly into the bloodstream. They affect physical strength, mental ability, stature, reproduction, hair growth, voice pitch, and behavior. How people think, act, and feel depends largely on these tiny secretions. Each gland produces one or more hormones. The glands include the thyroid, parathyroids, adrenals, ovaries, testes, islets of Langerhans in the pancreas, and the pituitary.

Reproductive System

The reproductive system of the male includes two testes, a duct system, accessory glands, and the penis. The reproductive system of the female consists of two ovaries, two fallopian tubes, the uterus, and vagina. (See Figure 4-15.)

ORGANS OF THE URINARY SYSTEM

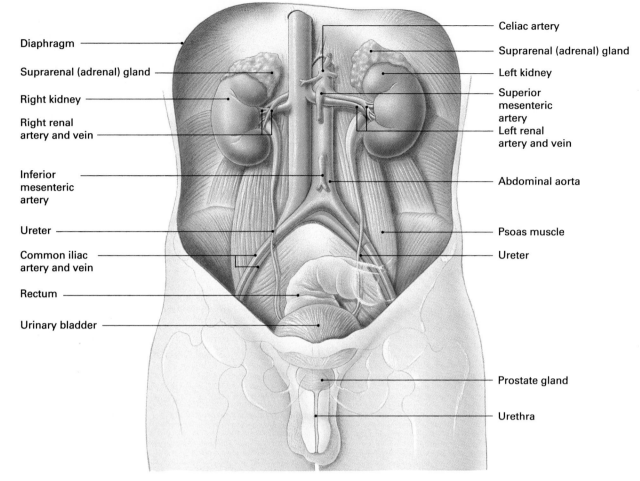

Diaphragm

Suprarenal (adrenal) gland

Right kidney

Right renal artery and vein

Inferior mesenteric artery

Ureter

Common iliac artery and vein

Rectum

Urinary bladder

Celiac artery

Suprarenal (adrenal) gland

Left kidney

Superior mesenteric artery

Left renal artery and vein

Abdominal aorta

Psoas muscle

Ureter

Prostate gland

Urethra

Figure 4-14   Urinary system.

Figure 4-15   Reproductive system.

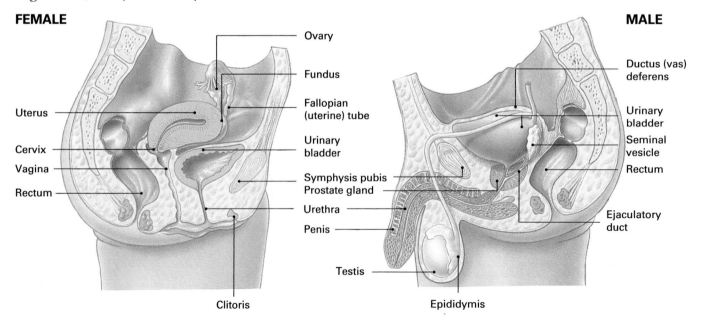

FEMALE

MALE

Ovary

Fundus

Fallopian (uterine) tube

Urinary bladder

Symphysis pubis
Prostate gland

Urethra

Penis

Uterus

Cervix

Vagina

Rectum

Clitoris

Testis

Epididymis

Ductus (vas) deferens

Urinary bladder

Seminal vesicle

Rectum

Ejaculatory duct

At the beginning of this chapter, you read that fire service First Responders were providing emergency care to a patient who "hurt his right leg." To see how chapter information applies to this emergency, read the following. It describes how the call was completed.

PATIENT HISTORY

I asked the patient to describe what happened. He said he mis-stepped as he was getting off the escalator. He didn't fall the entire length. He said that he felt a twisting and then a sudden pain in his lower right leg.

Denying any allergies, he also stated that he doesn't take any medications or have any medical problems. He ate a "couple" of tacos for lunch at the food court in the terminal.

ONGOING ASSESSMENT

We rechecked the pulse, movement, and sensation below the injury site. They were all present and the same as the first time we checked. We repeated his vital signs. Pulse was 84, strong, and regular. Respirations were 16, regular, and deep. Blood pressure was 114/78. We made sure he was comfortable and continued to monitor him carefully.

PATIENT HAND-OFF

When the EMTs arrived, my partner gave them the hand-off report:

"This is Allen Levine, who is 43 years old. He was getting off the escalator and mis-stepped. He felt a twisting in his right lower leg. He never lost consciousness and denies any other injury from the fall. He has pain, swelling, and deformity in the distal third of his tib/fib. There is adequate pulse, motor function, and sensation distal to the injury. The remainder of the physical exam was negative. We've held manual stabilization on the injured leg. Allen's vital signs are pulse 84 strong and regular, respirations 16 and adequate, blood pressure 114/78."

While the EMTs immobilized the patient's leg in a splint, we helped keep onlookers from invading the patient's privacy. It wasn't long before the EMTs and the patient were ready to head for the ambulance. We walked with them to make sure the way was clear. Then, at the ambulance, we helped to get the stretcher loaded.

It was an exciting night. We were glad that everyone turned out to be okay.

The human body is a complex network of bones, muscles, vital organs, and nerve pathways. Each of these body systems has its own individual functions, but all must work in concert. Understanding the anatomical terms and function of each system and the way they work together will help you to develop a solid baseline from which to build strong patient assessment skills.

Chapter Review

FIREFIGHTER FOCUS

Consider these important tasks: opening an airway, examining a patient for the ability to move a part of his or her body, updating incoming EMS personnel via radio. Each task requires a knowledge of the human body.

To open an airway or examine a patient for breathing, you must know how the body breathes and how to recognize when it is not doing so adequately. To examine for injuries to the spine or certain bones, you may wish to check the patient's ability to wiggle fingers and toes. Your knowledge of the human body will let you know why that is important.

When reporting a patient's condition, you must be able to describe it accurately to the incoming EMTs. Again, it is your knowledge of the body that will allow this to happen.

FIRE COMPANY REVIEW

Page references where answers may be found or supported are provided at the end of each question.

Section 1

1. How does a patient appear when he or she is in the anatomical position? The lateral recumbent position? Supine? Prone? (p. 52)

2. What are definitions of the terms anterior, medial, distal, superficial, and external? (pp. 52–53)

Section 2

3. What is the anatomy and function of the musculoskeletal system? Give a brief description. (pp. 56, 59)

4. What is the anatomy and function of the respiratory system? Give a brief description. (p. 59)

5. What is the anatomy and function of the circulatory system? Give a brief description. (pp. 59, 61)

RESOURCES TO LEARN MORE

Agur, A.M. *Grant's Atlas of Anatomy,* Ninth Edition. Baltimore: Williams & Wilkins, 1991.

Fremgen, B.F. *Medical Terminology: An Anatomy and Physiology Systems Approach*. Upper Saddle River: NJ: Brady/Prentice-Hall, 1997.

MAJOR BODY SYSTEMS

Musculoskeletal
- Bones
- Muscles
- Ligaments
- Tendons

Respiratory
- Nose and mouth
- Pharynx
- Epiglottis
- Larynx and trachea
- Bronchi
- Lungs
- Alveoli

Circulatory
- Heart
- Blood vessels
- Blood

Nervous
- Brain
- Spinal cord
- Nerves

Skin
- Epidermis
- Dermis
- Subcutaneous layer

Digestive
- Mouth
- Esophagus
- Stomach
- Pancreas
- Liver
- Gallbladder
- Intestines

Urinary
- Kidneys
- Bladder
- Ureters
- Urethra

Endocrine
- Glands
- Hormones

Reproductive
- Male: testes, duct system, accessory glands, penis
- Female: ovaries, fallopian tubes, uterus, vagina

Airway

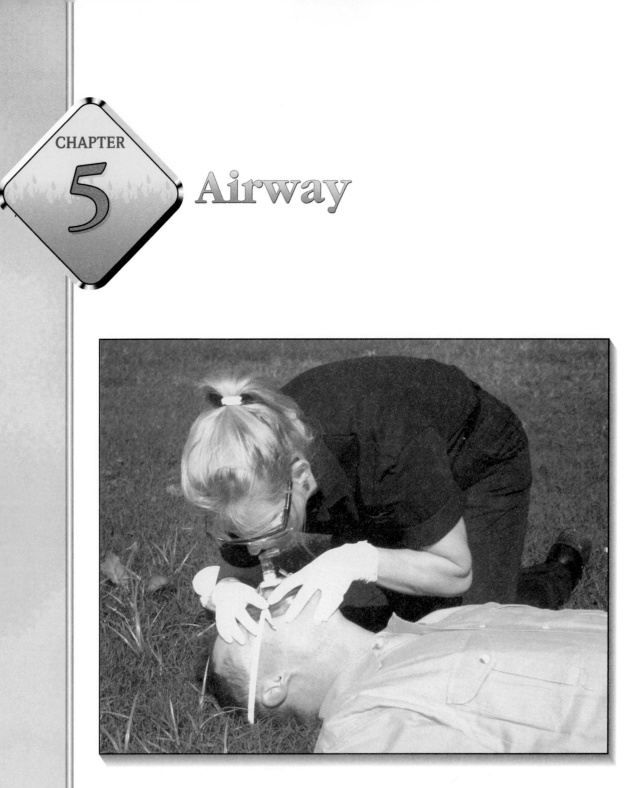

*I*NTRODUCTION   *The most important part of your job as a fire service First Responder involves a patient's airway. No matter what a patient's problem may be, you must determine whether or not the airway is open and clear first. It can spell the difference between life and death—for a business executive who chokes on a piece of steak, a child who falls into a swimming pool, or an unresponsive patient whose tongue is blocking her airway.*

Cognitive, affective, and psychomotor objectives are from the U.S. DOT's "First Responder: National Standard Curriculum." Enrichment objectives, if any, identify supplemental material.

Cognitive

2-1.1 Name and label the major structures of the respiratory system on a diagram. (p. 75)

2-1.2 List the signs of inadequate breathing. (pp. 87–88)

2-1.3 Describe the steps in the head-tilt chin-lift. (pp. 78–79)

2-1.4 Relate mechanism of injury to opening the airway. (pp. 78, 79)

2-1.5 Describe the steps in the jaw thrust. (pp. 79–80)

2-1.6 State the importance of having a suction unit ready for immediate use when providing emergency medical care. (pp. 85–86)

2-1.7 Describe the techniques of suctioning. (pp. 85–86)

2-1.8 Describe how to ventilate a patient with a resuscitation mask or barrier device. (pp. 88–89, 91)

2-1.9 Describe how ventilating an infant or child is different from an adult. (p. 93)

2-1.10 List the steps in providing mouth-to-mouth and mouth-to-stoma ventilation. (pp. 91–93)

2-1.11 Describe how to measure and insert an oropharyngeal (oral) airway. (pp. 81–82)

2-1.12 Describe how to measure and insert a nasopharyngeal (nasal) airway. (pp. 82–83)

2-1.13 Describe how to clear a foreign body airway obstruction in a responsive adult. (pp. 102–104)

2-1.14 Describe how to clear a foreign body airway obstruction in a responsive child with complete obstruction or partial airway obstruction and poor air exchange. (pp. 108–109)

2-1.15 Describe how to clear a foreign body airway obstruction in a responsive infant with complete obstruction or partial airway obstruction and poor air exchange. (pp. 107–108)

2-1.16 Describe how to clear a foreign-body airway obstruction in an unresponsive adult. (pp. 104–106)

2-1.17 Describe how to clear a foreign body airway obstruction in an unresponsive child. (p. 109)

2-1.18 Describe how to clear a foreign body airway obstruction in an unresponsive infant. (p. 108)

Affective

2-1.19 Explain why basic life support ventilation and airway protective skills take priority over most other basic life support skills. (pp. 74, 75, 110, 111)

2-1.20 Demonstrate a caring attitude towards patients with airway problems who request emergency medical services. (p. 77)

2-1.21 Place the interests of the patient with airway problems as the foremost consideration when making any and all patient care decisions. (pp. 110, 111)

2-1.22 Communicate with empathy to patients with airway problems, as well as with family members and friends of the patient. (p. 77)

Psychomotor

2-1.23 Demonstrate the steps in the head-tilt chin-lift. (pp. 78–79)

2-1.24 Demonstrate the steps in the jaw thrust. (pp. 79–80)

2-1.25 Demonstrate the techniques of suctioning. (pp. 85–86)

2-1.26 Demonstrate the steps in mouth-to-mouth ventilation with body substance isolation (barrier shields). (pp. 88–89, 91)

2-1.27 Demonstrate how to use a resuscitation mask to ventilate a patient. (pp. 94–95)

2-1.28 Demonstrate how to ventilate a patient with a stoma. (pp. 92–93)

(continued)

2-1.29 Demonstrate how to measure and insert an oropharyngeal (oral) airway. (pp. 81–82)

2-1.30 Demonstrate how to measure and insert a nasopharyngeal (nasal) airway. (pp. 82–83)

2-1.31 Demonstrate how to ventilate infant and child patients. (p. 93)

2-1.32 Demonstrate how to clear a foreign body airway obstruction in a responsive adult. (pp. 102–104)

2-1.33 Demonstrate how to clear a foreign body airway obstruction in a responsive child. (pp. 108–109)

2-1.34 Demonstrate how to clear a foreign body airway obstruction in a responsive infant. (pp. 107–108)

2-1.35 Demonstrate how to clear a foreign body airway obstruction in an unresponsive adult. (pp. 104–106)

2-1.36 Demonstrate how to clear a foreign body airway obstruction in an unresponsive child. (p. 109)

2-1.37 Demonstrate how to clear a foreign body airway obstruction in an unresponsive infant. (p. 108)

Enrichment

◆ Describe how to perform bag-valve-mask ventilation. (pp. 94–95)

◆ Describe the oxygen cylinders, oxygen delivery equipment, and oxygen administration guidelines. (pp. 95–100)

◆ Describe the special considerations related to administering oxygen to patients with chronic obstructive pulmonary diseases (COPD). (pp. 99–100)

ON SCENE

DISPATCH

Engine Six was dispatched to 426 Chapman Street, Apartment 14E, for an "unresponsive person."

SCENE SIZE-UP

As we approached the scene, we looked for crowds. Sometimes in the city, crowds can be more dangerous than the scenes themselves, but everything was quiet that morning. The call was on the fourteenth floor. We remained observant as we went up the elevator and approached the apartment. A man met us at the door. "Come quickly," he said. We saw that there was one patient, a female who was lying on the living room floor.

INITIAL ASSESSMENT

We already had on our gloves and eye protection when we reached the patient's side. She was unresponsive with snoring respirations.

The patient in this scenario requires immediate attention. No matter what the underlying reason for her condition, she will not survive without adequate respiration. Perhaps the most important part of her care will be the care she receives for her airway. Consider this patient as you read Chapter 5. What do you think the First Responders should do for her?

THE RESPIRATORY SYSTEM

The body can store food for weeks and water for days, but it can only store enough oxygen for a few minutes. When oxygen is cut off, brain cells begin to die in about five minutes.

Anatomy of the Respiratory System

The respiratory system supplies the body with the oxygen it needs. It also removes carbon dioxide. The major components of the respiratory system are the nose and mouth, pharynx (throat), epiglottis, trachea (windpipe), larynx (voice box), and the bronchi, lungs, and diaphragm. (See the

upper airway in Figure 5-1. You also may wish to review the diagrams in Chapter 4.)

Nose and Mouth

Air normally enters the body through the nose and mouth. There it is warmed, moistened, and filtered as it flows over the damp, sticky mucous membranes.

Pharynx

From the back of the nose and mouth, the air enters the pharynx (throat), the passageway for both food and air. Air from the mouth enters through the oral portion of the pharynx, or the oropharynx. Air from the nose enters through the nasal portion of the pharynx, or the nasopharynx. At its lower end, the pharynx divides in two. One

Figure 5-1 Anatomy of the upper airway.

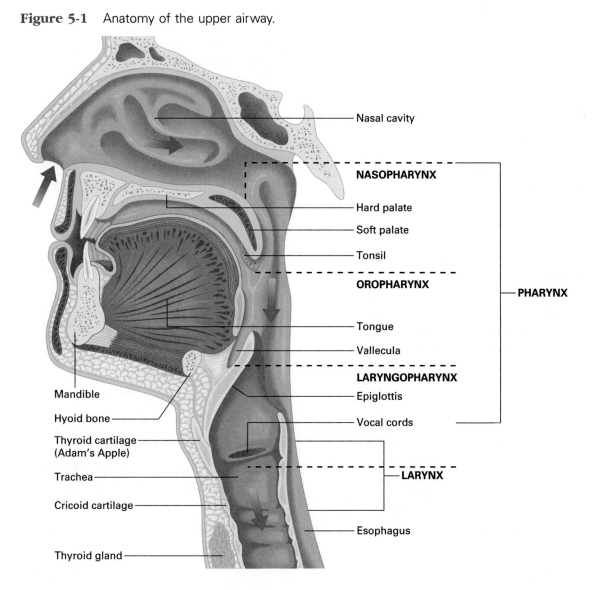

division is the esophagus, which leads to the stomach. The other is the trachea (windpipe), which leads to the lungs.

Epiglottis

The trachea is protected by a small leaf-shaped flap called the epiglottis. Normally, this flap covers the entrance of the larynx during swallowing so food and liquid cannot enter. However, with injury or illness, that reflex may not work properly. As a result, a patient could **aspirate** (inhale) liquid, blood, or vomit into the trachea and lungs, causing suffocation.

Trachea and Larynx

The trachea (windpipe) carries air from the nose and mouth to the lungs. Immediately above it is the larynx (voice box). The larynx can be easily felt with your fingertips at the front of the throat.

Bronchi and Lungs

The lower end of the trachea divides into two tubes called bronchi. The bronchi lead to the lungs. Each bronchus divides into the smaller bronchioles, somewhat like the branches of a tree. At the ends of the bronchioles are thousands of tiny air sacs called alveoli. Each alveolus is enclosed in a network of capillaries and is responsible for the exchange of oxygen and carbon dioxide.

The principal organs of respiration are the lungs. The lungs are two large lobed organs that house thousands of tiny alveoli.

Diaphragm

The diaphragm is a powerful dome-shaped muscle essential to breathing. It separates the thoracic cavity from the abdominal cavity. If it cannot contract effectively because of illness or injury, a patient will breathe inadequately and develop significant respiratory distress.

How Respiration Works

During inhalation, the diaphragm and the muscles between the ribs contract. This increases the size of the thoracic cavity, making it possible for the lungs to expand. The diaphragm moves down slightly, flaring the lower portion of the rib cage, which then moves upward and outward. This decreases pressure in the chest and causes air to flow into the lungs. (See Figure 5-2.)

In the lungs gases pass through the thin walls of the alveoli and capillaries. Oxygen enters the alveoli during inhalation and passes through the capillary walls into the bloodstream. Carbon dioxide passes from the blood through the capillary walls into the alveoli so it can be exhaled.

During exhalation, the diaphragm and the muscles between the ribs relax. This decreases the size of the thoracic cavity. The diaphragm moves up, the ribs move down and in, and air flows out of the lungs.

With some respiratory diseases, a patient has a hard time moving air out of the lungs. He or she has to use muscles not only to draw air in but also to force air out. As a result, both inhalation and exhalation require energy. Such patients tend to get exhausted quickly and can deteriorate rapidly.

Adequate breathing occurs at a normal rate (Table 5-1). For adults, that is 12 to 20 breaths per minute. For children, it is 15 to 30 breaths per minute. For infants, it is 25 to 50 breaths per minute. Adequate breathing is regular in rhythm and free of unusual sounds, such as wheezing or whistling. The chest should expand adequately and equally with each breath. The depth of the breaths should be adequate, too.

Breathing should be virtually effortless. It should be accomplished without the use of **accessory muscles** (additional muscles) in the neck, shoulders, and abdomen.

> ### PEDIATRIC NOTE
>
> When treating infants and children, remember the anatomical differences in their respiratory systems (described below). Knowing those differences can help you recognize airway problems and provide emergency care more effectively.

Infants and Children

The differences between the airways of infants and children and an adult's airway include the following (Figure 5-3):

◆ All airway structures, including the mouth and nose, are smaller than those of adults. They are more easily obstructed even by small

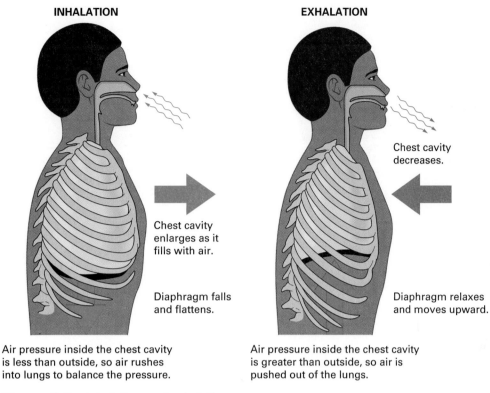

| INHALATION | EXHALATION |

Chest cavity enlarges as it fills with air.

Diaphragm falls and flattens.

Air pressure inside the chest cavity is less than outside, so air rushes into lungs to balance the pressure.

Chest cavity decreases.

Diaphragm relaxes and moves upward.

Air pressure inside the chest cavity is greater than outside, so air is pushed out of the lungs.

Figure 5-2   Inhalation and exhalation.

objects, blood, or swelling. Take extra care to keep the airway of infants and children open.

◆ The tongue takes up proportionally more space in the pharynx than the tongue of an adult. It can therefore block the airway more easily.

◆ The trachea is narrower than an adult's. It also is softer and more flexible. So tipping the head too far back or allowing it to fall forward can close the trachea. Because the head of an infant or young child is quite large relative to the body, a folded towel or similar item under the shoulders may be needed to keep the airway aligned and open.

◆ The chest wall is softer, so infants and children tend to rely on the diaphragm for breathing. Watch for excessive movement there. It can alert you to respiratory distress. Remember, the primary cause of cardiac arrest in infants and children is an uncorrected respiratory problem.

◆ Children have a higher oxygen need. Pound for pound, younger children need twice the amount of oxygen that adults do. Respiratory problems can cause them to deteriorate quickly.

TABLE 5-1	
NORMAL BREATHING RATES	
Adult	12–20 breaths per minute
Child	15–30 breaths per minute
Infant	25–50 breaths per minute

COMPANY OFFICER'S NOTE

When your patient is an infant or child, family members will need you, too. It will be a very stressful time for them. Be calm and caring as well as professional. You may be the one who is less involved in patient care and, therefore, more able to answer questions to comfort family members. If you are needed for patient care, that will be your priority. When possible, consider the family.

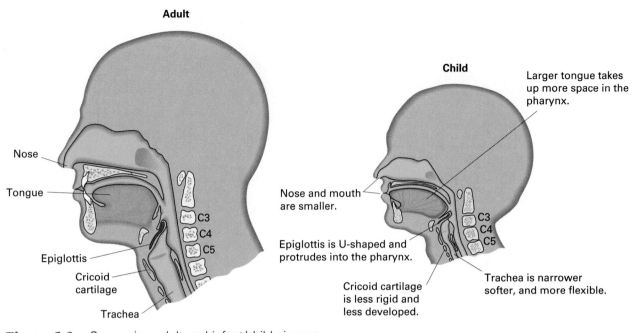

Figure 5-3 Comparing adult and infant/child airways.

PATIENT ASSESSMENT GUIDELINES

Opening the Airway

The tongue is the most common cause of airway obstruction in an unresponsive patient. A patient who loses consciousness will lose muscle tone. When that happens, the base of the tongue can fall back and **occlude** (block) the airway. The patient's effort to breathe then creates negative pressure, which pulls the tongue, epiglottis, or both into the throat.

In some cases, patients require **ventilation** (assisted breathing by forcing air into the patient's lungs). Before a patient who is breathing inadequately can receive it, he or she must have an open airway.

There are two maneuvers commonly used to open an airway: the **head-tilt/chin-lift maneuver** and the **jaw-thrust maneuver.** Both techniques move the tongue from the back of the throat and allow air to pass into the lungs. (The tongue is attached to the lower jaw. So moving the lower jaw forward will relieve an obstruction.)

Head-Tilt/Chin-Lift Maneuver

The head-tilt/chin-lift maneuver is the method of choice for opening the airway of an uninjured patient. The American Heart Association (AHA) recommends it for opening the airway of patients who do not have injuries to the head, neck, or spine. Use it first for unresponsive patients who are not injured.

To perform the head-tilt/chin-lift maneuver (Figures 5-4 and 5-5):

1. **Position your hand.** Place your hand on the patient's forehead. Use the hand that is closest to the patient's head.

2. **Apply firm, backward pressure** with the palm of your hand to tilt the head back.

3. **Position your other hand.** Place your fingertips under the bony part of the patient's lower jaw. (If the patient is an infant or child, place only your index finger under the jaw.)

4. **Lift the chin forward.** At the same time, support the jaw and tilt the head back as far as possible. The patient's teeth should be nearly together. (If the patient is an infant or child, tilt the head back only slightly, as if the child

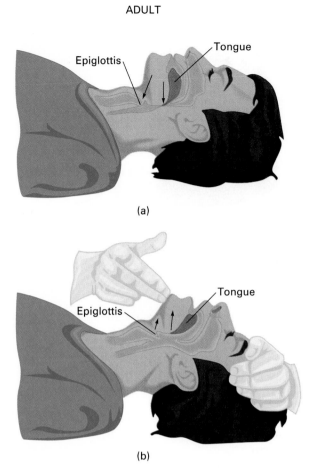

Figure 5-4 Head-tilt/chin-lift maneuver for an adult.

Figure 5-5 Head-tilt/chin-lift maneuver for an infant.

is "sniffing." Remember not to overextend the head.)

5. **Keep the head tilted back** by continuing to press your other hand on the patient's forehead.

There are several important precautions to remember when performing the head-tilt/chin-lift:

◆ Never let your fingers press deeply into the soft tissues under the chin. You could block the patient's airway.

◆ If necessary, use your thumb to press in the patient's lower lip, keeping the patient's mouth slightly open. Never use your thumb to lift the patient's chin.

◆ Do not let the patient's mouth close completely.

◆ If the patient has dentures (false teeth), try to hold them in place. It will help prevent the

patient's lips from interfering with breathing. If you are unable to manage the dentures, remove them.

Jaw-Thrust Maneuver

A jaw-thrust maneuver may be used instead of the head-tilt/chin-lift. Note, however, that it is tiring and technically difficult. It is the safest approach to opening the airway in a patient with suspected spine injury. With it, the patient's head and neck are brought into a neutral position. This means the head is not turned to the side, tilted forward, or tilted back.

Use the jaw-thrust maneuver on unresponsive patients who are injured or who have suspected spine injury. To perform it (Figure 5-6):

1. **Kneel above the patient's head.** Place your elbows on the surface where the patient is lying. Place one hand on each side of the head to hold it still.

2. **Grasp the angles of the patient's lower jaw on both sides.** (If the patient is an infant or child, place two or three fingers of each hand at the angle of the jaw.)

3. **Move the jaw forward,** using a lifting motion with both hands. This pulls the tongue away from the back of the throat. Be careful not to move or tilt the head.

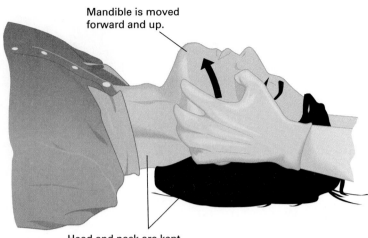

Figure 5-6 Jaw-thrust maneuver.

Mandible is moved forward and up.

Head and neck are kept in neutral in-line position.

4. Keep the patient's mouth slightly open. If necessary, pull back the lower lip with the thumb of your gloved hand.

If the jaw-thrust maneuver does not open the patient's airway, try again. Reposition the jaw and determine whether or not the airway is open. If repositioning does not work, insert an airway adjunct (described later in this chapter).

Inspecting the Airway

Assess the airway of each and every one of your patients. A clear and open airway, or **patent airway,** is absolutely necessary for adequate breathing.

You can determine that a patient's airway is patent if he or she is alert, responsive, and talking to you in a normal voice. If you see that the patient's mental status is altered, however, you have to look more closely. A patient who is drowsy, disoriented, confused or unresponsive may have blood, vomit, or excess saliva in the airway.

To inspect the airway of an unresponsive patient, first open the patient's mouth with a gloved hand. If necessary, use a **cross-finger technique** (Figure 5-7). That is, kneel above the patient. Then cross the thumb and forefinger of one hand. Place the thumb on the patient's lower incisors and the forefinger on the upper incisors. Then use a scissors-like motion to open the patient's mouth. When the mouth is open, look inside for fluids and solids, including broken teeth or dentures that may be blocking the airway. Finally, listen for unusual sounds.

Sounds that may indicate an airway obstruction include:

◆ *Snoring*—may indicate the upper airway is blocked by the tongue or by relaxed tissues in the throat.

◆ *Crowing*—like the cawing of a crow, may mean the muscles around the larynx are in spasm.

◆ *Gurgling*—blood, vomit, mucus, or another liquid may be in the airway.

◆ *Stridor*—a harsh, high-pitched sound during inhalation, may indicate the larynx is swollen and blocking the upper airway.

Figure 5-7 Cross-finger technique.

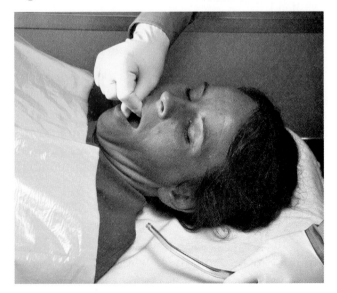

Airway Adjuncts

Once the airway is open, it may be necessary to insert an **airway adjunct** (an artificial airway). There are two kinds—the **oropharyngeal airway** and the **nasopharyngeal airway.** Both extend down towards, but do not pass through, the larynx. They often are used when patients are being artificially ventilated.

When using airway adjuncts, remember the following:

◆ The airway adjunct must be clean and clear of any obstructions.

◆ The proper size must be used to be effective and to prevent complications.

◆ A patient with an airway adjunct can still aspirate (inhale) secretions, blood, vomit, or other foreign substances into the lungs.

◆ The patient's mental status and gag reflex will tell you if an airway adjunct is appropriate. That is, in general, if the patient is unresponsive with no gag reflex, use the oropharyngeal airway. If the patient is responsive or has a gag reflex, use the nasopharyngeal airway.

◆ You must continually and carefully monitor the patient's mental status. If he or she becomes completely responsive or gags, you must remove the airway adjunct.

Oropharyngeal Airway

The oropharyngeal, or oral, airway is a semicircular device made of hard plastic or rubber (Figure 5-8). It is designed to hold the tongue away from the back of the throat at the level of the pharynx.

There are two common types. One is tubular, and the other has a channeled side. Both types are disposable and come in a variety of adult, child, and infant sizes.

The oral airway may be used to help maintain and open the airway of an unresponsive patient who has no gag reflex. Do not use this device on a patient who is responsive or has a gag reflex. If you do, it may cause vomiting or spasm of the vocal cords, which will further compromise the airway.

To insert an oropharyngeal airway, follow these steps (Figure 5-9).

1. **Select the proper size.** It should extend from the corner of the lip to the angle of the jaw or to the tip of the earlobe. Note that if the

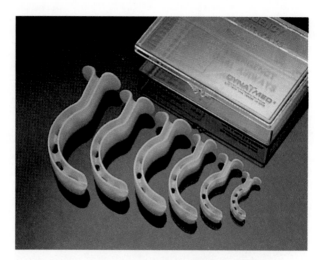

Figure 5-8 Oropharyngeal (oral) airways.

device is too long, it can push the epiglottis over the opening of the trachea, closing off the airway completely.

2. **Open the patient's mouth.** If necessary, use the cross-finger technique.

3. **Insert the adjunct upside down.** Be sure the tip is pointing toward the roof of the patient's mouth. (This helps to keep the tongue from being pushed back into the airway and causing an obstruction.)

4. **Advance the adjunct gently.** Stop when you encounter resistance. This will happen when the device comes in contact with the soft back of the roof of the mouth.

5. **Turn the airway 180°.** Do so while continuing to advance it until the flat flange at the top rests on the patient's front teeth. The airway follows the natural curve of the tongue and the oropharynx.

If the patient gags at any time during insertion, remove the oropharyngeal airway. It may then be necessary to use a nasopharyngeal (nasal) airway or no airway adjunct at all. If the patient tries to dislodge the device, remove it by gently pulling it out and down. Do not rotate the device during removal. Be prepared for vomiting. Have suction ready.

If your patient is an infant or child, a preferred alternative is to use a tongue depressor (blade) to help insert the device. For an infant or child, proceed as follows (Figure 5-10):

1. **Select the proper size oral airway.**

2. **Open the patient's mouth.**

Inserting an Oropharyngeal Airway

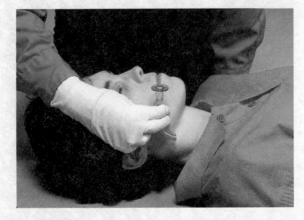

Figure 5-9a Measure to assure the correct size.

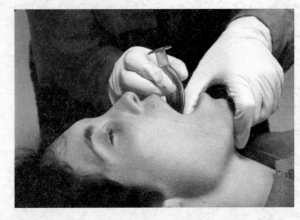

Figure 5-9b Insert the airway with its top pointing up toward the roof of the patient's mouth.

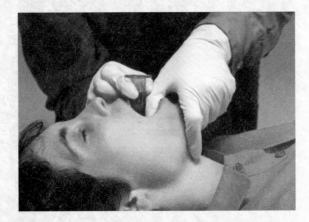

Figure 5-9c Gently rotate it until it reaches the proper position.

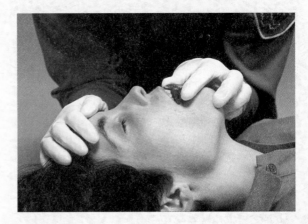

Figure 5-9d Continue until the flange rests on the patient's teeth.

3. **Insert a tongue depressor.** Stop when its tip is at the base of the tongue. Press the tongue depressor, and therefore the tongue, down towards the floor of the mouth and away from the opening of the throat.

4. **Insert the airway in its normal upright position.** Stop when the flange is seated on the patient's teeth. (Do not insert it upside down as you would in an adult. If you do, it could cause bleeding in the airway.)

Nasopharyngeal Airway

The nasopharyngeal, or nasal, airway is a curved hollow tube of soft plastic (Figure 5-11). It has a flange or flare at the top and a bevel at the bottom. It comes in a variety of sizes.

Nasal airways are less likely to cause vomiting than oral airways. This is because the soft tube moves and gives when the patient swallows. Use it to keep the tongue from blocking the airway in patients who are not fully responsive or who have a gag reflex. Use it also when the patient cannot take the oral airway or if the teeth are clenched tightly and will not open.

Even though a nasal airway is lubricated before insertion, it can be painful and may cause the lining of the nose to bleed into the airway.

To insert a nasal airway, follow these steps (Figure 5-12):

Figure 5-10 Inserting an oropharyngeal airway in an infant or child.

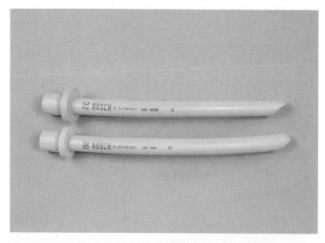

Figure 5-11 Nasopharyngeal (nasal) airway.

1. **Select the proper size.** It should extend from the tip of the patient's nose to the tip of the earlobe. Also, the diameter should fit inside the nostril without **blanching** (losing color from) the skin of the nose. If it is too long, it could send air into the stomach instead of the lungs, causing massive **gastric distention** (inflation of the stomach) and inadequate ventilation.

2. **Lubricate the device.** Use a sterile, water-soluble lubricant. This makes the airway easier to insert. It also reduces the chances of injuring the nasal lining. Do not use petroleum jelly, which can damage the lining of the nose and throat.

3. **Insert the airway posteriorly.** The bevel should point toward the **septum** when it is inserted into the right nostril. (The septum is the wall dividing the two nostrils.) Insert the device close to the midline, along the floor of the nostril, and straight back into the nasopharynx. When the airway is properly inserted, the flange should lie against the flare of the nostril.

4. **If the airway cannot be inserted in one nostril, try the other nostril.** Do not force a nasal airway into place. If you experience difficulty advancing the airway, remove it, turn it 180 degrees, and attempt to insert it in the other nostril.

After insertion, check to see that air is flowing through the airway as the patient breathes. If the patient is breathing spontaneously but no air movement is felt through the tube, remove it immediately and try inserting it in the other nostril.

Note that it is still necessary to maintain a head-tilt/chin-lift or jaw-thrust once the device is inserted.

Clearing the Airway

There are three ways a fire service First Responder can clear an airway of secretions—by the recovery position, finger sweeps, or suctioning. These techniques are not performed sequentially. The technique you choose depends on the patient's condition.

Recovery Position

If the patient is breathing adequately and has a pulse, the American Heart Association (AHA) recommends that you place him in the **recovery position.** It is the first step in maintaining an open airway.

This position uses gravity to keep the airway clear. It allows fluids to drain from the mouth instead of into the airway. The patient's airway is likely to remain open and airway obstructions are less likely to occur. Note that even though the patient is in a recovery position, you should continue to monitor him until the EMTs arrive and assume care.

Do not move the patient into the recovery position if you suspect trauma or spine injury. It

Inserting a Nasopharyngeal Airway

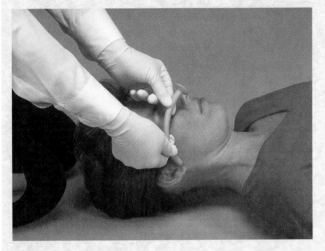

Figure 5-12a Measure the nasopharyngeal airway.

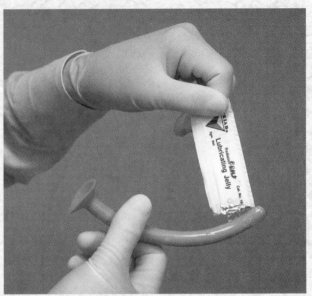

Figure 5-12b Lubricate it with a water-soluble lubricant.

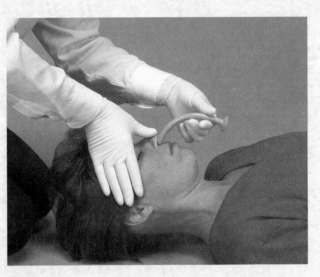

Figure 5-12c Insert the airway posteriorly.

should be used for an unresponsive, uninjured patient who is breathing adequately. He or she should stay in that position until the ambulance arrives. (If there is no breathing or breathing is inadequate, the patient must be supine so you can provide artificial ventilation.)

To move a patient into the recovery position, perform the following (Figure 5-13):

1. **Position the arms and legs.** Lift the patient's left arm above his head. Then cross the patient's right leg over the left leg.

2. **Support the patient's face** as you grasp his right shoulder.

3. **Roll the patient toward you onto his side** (the left side, preferably). Then place his right

Figure 5-13 Recovery position.

hand under the side of his face. If possible, move the patient's head, shoulders, and torso simultaneously as a unit without twisting. The head should be in as close to a midline position as possible.

4. Flex the patient's top leg at the knee.

After the patient is in position, continue to monitor the airway and breathing.

Finger Sweeps

A **finger sweep** is performed only on unresponsive patients. In a finger sweep, you use your index finger to remove solid objects from the airway. Always wear gloves when performing a finger sweep. Foreign material or vomit in the mouth should be removed quickly. Never probe deeply with your fingers in an infant's or child's mouth. (This is called a "blind" finger sweep.) If you do not see an object in the infant's or child's mouth, do not perform a finger sweep.

To perform a finger sweep on an uninjured unresponsive patient (Figure 5-14):

1. Roll the patient onto his left side. This position allows material to drain out of the mouth. It also helps to keep the tongue away from the back of the throat.

2. Open the patient's mouth and look inside. If you see liquids or semi-liquids, cover your gloved index and middle fingers with a cloth.

3. Wipe out the patient's mouth. Insert your index finger. Pass it along the inside of the cheek and into the throat at the base of the tongue. (Use your little finger for an infant or child.) Hook your finger to dislodge and remove any foreign object. Take extreme care

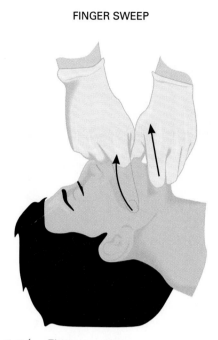

Figure 5-14 Finger sweep.

that you do not force an object deeper into the patient's throat.

Suctioning

Suction units use negative pressure to keep the airway clear. They remove blood, vomit, secretions, and other liquids from the mouth and airway. If you hear a gurgling sound during assessment or artificial ventilation, immediately suction the airway.

Most suction units cannot remove solid objects like teeth, particles of food, and other solid foreign bodies. Some cannot remove very thick vomit. In such situations, you may need to use an alternative piece of suction equipment or a finger sweep.

Suctioning Equipment. Portable suction units (Figure 5-15) can be manually or electrically powered. Some are oxygen- or air-powered. All produce a vacuum that can suction substances from the throat. Each should be inspected before a shift or on a regular basis.

Manual units do not require an energy source other than the person operating it. As a result, they lack some of the typical problems associated with electric- or oxygen-powered devices. They also can more effectively suction heavy substances, such as thick vomit.

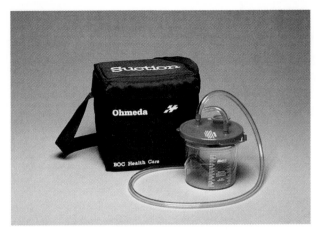

Figure 5-15 Portable suction unit.

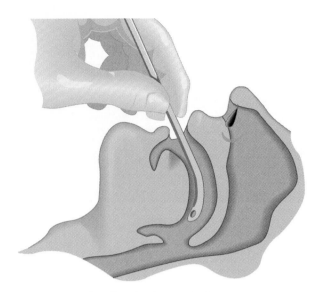

Figure 5-16 Suctioning technique.

Electric units must have fully charged batteries to function effectively. A low battery charge reduces the vacuum and the length of time the unit can be used. Some units allow for constant charging, so batteries remain full.

Any type of suction unit must have:

◆ Wide-bore, thick-walled, non-kinking tubing that fits a standard suction catheter.

◆ Several sterile disposable suction catheters. A "tonsil tip" (rigid plastic) is used to suction unresponsive patients. A soft and flexible one is generally used to suction the nose.

◆ An unbreakable collection bottle or container and a supply of water for rinsing and clearing tubes and catheters.

◆ Enough vacuum pressure and flow to suction substances from the throat effectively.

Principles of Suctioning. The procedure for suctioning varies, depending on the type of unit and catheter used. However, the following general principles apply (Figure 5-16):

1. **Be sure to take BSI precautions.** Suctioning involves removal of body fluids. The potential for coughing and splatters is high. So wear protective eye wear, a mask, and gloves. If you suspect tuberculosis (TB), wear an N-95 or HEPA respirator the entire time you are in contact with the patient.

2. **Use the correct type of catheter for your patient.** Use a "tonsil tip" or "tonsil sucker" catheter to suction the mouth and throat of an unresponsive patient, or an infant or child.

Use a flexible, or "French," catheter to suction the nose.

3. **Insert the catheter,** without suction, only to the base of tongue. Place the convex (bulging) side of the catheter against the roof of the patient's mouth.

4. **Apply suction** by moving the catheter from side to side. Stop after 15 seconds in an adult, 10 seconds in a child, and 5 seconds in an infant.

Do not exceed the maximum times noted above. Suctioning removes air and oxygen, which can result in a quick drop in blood oxygen levels and changes in heart rate. In an adult, watch for rapid, slow, or irregular heart rates. In an infant, watch for a decreased heart rate. If a decrease is noted, stop suctioning and reapply oxygen or ventilate for at least 30 seconds prior to suctioning again.

Assessing Breathing

After you establish an open airway, determine if the patient's breathing is adequate. The brain, heart, and liver are most sensitive to inadequate oxygen. Remember that brain cells start to die within minutes without an oxygen supply.

Determining the Presence of Breathing

Breathing should be effortless. Watch to determine whether or not the chest rises and falls as the

Like the size-up of a fire scene before the initial fire attack, patient care must also be done while looking at the big picture. As a company officer, it may be your responsibility to keep crew members focused on the important tasks. For example, it is easy, even for an experienced fire service First Responder, to focus on obvious injuries or deformities and lose track of more important initial tasks like airway control. It will be your job at the medical call, like the fire scene, to keep track of big picture items.

patient breathes. Also check to see if the patient is using accessory muscles to breathe. Look for excessive use of the neck muscles or pulling inward of the muscles between the patient's ribs.

Observe a responsive patient for the ability to speak. This ability means the air is moving past the vocal cords. Breathing may not be adequate, if the patient can only make sounds or can speak just a few words at a time before having to catch his breath. Patients who can speak full sentences without showing signs of distress or obstruction are breathing adequately.

In an unresponsive patient, use the cross-finger technique, if needed. Then open the airway with the head-tilt/chin-lift or jaw-thrust maneuver. Place your ear close to the patient's mouth and nose for 3 to 5 seconds, and:

◆ *Look* for the rise and fall of the patient's chest.

◆ *Listen* for air coming out of the patient's mouth and nose.

◆ *Feel* for air coming out of the patient's nose and mouth.

If the airway is obstructed, the patient's chest may still rise and fall. However, air will not be moving in and out of the patient's nose or mouth.

Note that **agonal respirations** (reflex gasping with no regular pattern or depth) may occur with cardiac arrest. They may also be a late sign of impending respiratory arrest. These reflex gasps should not be confused with breathing.

Signs of Inadequate Breathing

It is very important for you to recognize the signs of inadequate breathing. Some are subtle and require careful evaluation. If you are not sure that a patient needs breathing assistance, it is better to err on the side of safety and provide ventilations.

Inadequate breathing is characterized by the following signs (Figure 5-17). Note, however, that not all of the signs will be present at the same time. Any one of them may be reason enough to ventilate a patient without delay.

◆ *Too fast or too slow breathing rate.* Rates that are above or below normal indicate inadequate breathing.

◆ *Inadequate chest wall motion.* Adequate breathing is normally accompanied by the rise and fall of the chest. If the chest wall is not rising and falling as it should, or if the sides of the chest rise and fall unequally, breathing is inadequate.

◆ *Cyanosis.* This is a bluish discoloration of the skin and mucous membranes. It is a sign that body tissues are not receiving enough oxygen.

◆ *Mental status changes.* Remember that the mental status of a patient typically correlates with the status of his airway and breathing. A patient who becomes drowsy, disoriented, confused, or unresponsive may not be breathing adequately.

Figure 5-17 Signs and symptoms of inadequate breathing.

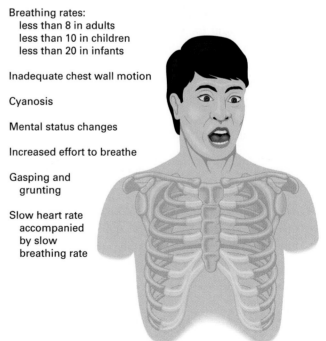

Breathing rates:
  less than 8 in adults
  less than 10 in children
  less than 20 in infants

Inadequate chest wall motion

Cyanosis

Mental status changes

Increased effort to breathe

Gasping and grunting

Slow heart rate accompanied by slow breathing rate

◆ *Increased effort to breathe.* Normal breathing is effortless. When you see a pronounced use of abdominal muscles to breathe, the patient is pushing on the diaphragm to force air out of the lungs. An infant also may develop a "seesaw" motion in which the abdomen and chest move in opposite directions. You may note retractions (pulling inward) between the ribs, above the collarbone, around the muscles of the neck, and below the rib cage as the patient inhales. Flaring of the nostrils during inhalation is another sign more commonly seen in infants and children.

◆ *Gasping and grunting.* These sounds mean the patient is having a difficult time moving air through the respiratory tract. Be alert for other abnormal sounds, such as snoring, crowing, gurgling, or stridor.

◆ *Slow heart rate accompanied by slow breathing rate,* especially in infants and children.

SECTION
3

GUIDELINES FOR EMERGENCY CARE

Artificial Ventilation

If a patient is breathing inadequately or is not breathing at all, that patient needs your immediate assistance. Artificial ventilation is a way of breathing for these patients.

The air we inhale contains about 21% oxygen. Only 5% is used by the body. The remaining 16% is exhaled. Since the air you breathe into a patient contains more than enough oxygen to keep him alive, artificial ventilation is sufficient to support life until high-concentration oxygen is available.

When performing artificial ventilation, monitor the patient continuously to make sure your breaths are adequate. Indications of adequate ventilations include:

◆ *Rate of respiration is adequate*—once every three seconds for infants and children, once every five seconds for adults.

◆ *Force of air is consistent.* It also is sufficient to cause the chest to rise during each ventilation.

◆ *Patient's heart rate returns to normal.* However, underlying medical conditions may pre-

vent this from happening even when ventilations are adequate.

◆ *Patient's color improves.*

Inadequate ventilation may occur because of problems with the patient's airway or because of improper use of a ventilation device. Indications of inadequate ventilations include:

◆ Chest does not rise and fall with each ventilation.

◆ Ventilation rate is too fast or too slow.

◆ Heart rate does not return to normal.

The risk of coming in contact with a patient's secretions, blood, or vomit while ventilating is high. Therefore, you must take BSI precautions. At a minimum, use gloves, eye wear, and a pocket face mask or other barrier device with a one-way filter.

There are many techniques for artificial ventilation. A fire service First Responder must be competent in three, listed here in order of preference: **mouth-to-mask, mouth-to-barrier device,** and **mouth-to-mouth.**

Mouth-to-Mask Ventilation

The most effective First Responder technique for ventilation is mouth-to-mask. A pocket face mask with a one-way valve is used to form a seal around the patient's nose and mouth. You blow into a port at the top of the mask to deliver the ventilation. The one-way valve diverts the patient's exhaled breath.

Mouth-to-mask is the preferred technique because it eliminates direct contact with the patient's nose, mouth, and body fluids. It prevents exposure to the patient's exhaled air. It also allows you to deliver ventilations of adequate force and volume. The mask you use should have the following characteristics.

◆ It should be transparent, so you can see vomit, blood, or other substances in the patient's mouth.

◆ It must fit snugly enough on the patient's face to form a good seal.

◆ It should be available in an average adult size and in additional sizes for infants and children.

◆ It must have a one-way valve, or it must be able to connect to a one-way valve at the ventilation port.

◆ If you have oxygen available, the mask must have an oxygen inlet port.

Mouth-to-mask ventilation is very effective because both your hands are used to create a seal around the mask. To perform the technique, position yourself at the top of the patient's head. Attach oxygen to the mask, if available. Then follow these steps (Figures 5-18 and 5-19).

1. **Position the mask on the patient.** The narrower top portion of the mask should be seated on the bridge of the nose. The broader portion should fit in the cleft of the chin. The position of the mask is critical. If it is wrong, it will leak and prevent you from delivering adequate ventilations.

2. **Seal the mask.** Place both thumbs on the top portion. Place the heels and palms of both hands along the sides. Compress the mask firmly around the edges to form a good seal.

3. **Open the patient's airway.** Place your index fingers on the part of the mask that covers the chin. Using your middle and ring fingers of both hands, grasp along the mandible (the bony part of the jaw). Pull upward to perform the head-tilt/chin-lift maneuver. Use a jaw-thrust maneuver if trauma is suspected.

4. **Deliver two slow initial breaths.** Place your mouth around the one-way valve and blow into the ventilation port. Each breath should be slow (delivered over 1.5 to 2.0 seconds for an adult and 1.0 to 1.5 seconds for an infant or child). It also should be steady and of sufficient

volume to make the chest rise (usually 800 to 1200 ml in an average adult). Make sure that you do not deliver too much air too fast, or you will force air into the patient's stomach.

5. **Determine if ventilations are adequate.** Watch the chest rise and fall. Listen and feel for air escaping when the patient exhales.

6. **Continue ventilations at the proper rate** (Table 5-2). For adults, deliver 10-12 breaths per minute with each lasting 1.5 to 2.0 seconds. For infants and children, deliver 20 breaths per minute with each lasting 1.0 to 1.5 seconds. For a newborn, deliver 40 breaths per minute with each breath lasting 1.0 to 1.5 seconds.

If you cannot ventilate the patient or if the chest does not rise adequately, position the patient's head and try again. (Improper head position is the most common cause of difficulty with ventilation.) If the second try also fails, assume the airway is blocked by a foreign object. Then follow the guidelines (later in this chapter) for removing it.

Mouth-to-Barrier Device Ventilation

A barrier device, such as a face shield, can be used during ventilation (Figure 5-20). It provides some of the same protection to the First Responder as a pocket face mask. A thin and flexible plastic face shield also can be folded and carried easily. Some are available in key-ring and belt-storage containers.

Barrier devices are thin enough to provide very low resistance to the ventilations you deliver to the patient. They provide some protection against contamination from body fluids. However, many do not have a one-way valve, which would divert the patient's exhaled air.

To provide ventilations using a barrier device, kneel at the patient's head. Then follow these steps.

1. **Position the device on the patient.** Pinch the nostrils closed if the device does not cover them.

2. **Open the patient's airway.** Use a head-tilt/chin-lift or a jaw-thrust maneuver.

3. **Deliver two slow initial breaths.** Place your mouth over the barrier device and blow into it. Each breath should be slow (delivered over 1.5 to 2.0 seconds for an adult and 1.0 to 1.5 seconds for an infant or child). It also

Figure 5-18 For a patient with suspected spine injury, be sure to maintain a jaw-thrust during mouth-to-mask ventilation.

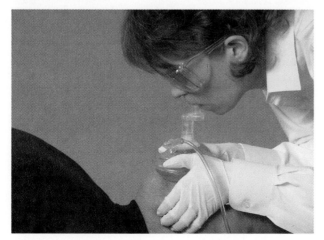

Artificial Ventilation

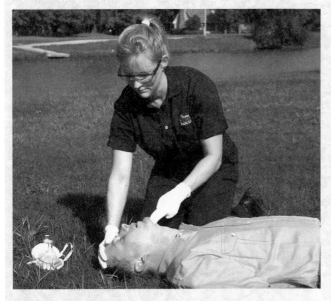

Figure 5-19a If no spine injury is suspected, open the airway with a head-tilt/chin-lift maneuver.

Figure 5-19b Listen, look, and feel to establish breathlessness.

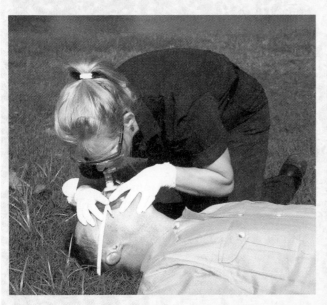

Figure 5-19c Deliver two slow breaths.

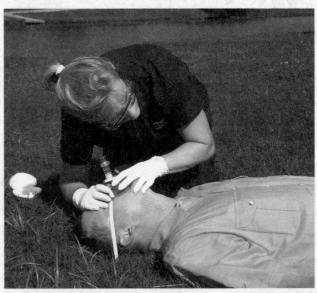

Figure 5-19d If successful, you will see the chest rise and feel exhaled air on your cheek after each breath.

TABLE 5-2

ARTIFICIAL VENTILATION RATES

Adult	10–12 breaths per minute at 1.5 to 2.0 seconds each
Child	20 breaths per minute at 1.0 to 1.5 seconds each
Infant	20 breaths per minute at 1.0 to 1.5 seconds each
Newborn	40 breaths per minute at 1.0 to 1.5 seconds each

should be steady and of sufficient volume to make the chest rise. Make sure that you do not deliver too much air too fast, or you will force air into the patient's stomach.

4. **Determine if ventilations are adequate.** Watch the chest rise and fall. Listen and feel for air escaping when the patient exhales.

5. **Continue ventilations at the proper rate.** For adults, deliver 10–12 breaths per minute, each lasting 1.5 to 2.0 seconds. For infants and children, deliver 20 breaths per minute, each lasting 1.0 to 1.5 seconds. For a newborn, deliver 40 breaths per minute, each breath lasting 1.0 to 1.5 seconds.

If you cannot ventilate the patient or if the chest does not rise adequately, position the patient's head and try again. If the second try also fails, assume the airway is blocked by a foreign object. Then follow the guidelines (later in this chapter) for removing a foreign body airway obstruction.

Mouth-to-Mouth Ventilation

The risk of contracting infectious diseases combined with the availability of protective devices

Figure 5-20 Barrier device.

makes mouth-to-mouth ventilation inadvisable for use by fire service First Responders. As described earlier, barrier devices and face masks with one-way valves are available. You should always use them as a BSI precaution. However, your decision is a personal one. Be sure to follow your local protocols. Mouth-to-mask and mouth-to-barrier device techniques should not replace training in mouth-to-mouth ventilation.

Mouth-to-mouth ventilation is a quick, effective method of delivering oxygen to a nonbreathing patient. It involves ventilating the patient with your exhaled breath while making mouth-to-mouth contact. Use the mouth-to-mouth technique only in emergency situations in which no protective devices are available. For example, you may find that you need to perform mouth-to-mouth on a family member at home where a barrier device is not available.

In mouth-to-mouth, you form a seal with your mouth around the patient's mouth. The obvious risk to you is exposure to body fluids and thus to infectious disease.

To perform the technique, kneel at the patient's head. Then follow these steps:

1. **Open the patient's airway.** Use a head-tilt/chin-lift or a jaw-thrust maneuver.

2. **Form an airtight seal.** Gently squeeze the patient's nostrils closed with the thumb and index finger of the hand that is holding the patient's head-tilt. Take a deep breath and form an airtight seal with your lips around the patient's mouth. If you are ventilating an infant or small child, cover both the nose and mouth with your lips.

3. **Deliver two slow initial breaths** (over 1.5 to 2.0 seconds for an adult and 1.0 to 1.5 seconds for an infant or child). Each should be

steady and of sufficient volume to make the chest rise. Make sure that you do not deliver too much air too fast, or you will force air into the patient's stomach.

4. **Determine if ventilations are adequate.** Watch the chest rise and fall. Listen and feel for air escaping when the patient exhales.

5. **Continue ventilations at the proper rate.** For adults, deliver 10-12 breaths per minute, each lasting 1.5 to 2.0 seconds. For infants and children, deliver 20 breaths per minute, each lasting 1.0 to 1.5 seconds. For a newborn, deliver 40 breaths per minute, each breath lasting 1.0 to 1.5 seconds.

If you cannot ventilate the patient or if the chest does not rise adequately, position the patient's head and try again. If the second try also fails, assume the airway is blocked by a foreign object. Then follow the guidelines (later in this chapter) for removing a foreign body airway obstruction.

Mouth-to-Stoma Ventilation

A patient who has had all or part of the larynx surgically removed has had a laryngectomy (Figure 5-21). This patient will have a **stoma,** a permanent opening that connects the trachea directly to the front of the neck. These patients breathe only through the stoma.

To perform artificial ventilation on a patient with a stoma, remove all coverings such as scarves or ties from the stoma area. Then follow these steps (Figure 5-22):

1. **Clear the stoma of any foreign matter.** Use a gauze pad or handkerchief. Do not use tissue, which can shred and cling.

2. **Form an airtight seal around the stoma.** Whenever possible, use a barrier device such as a pocket face mask or shield.

3. **Blow slowly through the stoma** for 1.5 to 2.0 seconds. Use just enough force to make the patient's chest rise.

4. **Determine if ventilations are adequate.** Allow time for exhalation. Watch for the patient's chest to fall. Feel to make sure air is escaping back through the stoma as the patient exhales.

5. **Continue ventilations at the proper rate.** For adults, deliver 10–12 breaths per minute

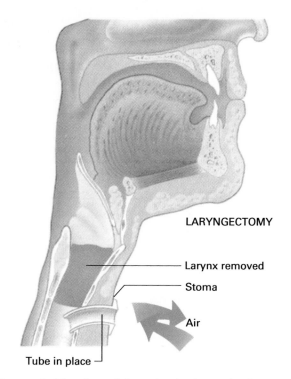

LARYNGECTOMY

Larynx removed

Stoma

Air

Tube in place

Figure 5-21 A neck breather's airway is changed by surgery.

with each lasting 1.5 to 2.0 seconds. For infants and children, deliver 20 breaths per minute with each lasting 1.0 to 1.5 seconds. For a newborn, deliver 40 breaths per minute with each breath lasting 1.0 to 1.5 seconds.

If the chest does not rise, the patient may be a "partial neck breather." This patient has had only part of the larynx removed. He or she can breathe through both the stoma and the mouth

Figure 5-22 Mask-to-stoma ventilation.

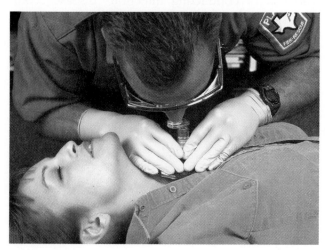

and nose. Seal the patient's nose and mouth with one hand. Pinch off the nose between your third and fourth fingers. Seal the lips with the palm of the same gloved hand. Hook your thumb under the patient's chin, and press up and back. Continue ventilations through the stoma.

Infants and Children

Many of the steps involved in managing the airway of an infant or child are the same as those for an adult. However, remember there are important differences.

You must position the head carefully when preparing for artificial ventilation. Keep an infant's head in a neutral ("sniffing") position. You can extend the head slightly beyond neutral if the patient is older than one year. But do not extend the infant's or child's head too far. The airway is more flexible than an adult's and can be easily overextended, which can in itself block the airway.

Consider an oral airway for an infant or child. The primary cause of blocked airway in these patients is the tongue. Try positioning the head and pulling the jaw forward to move the tongue away from the back of the throat. If that is unsuccessful, use an oropharyngeal airway to keep the tongue away from the throat and the airway open.

Depending on the age and size of an infant, you may be able to form a seal with your mouth over the infant's mouth and nose. If an adult-size pocket face mask must be used, position the mask upside down (Figure 5-23).

Guard against gastric distention (inflation of the stomach). It is common in infants and children who are being ventilated. During artificial ventilation, air may get into the esophagus and stomach if ventilations are too forceful. Gastric distention may significantly impair your attempts at ventilation, because it forces the diaphragm up, limiting the amount of air that can enter the lungs. Monitor the infant or child carefully to make sure the chest is rising and falling with ventilations. Listen and feel for exhaled air. Watch carefully to make sure the abdomen does not start to extend.

The American Heart Association (AHA) recommends against pressing on the abdomen to relieve gastric distention. To help avoid gastric distention keep the following points in mind:

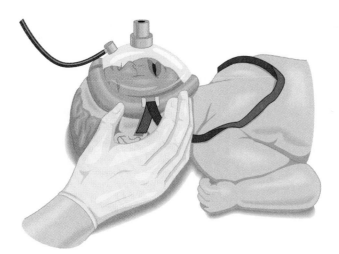

Figure 5-23   For an infant, the pocket face mask is reversed.

- Breathe slowly and with just enough force to make the chest rise. If you notice that the abdomen is starting to distend, reduce the force of your ventilations.
- Keep the infant's or child's head in a neutral position.
- Allow the infant or child to exhale between ventilations.
- Be alert for vomiting. If it looks like the infant or child is about to vomit, stop ventilating immediately and roll him onto his side. This position allows the vomit to flow from the mouth rather than into the lungs. If the infant or child vomits, suction or quickly wipe the mouth out with gauze pads. Wipe off the face, and return to ventilation.

GERIATRIC NOTE

Many patients have dentures or other dental appliances. If the dentures are secure in the mouth, leave them in place. It is much easier to create an airtight seal with them there. If the dentures are extremely loose, remove them so they do not block the airway. Partial dentures—plates and bridges—may become dislodged, too. If they are loose, remove them.

Reassess the mouth frequently in patients who have dental appliances to make sure they have not come loose.

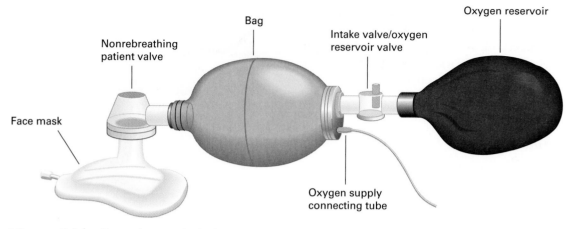

Nonrebreathing
patient valve

Bag

Intake valve/oxygen
reservoir valve

Oxygen reservoir

Face mask

Oxygen supply
connecting tube

Figure 5-24   Bag-valve-mask device.

Bag-Valve-Mask Ventilation

The bag-valve-mask (BVM) is a hand-operated device (Figure 5-24). It consists of a self-inflating bag, one-way valve, face mask, and oxygen reservoir. The BVM device has a volume of about 1600 milliliters. When used with oxygen, it can deliver almost 100% oxygen to the patient.

Note: it is highly recommended that two rescuers operate a BVM. It is too difficult and tiring for one rescuer to use. If you are alone, we recommend that you use a pocket face mask, which makes it easier for one rescuer to create a seal on the patient's face.

The BVM is available in infant, child, and adult sizes. Whatever the size, it should have the following features:

◆ Self-refilling bag that is disposable or easily cleaned and sterilized.

◆ Non-jam valve that allows a maximum oxygen inlet flow of 15 liters per minute or greater.

◆ No pop-off valve. If there is one, it should be disabled. Failure to do so may result in inadequate ventilations.

◆ Standardized 15/22 mm fittings.

◆ Oxygen inlet and reservoir that allows for a high concentration of oxygen.

◆ True nonrebreather valve.

◆ Ability to perform in all environments and temperature extremes.

To provide artificial ventilation by way of a BVM, follow these steps (Figure 5-25):

1. **Open the patient's airway.** An oral or nasal airway may be necessary in conjunction with the BVM. Note, if you suspect injury to the head or spine, stabilize the patient's head between your knees or have an assistant manually stabilize it.

2. **Select the correct size mask**—adult, child, or infant size.

3. **Attach oxygen,** if the BVM device has an oxygen port.

4. **Position the mask.** Place your thumbs over the top half of the mask. Index and middle fingers should be over the bottom half. Then put the narrow end (apex) of the mask over the bridge of the patient's nose. Lower the mask over the mouth and upper chin. If the mask has a large round cuff around a ventila-

Figure 5-25   Two rescuers operating a bag-valve-mask device.

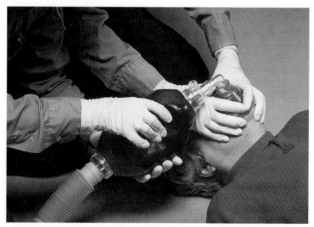

tion port, center the port over the patient's mouth. Use your ring and little fingers to bring the jaw up to the mask. Use the jaw-thrust, if you suspect head or spine injury. Be sure to avoid tilting the head or neck.

5. **Connect the mask to the bag,** if it has not already been done.

6. **Operate the bag.** Your partner should squeeze the bag with two hands until the patient's chest rises. For adults, deliver 10-12 breaths per minute with each lasting 1.5 to 2.0 seconds. For infants and children, deliver 20 breaths per minute with each lasting 1.0 to 1.5 seconds.

If you are alone, form a "C" around the ventilation port with your thumb and index fingers. Use your middle, ring, and little fingers under the jaw to maintain a chin lift and complete the seal. Squeeze the bag with your other hand while observing the chest rise and fall.

If the chest does not rise and fall with your ventilations, reposition the patient's head or jaw. Check again for an airway obstruction. If air is escaping from under the mask, reposition your fingers and check the position of the mask. If the patient's chest still does not rise, use an alternative method such as mask-to-mouth ventilation.

Assisting Inadequate Breathing

If your patient is breathing but breathing inadequately, you must assist. Provide ventilations while the patient is inhaling or trying to breathe on his or her own.

Since inadequate breathing is often slower than usual, you will need to provide additional ventilations in between the patient's own attempts to breathe. If breathing is rapid and very shallow, and therefore inadequate, provide assisted ventilations when the patient begins a respiration. You will find that providing a ventilation while the patient is exhaling can create resistance or an unusual noise.

Simply continue to time your assisted ventilations as best you can with the patient's own respiratory effort. Recognizing inadequate breathing and then assisting the patient with ventilations are among the best things you can do for your patient. Your intervention may prevent the patient from lapsing into complete respiratory and cardiac arrest.

Oxygen Therapy

Oxygen equipment can be an excellent tool for a well-trained fire service First Responder. It allows you to deliver oxygen to patients who desperately need it. However, if you are not allowed to administer oxygen in your EMS system, do not wait for it to arrive before providing emergency care.

Indications

Conditions that may require oxygen therapy include injury, heart or breathing problems, shock, and any other condition that prevents the efficient flow of oxygen throughout the body. Signs and symptoms that indicate the need for oxygen are:

◆ Poor skin color (blue, gray, or pale).

◆ Unresponsiveness.

◆ Cool, clammy skin.

◆ Respiratory distress.

◆ Blood loss.

◆ Chest pain.

◆ Trauma (injury).

Note that when you provide artificial ventilation, you can give a higher concentration of oxygen to the patient by connecting supplemental oxygen to your pocket face mask.

Oxygen Cylinders

All oxygen cylinders are manufactured according to strict governmental (U.S. Department of Transportation) regulations. According to those regulations, cylinders must be checked for safety at least once in 5 years. New cylinders should be checked every 10 years. (See Figure 5-26 for an example of a portable oxygen cylinder.)

A number of different types of oxygen cylinders are available. They vary in size and volume. Even though the volume of oxygen may vary, all of the cylinders when full are at the same pressure, about 2000 psi. Cylinder sizes are identified by letter. The following are sizes used in emergency medical care:

◆ D cylinder — 350 liters.

◆ E cylinder — 625 liters.

◆ M cylinder — 3000 liters.

◆ G cylinder — 5300 liters.

◆ H cylinder — 6900 liters.

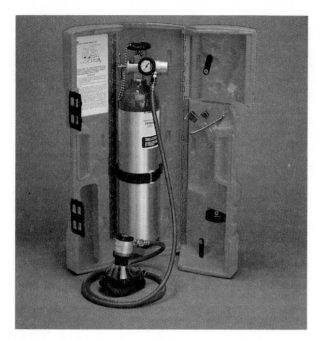

Figure 5-26 Basic portable oxygen cylinder.

Gas flow from an oxygen cylinder is controlled by regulators. They reduce pressure in the cylinder to a safe range of about 50 psi and control the flow to 15–25 liters per minute. Regulators are attached to the cylinder by a yoke. Each yoke fits only the cylinders made for one type of gas. In addition, all gas cylinders are color coded according to contents. Oxygen cylinders in the U.S. are generally steel green or aluminum gray.

Two types of regulators may be attached to oxygen cylinders. They are high-pressure regulators and therapy regulators.

The high-pressure regulator can provide 50 psi to power a demand-valve type resuscitator (flow-restricted oxygen-powered ventilation device) or a suction device. It has a threaded outlet and one gauge, which registers cylinder contents. It cannot be used interchangeably with the therapy regulator. It has no mechanism to adjust flow rate, and it is designed specifically for use with other equipment. To use a high-pressure regulator, attach the equipment supply line to the threaded outlet and open the cylinder valve fully. Then back off one-half turn for safety.

The therapy regulator can administer up to 15–25 liters of oxygen per minute. It has two controls. One shows cylinder contents and the other allows you to provide a metered flow of oxygen to the patient. The cylinder is full when the pressure is 2000 psi or greater. This pressure drops in direct proportion to the contents. For example, if the pressure is 1000 psi, the cylinder is half full. Adjust the flow meter to provide oxygen appropriate to the device used and the condition of the patient. Follow local protocol.

Safety Precautions. Observe the following safety precautions when you handle oxygen cylinders:

◆ Never allow combustible materials, such as oil or grease, to touch the cylinder, regulator, fittings, valves, or hoses.

◆ Never smoke or allow others to smoke in any area where oxygen cylinders are in use or on standby.

◆ Store the cylinders below 125° F (51.7° C).

◆ Never use an oxygen cylinder without a safe, properly fitting regulator valve.

◆ Never use a valve made for another gas, even if it has been modified.

◆ Keep all valves closed when the oxygen cylinder is not in use, even when a tank is empty.

◆ Keep oxygen cylinders secure to prevent them from toppling over. In transit, they should be in a carrier rack.

◆ Never place any part of your body over the cylinder valve. A loosely fitting regulator can be blown off with sufficient force to cause serious injury.

◆ Never stand an unsecured oxygen tank near the patient. If the tank is not in a commercial pack, lie it on its side by the patient.

Using the Cylinders. Prepare the tank if it is not used every day by following these steps (Figure 5-27):

1. **Check the cylinder.** Place the cylinder securely upright, and place yourself to the side. Then identify the cylinder as oxygen and remove the protective seal.

2. **"Crack" the tank.** Use the wrench supplied. Slowly open and rapidly close the cylinder valve to clear it of debris.

3. **Inspect the regulator valve** to be certain that it is the right type of oxygen cylinder. Be sure it has an intact washer. Place the yoke of the regulator over the cylinder valve and align the pins.

4. **Hand-tighten the T-screw** on the regulator.

5. **Open the main cylinder valve to check the pressure.** Make one-half turn beyond the point where the regulator valve becomes pressurized. Read the gauge to be sure that the tank has an adequate amount of oxygen. Most tanks will not function with less than 200 psi. Many organizations have a policy to replace or refill oxygen tanks that get below 500 psi.

6. **Attach the oxygen-delivery device to the regulator.**

7. **Adjust the flow meter** to the appropriate liter flow.

8. **Apply the oxygen delivery device to the patient.**

When you are ready to stop oxygen therapy, detach the mask from the patient. Then shut off the control valve until the liter flow is at zero. Shut off the main cylinder valve. Then bleed the valves by leaving the control valve open until the needle or ball indicator returns to zero. Shut the control valve on all tanks you carry.

It should be part of your daily routine to check the oxygen tank carried in your vehicle. You should open the main cylinder valve and check the pressure remaining on the cylinder. It is very discouraging, not to mention negligent, to arrive at the scene of a crash with lights, sirens, and other fanfare only to find that you have an empty oxygen cylinder.

Always replace a cylinder when the pressure is low. Follow local guidelines. Have backup portable oxygen cylinders in your vehicle. Note that oxygen itself does not burn. It does, however, feed and support combustion, making things burn that normally would not, especially when the oxygen is pressurized. Make absolutely sure that there are no open flames in the area when you are using oxygen.

Oxygen Delivery Equipment

A variety of devices are available to deliver oxygen to the patient. Proper training in their use is essential. Follow all local protocols.

Oxygen equipment either delivers low- or high-flow oxygen. Use high-flow oxygen through a **nonrebreather mask.** Use low-flow oxygen through a **nasal cannula.** Note that the patient must be breathing in order for you to use either of these devices. If the patient is not breathing or is breathing inadequately, begin artificial ventilation with supplemental oxygen.

Nonrebreather Mask. A nonrebreather mask has an oxygen reservoir bag and a one-way valve. The one-way valve allows the patient to inhale from the bag and exhale through the valve. Adjust the oxygen flow to prevent the bag from collapsing during inhalation, using about 10 to 15 liters per minute.

A nonrebreather mask requires a tight seal. If fitted properly to the face, it can deliver oxygen concentrations up to 90%. So it is ideally suited for patients who have severe hypoxemia (deficiency of oxygen in the blood), such as those with chest pain or injuries. Remember that flow rate must be adequate to keep the bag inflated as the patient breathes. If the bag collapses, the patient will not receive oxygen and may suffocate. Be ready to remove the mask if the patient vomits. Note also that caution must be exercised when using this device on patients with some chronic lung diseases. (See "Special Considerations" below.)

To use the nonrebreather mask (Figure 5-28):

1. **Select a mask.** It should have the oxygen supply tube preattached. The other end attaches to the oxygen source.

2. **Fill the bag.** Turn on the oxygen at 10 to 15 liters per minute to fill the bag. Then set the flow at the prescribed level. Make sure the bag is full before using it. Briefly cover the gasket where oxygen from the reservoir enters the mask to help fill the reservoir quickly.

3. **Apply the mask.** Gently place it over the patient's face. Slip the loosened elastic strap over the head so that it is positioned below or above the ears. Then pull the ends of the elastic until the mask fits the patient's face. Note that you should not lift the head of a trauma patient. Such movement may worsen a spine injury. Instead, gently tape the mask to the patient's cheeks. Then, monitor the airway.

Some masks have a thin piece of metal where the mask covers the bridge of the patient's nose. To ensure a good seal, that metal should be pinched so that the mask conforms to the shape of the nose.

Patients who have trouble breathing or who are in shock may get anxious when you try to place a mask on them. To these patients, it feels

Oxygen Administration

Figure 5-27a Identify the cylinder as oxygen and remove the protective seal.

Figure 5-27b Crack the main cylinder for one second to remove dust and debris.

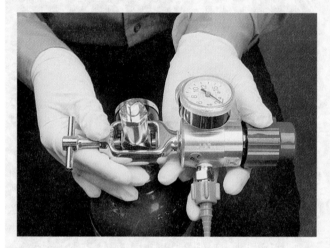

Figure 5-27c Place the yoke of the regulator over the cylinder valve and align the pins.

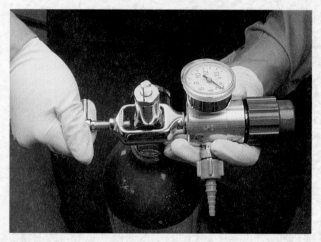

Figure 5-27d Hand-tighten the T-screw on the regulator.

as if they are being suffocated. Try to convince them to accept the mask. Explain to the patient what the mask is and why you are using it. You might tell them that the mask can feel confining, but that it also provides the high concentration of oxygen they need. But remember, a competent patient has a right to refuse treatment. Respect it.

As a last resort, use a nasal cannula to provide some oxygen. If that too is refused, the patient may allow you to hold the nasal cannula near his nose. This is referred to as the "blow-by" method.

Nasal Cannula. One of the most common oxygen devices is the nasal cannula. Its two soft plastic tips are inserted a short distance into the nostrils. The tips are attached to the oxygen source with thin tubing. It is comfortable and convenient. Most patients are able to tolerate it with ease. It is good for patients who are anxious about a mask, for those who are nauseated or vomiting, and in situations in which you need to communicate with the patient.

The nasal cannula provides safe, comfortable, low-flow oxygen in concentrations of 24% to 44% with a one- to six-liter flow. It should be used at

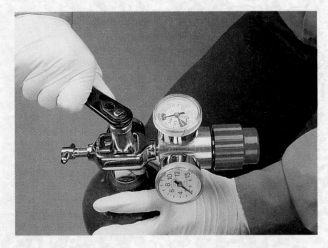

Figure 5-27e Open the main cylinder valve to check the pressure.

Figure 5-27f Attach the oxygen-delivery device to the regulator.

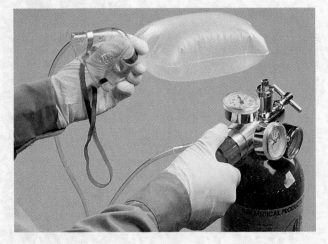

Figure 5-27g Adjust the flow meter to the appropriate liter flow.

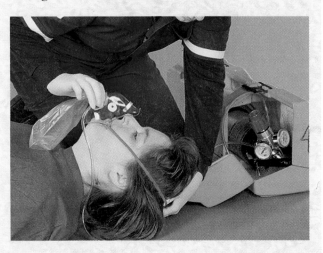

Figure 5-27h Apply the oxygen delivery device to the patient.

low rates of less than six liters per minute. Higher flows can cause headaches, drying of the membranes in the nose, and nosebleeds.

To use a nasal cannula (Figure 5-29):

1. **Set liter flow** to the desired rate before inserting the device. Make sure oxygen flows from the cannula tips.

2. **Insert the two tips.** Put them into the nostrils with the tab facing out.

3. **Position the tubing.** It should be over and behind each ear. Gently secure it by sliding the adjuster underneath the chin.

Do not adjust the tubing too tightly. If an elastic strap is used, adjust it so that it is secure but comfortable. If the tubing causes irritation, pad the patient's cheeks and behind the ears with 2″ × 2″ gauze pads. Be sure to check placement often. The cannula can be dislodged easily.

Special Considerations

WARNING! The term "respiratory depression" refers to a slow breathing rate of less than eight breaths per minute. It can occur—though rarely—when oxygen is applied to patients who have a

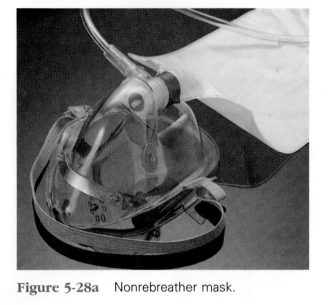

Figure 5-28a   Nonrebreather mask.

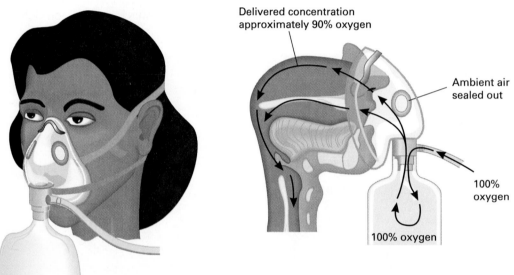

Delivered concentration
approximately 90% oxygen

Ambient air
sealed out

100%
oxygen

100% oxygen

Figure 5-28b   Nonrebreather applied to a patient.

chronic obstruction pulmonary disease (COPD) such as emphysema or chronic bronchitis. Beware of this rare complication.

Patients with COPD may, over time, lose the normal ability to use increased carbon dioxide levels in the blood as a stimulus to breathe. When this occurs, the COPD patient's body may use low blood oxygen as the factor that stimulates breathing.

Because of this so-called hypoxic drive, EMS personnel have for years been trained to administer only low concentrations of oxygen to these patients. However, more harm is done by withholding high-concentration oxygen from these patients than could be done by administering it.

As a First Responder, you will probably never see adverse conditions that result from oxygen administration. The time required for such conditions to develop is usually too long to cause any problem during emergency care in the field. The bottom line is this: Never withhold high-concentration oxygen from a patient who needs it.

Foreign Body Airway Obstruction (FBAO)

An upper airway obstruction is anything that blocks the nasal passages, the back of the mouth, or the throat. A lower airway obstruction can be caused by breathing in a foreign body or by

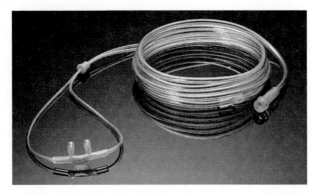

Figure 5-29a Nasal cannula.

Figure 5-29b Nasal cannula applied to a patient.

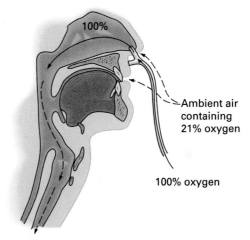

100%

Ambient air
containing
21% oxygen

100% oxygen

24% to 44% oxygen
concentration delivered

severe spasm of the bronchial passages. A **foreign body airway obstruction (FBAO)** is a true emergency. It must be cleared from the airway before the patient can breathe and before you can give artificial ventilation.

Airway obstruction in a responsive patient can be the cause of cardiac arrest. The most common FBAO is food. If your patient was eating prior to collapsing, suspect he or she choked on food. Elderly people are at risk for choking because they have a weaker gag reflex, as do people who are intoxicated, stroke victims, and handicapped patients. As a result, they are more frequently misdiagnosed as having heart disease if they collapse. Other common causes of airway obstruction in a responsive patient are bleeding into the airway and aspirated vomit.

Airway obstruction can also be the result of cardiac arrest. In an unresponsive patient, an FBAO may be caused by vomiting, loose or broken dentures or bridges, or injury to the face or jaw. The most common source of upper airway obstruction in an unresponsive patient is the tongue.

Other causes of airway obstruction include secretions, blood clots, cancerous conditions of the mouth or throat, enlarged tonsils, and acute epiglottitis.

Types of FBAO

There are two types of foreign body airway obstruction (FBAO)—partial and complete. A **partial FBAO** means an object is caught in the throat but it does not totally occlude (block) breathing. Even if the patient has good air exchange, you should never leave him alone with a partial airway obstruction. The obstruction can shift and become complete.

A patient with a partial FBAO but with good air exchange may:

◆ Remain responsive.

◆ Be able to speak.

◆ Cough forcefully.

◆ Wheeze between coughs.

A patient with a partial FBAO and poor air exchange may have:

◆ Weak, ineffective cough.

◆ High-pitched noise when inhaling.

◆ Increased respiratory difficulty and may clutch at the throat.

◆ Cyanosis (bluish discoloration of the skin and mucous membranes).

In a **complete FBAO,** all air exchange has stopped because an object fully occludes the patient's airway. The patient may be either responsive or unresponsive, depending in part on how long the airway has been blocked.

A patient with a complete FBAO will be unable to breathe, cough, or speak. He or she may clutch at the neck with thumb and fingers (the universal signal for choking). The amount of oxygen in the blood will decrease rapidly when air cannot enter the lungs. This will result in unresponsiveness. Death also will occur rapidly if the obstruction is not relieved.

The way you manage an obstructed airway depends on whether the obstruction is partial or complete.

Partial FBAO with Good Air Exchange

A patient with a partial obstruction and good air exchange is responsive and able to cough forcefully. In this case:

◆ Do not interfere with the patient's own attempts to dislodge the obstruction by coughing.

◆ Encourage the patient to "cough up" the foreign body.

◆ Do not make any specific attempts to relieve the obstruction.

◆ Never leave the patient until you are certain the airway is clear and there are no other problems that threaten the airway.

◆ If the patient cannot dislodge the object on his own, even if good air exchange continues, activate the EMS system.

◆ Place the patient in a position of comfort, where it is easiest for him or her to breathe.

Partial FBAO with Poor Air Exchange or Complete Obstruction

The American Heart Association (AHA) recommends the **Heimlich maneuver** in cases of partial airway obstruction with poor air exchange and in cases of complete airway obstruction. Also called "subdiaphragmatic abdominal thrusts," or simply "abdominal thrusts," the Heimlich maneuver pushes the diaphragm quickly upward (Figure 5-30). This action forces enough air from the lungs to dislodge and expel the foreign object. The AHA recommends against the use of back blows in an adult.

Each individual abdominal thrust must be delivered with enough force and pressure to dislodge the foreign object. You must keep trying if the first thrust is unsuccessful. Deliver each thrust with the intent of relieving the obstruction. It may take as many as five or more thrusts to succeed.

Responsive Adult. If the patient is responsive, perform the Heimlich maneuver as follows (Figure 5-31):

1. Get in position. Stand behind the patient. Wrap your arms around his waist. Keep your elbows out, away from his ribs.

Figure 5-30 Abdominal thrusts push the diaphragm up, forcing air to expel the foreign object.

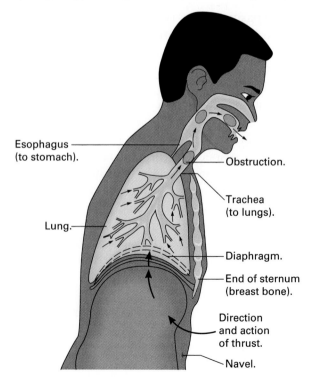

Esophagus (to stomach).

Obstruction.

Trachea (to lungs).

Lung.

Diaphragm.

End of sternum (breast bone).

Direction and action of thrust.

Navel.

Foreign Body Airway Obstruction— Responsive Adult

Figure 5-31a Universal sign of choking.

Figure 5-31b Determine if the patient can speak or cough by asking, "Are you choking?"

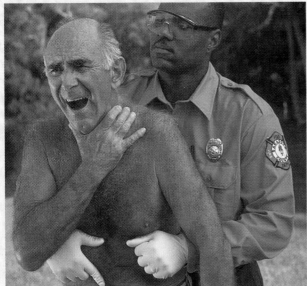

Figure 5-31c If so, perform the Heimlich maneuver to dislodge the object.

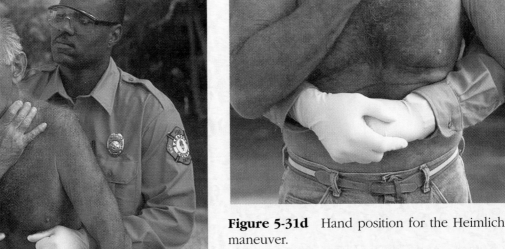

Figure 5-31d Hand position for the Heimlich maneuver.

2. **Position your hands.** Make a fist with one hand. Place the thumb side of the fist on the middle of the abdomen slightly above the navel and well below the xiphoid process.

3. **Perform an abdominal thrust.** First, grasp your fist with your other hand, thumbs toward the patient. Then press your fist into the patient's abdomen with a quick inward and upward thrust.

4. If the first thrust does not dislodge the foreign body, make each new thrust separate and distinct. Continue until the object is expelled or the patient becomes unresponsive.

Beware of certain dangers. First, if you are improperly positioned or if you perform the thrusts too rapidly or too forcefully, you can lose your balance and fall into the patient. If your hands are too high, you could cause internal injury. Finally, the Heimlich maneuver can cause vomiting. Correct hand placement and use of appropriate force minimizes this risk.

Pregnant or Obese Responsive Adult. If the patient is in the advanced stages of pregnancy or is markedly obese, there may be no room between the rib cage and the abdomen to perform abdominal thrusts, or you may be unable to reach around the patient. In these cases, perform chest thrusts as follows (Figure 5-32):

1. Get in position. Stand behind the patient. Place your arms directly under the patient's armpits. Wrap your arms around the patient's chest.

2. Position your hands. Make a fist with one hand. Place the thumb side of your fist on the middle of the patient's sternum. If you are near the margins of the rib cage, your hand is too low.

3. Perform a chest thrust. First, seize your fist firmly with your other hand. Then thrust backward sharply.

4. If the first thrust does not dislodge the foreign body, repeat thrusts until the object is expelled or the patient becomes unresponsive.

Unresponsive Adult. If your patient is unresponsive when you find him, activate the EMS system. Then place the patient in a supine position and proceed with the following (Figure 5-33):

1. Attempt to ventilate the patient. First open the airway. Then try to ventilate using the mouth-to-mask, mouth-to-barrier device, or mouth-to-mouth technique.

2. If ventilation is unsuccessful, reposition the patient's head and try again. If ventilation is still unsuccessful, proceed with the following.

3. Get in position. Kneel astride the patient's thighs.

Figure 5-32 Chest thrusts on a standing obese patient with an FBAO.

4. Position your hands. Place the heel of one hand on the midline of the patient's abdomen, slightly above the navel and well below the xiphoid process. Place your second hand on top of your first.

5. Perform up to 5 abdominal thrusts. Press into the abdomen with quick upward thrusts. If you are in the right position, the thrusts will stay in the center of the abdomen and will not veer to the left or right.

6. Perform a tongue-jaw lift. That is, move to the head, grasp both the tongue and the lower jaw between your thumb and fingers and lift the jaw.

7. Perform a finger sweep. Use the hooked index finger of your gloved hand to sweep deeply into the mouth along the inner cheeks and to the base of the tongue. Use a hooking motion to dislodge the foreign object, move it to the mouth, and remove it. Remember, use this maneuver only on an unresponsive patient.

8. If the foreign body is not dislodged, repeat the sequence. That is, alternate these maneu-

Foreign Body Airway Obstruction— Unresponsive Adult

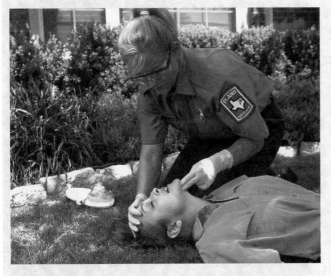

Figure 5-33a If the patient is unresponsive when you find him, open the airway.

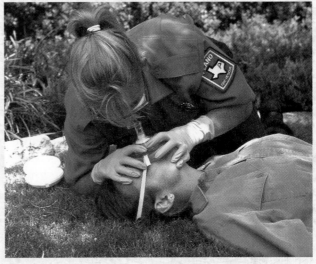

Figure 5-33b Try to ventilate. If unsuccessful, reposition the head and try again.

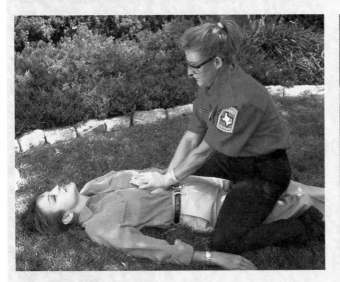

Figure 5-33c Deliver up to five abdominal thrusts.

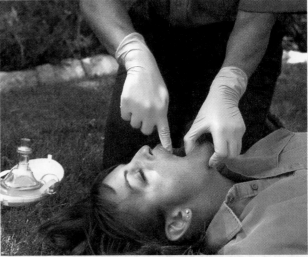

Figure 5-33d Remove the object, or repeat attempted ventilation, abdominal thrusts, tongue-jaw lift, and finger sweep until effective.

vers in rapid sequence: attempted ventilation, repeated abdominal thrusts, and tongue-jaw lift with finger sweep. Continue until the foreign body is expelled and ventilation is successful or until you are relieved by other EMS personnel.

If a choking patient becomes unresponsive while you are attempting to dislodge an FBAO, then follow these steps:

1. **Perform a tongue-jaw lift.** This action will draw the tongue away from the back of the

throat and a foreign body that may be lodged there. In fact, the maneuver itself may partially relieve the obstruction.

2. **Perform a finger sweep** to remove the object.

3. **If necessary, continue with airway care** as outlined above for an unresponsive adult.

Pregnant or Obese Unresponsive Adult. If you find an unresponsive patient who is in the late stages of pregnancy or who is markedly obese, attempt to ventilate. If you are unsuccessful, reposition the head and try again. If still unsuccessful, you must use chest thrusts (Figure 5-34). That is, kneel close to the patient's side. Place the heel of one hand directly over the lower half of the sternum, above the tip of the xiphoid process. Place your second hand on top of the first. Give distinct, separate thrusts downward.

If the first few thrusts do not dislodge the foreign body, repeat the sequence of attempted ventilation, repeated thrusts, tongue-jaw lift, and finger sweep in rapid sequence. Continue until the foreign body is expelled and ventilation is successful or until you are relieved by other EMS personnel.

FBAO in Infants and Children

Manage a complete airway obstruction in children older than eight years the same way you would for adults. Treatment for infants (birth to one year) and children (one to eight years) is different.

More than 90% of childhood deaths for FBAO are in children younger than five years old. Nearly 65% of those who die are infants. The most common causes of FBAO in infants and small children are toys, balloons, small objects such as plastic lids, and food such as hot dogs, round candies, nuts, and grapes.

Airway obstruction in an infant or small child also can be caused by swelling and infection. Both narrow the airway. Croup and epiglottitis can cause complete blockage of the airway.

Suspect that obstruction is caused by infection instead of by a foreign object if you detect:

◆ Fever, especially if accompanied by congestion.

◆ Hoarseness.

◆ Drooling.

◆ Lethargy or limpness.

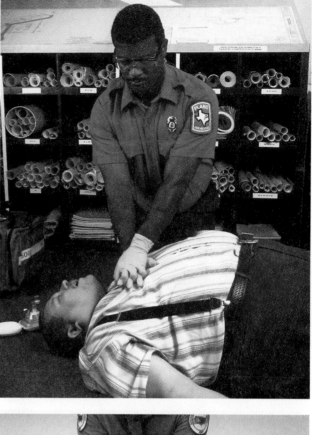

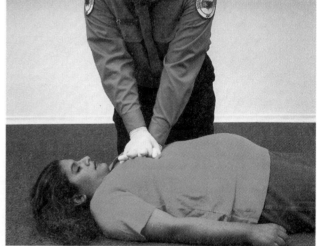

Figure 5-34 Chest thrusts for an FBAO in a supine obese patient (top) and a supine pregnant patient (bottom).

If you suspect the obstruction is caused by infection, arrange for immediate transport to a medical facility.

Suspect an FBAO in an infant or child who has sudden onset of respiratory distress associated with coughing, gagging, stridor, or wheezing, especially when food or small items are found near the child. Try to clear only a complete airway obstruction or a partial airway obstruction with poor air exchange.

Never perform a blind finger sweep on an infant or child. In a child age one or older, perform a tongue-jaw lift, look into the airway, and use your finger to sweep the foreign body out only if you can actually see it.

Responsive Infant. If the infant has a partial airway obstruction but still has good air exchange, activate the EMS system. Let the infant try to expel the object by coughing. Place the patient in a position of comfort (in the parent's arms if possible) so that secretions and vomit will drain out of the mouth. The jaw will also fall forward, bringing the tongue and epiglottis away from the back of the throat.

If the infant has serious difficulty breathing, an ineffective cough, and no strong cry, he or she has a partial obstruction with poor air exchange.

According to the AHA, you should perform the following procedure only if the infant has a complete FBAO or a partial obstruction with poor air exchange and only if the obstruction is due to a witnessed or a strongly suspected foreign object. Do not perform the following procedure if you suspect the obstruction is caused by infection. Instead, arrange for immediate transport to a medical facility.

To relieve an FBAO in a responsive infant:

1. **Call out "Help!"** When someone responds, have that person activate the EMS system. If you are alone, activate the EMS system within one minute, if you are unable to clear the airway.

2. **Get in position.** Straddle the infant over one of your arms, face down with his head lower than the rest of the body. Rest your arm on your thigh for support. Support the infant's head by firmly holding the jaw with your hand.

3. **Deliver up to five back blows.** Use the heel of your hand between the shoulder blades (Figure 5-35).

4. **If the FBAO is not expelled,** turn the infant face up. Support him on your arm, with his head lower than the rest of his body.

5. **Position your hand.** Place your middle and ring finger over the middle of the infant's sternum. They should be one finger-width below

Figure 5-35   Back blows for an infant with an FBAO.

an imaginary line drawn between the infant's nipples (Figure 5-36).

6. **Deliver up to five chest thrusts.** Use a quick downward motion.

7. **If the first set of thrusts do not dislodge the foreign body,** continue alternating sets of five back blows and five chest thrusts until the FBAO is expelled or the infant becomes unresponsive.

Figure 5-36   Locating finger position for chest thrusts on an infant with an FBAO.

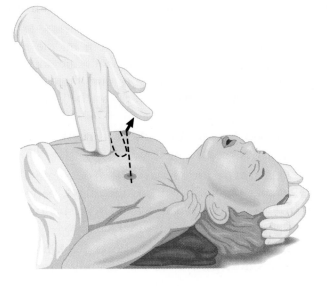

When an FBAO is expelled, a responsive infant will usually start to cry and make noise. An unresponsive infant may not make noise or breathe spontaneously. You will need to try to ventilate the infant to determine if air is reaching the lungs.

Unresponsive Infant. If you find an infant who is unresponsive:

1. **Call out "Help!"** When someone responds, have that person activate the EMS system. If you are alone, activate the EMS system within one minute, if you are unable to clear the airway.

2. **Position the patient.** Place the infant on his back on a firm, hard surface. Support his head and neck. Note that the infant's head should be in a neutral position. Overextension of the infant's neck can obstruct the airway.

3. **Attempt to ventilate the patient.** First open the airway. Then seal your mouth over the infant's mouth and nose. Deliver slow, gentle breaths until you see the chest rise.

4. **If ventilation is unsuccessful,** reposition the infant's head and try again. If ventilation is still unsuccessful, proceed with the following.

5. **Get in position.** Straddle the infant over one of your arms, face down with his head lower than the rest of the body. Rest your arm on your thigh for support. Support the infant's head by firmly holding the jaw with your hand.

6. **Deliver up to five back blows.** Use the heel of your hand between the infant's shoulder blades.

7. **If the FBAO is not expelled,** turn the infant face up. Support him on your arm, with his head lower than the rest of his body.

8. **Position your hand.** Place your middle and ring fingers over the middle of the infant's sternum. They should be one finger-width below an imaginary line drawn between the infant's nipples.

9. **Deliver up to five chest thrusts.** Use a quick downward motion.

10. **Perform a tongue-jaw lift.** Look into the infant's mouth. If you see the object, perform a finger sweep to remove it. (Never perform a blind finger sweep in an infant.)

11. **If the first set of ventilations, back blows, and chest thrusts do not dislodge the for-**

eign body, repeat the procedure. Continue with alternating sets of attempts to ventilate, five back blows, five chest thrusts, and tongue-jaw lift until the FBAO is expelled or until you are relieved by another EMS worker.

If a choking infant becomes unresponsive while you are attempting to dislodge an FBAO, have a second person activate EMS. Perform a tongue-jaw lift and, if you can see the object, perform a finger sweep to remove it. If necessary, continue with airway care as described above for an unresponsive infant.

Once the foreign body is removed, check the infant for breathing and pulse. If there is no pulse, begin infant CPR as described in Chapter 6. If the infant is breathing and has a pulse, place him or her in a recovery position. Monitor breathing and pulse while you maintain an open airway.

Responsive Child. A patient one to eight years old is considered a child. If a child has a partial obstruction with good air exchange, encourage him to expel it himself by coughing. If there is a partial obstruction with poor air exchange or a complete obstruction, then follow the steps outlined below (Figure 5-37):

1. **Get in position.** Stand behind the child. Wrap your arms around his waist. Keep your elbows out, away from his ribs.

2. **Position your hands.** Make a fist with one hand. Place the thumb side of the fist on the

Figure 5-37 Performing the Heimlich on a standing or sitting responsive child with an FBAO.

middle of the abdomen slightly above the navel and well below the xiphoid process. Never place your hands on the xiphoid process or on the lower edge of the ribs. Keep in mind a child's smaller size and proportions as you determine hand placement.

3. **Perform an abdominal thrust.** First, grasp your fist with your other hand, thumbs toward the patient. Then press your fist into the patient's abdomen with a quick inward and upward thrust.

4. **If the first thrust does not dislodge the foreign body,** make each new thrust separate and distinct. Continue until the object is expelled or the patient becomes unresponsive.

Unresponsive Child. If you find an unresponsive child, have a second person activate the EMS system. Then:

1. **Position the patient.** Place the child in a supine position.

2. **Attempt to ventilate the patient.** First open the airway. Then attempt to ventilate using the mouth-to-mask, mouth-to-barrier device, or mouth-to-mouth technique.

3. **If ventilation is unsuccessful,** reposition the patient's head and try again. If ventilation is still unsuccessful, proceed with the following.

4. **Get in position.** Kneel astride the child's thighs.

5. **Position your hands.** Place the heel of one hand on the midline of the child's abdomen, slightly above the navel and well below the xiphoid process. Place your second hand on top of your first.

6. **Perform up to five abdominal thrusts.** Press into the abdomen with quick upward thrusts.

7. **Perform a tongue-jaw lift.** If you see the object, perform a finger sweep to remove it. Do not perform blind finger sweeps on a child.

8. **If the first five thrusts and tongue-jaw lift do not dislodge the foreign body,** repeat the sequence. That is, alternate these maneuvers in rapid sequence: attempted ventilation, abdominal thrusts, and tongue-jaw lift. Continue until the foreign body is expelled and ventilation is successful. If you are not successful after one minute, activate the EMS if it has not already been done.

If a choking child becomes unresponsive while you are attempting to dislodge an FBAO, have a second person activate EMS. Perform a tongue-jaw lift and, if you can see the object, remove it with a finger sweep. Then, if necessary, continue with airway care as outlined above for an unresponsive child.

 **ON SCENE FOLLOW-UP**

> At the beginning of this chapter, you read that fire service First Responders were called to help an unresponsive adult with snoring respirations. To see how chapter skills apply to this emergency, read the following. It describes how the call was completed.

INITIAL ASSESSMENT (CONTINUED)

Pete, my partner, immediately performed a maneuver to open the patient's airway, which eliminated the snoring sounds in her throat. Pete suctioned the airway to remove the build up of secretions. Mrs. Franklin, the patient, didn't have a gag reflex, so she accepted the oropharyngeal airway well.

Our assessment of her breathing revealed that there was minimal chest movement and slow respirations. She also was beginning to show signs of blue coloring around her lips.

Realizing that Mrs. Franklin was breathing inadequately and her pulse was rapid and weak, we ventilated her using a pocket face mask with one-way valve and supplemental oxygen. Johnny, the lieutenant and officer in charge, radioed the ambulance. He updated them on the patient's condition and made sure ALS was en route.

PHYSICAL EXAMINATION

Mr. Franklin told us his wife had a medical problem, not a traumatic condition. I did a quick physical exam, while Pete continued to assist ventilations. There were no signs of injury on Mrs. Franklin's head, neck, chest, abdomen, or extremities. Her pulse was 96 and weak. The respiratory rate was 8 and shallow.

PATIENT HISTORY

Mr. Franklin went on to tell us that his wife had a heart attack several years ago. She takes medications for her heart and high blood pressure. She complained of a headache about an hour before she became unresponsive. She ate breakfast earlier. She has no allergies.

ONGOING ASSESSMENT

Our primary focus was on making sure Mrs. Franklin was ventilated properly. We also checked her pulse frequently. On our second check, her pulse was 104, bounding, and regular. Her respirations were about 6 and shallow, being assisted.

PATIENT HAND-OFF

We had the airway under control when the ambulance arrived. The patient's color had improved. Her pulse also had slowed a bit. Johnny, our lieutenant, gave the EMTs the hand-off report:

"This is Mrs. Franklin. She is 74 years old. She had a headache about an hour ago. She was found by her husband slumped over in a chair. He moved her to the floor. We found that she had inadequate ventilations and began assisting with a pocket mask and oxygen. She will groan with loud verbal stimulus. Her respiratory rate is about 6, pulse 104 and bounding. She has a history of heart attack and high blood pressure. She ate breakfast. She has no allergies."

I saw Mr. Franklin in the grocery store recently. He told me that his wife had had a severe stroke. She remained in the hospital for some time and was eventually moved to a rehabilitation center. He thanked me again and told me that he hoped his wife would be home soon.

It has been said that the priorities in patient care are: airway, airway, and airway! Without a clear and open airway plus adequate ventilations, no patient can survive. So even as you approach your patient's side, the first questions in your mind should be "Is she breathing? Is she breathing adequately?" Remember, without an airway, there is no chance of survival.

Chapter Review

Without adequate respiration, we die. The body requires oxygen to live. These statements alone provide your focus. As a fire service First Responder, you must evaluate respiration. Then do everything possible to make sure respirations continue adequately and without obstruction.

Consider this scenario. You come upon the scene of a motor-vehicle crash. You find a patient who has been ejected from the vehicle. While he appears to have multiple injuries, you immediately notice bleeding from the mouth and nose. The patient is unresponsive and is making gurgling sounds from his airway.

After performing a scene size-up and taking BSI precautions, your first priority is the airway. The patient needs suctioning. Depending on the amount of bleeding, he may need frequent suctioning. If you leave the airway, even briefly to perform more of your assessment, the patient will lose his airway and die.

If all you do is take care of the airway, and you do it well, you have done everything possible to save a life. A nicely assessed and packaged patient who does not have an airway has not been properly cared for. In the ABCs and in the care you provide, the airway comes first.

FIRE COMPANY REVIEW

Page references where answers may be found or supported are provided at the end of each question.

Section 1

1. What are the nine major components of the respiratory system? (pp. 75–76)

Section 2

2. What are two maneuvers that you can use to open a patient's airway? Describe when you should use each one. (pp. 78–80)

3. What are two types of airway adjunct? Describe when you should use each one. (p. 81)

4. How can you find out if a patient is breathing? (pp. 86–88)

5. What are the signs of inadequate breathing? (pp. 87–88)

Section 3

6. What are three techniques used by a fire service First Responder to artificially ventilate a patient? Name them in order of preference. (p. 88)

7. How do you perform artificial ventilation? Briefly describe each step, including rates for infants, children, and adults. (pp. 88–95)

8. What are the signs that show your ventilations are adequate? (p. 88)

9. How do you relieve a foreign body airway obstruction in a responsive adult? Briefly, describe each step. (pp. 102–104)

10. How do you relieve a foreign body airway obstruction in an unresponsive adult? Briefly describe each step. (pp. 104–106)

RESOURCES TO LEARN MORE

Roberts, J.R., and J.R. Hedges. *Clinical Practices in Emergency Medicine,* Third Edition. Philadelphia: W.B. Saunders, 1997.

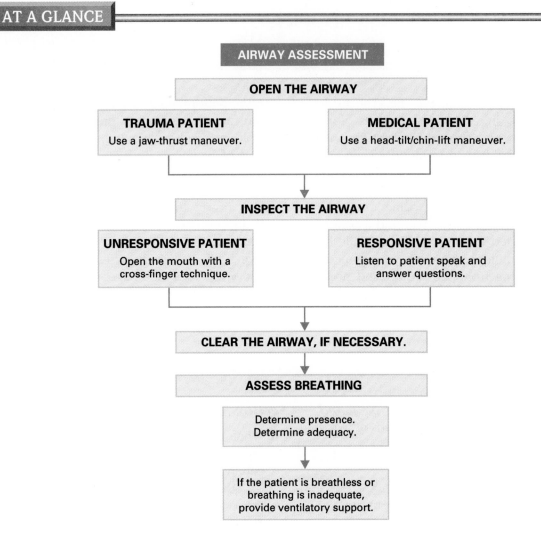

AIRWAY ASSESSMENT

OPEN THE AIRWAY

TRAUMA PATIENT
Use a jaw-thrust maneuver.

MEDICAL PATIENT
Use a head-tilt/chin-lift maneuver.

INSPECT THE AIRWAY

UNRESPONSIVE PATIENT
Open the mouth with a
cross-finger technique.

RESPONSIVE PATIENT
Listen to patient speak and
answer questions.

CLEAR THE AIRWAY, IF NECESSARY.

ASSESS BREATHING

Determine presence.
Determine adequacy.

If the patient is breathless or
breathing is inadequate,
provide ventilatory support.

Circulation

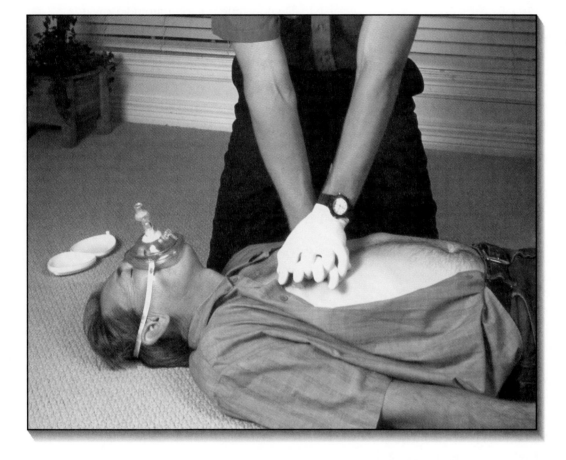

*I*NTRODUCTION   *Heart disease takes the lives of about 600,000 Americans each year. Included in that total are the 350,000 people a year—about 1,000 a day—that suffer from sudden cardiac arrest. Only one in 20 survive. The people who can be saved need* immediate *CPR followed by advanced medical care, including defibrillation, within 8 to 10 minutes of collapse.*

But even with CPR, many patients will not live. They may have been without a pulse or breathing for too long or cardiac arrest may have caused irreversible damage to the heart. Please do not let that

113

discourage you. Emergency care in the field is still critical to saving many lives. And today's new hospital techniques often help to reverse the crippling effects of heart attack or cardiac arrest.

Note that several studies have shown that you can lose your CPR skills unless you have frequent practice and retraining. Retraining should occur every one to two years.

OBJECTIVES

Cognitive, affective, and psychomotor objectives are from the U.S. DOT's "First Responder: National Standard Curriculum." Enrichment objectives, if any, identify supplemental material.

Cognitive

4-1.1   List the reasons for the heart to stop beating. (p. 116)

4-1.2   Define the components of cardiopulmonary resuscitation. (p. 117)

4-1.3   Describe each link in the chain of survival and how it relates to the EMS system. (pp. 116–117)

4-1.4   List the steps of one-rescuer adult CPR. (pp. 121–123)

4-1.5   Describe the technique of external chest compressions on an adult patient. (pp. 119–121)

4-1.6   Describe the technique of external chest compressions on an infant. (pp. 130–132)

4-1.7   Describe the technique of external chest compressions on a child. (pp. 130–132)

4-1.8   Explain when the First Responder is able to stop CPR. (p. 117)

4-1.9   List the steps of two-rescuer adult CPR. (p. 125)

4-1.10 List the steps of infant CPR. (pp. 130–132)

4-1.11 List the steps of child CPR. (pp. 130–132)

Affective

4-1.12 Respond to the feelings that the family of a patient may be having during a cardiac event. (p. 117)

4-1.13 Demonstrate a caring attitude towards patients with cardiac events who request emergency medical services. (p. 117)

4-1.14 Place the interests of the patient with a cardiac event as the foremost consideration when making any and all patient care decisions. (p. 117)

4-1.15 Communicate with empathy with family members and friends of the patient with a cardiac event. (p. 117)

Psychomotor

4-1.16 Demonstrate the proper technique of chest compressions on an adult. (pp. 119–121)

4-1.17 Demonstrate the proper technique of chest compressions on a child. (pp. 130–132)

4-1.18 Demonstrate the proper technique of chest compressions on an infant. (pp. 130–132)

4-1.19 Demonstrate the steps of adult one rescuer CPR. (pp. 121–123)

4-1.20 Demonstrate the steps of adult two rescuer CPR. (p. 125)

4-1.21 Demonstrate child CPR. (pp. 130–132)

4-1.22 Demonstrate infant CPR. (pp. 130–132)

ON SCENE

DISPATCH

Our F.A.S.T. (First Aid Stabilization Team) unit was dispatched on an EMS assist because the closest ambulance was unavailable. The call was for a possible heart attack. As I climbed into our vehicle, I realized that time would really count.

SCENE SIZE-UP

There were three of us assigned to the F.A.S.T unit. Being the senior officer, I sized up the scene carefully from the cab before we got out. I reminded the others to put on their BSI equipment. An elderly man met us half-way to the door and stated, "You better hurry up. They're doing CPR on my brother."

INITIAL ASSESSMENT

Inside the home, we observed a woman performing CPR on an older gentleman. The woman seemed to be doing pretty well but she was getting tired. My crew approached and took over. Checking the patient's ABCs (airway, breathing, and circulation), we found no pulse or respirations. We continued with CPR.

> *CPR is an important skill for the First Responder. Though needed in only a small percentage of calls, when it is used, it is vitally important. In the scenario described above, the firefighters have begun CPR. How long do you think they should continue? Could CPR injure the patient? Will the patient live? Consider these questions as you read Chapter 6.*

SECTION 1
THE CIRCULATORY SYSTEM

The circulatory system is responsible for delivering oxygen and nutrients to the body's tissues. It also is responsible for removing waste from the tissues. Its basic components are the heart, arteries, veins, capillaries, and blood.

The heart is a hollow, muscular organ about the size of a fist. It lies in the lower left central region of the chest between the lungs. It is protected in the front by the ribs and sternum (breastbone). In the back it is protected by the spinal column. The heart contains four chambers. The two upper ones are the left and right atria. The two lower ones are the left and right ventricles. The septum (a wall) divides the right side of the heart from the left side. The heart also contains several one-way valves that keep blood flowing in the correct direction.

The circulatory system contains blood vessels. These vessels transport blood throughout the body. Arteries transport blood away from the heart. Veins carry blood back to the heart. The tiny capillaries allow for the exchange of gases and nutrients between the blood and the cells of the body.

How the Heart Works

The heart is like a two-sided pump (Figure 6-1). The left side receives oxygenated blood from the lungs and pumps it to all parts of the body. The right side receives blood from the body and then pumps it to the lungs to be reoxygenated.

The blood is kept under pressure and in constant circulation by the heart's pumping action. In a healthy adult at rest, the heart contracts between 60 and 80 times per minute. The pulse is a sign of the pressure exerted during each contraction of the heart. Every time the heart pumps, a wave of blood is sent through the arteries. That wave is

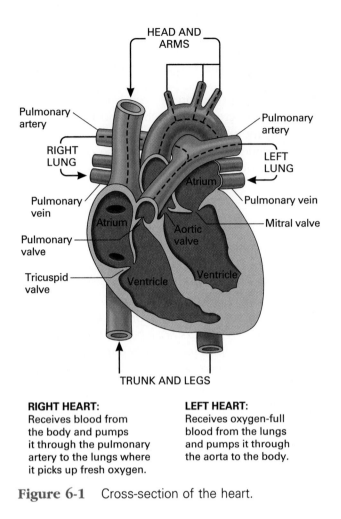

RIGHT HEART:
Receives blood from the body and pumps it through the pulmonary artery to the lungs where it picks up fresh oxygen.

LEFT HEART:
Receives oxygen-full blood from the lungs and pumps it through the aorta to the body.

Figure 6-1   Cross-section of the heart.

felt as a pulse. It can be palpated, or felt, most easily where a large artery lies over a bone close to the skin. Sites include: the carotid pulse in the neck, the brachial pulse in underside of the upper arm, the radial pulse in the thumb side of the wrist, and the femoral pulse in the upper thigh.

The pulse is felt most easily over the carotid artery on either side of the neck. The carotid artery should be palpated (felt) first when a patient is unresponsive. The radial artery on the thumb side of the inner surface of the wrist is also easy to feel. It should be palpated first when the patient is responsive.

The heart, lungs, and brain work together closely to sustain life. The smooth functioning of each is critical to the others. When one organ cannot perform properly, the other two are handicapped. If one fails, the other two soon will follow.

When the Heart Stops

Clinical death occurs when a patient is in **respiratory arrest** and **cardiac arrest.** Immediate

CPR slows the death process and is critical for patient survival. However, if a patient is clinically dead for 4 to 6 minutes, brain cells begin to die. After 8 to 10 minutes without a pulse, irreversible damage occurs to the brain, regardless of how well CPR is performed.

There are many reasons why a heart will stop. They include heart disease, stroke, allergic reaction, diabetes, prolonged seizures, and other medical conditions. The heart also may stop because of a serious injury. In infants and children, respiratory problems are the most common cause of cardiac arrest. This is why airway care is so important in young patients.

> **COMPANY OFFICER'S NOTE**
>
> Remember: A heart attack and cardiac arrest are not the same! A person having a heart attack will most likely be conscious and talking to you when you arrive on scene. However, it is important for you to know that 30% of heart-attack victims will be in cardiac arrest when the fire department First Responder arrives on scene. Therefore, hope for the best (a conscious, alive, and talking patient), but be prepared for the worst (an unconscious, non-breathing, and pulseless patient in cardiac arrest).

The patient in respiratory and cardiac arrest has the best chance of surviving if all of the links in the **chain of survival** come together. This "chain" as identified by the American Heart Association (AHA) contains four links:

◆ *Early access.* Lay people must activate the EMS system immediately. The use of a universal access number (9-1-1, for example) helps to speed system access.

◆ *Early CPR.* Family members, citizens, and First Responders must be trained in CPR and begin as soon as possible. CPR will help to sustain life until the next step.

◆ *Early defibrillation.* Defibrillation is the process by which an electrical current is sent to the heart to correct fatal heart rhythms. The earlier defibrillation can be performed, the better.

Early advanced care. Advanced care, or the administration of medications and other advanced therapies, must start as soon as possible. This can be done by paramedics at the scene or by prompt transportation to the emergency department.

As a First Responder, you have an important role. You can provide early CPR and, if permitted in your area, defibrillation.

The principle of CPR is to oxygenate and circulate the blood of the patient until defibrillation and advanced care can be given. Any delay in starting CPR increases the chances of nervous system damage and death. The faster the response, the better the patient's chances are. Survival rates improve when the time between arrest and the delivery of defibrillation and other advanced measures is short.

FIRE DRILL

When responding to a structure fire, it is important to get equipment and manpower on scene as quickly as possible. In fact, the time of ignition of a fire to flashover is about eight minutes (time-temperature curve). Therefore, it is critical to assess, perform rescue, and ventilate as soon as possible in order to avoid "irreversible" damage to the structure. Similarly, it is just as critical to assess, perform rescue breathing and chest compressions, and defibrillate within eight minutes to avoid irreversible damage to the structure of the brain.

SECTION

2 CARDIOPULMONARY RESUSCITATION (CPR)

According to the American Heart Association (AHA), proper assessment of the patient's airway, breathing, and circulation is critical to successful CPR. The AHA also states that no patient should undergo the intrusive procedures of CPR until need is clearly established.

You can establish the need for CPR by determining that the patient is unresponsive, breathless, and pulseless.

To perform CPR, you must maintain an open airway, provide artificial ventilation, and provide artificial circulation by means of chest compressions. (For a detailed discussion of airway maintenance and artificial ventilation, see Chapter 5.)

CPR must begin as soon as possible and continue until the:

◆ First Responder is exhausted and is unable to continue.

◆ Patient is turned over to another trained rescuer or the hospital staff.

◆ Patient is resuscitated.

◆ Patient has been declared dead by a proper authority.

A cardiac event is very serious. Without your interventions, the patient may not survive. So remember to place his or her interests first. And be sure to demonstrate a caring attitude. When possible, respond to the feelings of the patient's family and friends with empathy.

Steps Preceding CPR

Before providing CPR to a patient, you must first (Figure 6-2):

◆ Determine unresponsiveness.

◆ Determine breathlessness.

◆ Determine pulselessness.

To determine unresponsiveness, tap or gently shake the patient and shout, "Are you okay?" If the patient does not respond, and if you are alone, immediately activate the EMS system for an adult (after one minute of care for an infant or child). This increases the patient's chances of early defibrillation and early advanced care. Then continue with your assessment.

If, after opening the patient's airway, you determine breathlessness, provide artificial ventilation. (Follow the procedures described in Chapter 5.) Provide supplemental oxygen, if you are allowed, by attaching it to your pocket face mask. The oxygen should flow at 15–25 liters per minute.

To determine pulselessness, find the carotid artery pulse point (Figure 6-3):

1. Place two fingers on the larynx ("Adam's apple").

Steps Preceding CPR

Figure 6-2a Determine unresponsiveness.

Figure 6-2b Be sure EMS has been activated.

Figure 6-2c Position the patient. He or she should be supine on a firm, flat surface.

Figure 6-2d Open the airway.

2. **Slide your fingers to the side.** Stop in the groove between the larynx and the large neck muscle.

3. **Feel for the pulse.** Press for 5 to 10 seconds, gently enough to avoid compressing the artery. Do not use your thumb. Do not rest your hand across the patient's throat.

If the patient has a pulse—even a weak or irregular one, do not begin chest compressions.

You could cause serious problems. Monitor the pulse frequently. If you find the patient has no pulse, assume that he or she is in cardiac arrest. Begin CPR immediately.

Note that to perform CPR correctly, your patient must be in a supine position on a firm, flat surface such as the floor or a backboard.

You may wish to refer to the CPR summary, Figure 6-4, as you read the rest of this chapter.

Figure 6-2e Determine breathlessness.

Figure 6-2f Provide artificial ventilation.

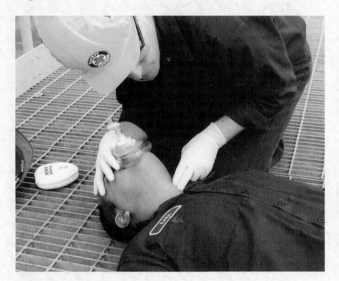

Figure 6-2g Determine pulselessness.

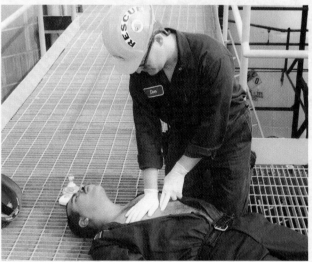

Figure 6-2h Bare the chest, locate the proper hand position, and begin CPR.

CPR for Adults

CPR involves a combination of skills. When a patient's heart has stopped, artificial ventilation alone cannot help the patient. Chest compressions also must be used to circulate the oxygen in the blood.

Chest compressions consist of rhythmic, repeated pressure over the lower half of the sternum. They cause blood to circulate as a result of the build-up of pressure in the chest cavity. When combined with artificial ventilation, they provide enough blood circulation to maintain life for a short time.

To perform chest compressions, take BSI precautions. Then follow these steps:

1. **Position the patient.** He or she must be supine on a firm, flat surface such as the floor.

2. **Uncover the chest.** Remove the patient's shirt or blouse. Do not waste time unbuttoning it.

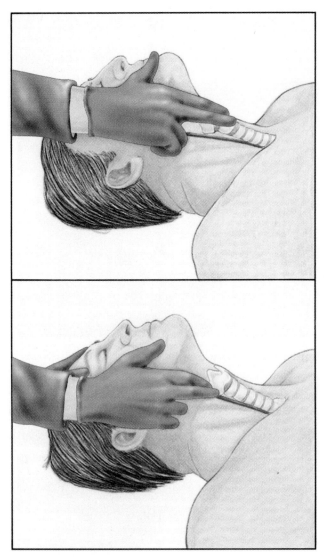

Figure 6-3 Locating the carotid artery.

process (the notch). Put your index finger of the same hand on the lower end of the patient's sternum. Then place the heel of your other hand alongside your fingers. There should be two finger widths between the tip of the sternum and the place where you rest the heel of your hand. When you apply pressure at this point, the sternum is flexible enough to be compressed without breaking. (A fracture of the sternum or the ribs can cut the heart, lungs, or liver.)

6. **Position your hands.** Put your free hand on top of the hand that is on the sternum. Your hands should be parallel. Extend or interlace your fingers to hold them off the chest wall. If your fingers rest against the chest wall during compressions, you increase the chance of separating and injuring the patient's ribs. (An alternative position for large hands and hands or wrists with arthritis is to use your free hand to grasp the wrist of the hand on the patient's sternum.)

7. **Position your shoulders.** Put them directly over your hands.

8. **Perform chest compressions.** Keeping your arms straight and your elbows locked. Your shoulders should be directly over the patient's chest. Compress the chest while bending or pivoting from the hips. Apply firm, heavy pressure. Depress the sternum 1.5 to 2 inches (40 mm to 55 mm) on an adult. Be sure the thrust is straight down into the sternum. If it is not, compressions may not be centered properly on the chest and part of the force of the thrust will be lost.

Using the weight of your body, deliver smooth compressions without jerking or jabbing. If necessary, add force to the thrusts with your shoulders. Never add force with your arms—the force is too great and could fracture the sternum. Compressions should be 50% of the cycle. That is, the compression and release time should be about equal.

9. **Completely release pressure after each compression.** Let the sternum return to its normal position, and allow blood to flow back into the chest and heart. If you do not release all pressure, blood will not circulate properly. Do not lift or move your hands in any way.

Rip it open or pull it up. Cut a woman's bra in two or slip it up to her neck.

3. **Get in position.** Kneel close to the patient's side. Have your knees about as wide as your shoulders.

4. **Locate the xiphoid process** (Figure 6-5). First feel the lower margin of the rib cage on the side nearest you. Use the middle and index fingers of your hand, the one closest to the patient's feet. Then run your fingers along the rib cage to the notch where the ribs meet the sternum in the center of the lower chest.

5. **Locate the compression site** (Figure 6-6). Place your middle finger on the xiphoid

	Adult (over 8 years)	Child (1 to 8 years)	Infant (under 1 year)
Hand Position	Two hands on lower half of sternum	Heel of one hand on lower half of sternum	Two or three fingers on lower half of sternum (one finger width below nipple line)
Compressions	Approximately 1.5 to 2 inches in depth (40 to 55 mm)	Approximately 1 to 1.5 inches in depth (25 to 40 mm)	Approximately ½ to 1 inch in depth (15 to 25 mm)
Breaths	Slowly, until chest gently rises (about 1.5 to 2 seconds per breath)	Slowly, until chest gently rises (about 1 to 1.5 seconds per breath)	Slowly, until chest gently rises (about 1 to 1.5 seconds per breath)
Cycle	15 compressions, 2 breaths (one rescuer) 5 compressions, 1 breath (two rescuers)	5 compressions, 1 breath	5 compressions, 1 breath
Rate	15 compressions in about 10 seconds or 80–100 per minute	5 compressions in about 3 seconds or 100 per minute	5 compressions in about 3 seconds or at least 100 per minute

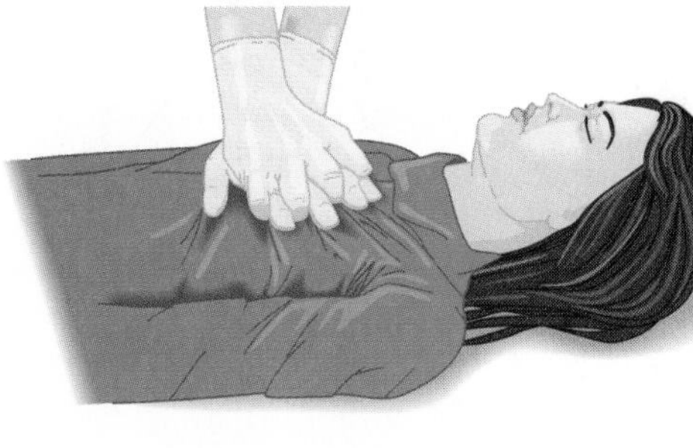

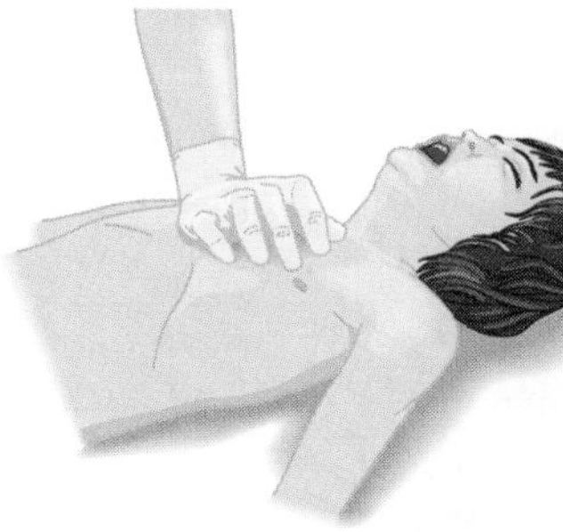

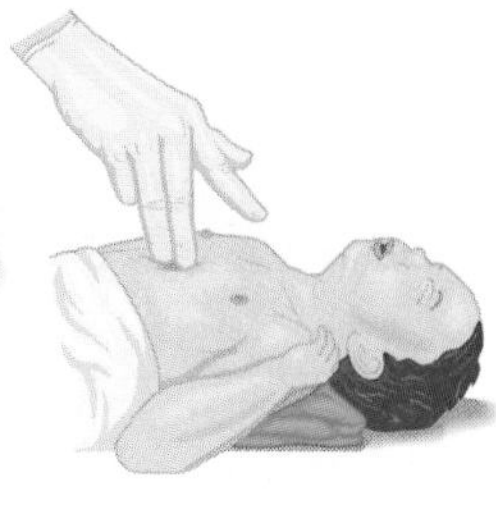

Figure 6-4 CPR Summary.

You could lose proper positioning. Avoid sudden jerky movements.

10. **Count as you administer compressions.** You should be able to say (and do) the following in a bit less than two seconds:

   ◆ One — push down.
   ◆ and — let up.
   ◆ Two — push down.
   ◆ and — let up.

This procedure should let you administer 80 to 100 compressions per minute to an adult. Practice until you can perform 15 complete compressions in 9 to 11 seconds. Beware of becoming hyperventilated. If you find you are, continue breathing at a regular tempo, but not at the same rhythm you used before.

One-Rescuer Adult CPR

To perform CPR alone, you must do the following: Determine unresponsiveness. Activate the EMS system, if it has not already been done. Open the airway, and determine breathlessness. Perform artificial ventilation, and remove foreign body airway obstructions as needed. If breathing is restored and the patient has a pulse, then place him or her in the recovery position. Do not begin CPR.

If the patient's pulse is absent, begin CPR as follows (Figure 6-7). Be certain that you have taken all proper BSI precautions. Then:

1. **Get in position,** and locate the proper hand position (described earlier).

2. **Perform chest compressions.** Perform 15 chest compressions at a rate of 80 to 100 per

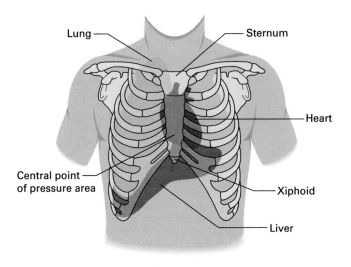

Lung — — Sternum

— Heart

Central point — of pressure area

— Xiphoid

— Liver

Posterior movement of xiphoid may lacerate liver. Lowest point of pressure on sternum must be above, not on, the xiphoid.

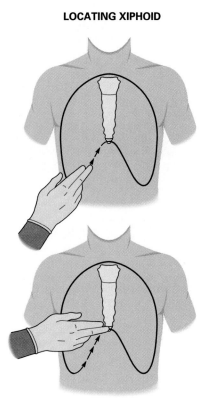

Figure 6-5 Locating the xiphoid process.

minute. Either count out loud or use some other way to keep track of the number of compressions you deliver.

3. **Deliver two slow breaths.** After completing 15 compressions, open the airway. Then deliver two breaths, each lasting 1.5 to 2 seconds. Be sure you inhale deeply between breaths. Continue until you have completed four cycles of 15 compressions and 2 ventilations.

4. **Then check the patient's pulse.** Check for 5 to 10 seconds at the carotid artery. If the pulse has returned, monitor the patient's pulse and breathing closely until the EMTs arrive. If the pulse has returned but the patient is not breathing, provide artificial ventilation. If there is still no pulse, resume CPR. Check again for pulse and breathing every two or three minutes.

If another First Responder trained in CPR arrives at the scene, he or she should do two things. First, the rescuer should verify that the EMS system has been activated and is responding. He should activate the EMS system, if necessary.

Second, the rescuer may take over CPR when the first rescuer gets tired.

To relieve the first rescuer with as little interruption as possible, follow these steps:

◆ If the first rescuer is currently performing chest compressions, the second rescuer should take a position at the patient's head. The second rescuer may then attempt to check the pulse while the first rescuer compresses the chest. Adequate CPR will usually create a carotid pulse. When the first rescuer completes the compressions, the second rescuer should provide two ventilations and check the pulse. The second rescuer can then resume CPR.

◆ If the first rescuer is performing ventilations when the second rescuer arrives, the second rescuer should prepare to perform compressions. After the first rescuer completes two ventilations and checks the pulse, the second rescuer should begin compressions.

There is no exact sequence to cover all situations. The examples above are efficient ways to change rescuers when performing CPR. The main

Correct Position for CPR

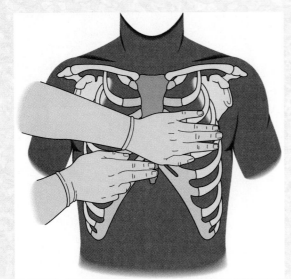

Figure 6-6a Place the heel of your hand on the patient's sternum.

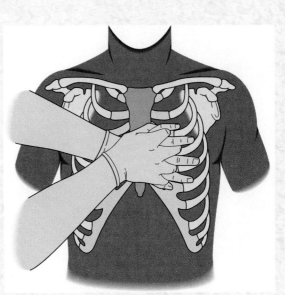

Figure 6-6b Interlace your fingers.

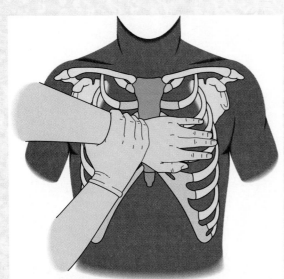

Figure 6-6c Alternative hand placement for large hands and hands or wrists with arthritis.

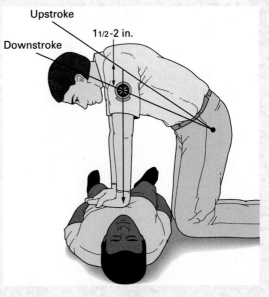

Figure 6-6d Position your shoulders, and then perform chest compressions.

objective is to minimize the amount of time the patient goes without CPR. It is usually convenient to incorporate pulse and breathing checks into the changes.

Any rescuers who are not currently performing CPR can help prepare the scene for the arrival of the ambulance. EMTs and paramedics require space for stretchers, equipment, and additional personnel. Moving furniture away from the patient may help to create extra space. Directing the ambulance crew to the patient is also valuable. If time permits, find out from family and bystanders the exact sequence of events leading to the time the patient's heart stopped.

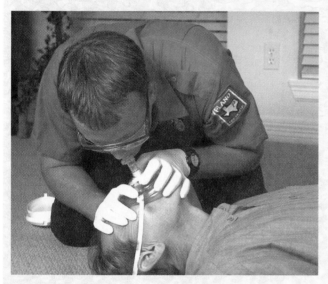

Figure 6-7a After determining unresponsiveness and breathlessness, provide artificial ventilation.

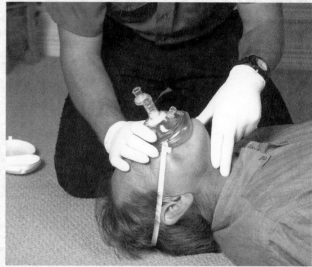

Figure 6-7b Determine pulselessness.

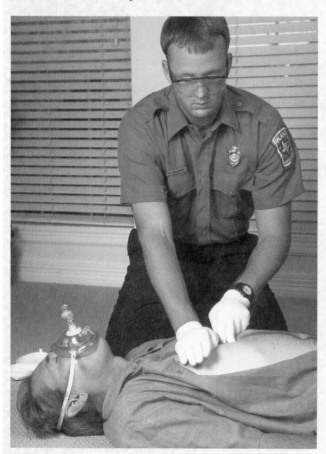

Figure 6-7c Expose chest, and locate proper hand position.

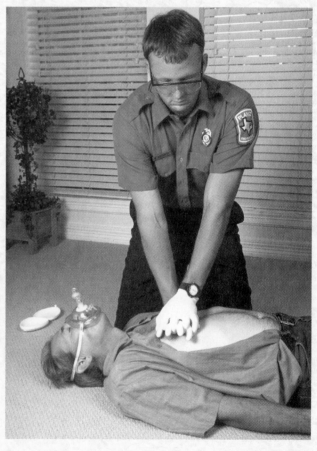

Figure 6-7d Perform chest compressions and ventilations at the proper rate.

Two-Rescuer Adult CPR

All First Responders should learn both the one-rescuer and two-rescuer techniques. The two-rescuer coordinated technique is less tiring. When possible, use an oral airway and a pocket face mask.

Before performing CPR, you and your partner must first determine that the patient is unresponsive, breathless, and pulseless. One rescuer may determine unresponsiveness, provide initial ventilations, and check the pulse. At the same time, the second rescuer can activate the EMS system and prepare to do compressions.

Note that in two-rescuer CPR, the ratio of compressions to ventilations is 5:1 (five compressions and then one ventilation). The ventilation should be delivered during a 1.5- to 2-second pause after every fifth chest compression.

If the patient remains unresponsive, breathless, and pulseless after your initial breaths, proceed as follows (Figure 6-8):

1. **Get in position.** The rescuers, if possible, should take position on opposite sides of the patient. One rescuer kneels by the patient's side for compressions. The other kneels at the patient's head and provides ventilations.

2. **Perform five chest compressions.** Perform them at a rate of 80 to 100 per minute. The compression rescuer should count the sequence out loud. Use an audible count of "one and two and three and four and five and pause," so that five compressions can be achieved every three to five seconds.

3. **Deliver one slow breath.** The ventilation rescuer should take a deep breath on "three and," get into position to ventilate on "four and," and begin breathing into the patient after "five." The compression rescuer pauses for 1.5 to 2 seconds so that the patient receives a slow, full breath. If you have the proper equipment and training, ventilate with 100% oxygen.

4. **After one minute of CPR, check the patient's pulse.** Check for 5 to 10 seconds at the carotid artery. If the pulse has returned, monitor the patient's pulse and breathing closely until the EMTs arrive. If the pulse has returned but the patient is not breathing, provide artificial ventilation. If there is still no pulse, resume CPR. Check again for pulse and breathing every two or three minutes.

When the compression rescuer gets tired, he or she should switch with the ventilation rescuer. Here is the seven-second method (Figure 6-9):

1. The tired compression rescuer calls for a switch at the beginning of the compression cycle by substituting "change" for "one." The audible count remains the same for the remaining four compressions. (Any mnemonic that satisfactorily accomplishes the change is acceptable. Another popular technique uses "Change, on, the, next, breath.") A similar phrase can also be used to call for the move of the patient or for a pulse check. It simply involves substituting. For example:

   ◆ "1 and 2 and 3 and 4 and 5 and pause."
   ◆ "Change and 2 and 3 and 4 and 5 and pause."
   ◆ "Lift and 2 and 3 and 4 and 5 and pause."
   ◆ "Pulse and 2 and 3 and 4 and 5 and pause."

2. After the fifth compression, the ventilation rescuer should give a full breath. Then he or she should move to the chest, locate the xiphoid process, and get hands in position for compressions.

3. At the same time, the compression rescuer should move quickly to the patient's head. Then he or she checks the carotid pulse and breathing for three to five seconds.

4. If no pulse is found, the rescuer at the head gives a breath and announces, "No pulse. Continue CPR."

5. The rescuer at the chest is in position and begins compression. If shortness of breath prevents the rescuer from giving a full count out loud, he or she should at least say the "four and five and" count so that the ventilation rescuer will know when to breathe.

Monitoring the Patient

The patient's condition needs to be monitored throughout CPR. This will ensure that rescue efforts are effective. It also lets you know when spontaneous breathing and the pulse returns.

In two-rescuer CPR, there is a ventilation rescuer and a compression rescuer. To monitor the effectiveness of chest compressions, the ventilation rescuer should feel for a pulse at the carotid artery during compressions. To determine if a spontaneous pulse has returned, the ventilation rescuer should check the carotid artery for three

Two-Rescuer Adult CPR

Figure 6-8a The ventilation rescuer provides artificial ventilation while the compression rescuer bares the patient's chest.

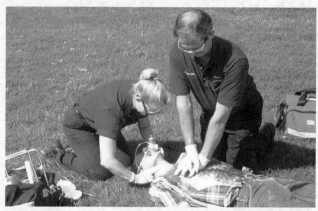

Figure 6-8b The ventilation rescuer determines pulselessness as the compression rescuer gets into position.

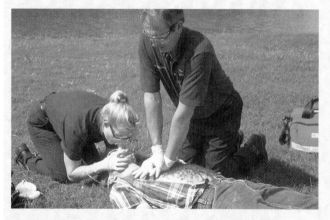

Figure 6-8c Both work together to perform compressions and ventilations at the correct ratio and rate.

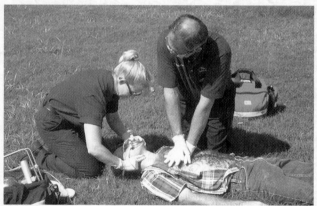

Figure 6-8d After the first minute and every few minutes thereafter, stop CPR to assess the carotid pulse.

to five seconds at the end of the first minute of CPR and every few minutes thereafter. Note that the pulse must be checked when CPR is not in progress.

In general, CPR should not be interrupted for more than five seconds. One of the few exceptions to this rule applies to moving a patient. It may not be possible to perform CPR in a cramped bedroom or other small area. In this case, it is acceptable to move the patient so proper CPR can be performed. These actions must be kept as close to five seconds as possible.

Signs of Successful CPR

Signs of successful CPR include the following:

◆ Each time the sternum is compressed, you should feel a pulse in the carotid artery. It may feel like a flutter.

◆ Chest should rise and fall with each ventilation.

◆ Patient's skin color may improve or return to normal.

Other possible but less likely occurrences include:

Changing Positions

Figure 6-9a The tired compression rescuer calls for a switch.

Figure 6-9b The ventilation rescuer delivers a breath as usual, then moves to the patient's side. The second rescuer moves to the patient's head.

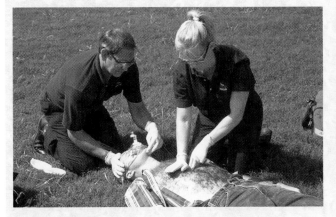

Figure 6-9c The rescuer at the head opens the airway and checks respirations and pulse for five seconds. The second rescuer prepares for compressions.

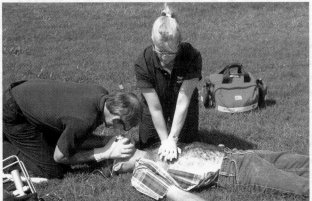

Figure 6-9d If the patient is still unresponsive, breathless, and pulseless, continue CPR.

◆ Pupils may react or appear to be normal. (Pupils should constrict when exposed to light.)

◆ Heartbeat may return.

◆ Spontaneous gasp of breathing may occur.

◆ Patient may move his or her own arms or legs.

◆ Patient may try to swallow.

Remember that "successful" CPR does not mean that the patient lives. "Successful" only means that you performed CPR correctly. Very few patients will survive if they do not receive early defibrillation and advanced cardiac life support (ACLS). The goal of CPR is to extend the window of survival. Hopefully, advanced providers will arrive in time.

Mistakes in Performing CPR

The most common ventilation mistakes are as follows:

◆ Failing to maintain an adequate head tilt.

◆ Failing to maintain an adequate seal around the patient's mouth, nose, or both with a pocket face mask or face shield. The seal should be released when the patient exhales.

- Completing a two-rescuer cycle in fewer than five seconds.
- Failing to watch and listen for exhalation.
- Not giving full breaths.
- Providing breaths too rapidly.

Some common chest compression mistakes include the following:

- Bending your elbows instead of keeping them straight.
- Not aligning your shoulders directly above the patient's sternum.
- Placing the heel of your bottom hand too low or not in line with the sternum (Figure 6-10).
- Not depressing the sternum to proper depth.
- Not extending the fingers of your hands, touching the patient's chest.
- Pivoting at the knees instead of at your hips.
- Compressing at an incorrect rate.
- Moving your hands from the compression site between compressions.

COMPANY OFFICER'S NOTE

The safety officer on the fireground provides an important function: ensuring operations are performed safely. Similarly, when operating at a cardiac-arrest call—and when staffing permits—one person should ensure that proper CPR is being performed. Simultaneously, the EMS safety officer can coordinate the arrival and activities of the ALS ambulance squad.

Complications Caused by CPR

Even properly performed, CPR may cause rib fractures in some patients. Other complications that can occur with proper CPR include:

- Fracture of the sternum, which is common in older patients.
- Pneumothorax (collapse of the lungs caused by air in the chest).
- Hemothorax (collapse of the lungs caused by bleeding in the chest).

- Cuts and bruises to the lungs.
- Lacerations (cuts) to the liver.

These complications are rare. However, you can help minimize the risk by giving careful attention to your performance. Remember that effective CPR is necessary, even if it results in complications. After all, the alternative is death.

In addition, the rib cartilage in elderly patients separates easily. You will hear it crunch as you compress. Be sure that your hand is positioned correctly and that you are compressing to the correct depth, but do not stop.

FIRE DRILL

On the fireground, attention to proper procedures during operations can minimize the damage to the structure. For example, proper hose placement and proper water flow will extinguish the fire and minimize damage to the contents and structure itself. Similarly, when performing CPR, attention to proper hand placement and proper depth of compression will minimize any injuries that CPR may cause to the patient.

CPR for Infants and Children

Infants (up to one year old) and children (one to eight years old) need slightly different care. Cardiac arrest in them is rarely caused by heart problems. The heart nearly always stops beating because of too little oxygen due to injuries, suffocation, smoke inhalation, **sudden infant death syndrome (SIDS),** or infection.

See Chapter 5 for how to determine if your infant or child patient is unresponsive and breathless. Follow the directions there, too, for providing emergency care to an infant or child who is not breathing. Remember that according to AHA guidelines, if you are alone, you should resuscitate an infant or child for one minute before you activate the EMS system.

Determining that your infant or child patient is pulseless is important. For an infant, check the brachial pulse on the inside of the upper arm between the elbow and shoulder. Press the artery gently with your index and middle fingers. Never use your thumb. In a child, check the pulse at the carotid artery.

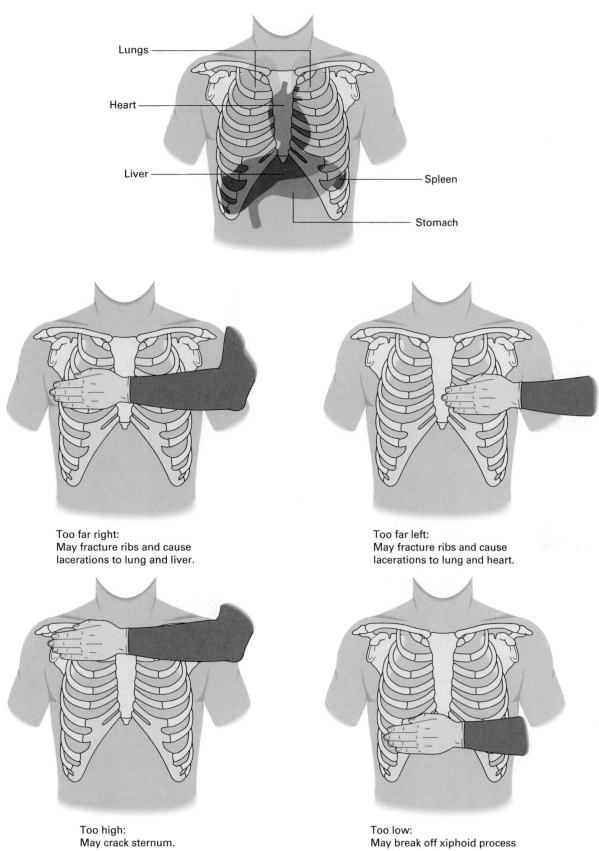

Too far right:
May fracture ribs and cause lacerations to lung and liver.

Too far left:
May fracture ribs and cause lacerations to lung and heart.

Too high:
May crack sternum.

Too low:
May break off xiphoid process and lacerate the liver.

Figure 6-10   Consequences of improper hand placement.

Infant and Child CPR

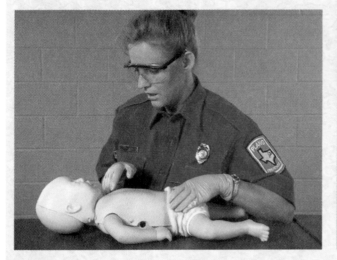

Figure 6-11a Determine unresponsiveness by tapping the infant and speaking loudly.

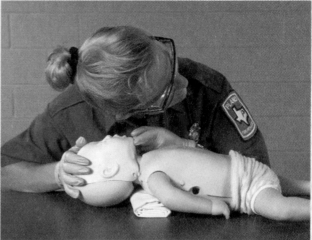

Figure 6-11b Gently open the airway, and determine breathlessness.

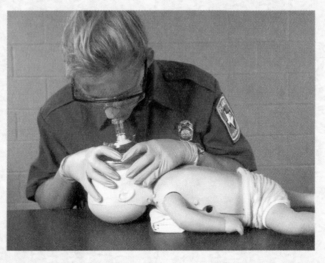

Figure 6-11c Cover the infant's mouth and nose with a pocket mask. Then ventilate.

It can be difficult to find a pulse in an infant or child. *So, do not spend too much time trying to locate one.* According to the AHA, if the infant or child is not breathing, heart rate is probably inadequate and chest compressions are usually necessary.

Performing Infant or Child CPR

If the infant or child is unresponsive, breathless, and pulseless, begin CPR. Be certain you have taken BSI precautions. Then follow these guidelines (Figures 6-11 and 6-12):

1. **Position the patient.** Make sure the patient is lying on a firm, flat surface. If the patient is an infant, put him or her in your lap with head tilted back slightly. Use your palm to support the baby's back. Make sure his or her head is not higher than the rest of the body.

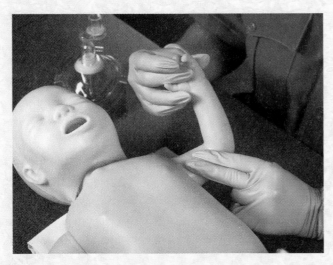

Figure 6-11d Determine pulselessness at the brachial artery.

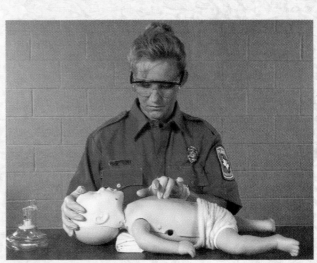

Figure 6-11e Locate the correct hand position, and begin chest compressions.

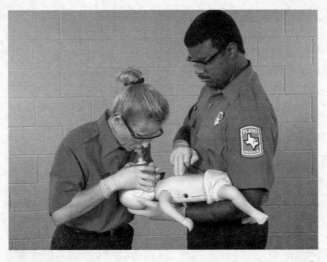

Figure 6-11f Performing CPR while carrying the baby.

2. **Locate the compression site.** For an infant, it is one finger-width below an imaginary line between the nipples.

   For a child, locate the lower margin of the rib cage on the side next to you. Use your middle and index fingers, while your hand nearest the child's head maintains head tilt. Follow the rib cage to the xiphoid process, where the ribs and sternum meet. Then place your index finger next to the middle finger.

   While looking at the position of the index finger, lift that hand and place its heel next to where the index finger was.

3. **Perform chest compressions.** For an infant, use the flat part of your middle and ring fingers to compress the infant's sternum one-half to one inch (15 mm to 25 mm), or one-third to one-half of the depth of the chest. The compression rate for an infant is at least 100 per minute.

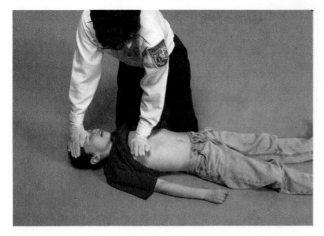

Figure 6-12 Chest compressions on a larger child.

For a child, compress the sternum 1.0 inch to 1.5 inches (25 mm to 40 mm), or one-third to one-half of the depth of the chest, with the heel of one hand. The compression rate for a child is 100 compressions per minute.

The ratio of compressions to ventilations in both infants and children is 5:1 (five compressions and then one ventilation). In one-rescuer CPR for an infant or child, ventilate during a pause after each fifth compression. Count compressions at this rhythm:

◆ Infant—1, 2, 3, 4, 5, breathe.
◆ Child—1 and 2 and 3 and 4 and 5 and breathe.

After one minute, or 20 cycles, check for the return of a spontaneous pulse.

Signs of Successful Infant and Child CPR

The methods of checking for successful CPR in infants and children are almost the same as for adults.

◆ Check the patient's pulse periodically. In the infant, check the brachial pulse. In the child, check the carotid pulse.
◆ Check the pupils. CPR is successful if they are reacting normally or appear to be normal.
◆ Watch for a spontaneous heartbeat, spontaneous breathing, and responsiveness.

Complications of Infant and Child CPR

One of the most common complications with injury and sudden illness in children is hypothermia (a below-normal body temperature). So keep the infant or child warm.

 ## ON SCENE FOLLOW-UP

At the beginning of this chapter, you read that First Responders took over CPR from a bystander. To see how chapter skills apply to this emergency, read the following. It describes how the call was completed.

PHYSICAL EXAMINATION

Our first concern was providing good CPR. Two rescuers in my crew were doing that. A thorough physical exam would have to wait. The patient was on the living room floor and didn't appear to have any injuries from falling to the ground.

PATIENT HISTORY

I talked to the bystander who had identified himself earlier as the patient's brother. He told

me that the patient had been mowing the lawn when he began having chest pain. He stated that his brother had entered the living room, clutching his chest and complaining of pressure beneath his breast bone. A minute later, he witnessed the collapse to the floor. He stated he thought the patient had collapsed about five minutes before we arrived.

The patient's wife was also in the house. She told me that her husband was 71 and had bypass surgery two years ago after a heart attack. He was on medication for his heart and high blood pressure. She went to get it, as I radioed the incoming EMS units with an update.

ONGOING ASSESSMENT

I assumed the role of "safety officer" and monitored the effectiveness of the ongoing CPR. I also maintained contact with the responding ambulance crew. I asked my crew members if either of them needed a break. They stated they were okay. All we could do at this point was await the EMT's arrival. It seemed like it took them forever, yet within four minutes they were on scene.

PATIENT HAND-OFF

When the EMTs arrived, I told them that the patient is a 71-year-old male. He was mowing the lawn and then entered the house complaining of chest pain. He collapsed to the floor while talking to his brother. While his brother called 9-1-1, the wife began CPR. The patient was unresponsive with no pulse or respirations when CPR was continued by our crew. I told them I didn't believe the patient had any injuries. He had a history of bypass surgery and high blood pressure, and took medication. I gave them the patient's medication vials.

The EMTs continued emergency care as we watched. It took three shocks from the AED to get the patient's heart started again. But he still had no respirations. So one of them continued to ventilate the patient while the others put the patient on a backboard. The backboard would give them a hard surface to compress against in case they had to start CPR again.

Later, the EMTs told me that the patient had improved slightly in the ambulance and was transferred to the cardiac unit at the hospital. Not all patients survive. I was happy we helped one who did.

> *Heart disease is still the number one killer in the U.S. Be prepared to provide CPR to any patient who needs it, and remember to take refresher courses frequently.*

Chapter Review

Circulation, like respiration, is essential to life. If a patient's heart fails to beat, he or she will surely die unless actions are taken to restore the heartbeat. As a First Responder, you will help to take these actions.

Cardiopulmonary resuscitation (CPR) is the first step in saving the life of a patient whose heart has stopped beating. The sooner CPR is started, the better, because permanent brain damage may occur after as few as four minutes.

The next chapter covers automated external defibrillation. The AED is a device that applies an electric shock to the patient's chest, which can restore a heartbeat. You will recall that the chain of survival requires early access, early CPR, and early defibrillation together with advanced care for an optimal chance of survival.

Perform CPR in accordance with AHA standards when you are called to do so. Be sure you have a barrier device with you at all times.

FIRE COMPANY REVIEW

Page references where answers may be found or supported are provided at the end of each question.

Section 1

1. What is the name and location of each of the four chambers of the heart? (p. 115)

2. What are the links in the "chain of survival"? (pp. 116–117)

3. What is the pulse? Where can you best palpate it? (pp. 115–116)

Section 2

4. Before performing CPR on your patient, what must your assessment of his or her condition reveal? (p. 117)

5. How can you find the correct CPR compression site on an infant, child, and adult? (pp. 119, 131)

6. What are the appropriate compression depths for an infant, child, and adult? (pp. 120, 131–132)

7. Why is it essential to perform CPR in spite of the problems it may cause? (p. 128)

8. When should you activate EMS if your patient is an unresponsive adult? An unresponsive infant? (pp. 117, 128)

RESOURCES TO LEARN MORE

Basic Life Support for Healthcare Providers: 1997–99 Emergency Cardiovascular Care Programs. Dallas: American Heart Association, 1997.

Heartsaver Plus: 1997–99 Emergency Cardiovascular Care Programs. Dallas: American Heart Association, 1997.

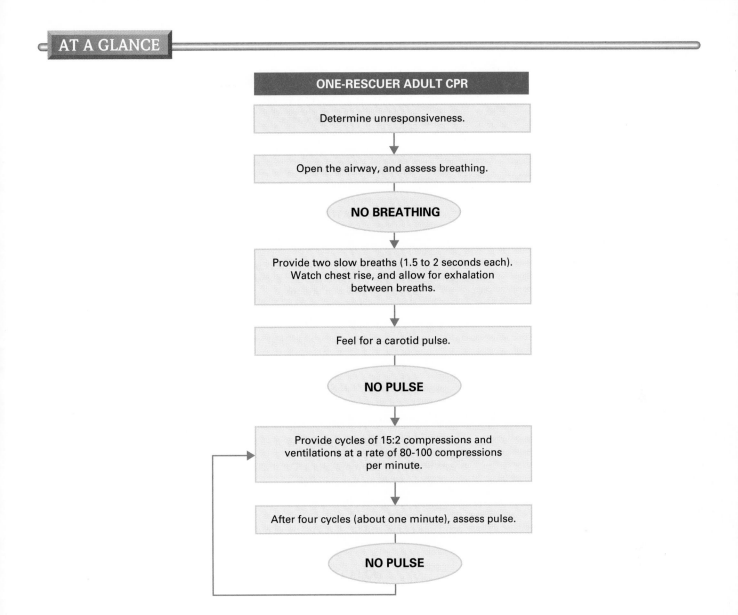

ONE-RESCUER ADULT CPR

Determine unresponsiveness.

Open the airway, and assess breathing.

NO BREATHING

Provide two slow breaths (1.5 to 2 seconds each). Watch chest rise, and allow for exhalation between breaths.

Feel for a carotid pulse.

NO PULSE

Provide cycles of 15:2 compressions and ventilations at a rate of 80-100 compressions per minute.

After four cycles (about one minute), assess pulse.

NO PULSE

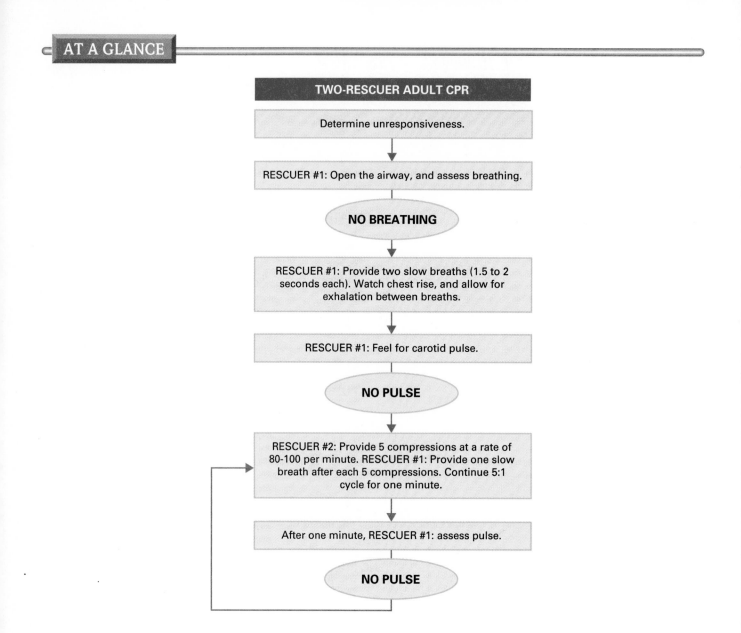

TWO-RESCUER ADULT CPR

Determine unresponsiveness.

RESCUER #1: Open the airway, and assess breathing.

NO BREATHING

RESCUER #1: Provide two slow breaths (1.5 to 2 seconds each). Watch chest rise, and allow for exhalation between breaths.

RESCUER #1: Feel for carotid pulse.

NO PULSE

RESCUER #2: Provide 5 compressions at a rate of 80-100 per minute. RESCUER #1: Provide one slow breath after each 5 compressions. Continue 5:1 cycle for one minute.

After one minute, RESCUER #1: assess pulse.

NO PULSE

Automated External Defibrillation

*I*NTRODUCTION   *As you will recall from Chapter 6, early defibrillation is part of the "chain of survival." In order to get defibrillation to patients early enough, as many trained people as possible—not just physicians and paramedics—must be able to perform this life-saving skill. In fact, now that members of the public have access to automated external defibrillators (AEDs), you may find defibrillation in progress when you arrive on scene. Do not be*

surprised. Document care provided upon arrival. Follow local
protocol thereafter. Automated external defibrillators have made that
possible. Now First Responders, EMT-Basics, and even members of the
public are able to defibrillate a patient when it is needed.

OBJECTIVES

Cognitive, affective, and psychomotor objectives are from the U.S. DOT's
"First Responder: National Standard Curriculum." Enrichment objectives, if
any, identify supplemental material.

Cognitive

No objectives are identified.

Affective

No objectives are identified.

Psychomotor

No objectives are identified.

Enrichment

◆ List the indications and contraindications for
automated external defibrillation. (pp. 140–141)

◆ Differentiate between fully automated and
semi-automated external defibrillators. (p. 139)

◆ List the steps in the operation of an automated
external defibrillator. (pp. 143–144)

◆ Describe the care for a patient whose pulse has
returned after automated external defibrillation.
(pp. 144–145)

◆ Discuss the need to complete the automated
defibrillator operator's shift checklist. (p. 145)

◆ Explain the role medical direction plays in the
use of automated external defibrillation.
(pp. 139, 145)

◆ Describe the maintenance of an automated
external defibrillator. (p. 145)

ON SCENE

DISPATCH

Our engine company was sent on a call to an
unresponsive woman at 326 Riverview Lane.

SCENE SIZE-UP

We approached the scene carefully and saw a
man frantically waving to us. We were still alert
to the possibility of danger when we got to the
house. The man told us that his neighbor, a

nurse, was performing CPR. We saw her working
on an elderly woman who was lying on the
lawn. "My wife collapsed while she was carrying
in the groceries," the man said. "Please help her.
Please!"

There was only one patient. We decided to
call for paramedic assistance right away. After
putting on our gloves and grabbing a pocket
face mask, we joined the nurse at CPR. She was
doing a good job.

INITIAL ASSESSMENT

We checked the patient's pulse and respirations. There were none.

Cardiac arrest is literally a life-or-death situation. Some patients may be saved through the use of an automated external defibrillator. Consider this patient as you read Chapter 7. See if you can decide if she needs defibrillation. If she does, then how and when would you apply it?

ABOUT DEFIBRILLATORS

Defibrillation is the application of an electric shock to the chest of a patient who is in cardiac arrest. Most physicians and paramedics use **manual defibrillation.** That is, they interpret a machine's report and decide if shocks are indicated. If so, they apply the shocks themselves with "paddles" to the patient's chest.

The **automated external defibrillator (AED)** can perform all those tasks. It has a microprocessor that actually interprets a heart rhythm, just as a doctor would. When necessary, shocks are then delivered by the device directly to the patient.

Actual shocks are delivered by way of adhesive pads. They are connected to the AED through cables, which can transmit a shock that is powerful enough to correct a lethal heart rhythm. The pads make defibrillation safer, since no one needs to touch the patient during analysis or shocks.

EMS systems that allow fire service First Responders to use AEDs should have all of the following in place:

◆ *All of the links of the "chain of survival."* Without early access, early CPR, and early advanced care, the patient will not have the best chance of survival. (See Figure 7-1.)

◆ *Medical direction.* A physician must issue standing orders for First Responders to use an AED.

◆ *Quality improvement programs.* These monitor the use of AEDs in the field.

◆ *Mandatory continuing education.*

Types of AEDs

There are two types of AED (Figure 7-2). One type is fully automated. The other is semi-automated. An operator attaches a fully automated one to the patient, turns it on, and the AED does the rest. The **semi-automated external defibrillator** performs the same tasks, but the operator must push a button to analyze the patient's heart rhythm and deliver a shock.

There are many brands of AED. Many of them include:

◆ *On/off button.* This button controls power to the AED.

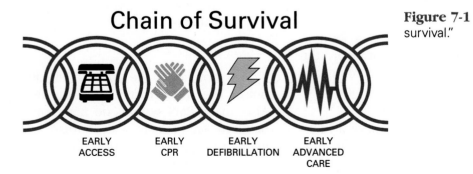

Chain of Survival

EARLY ACCESS   EARLY CPR   EARLY DEFIBRILLATION   EARLY ADVANCED CARE

Figure 7-1 "Chain of survival."

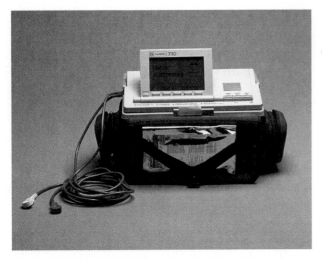

Figure 7-2a Example of a fully automated defibrillator.

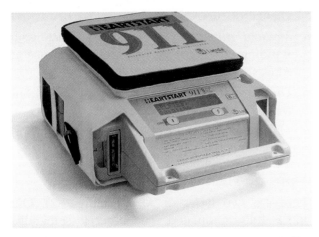

Figure 7-2b Example of a semi-automated defibrillator. *(Source: Laerdal Medical Corporation)*

- *Analyze button.* Press this button to have the AED analyze heart rhythm.

- *Shock button.* Push this button to deliver a shock to the patient.

- *Voice synthesizer.* Some units have an electronic voice that prompts you to perform specific actions. It may direct you to "analyze rhythm" or "push to shock" at the proper time.

- *Tape recorder.* Some units have a built-in microphone and tape recorder that record the events of a cardiac arrest. The recording can then be used for training or quality improvement at a later date.

- *ECG screen or light.* Some AEDs have a built-in screen or light that shows the electrical activity of the heart.

OPERATING A DEFIBRILLATOR

AEDs are very safe and accurate. Even so, follow operating guidelines carefully. They will assure safe and proper use of the machine. Also:

- Become familiar with the AED you are using.

- Make sure the AED batteries are fully charged. Carry extra ones, too.

- Carefully follow local protocols on AED use in your area.

Make sure no one touches the patient while the AED is analyzing heart rhythm or delivering shocks. If someone is in contact, the shock may be transferred to that person. It also may cause interference with the accuracy of the AED.

COMPANY OFFICER'S NOTE

When operating an AED, follow these safety guidelines:

- Do not shock a patient who is wet. Water conducts electricity. So either dry the patient's chest or move him or her to a dry environment (inside, if it is raining).

- Do not shock a patient if he or she is in contact with a metal surface, especially if it is wet or if other people are touching that surface. That includes your stretcher! Make sure no one is in contact with your stretcher if an AED is about to deliver a shock.

- Do not ventilate a patient when a shock is about to be delivered. Instead, wait for the shock to occur and then deliver ventilations between shocks.

- Do not use an AED in a moving vehicle. The device will not properly analyze the rhythm. You must completely stop the vehicle in order to determine if more shocks are indicated.

Do not apply the AED to a patient with a pulse. The shock could cause the heart to stop. Also, many EMS systems do not allow an AED to be used on patients with cardiac arrest due to injury or hypothermia (low body temperature). Learn your local protocols.

Heart Rhythms

The normal electrical impulses of the heart occur in an orderly, rhythmic fashion. When you place an AED on a patient, it evaluates your patient's heart rhythm (Figure 7-3). There are two specific ones that require an AED shock. They are:

◆ *Ventricular fibrillation.* This is a chaotic, unorganized heart rhythm. It cannot produce a pulse or circulate blood.

◆ *Ventricular tachycardia.* This is a rhythm that is more organized but very rapid and inefficient. It can produce a pulse. (Note that AED shocks must only be delivered to patients with no pulse.)

Heart rhythms that do not require an AED shock are:

◆ *Pulseless electrical activity (PEA).* If you were to look at an electrocardiogram (ECG) of this rhythm, you might think nothing was wrong. You would find electrical activity on the display, but you would not find a pulse.

◆ *Asystole.* Also known as "flat line," this is a condition where there are no electrical impulses present and, therefore, no pulse. The ECG would show a flat line.

A saying common to doctors and paramedics applies to First Responders who use AEDs: "Treat the patient, not the machine." For example, if the AED reads a "flat line," it may only mean that one of the electrodes or cables is loose. Other times an AED might read "normal" electrical activity. If you were to pay attention only to the monitor,

you would not know that a PEA patient needs immediate CPR. Remember, the AED only analyzes heart rhythms. It does not check pulse.

Always follow local protocols to determine whether or not defibrillation is needed. If you have questions, contact medical direction. Practice using an AED often, and attend continuing education classes.

Operation Guidelines

Automated defibrillation is easy to learn. An AED may be applied to deliver a shock to a patient in less than a minute. A sequence of shocks may be delivered in about 90 seconds.

Always remember that using an AED is a definitive step toward returning the heart to a normal rhythm and function. CPR is vital, but its main purpose is to prolong life until defibrillation can be performed. Where there is a pulseless patient and a defibrillator ready to go, use the defibrillator first!

See ideal positioning in Figure 7-4.

Applying Adhesive Pads

It is through the adhesive pads that the AED monitors heart rhythm and delivers shocks. So the pads must be placed in very specific locations (Figure 7-5). Remember, all directions refer to the patient's right and left, not yours.

Place one pad just below the patient's right clavicle and to the right of the sternum. Then attach the cable with the white tip to it. Place the other pad over the patient's left lower ribs, and

Heart Rhythms

Chaotic electrical discharge as occurs
in heart muscle wall.

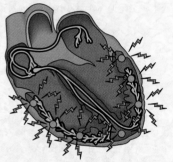

Chaotic electrical discharge as seen on an ECG tracing.

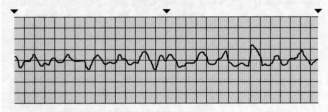

Figure 7-3a   Ventricular fibrillation.

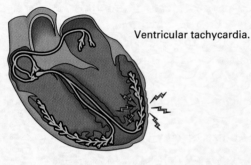

Ventricular tachycardia.

ECG tracing of ventricular tachycardia.

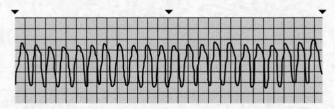

Figure 7-3b   Ventricular tachycardia.

Asystole

ECG tracing of asystole

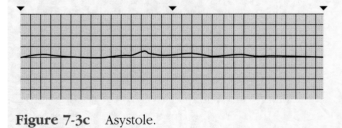

Figure 7-3c   Asystole.

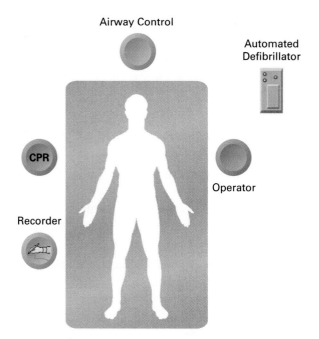

Figure 7-4   Ideal positioning. (Alternatives may be necessary.)

attach the red electrode to it. The mnemonic "white to right, red to ribs" may help you remember where the pads and cables are to be placed.

Occasionally, you will find patients who have nitroglycerin patches on their chests. Do not place pads on or near them. Arcing or burning of the patient's skin could occur. Instead, remove the patches and use a towel to wipe off any paste that remains.

Adhesive pads should be securely applied to the chest. This may not be possible if the patient is extremely hairy. Be sure to carry a razor with your AED. Use it to quickly but safely shave the areas where the pads are to adhere.

Procedure for Unresponsive Patients and Those with CPR in Progress

After sizing up the scene and taking BSI precautions, perform an initial assessment. Stop CPR, if it is in progress, to check the patient's pulse and respirations (Figure 7-7). Then:

1. **Attach the AED,** if the patient is pulseless. If there will be a delay in administering a shock, have trained bystanders or responders resume CPR.

2. **Turn on the AED power.**

3. **Stop CPR,** and instruct everyone to clear the patient.

4. **Press the "analyze" button.**

   If the AED advises to administer a shock, then make sure everyone is still clear and:

1. **Deliver the shock.**

Figure 7-5   Defibrillator pad placement.

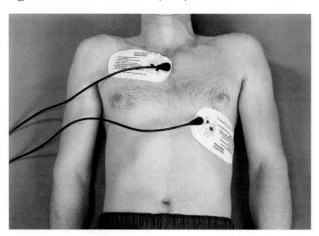

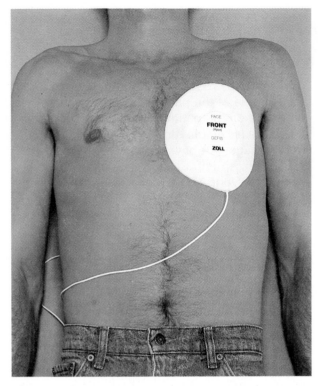

Figure 7-6a Alternative pad placement, front view.

Figure 7-6b Alternative pad placement, back view.

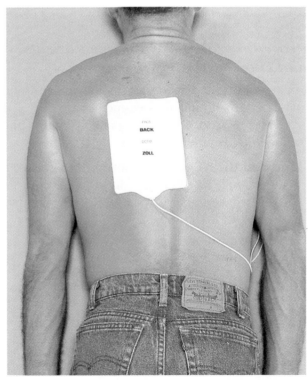

2. **Reanalyze the heart rhythm.** Some AEDs may do this automatically. Others may require you to push a button.

3. **Deliver a second shock,** if you are advised to do so by the AED. Again the AED will reanalyze the heart rhythm.

4. **Deliver a third shock,** if you are advised to do so by the AED.

5. **Check the patient's pulse** after the third shock. Note that it is not necessary to check the pulse between shocks, unless the AED prompts you to do so. It can waste valuable time.

6. **If a pulse is present, check respirations.** Remember, the patient most likely will have inadequate breathing at this point if he or she is breathing at all. If breathing is inadequate, assist ventilations.

   If there is no pulse, begin CPR for one minute. After one minute, repeat the cycle described above of up to three shocks.

7. Many EMS systems recommend that after two sets of three stacked shocks, transport should occur. Follow local protocols.

   If the AED advises NOT to administer a shock, then:

1. **Recheck the patient's pulse.**

2. **If a pulse is present, check respirations.** If breathing is inadequate, assist with ventilations. **If the pulse is not present,** perform CPR for one minute and reanalyze the heart rhythm.

   If you are alone and find an unresponsive patient, then:

1. **Assess the patient's breathing and pulse.** If they are absent and your AED is immediately available, proceed with the following.

2. **Apply the AED, and analyze the heart rhythm immediately.** Do not begin CPR first. Defibrillation is more important at this point and better for the patient.

3. **Follow the instructions** outlined above in response to the AED prompts.

Post-Resuscitation Care

In some cases, not all, your patient will regain a pulse. This is exciting, but there is much more to be done. Your patient is still in serious condition.

Remember, when caring for a patient whose pulse has returned:

◆ Monitor the pulse carefully. It may disappear.

◆ Many patients will not breathe, even though a pulse has returned. This is common. Ventilate the patient or assist ventilations as necessary.

◆ If your patient has regained pulse and adequate respirations—and has not been injured, place him or her in a recovery position.

◆ Apply high-concentration oxygen, if you are trained and allowed to do so.

◆ Keep the AED attached to the patient. EMS personnel who take over care will want it attached during transport.

◆ Perform ongoing assessments of the patient until you turn over care.

Advanced care also will help the patient. It includes medications and other interventions that help to stabilize the patient and prevent another cardiac arrest. So, if you have not done so already, arrange for the paramedics to respond to the scene. If they cannot respond in a reasonable amount of time, an ambulance should transport the patient promptly to the nearest hospital emergency department.

Call Review

After using an AED, the call should be reviewed by a quality improvement committee. (Your medical director will be involved.) The committee will determine if protocols were followed. This review uses the run report and data module of the AED and, if your AED was equipped, the tape recording of the call.

While the call review looks for problem areas, it is not designed to get people "into trouble." If needed, the committee may recommend further training for some or all rescuers who use AEDs. The goal of the review is to have trained fire service First Responders providing quality care to patients.

AED Maintenance

The AED is a vital piece of equipment. Do not arrive at the side of a patient with one that does not work. That would be devastating. It also may be a cause of liability against you and your agency.

Complete the operator's checklist at the beginning of every shift or, in the case of a volunteer agency, on a regular basis. (See Figure 7-8.) Make sure that the AED battery—and spares—are fully charged. Make sure the leads are with the unit. Several sets of adhesive pads also should be available. Always treat the AED carefully. Do not jar it or handle it roughly. Finally, follow all manufacturer's guidelines.

Using a Semi-Automated Defibrillator

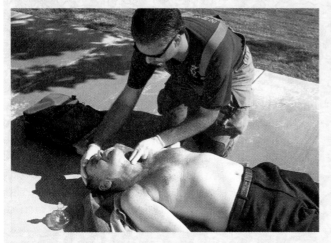

Figure 7-7a Determine breathlessness and pulselessness.

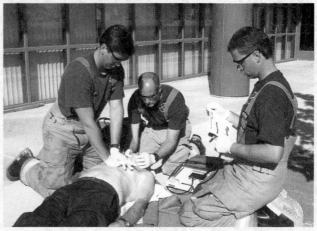

Figure 7-7b One rescuer initiates CPR, while the other prepares the AED.

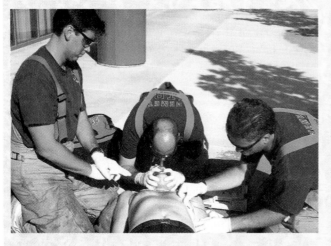

Figure 7-7c Place electrodes on the patient's chest.

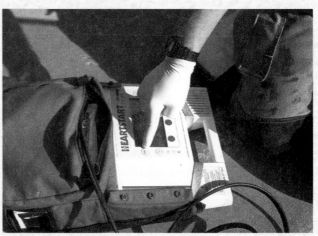

Figure 7-7d Turn on the AED.

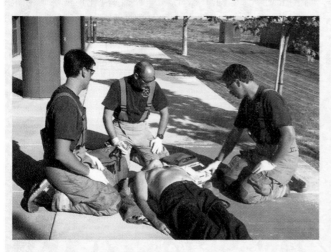

Figure 7-7e Stop CPR and get clear of the AED as it analyzes heart rhythm.

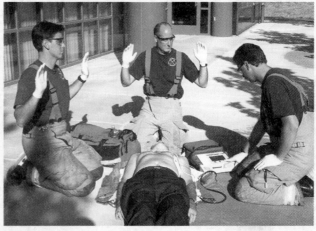

Figure 7-7f If a shock is advised, clear all others from the patient and deliver the shock.

AUTOMATED DEFIBRILLATORS: OPERATOR'S SHIFT CHECKLIST

Date: _____ Shift: _____ Location: _____

Mfr/Model No.: _____ Serial No. or Facility ID No.: _____

At the beginning of each shift, inspect the unit. Indicate whether all requirements have been met. Note any corrective actions taken. Sign the form.

	Okay as found	Corrective Action/Remarks
1. Defibrillator Unit		
Clean no spills, clear of objects on top, casing intact		
2. Cables/Connectors		
a. Inspect for cracks, broken wire, or damage b. Connectors engage securely		
3. Supplies		
a. Two sets of pads in sealed packages, within expiration date *g. Spare charged battery b. Hand towel *h. Adequate ECG paper c. Scissors *i. Manual override module, key or card d. Razor * e. Alcohol wipes *j. Cassette tape, memory module, and/or event card plus spares * f. Monitoring electrodes		
4. Power Supply		

a. Battery-powered units
  (1) Verify fully charged battery in place
  (2) Spare charged battery available
  (3) Follow appropriate battery rotation schedule per manufacturer's recommendations
b. AC/Battery backup units
  (1) Plugged into live outlet to maintain battery charge
  (2) Test on battery power and reconnect to line power		

5. Indicators/*ECG Display		
* a. Remove cassette tape, memory module, and/or event card *e. "Service" message display off b. Power on display *f. Battery charging; low battery light off c. Self-test ok g. Correct time displayed — set with dispatch center * d. Monitor display functional		
6. ECG Recorder		
a. Adequate ECG paper b. Recorder prints		
7. Charge/Display Cycle		
* a. Disconnect AC plug — battery backup units *e. Manual override functional b. Attach to simulator f. Detach from simulator c. Detects, charges and delivers shock for "VF" *g. Replace cassette tape, module, and/or memory card d. Responds correctly to non-shockable rhythms		
8. *Pacemaker		
a. Pacer output cable intact c. Inspect per manufacturer's operational guidelines b. Pacer pads present (set of two)		
☐ **Major problem(s) identified** **(OUT OF SERVICE)**		

*Applicable only if the unit has this supply or capability

Signature: _____

Figure 7-8 Operator's shift checklist for AEDs, courtesy of Laerdal.

At the beginning of this chapter, you read that fire service First Responders were at the scene of a patient in respiratory and cardiac arrest. To see how chapter skills apply to this emergency, read the following. It describes how the call was completed.

INITIAL ASSESSMENT (CONTINUED)

I had the AED. It took only a few seconds to hook up the electrodes and cables while CPR continued. I turned on the AED and instructed everyone to clear the patient while it analyzed the heart rhythm. The machine advised me to press the shock button.

After making sure everyone was clear, I shocked the patient. The machine automatically went back to "analyze" mode. It instructed me to shock again. When everyone was clear, I shocked again. The AED analyzed. It then told me to check the pulse. I couldn't believe it. There was one.

We immediately checked the patient's respirations. There were none. We hooked oxygen into the pocket face mask and ventilated her.

PATIENT HISTORY

Among the facts the patient's husband, Mr. Jones, told us was that his wife was in good health. She hadn't been to a doctor in years. She took no medications. She had no complaints before she went down. They had sandwiches for dinner two hours ago. She had no allergies. "She was always so healthy. I just don't understand," he said.

PHYSICAL EXAMINATION

We checked quickly for any obvious signs of injury and found none.

ONGOING ASSESSMENT

We never got to the ongoing assessment. The medics were already on scene.

PATIENT HAND-OFF

The hand-off report to the medics was as follows:

"This is Georgia Jones. She is 66 and was carrying groceries into the house when she collapsed on the grass. A nurse who lives next door started CPR. She was doing a good job. We hooked up the AED and gave two shocks, which caused the pulse to return, but not respirations. We ventilated the patient and monitored her pulse until you arrived. Georgia hasn't been to the doctor recently, has no meds, and no allergies. Her husband says she is in good health, and there were no problems before she collapsed. There appear to be no injuries from the fall. She had a sandwich two hours ago for dinner."

The medics began their advanced life support. They started intravenous lines (IVs), put in airway adjuncts, and administered medications. The patient went back into cardiac arrest once while they were caring for her. They used their manual defibrillator and got her pulse back with one shock. Georgia made it to the hospital with a pulse, but the doctors said her heart and brain had suffered too much damage to survive.

After a day in the cardiac unit, Georgia passed away. We were sad to hear it and disappointed. We realized that we had done the best we could. Many patients don't live. I know that. The next time maybe we'll get to shake the hand of the person we help when they walk out of the hospital.

SPECIAL NOTE: The AED is a device that has tremendous potential to save lives. But many patients will not be saved. Some will have been "down" too long or suffered heart damage so great that they could not be saved even if they were in a hospital. Do your best for all patients. If a patient does not survive, even with your best efforts, it is not your fault.

Losing a patient is a difficult experience. Remember to focus on the good you have done for prior patients and the good you will do for future ones. Also seek out experienced members of your agency, because they undoubtedly have felt the same at some point in their careers. If necessary, speak to your medical director about the call.

Chapter Review

FIREFIGHTER FOCUS

Defibrillation saves lives. It is a skill that the American Heart Association (AHA) believes should be taught to every First Responder.

Remember that CPR will sustain life only for a period of time. An AED can actually shock certain abnormal heart rhythms back to normal. Never delay defibrillation. If you have a choice between it and CPR, choose defibrillation. The earlier the shock is applied, the better chance you have of restoring normal heart patterns in your patient.

Finally, you may be under a lot of stress at the scene of a cardiac arrest, especially if you have not been on many such calls. The old adage "practice makes perfect" applies. If you practice with your AED often and you know your local protocols, you will perform better under the stress of the situation.

FIRE COMPANY REVIEW

Page references where answers may be found or supported are provided at the end of each question.

Section 1

1. What is the difference between automated and semi-automated defibrillators? (p. 139)

Section 2

2. What role does medical direction play in the use of an AED? (pp. 139, 145)

3. When should the AED not be used? (pp. 140–141)

4. Why is it so important to make sure no one is touching the patient while the AED analyzes or shocks the patient? (p. 140)

5. If given a choice between performing CPR and using an AED, which should be performed first? Explain your answer. (p. 141)

6. Where are the AED's adhesive pads placed on the patient? (pp. 141, 143)

7. When should the patient's pulse be checked—before or after the first shock? When should it be checked again? (pp. 143–145)

RESOURCES TO LEARN MORE

Graves, J.R., et al. *RapidZap: Automated Defibrillation*. Upper Saddle River, NJ: Brady/Prentice-Hall, 1989.

Weigel, A., et al. *Automated Defibrillation*. Upper Saddle River, NJ: Brady/Prentice-Hall, 1988.

AED CARDIAC ARREST TREATMENT SEQUENCE

Verify patient is unresponsive, not breathing, and pulseless.

↓

Have partner start CPR.

↓

Prepare AED and apply pads to patient.

↓

Turn AED on and begin narrative.

↓

Clear patient and press *Analyze* button.

"Deliver shock" message
- Say "Clear!" and assure no one is touching the patient.
- Deliver 3 shocks in succession as long as AED gives "Deliver shock" message.
- Check pulse.
- If no pulse, resume CPR for 1 minute.
- Press *Analyze* button.
- If AED gives "Deliver shock" message, give 3 more shocks in succession as long as AED gives "Deliver shock" message.
- After 6 shocks, stop defibrillation.

"No shock" message
- Check pulse. If none, give CPR for 1 minute.
- Press *Analyze* button.
- "No shock" message.
- Check pulse. If none, give CPR for 1 minute.
- Press *Analyze* button.
- "No shock" message.

Check pulse. If none, resume CPR.

Notes: Whenever there is a "No shock" message, check for a pulse. If the patient has a pulse, provide oxygen via nonrebreather mask or ventilations with high-concentration oxygen as needed.

If you initially shock the patient and then receive a "No shock" message before giving 6 shocks, follow the steps in the right-hand column above.

If you initially receive a "No shock" message and then on a subsequent analysis receive a "Deliver shock" message, follow the steps in the left-hand column above.

Occasionally, you many need to shift back and forth between the two columns. If this happens, follow the steps until you receive one of the following indications for transport:

- You have administered 6 shocks.
- You have received 3 consecutive "No shock" messages (separated by 1 minute of CPR).
- The patient regains a pulse.

If the patient is resuscitated and rearrests, start the sequence of shocks from the beginning.

Scene Size-Up

*I*NTRODUCTION   *There are three important reasons to perform a thorough scene size-up. First, scene safety must be assured. Second, it is necessary to identify the mechanism of injury or nature of illness. Finally, you will need to determine if additional resources are needed. Most likely, you will not have contact with your patient during scene size-up. Even so, your observations, decisions, and actions set the foundation for the success of the entire call.*

Cognitive, affective, and psychomotor objectives are from the U.S. DOT's "First Responder: National Standard Curriculum." Enrichment objectives, if any, identify supplemental material.

Cognitive

3-1.1 Discuss the components of scene size-up. (pp. 151, 152, 155, 163, 166)

3-1.2 Describe common hazards found at the scene of a trauma and a medical patient. (p. 153)

3-1.3 Determine if the scene is safe to enter. (pp. 152–155)

3-1.4 Discuss common mechanisms of injury/nature of illness. (pp. 155–163)

3-1.5 Discuss the reason for identifying the total number of patients at the scene. (p. 163)

3-1.6 Explain the reason for identifying the need for additional help or assistance. (p. 163)

Affective

3-1.22 Explain the rationale for crew members to evaluate scene safety prior to entering. (p. 154)

3-1.23 Serve as a model for others by explaining how patient situations affect your evaluation of the mechanism of injury or illness. (pp. 155–163)

Psychomotor

3-1.33 Demonstrate the ability to differentiate various scenarios and identify potential hazards. (pp. 152–154, 163)

ON SCENE

DISPATCH

I am a First Responder for the Fire Department. My partner and I were dispatched to a car crash at the corner of Fry Boulevard and Garden Avenue. The caller indicated that the crash seemed "pretty bad."

SCENE SIZE-UP

When we approached the scene, we saw the caller was correct. It was a head-on into a telephone pole. I asked my partner to alert the trauma center, determine if a helicopter was available, and, if it was, to have it put on standby.

> *The scene size-up on this call will be very important. As you read this chapter, consider what the First Responders should do to size up the scene, as well as handle other parts of the call.*

1 ASSURING PERSONAL SAFETY

BSI Equipment

Body substance isolation (BSI) precautions must be taken on every call. See Chapter 2 for a full discussion. What follows is a brief review.

BSI equipment includes:

- *Gloves.* Wear them routinely on every call.
- *Eye protection.* Like gloves, eye protection should be used on every call. After assessing the scene, it may not be necessary to continue wearing them unless there is a chance of blood or body fluids spraying or splashing into your eyes.

◆ *Mask.* Wear one when there is any chance of blood or body fluids spraying or splashing into your nose or mouth.

◆ *Gown.* Wear one when there is any chance of clothing becoming soiled with blood or other body fluids.

Remember that you should always have personal protective equipment available. When you approach the scene, anticipate which items may be needed and then put them on. Waiting too long may cause you to become so involved in patient care that you forget to protect yourself.

Figure 8-1a Plan for the possibility of a dangerous scene.

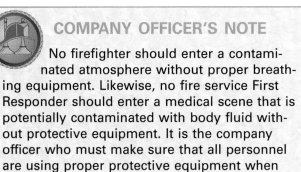

> **COMPANY OFFICER'S NOTE**
>
> No firefighter should enter a contaminated atmosphere without proper breathing equipment. Likewise, no fire service First Responder should enter a medical scene that is potentially contaminated with body fluid without protective equipment. It is the company officer who must make sure that all personnel are using proper protective equipment when necessary.

Figure 8-1b Observe the scene for signs of potential danger.

Nothing is more important at the emergency scene than your safety. Hazards may include obvious situations such as violence, downed power lines, or hazardous materials. Do not overlook the dangers at other scenes such as car crashes, unstable vehicles, unstable surfaces (slopes, ice, etc.), and dangerous pets. Place your safety first. If you do not, you may become a patient yourself and quite possibly prevent others from caring for the patient you were sent to help.

The vast majority of calls go by uneventfully. When there is danger, three words sum up the actions required to respond appropriately: plan, observe, and react (Figure 8-1).

Figure 8-1c React to danger appropriately: retreat, radio, and reevaluate.

> **FIRE DRILL**
>
> If someone dies by chemical hazard, electric shock, or other on-scene danger, it should be the patient, not you.

Plan

Many First Responders work together to prevent danger and know what do to when danger strikes. However, scene safety begins long before the actual emergency. For example, you should:

◆ *Wear safe clothing.* Nonslip shoes and other practical clothing will help you to respond to danger without restriction.

◆ *Prepare your equipment properly.* Make sure it is not cumbersome. Remember that you will be carrying it into emergencies. If your first response kit is too heavy or too large, it will distract your attention from where it should be—observing the scene.

◆ *Carry a portable radio.* A radio allows you to call for help if you are separated from your vehicle.

◆ *Plan safety roles.* If there will be more than one rescuer on any call, tasks can be split. For example, one rescuer can care for the patient, while the other observes for safety. The "observer" could look for nearby weapons or other threats, for example, and for clues to the patient's condition, such as prescription medications.

Observe

Remember that it is always better to prevent danger than it is to deal with it. Observation and awareness are the best ways to accomplish this goal.

Observation begins early in the call. As you approach the scene, turn off your lights and sirens to avoid broadcasting your arrival and attracting a crowd. Observe the neighborhood as you look for house numbers. If possible, do not park directly in front of the call. This provides two benefits. First, you may be able to approach the scene unnoticed, which allows you to size it up without distraction. Second, since many first response units do not transport, the area directly in front of the call is left open for the ambulance.

As you approach an emergency scene, look for the following signs of potential danger:

◆ *Violence.* Any indication that violence has or may take place is significant. These signs include arguing, threats, or other violent behavior. Also notice any broken glass, overturned furniture, or the like.

◆ *Weapons of any kind.* If a weapon is on scene, it is a serious potential danger.

◆ *Signs of intoxication or drug use.* When people are under the influence of alcohol or drugs, their behavior is unpredictable. In addition, even though you see yourself as there to help, other people may not. If you are in uniform, you may be mistaken for the police, especially if you drove up in a vehicle with lights and sirens.

◆ *Anything unusual.* Even an awkward silence should cause you to be wary. Emergencies are usually very active events. In situations where you observe an unusual silence, a certain amount of caution is advisable.

Note that nothing in this textbook is meant to create fear or unwarranted suspicion. Remember that the vast majority of EMS calls will go by uneventfully. Some calls, however, do pose a threat. Those calls usually provide subtle clues that may be picked up before the danger strikes. Use your observation skills on every call to determine important safety information.

Remember, the general rule is: *If the scene is unsafe, make it safe if you are trained to do so. If not, do not enter and call for the appropriate teams to handle the situation.*

React

If you find danger at the scene, there are three "Rs" of reacting: *retreat, radio,* and *reevaluate.*

Retreat

It is not a fire service First Responder's responsibility to subdue violent persons or wrestle weapons away. A clear and justified course is to retreat from danger.

There are safer ways to retreat than others. When leaving the scene of danger, remember the following points:

◆ *Flee far enough away that danger will not threaten you again.* Retreating only a short distance keeps you in danger. In addition, when fleeing a scene of danger, place two major obstacles between you and it. If a dangerous person moves in your direction and gets through one obstacle, the second obstacle acts as a built-in buffer.

◆ *Take cover.* Cover and concealment are important. Find a position that hides your body and protects it from projectiles (getting behind a brick wall, for example). This is preferred over concealment, which only hides your body but offers no protection (like getting behind a shrub). When fleeing danger, moving a considerable distance from the scene and taking cover are usually the best options.

◆ *Discard your equipment.* Do not get bogged down. The equipment you carry can be thrown at the subject's feet to give you additional time to retreat.

Radio

The portable radio is an important piece of safety equipment. Its main function is to call for police assistance and to warn other rescuers of impending danger. When using the radio, speak clearly and slowly. Advise the dispatcher of the exact nature and location of the problem. Specify how many people are involved and whether or not weapons were observed.

Remember that the information you have must be shared as soon as possible to prevent others from coming up against the same danger.

Reevaluate

Do not reenter the scene until the police have secured it. Even then, keep in mind that violence may begin again. Emergencies are situations packed with stress for families, victims, rescuers, and bystanders. Maintain a level of observation throughout the call. Occasionally, weapons or illegal drugs are found while you are assessing your patient. Notify the police immediately.

After the call, document the situation on your run report. Occasionally, the danger may cause delays in reaching the patient. Courts have held this acceptable, provided that there has been a real and documented danger.

SECTION 2

MECHANISM OF INJURY OR NATURE OF ILLNESS

During scene size-up, you must determine the nature of the patient's problem. You may already have some idea from dispatch whether your patient is a **medical patient** (ill) or a **trauma**

patient (injured). So when you scan the scene for safety factors, also try to determine if the patient is ill or injured.

A medical patient's condition is caused by some internal factor such as a heart or breathing problem. There is nothing at the scene that suggests injury. In this case, speak to the patient, family, or bystanders to determine why EMS was called and what the **nature of illness** might be.

When you scan the trauma scene, note the **mechanism of injury** (forces that caused the injury). For example, if your patient fell from a ladder, it would be important to note how far the patient fell. The greater the distance, the more serious and extensive the injuries may be.

Occasionally, a patient may have a combination of illness and injury. Consider the patient who fell from a ladder, for instance. What if he passed out from a medical problem and then fell to the ground? As you approach the scene, the mechanism of injury may be obvious. The illness may not be. It will be your examination of the scene, as well as the eventual patient history, that will make a difference. (See Chapter 9 for instructions on how to gather a patient history.)

FIRE DRILL

A scene size-up at an EMS call accomplishes the same objectives as one performed on the fireground: a rapid, yet deliberate consideration of critical factors and the development of a plan based on those factors. Patient care will improve because of this process.

Remember that the mechanism of injury is an important part of your assessment of a trauma patient. It can suggest which body parts are injured and how severe the injuries might be. Whenever you care for a trauma patient, maintain a high index of suspicion and take note of:

◆ Body position at the time of impact.

◆ Part of the body impacted.

◆ Object that penetrated the body, or the surface the body landed on.

◆ Distance involved, if any.

Common mechanisms of injury are falls, vehicular collisions, penetrating objects such as bullets and knives, fire, and explosions.

Car Crashes

There are five basic types of car crashes: head-on impact, rear impact, side impact, rotational impact, and rollover. Each one has its own predictable pattern of injury.

Head-on Impact

A head-on impact occurs when a car hits another vehicle or an immovable object, such as a tree (Figure 8-2). The greater the car's speed, the greater the energy, the greater the damage. When the car stops, its occupants continue to travel forward. They take one of two possible pathways of motion—up-and-over or down-and-under (Figure 8-3). Each pathway has a distinctive pattern of injury, which can be affected by the use of a seat belt.

In the up-and-over pathway, the torso may be thrown over the steering wheel. The face, head, and neck strike the windshield. The chest and abdomen may then strike the steering wheel. Note the following patterns of injury:

◆ *Face, head, and neck injuries.* Look for obvious clues like hair, tissue, or blood on the windshield or rear-view mirror. The windshield also may bulge out in a classic bull's-eye or spider-web pattern.

   The face can sustain extensive soft-tissue damage. However, bleeding from its rich supply of blood vessels may not be as serious as it looks. Watch for airway problems if there is bleeding from the mouth, nose, or face.

Skull fracture may occur. Almost all head injuries have the potential to cause damage to the brain. Brain tissue can compress, rebound against opposite sides of the skull, and bruise. Brain tissue also can be cut or bruised on the floor of the skull, which is very rough or jagged.

Energy can travel down the neck, causing the potential for cervical-spine injury. The neck may be flexed or extended too far, resulting in whiplash injuries or fractures. An impact at the top of the head can cause compression fractures of the cervical spine. The anterior neck also can be injured by hitting the steering wheel or dashboard. Cartilage rings in the trachea (windpipe) can be separated, which would impair breathing.

◆ *Chest injuries.* When the chest strikes the steering wheel, the ribs and sternum may break. The heart may be compressed and bruised, making it unable to pump blood effectively. The aorta may be torn, resulting in life-threatening bleeding. As the lungs are compressed, they can be bruised or ruptured. Remember that broken ribs can also injure the lungs and heart.

◆ *Abdominal injuries.* When the abdomen strikes the steering wheel, the liver, spleen, and other organs are compressed. Sometimes they are cut. The liver may be cut in half as it is forced against the ligament that holds it in place. The spleen may be torn from its attachment, resulting in severe internal bleeding.

In the down-and-under pathway of motion, a body slides under the steering wheel. The knees strike the dashboard. Energy travels up the legs.

Figure 8-2 Head-on impact.

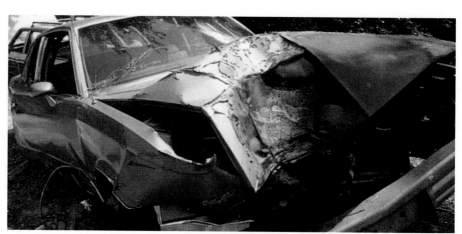

a.

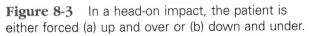

b.

Figure 8-3 In a head-on impact, the patient is either forced (a) up and over or (b) down and under.

The abdomen and then the chest strike the steering wheel. Classic injuries include dislocated hip and broken patella (kneecap), femur (thigh bone), and pelvis. In cases where an airbag is deployed, unrestrained drivers tend to go under the bag, sustaining serious lower extremity injuries.

Rear Impact

Rear impact occurs when a car is struck from behind by another vehicle traveling at greater speed (Figure 8-4). The car that is hit accelerates suddenly, and the occupant's body is slammed backward and then forward (Figure 8-5). Suspect the same kinds of injuries as discussed for head-on collisions. If positioned properly, a headrest will prevent the head from whipping back. If the headrest is not in place, suspect soft-tissue injury to the neck, compression of the cervical spine, and cervical-spine fractures.

Side Impact

The side impact is often called a "broadside" or "T-bone" collision. The person closest to the impact absorbs more energy than a person on the opposite side. As the energy of the impact is absorbed, the body is pushed sideways and the head moves in the opposite direction. The following injuries commonly occur:

◆ *Head and neck injuries.* The head often impacts the door post. This can result in skull injury, brain injury, and tears in neck muscles and ligaments. Cervical-spine fractures are common, since the vertebrae are not designed for extreme lateral movement.

◆ *Chest injuries.* If the door slams against the shoulder, the clavicle may probably break. If the arm is caught between the door and the chest, or if the door impacts against the chest directly, suspect broken ribs and possible breathing problems. Fractures low in the rib cage can injure the liver and spleen.

Figure 8-4 Rear impact.

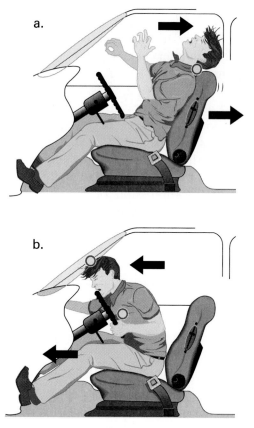

a.

b.

Figure 8-5 Rear impact forces the patient (a) back and then (b) forward.

◆ *Pelvic injuries.* Lateral impact to the pelvis often causes fractures of the pelvis and femur.

The person on the opposite side of the car is subject to similar kinds of head and neck injuries.

In addition, if there is more than one person sitting on a seat, heads often collide.

Rotational Impact

A rotational impact is one that occurs off center. The car strikes an object and rotates around it until the car either loses speed or strikes another object. The sturdiest structures in the car (such as the steering wheel, dashboard, door posts, and windows) are the ones that cause the most serious injuries. A variety of injury patterns may occur due to the initial strike and subsequent striking of stationary objects. Look for the same kinds of injuries that occur with head-on and side impacts.

Rollover

During a rollover (Figure 8-6), car occupants change direction every time the car does (Figure 8-7). Every fixture inside the car becomes potentially lethal. A specific pattern of injury is impossible to predict, but rollovers usually cause injuries to more than one body system. In this type of car crash, and all others, note the location of objects in the car that may have become missiles upon impact.

If car occupants are not wearing seat belts, they have a much greater chance of being thrown from the car, either partially or fully. Common injuries include severe soft-tissue injuries, multiple broken bones, and crushing injuries resulting from the car rolling over the occupant. Ejected patients also have a much higher chance of spine injury or death.

Figure 8-6 Rollover impact.

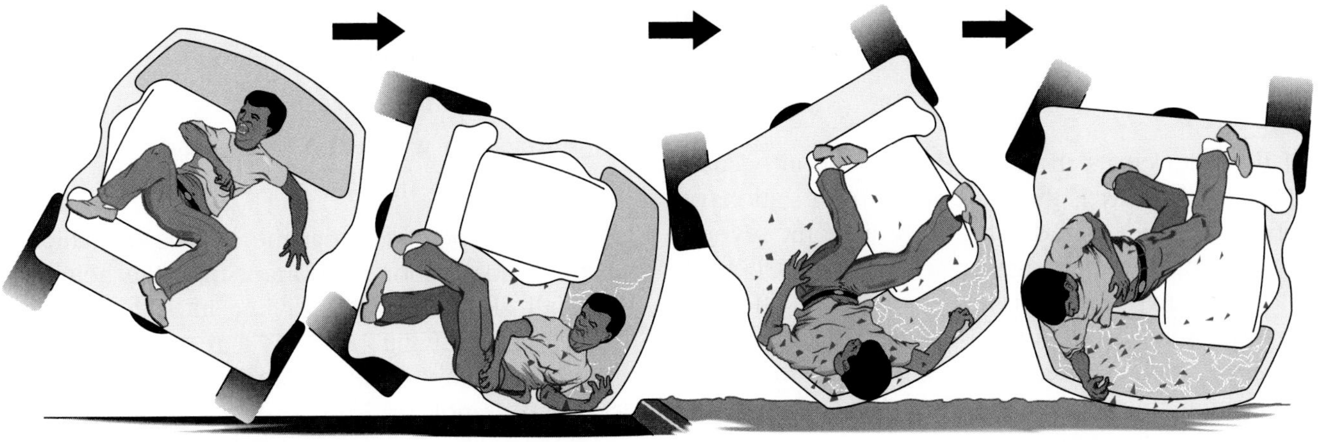

Figure 8-7 In a rollover, the unrestrained occupant changes direction every time the car does.

Restraints

Restraints help to reduce the severity of injuries. However, occupants of a car can still be injured, especially if restraints are not used properly.

◆ *Lap belt.* When worn properly, a lap belt can prevent the occupant from being thrown out of the car. But it does not prevent the head, neck, and chest from striking the steering wheel or dashboard. When the torso is thrown forward, compression fractures of the lower back can occur. If the belt is worn across the upper abdomen instead of over the pelvis, the spleen, liver, intestines, or pancreas can be injured. There also may be enough pressure in the abdomen to injure the diaphragm and force abdominal contents into the chest cavity.

◆ *Lap-and-shoulder belt.* This type of restraint can prevent the occupant from striking the steering wheel or dashboard. A severe impact can cause the shoulder belt to break the clavicle. Even properly used, it does not prevent the head and neck from moving sideways or forward and back. If a headrest is not in place, suspect cervical-spine injury.

◆ *Air bag.* It cushions the occupant and absorbs energy in a head-on crash. Without a seat belt, however, it is not effective in rollovers, rear-end or side collisions, or collisions involving multiple vehicles and repeated impact. Serious, even fatal injuries have occurred in children in the passenger seat when airbags deploy.

◆ *Car seat.* An infant car seat that faces backward in an upright position minimizes risk in a head-on collision. In a crash, all unrestrained parts of an infant's body continue to move forward. The greatest danger is to the neck. An infant's head is large for its body, and it tends to snap forward with great force. Suspect injury to the neck any time the mechanism of injury indicates it may be possible, even if the patient appears to be uninjured.

Motorcycle Crashes

Motorcycle crash injuries are greatly reduced when the rider wears a helmet. With no helmet, chances of severe head injury and death increase 340%. There are three types of impact in motorcycle accidents. They are head-on, angular, and ejection. "Laying the bike down," an evasive action, often prevents serious injuries but can cause extensive scrapes, bruises, and burns.

Head-on Impact

In a head-on impact, the rider generally impacts the handlebars at the same speed the bike was traveling. A variety of injuries can be expected. For example, if the rider's feet get caught, the thighs or pelvis will strike the handlebars, resulting in fractures.

Angular Impact

In this type of impact, the rider strikes an object at an angle. The object then usually collapses on the rider. Common objects are the edges of signs, outside mirrors on cars, or fence posts. Severe amputations can result.

Ejection

If the rider clears the handlebars, ejection occurs. The rider's body is thrown until it hits a stationary object or the ground. The body may hit several objects or strike the ground many times before stopping. Expect severe head and facial injuries if the rider is not wearing a helmet. Fractures and internal injuries are likely. Expect severe soft-tissue damage if the rider is not wearing boots and leather clothing.

Laying the Bike Down

A rider who anticipates a crash may try to "lay the bike down." That is, the rider may turn the motorcycle sideways and drag a leg on the ground to lose enough speed to get off. If successful, the rider slides along the ground, clearing the bike and the object it hits. Expect severe abrasions (scrapes) from contact with the pavement. Many victims are also burned by contact with the motorcycle's hot exhaust pipe.

Recreational Vehicle Crashes

Injuries caused by recreational vehicle crashes are similar to those caused by motorcycles. However, since ATVs (all-terrain vehicles) are often used on fields and hilly terrain, patients can be harder to reach. One of the most dangerous ATV crashes occurs when a rider runs into an unseen wire fence. Cut neck vessels, severed windpipes, and even decapitations have resulted.

ATVs are prone to collision with other vehicles. Expect head, neck, and extremity injuries. The three-wheel ATV is especially unstable. A simple turn can cause a rollover, resulting in head and crush injuries. Unfortunately, many of those who ride ATVs are children. Their lesser body weights make ATVs even more unstable.

Another type of recreational vehicle is the snowmobile. It is often driven at high speed. When there is a crash, riders sustain severe head and neck injuries. Rollovers are also common.

Falls

The most common mechanism of injury is a fall. Falls account for more than half of all trauma-related accidents. The severity of injury depends on the following:

- Distance of the fall.
- Anything that interrupts the fall.
- Body part that impacts first.
- Surface on which the victim lands.
- Cause of the fall.

Some experts say that the surface on which a victim lands is more important than the height of the fall. For example, diving into deep water from a high diving board is a recreational activity. Diving the same distance onto a concrete sidewalk is not.

Generally, a fall of three times a patient's height onto an unyielding surface is considered severe. The U.S. Department of Transportation suggests that any fall of 15 feet (5 m) or greater onto an unyielding surface is severe. Other sources say that falls from greater than 20 feet (6 m) should be considered severe in adults. Falls from 10 feet (3 m) or more should be considered severe in children. Whatever the case, you should have a high degree of suspicion about internal injuries no matter how the patient looks.

Feet-First Fall

A feet-first landing causes energy to travel up the skeleton (Figure 8-8). If the knees are flexed when the person lands, injury to the bones will be less severe. Common injuries include fractures of the spine, hip socket, femur, heel, and ankle. Head, back, and pelvis injuries are common if the victim falls backward. If the victim extends arms to break a forward fall, expect broken wrists. A broken shoulder and clavicle are common, too.

In falls of 15 feet (5 m) or more, internal organs are likely to be severely injured from sudden deceleration. The liver may be sliced in two. The spleen or kidneys may be torn from their attachments. The heart may be torn from the aorta.

Head-First Fall

In head-first falls, the pattern of injury begins with the arms and extends up to the shoulders. Head and spine injuries are very common. There is usually extensive damage to the neck. When the body is falling, the torso and legs are thrown either forward or backward, commonly causing chest, lower spine, and pelvis injuries.

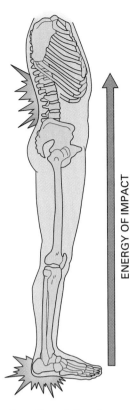

Figure 8-8 In feet-first falls, the energy of impact is transmitted up the skeletal system.

Penetrating Trauma

This kind of injury occurs when an object penetrates the surface of the body. Hand-powered weapons, such as knives or arrows, generally cause low-velocity injuries. These are limited to the immediate site of impact. Projectiles powered by another source, such as bullets from a handgun, cause medium-velocity or high-velocity injuries. These affect tissues far from the site of impact.

When you encounter a scene of violence, it is absolutely essential that you make sure the scene is safe before you try to reach a patient. Follow local protocol.

Low-Velocity Injuries

Among the factors that help determine the severity of a low-velocity injury are the sex of the offender, the position of the victim when stabbed, and the length of the object used to penetrate the body. The sex of the offender can give you important clues. Women generally have less upper-body strength. They usually stab overhand, or downward. Men have more upper-body strength, so they usually stab up and out. For example, if the victim was stabbed on the right side by a man, the most likely injuries would be to the liver, kidney, and intestine. If the victim was stabbed by a woman, the most likely injury would be to the lungs.

The length of the weapon also gives valuable clues. For example, a person stabbed in the chest with a short paring knife would probably suffer a pneumothorax (collapse of the lungs due to air in the chest). The same stab wound inflicted by a long carving knife could cut the pulmonary veins, the aorta, and the heart muscle itself.

Medium- and High-Velocity Injuries

In general, medium-velocity weapons include shotguns and hand guns. High-velocity weapons include high-power rifles. Knowing a weapon's velocity helps to determine how severe an injury might be. Other factors include:

- Trajectory, or the path the bullet travels after it enters the body.
- Drag, or the factors that slow a bullet down.
- Impact point, or the bullet's point of entry into the body.
- Whether or not the bullet fragments, or breaks apart.
- Kind of pressure wave caused by the energy of the bullet as it travels through body tissue.

The injury can also be complicated by clothing, gun powder, bacteria, and other foreign matter that is pulled into the wound. A soft-nose, high-velocity bullet is especially destructive. It can cause a wave of energy through the body that is 30 times the diameter of the bullet (Figure 8-9).

As you care for a gunshot victim, remember that tissue damage can be much more widespread than the surface wound indicates. A bullet wound that bleeds very little can be accompanied by a devastating internal injury.

If all the energy of a bullet is absorbed by the body, the bullet will remain there. If all is not absorbed, the bullet will exit the body. You need to assess the victim carefully. Look for both an entry and an exit wound. Note that an exit wound can be much larger than the entrance wound.

Blast Injuries

The most common explosions involve natural gas, gasoline, fireworks, or grain elevators. They occur

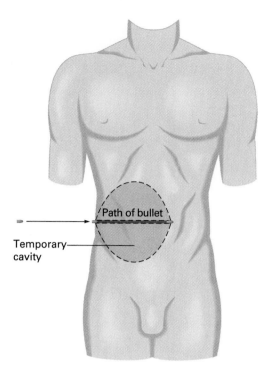

Figure 8-9 A gunshot wound can cause devastating damage, much more than a surface wound might indicate.

Path of bullet

Temporary cavity

Figure 8-10 Blast forces.

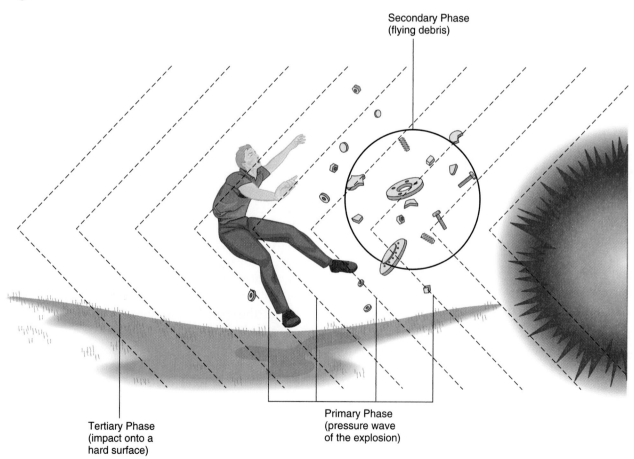

Secondary Phase (flying debris)

Tertiary Phase (impact onto a hard surface)

Primary Phase (pressure wave of the explosion)

in three phases, each with a typical pattern of injury (Figure 8-10):

- *Primary.* The pressure wave of an explosion affects gas-containing organs such as the lungs, stomach, intestines, inner ears, and sinuses. Blood vessels and membranes of the organs can be ruptured. Death can occur without any obvious external injury. Common primary blast injuries include pneumothorax, pulmonary contusion (bruising of the lungs), and perforation of the stomach and intestines.

- *Secondary.* Injuries result from flying debris created by the force of the blast. Unlike primary blast injuries, these are obvious. They most commonly include open wounds, impaled objects, and broken bones.

- *Tertiary.* Injuries in this phase occur when the victim is thrown away from the source of the explosion. They are basically the same as those of a victim who is ejected from a car during a collision.

SECTION 3 RESOURCE DETERMINATION

Once you have assured scene safety and determined the mechanism of injury or nature of illness, you must make sure you have the appropriate resources. For example, if the scene involves hazardous materials, you need to call a specialized hazmat team. If you have two or more patients, you may need to call for extra EMS personnel and ambulances. Consider implementing the incident management system (IMS) used in your area.

If special teams or extra units will be required, call dispatch to request them. Do this before you begin patient care. Experience has shown that getting immersed in patient care can cause any fire service First Responder to forget to call for additional help. Call first. It will help to prevent inefficiency and confusion later.

Never be too proud to ask for help when you need it! Situations in which you may need help include the following:

- When there are more patients than you can properly care for, call for additional personnel and ambulances. Until help arrives, you will need to make decisions about who must be treated first based on the severity of injuries. This is called "triage." (See Chapter 22 for details.)

- Hazardous materials can cause complicated emergencies. And they are everywhere. They are stored in homes and businesses. They are transported by ground, sea, and air. Hazmat incidents require specialized training to manage and decontaminate. Be alert and call for a hazmat team as soon as an incident is suspected. (See Chapter 24 for details.)

- It may be necessary to call for law enforcement if violence or the potential for violence exists. Situations such as problems with traffic, bystanders and crowds, or violations of the law will also require the police.

Other situations may require the power company, or electric utility, for downed wires or gas company for a natural gas shutdown. Confined-space rescue, high- and low-angle rescue, and helicopter rescue or evacuation also may be needed. Call for the resources you need immediately. If later you find they are not needed, they may be canceled. Time is of the essence.

COMPANY OFFICER'S NOTE

Hope for the best, but expect the worst. En route to a car crash, for example, ask yourself: "What is the speed limit where the collision occurred?" If it is 65 mph (105 km/h), you can almost be certain that there will be multiple patients and severe injuries. Therefore, ask for additional resources while en route to the scene, instead of upon arrival. If you find you do not need them, you can always withdraw your request.

CHAPTER 8 *Scene Size-Up* **163**

At the beginning of this chapter, you read that fire service First Responders were at the scene of an auto crash. To see how chapter skills apply to this emergency, read the following. It describes how the call was completed.

SCENE SIZE-UP (CONTINUED)

We kept a safe distance from the car. We could see that the pole seemed intact, and there were no wires down. There was fluid leaking from somewhere underneath the vehicle. However, we determined it was from the radiator. I saw that there were three patients in the car. I immediately called the dispatcher to have two ambulances respond to the scene. I also called for an extrication truck.

Two of the patients began climbing out of the car before my partner or I could get to them. The driver remained in the car. He had blood on his face. Looking at the mechanism of injury, I felt that he could be seriously injured. There was a lot of damage to the front of the car, and a large star in the windshield where his head hit.

We put on turnout gear and eye protection. Over our latex gloves, we put on heavy duty gloves to protect ourselves from all the broken glass. Approaching the two adult patients who had climbed out of the car, I noticed no obvious injuries. I asked them if they were okay, and both stated they were fine. I asked if they had any neck or back pain (no) and if we could evaluate them. Neither one wished to be examined. I decided to focus on the patient remaining in the vehicle.

INITIAL ASSESSMENT

The car was stable and safe, so I climbed into the back seat and held the patient's head and neck in a neutral in-line position. I asked him some quick questions, and saw that he was alert.

The blood on the patient's face wasn't causing any airway problems. He was breathing deeply and at a good rate. This was a 30-year-old man who could be seriously injured due to the mechanism of injury.

My partner went to the window nearest the patient. She talked to him to calm him down. We saw no severe bleeding, and his pulse was good. My partner placed him on oxygen and then updated the incoming units.

PHYSICAL EXAMINATION

The man had cuts and bruises to his forehead, which was oozing blood. He admitted that he was not wearing a seat belt at the time of the crash. He complained of pain in his neck and chest, and his neck was tender upon palpation. The left side of his chest was also tender, but there were no signs of broken ribs or open wounds. My partner listened to the patient's lungs. Air was moving in and out of both, she said. The man had no problems with his abdomen or hips. We could not get to his legs.

PATIENT HISTORY

We had not gotten far into the patient history when the extrication vehicle arrived on-scene and started extrication. It was more important to get the patient out of the car, and we could not hear or do much with the tools in operation anyway. We told our patient that it was going to be noisy for a minute. We explained what was going on. We also put a blanket over him for protection from any broken glass or debris.

ONGOING ASSESSMENT

We continued to monitor the patient's breathing and pulse during extrication. We also continued to talk with him. We found out that he was concerned that his wife would worry about him not showing up at home. We told him we would call her on our cell phone, so he gave us her number. I also found out he was a Chicago Cubs fan. I kidded him about that, saying that the pain he was now feeling was probably nothing compared to the years of pain the Cubs had put him through. He laughed. It was an encouraging sign.

PATIENT HAND-OFF

When the patient was extricated, the paramedics took over care. We told them what we had found so far: respiratory rate and pulse, the patient's complaints, our physical findings, and that we had him on oxygen (and that he was a Cubs fan). We also asked if we should cancel the helicopter from the trauma center. They told us to go ahead, and thanked us.

We stuck around and helped with the backboarding. Later we found out that the man had trouble breathing in the ambulance. He had a hemothorax (blood accumulated in his left lung). Fortunately, the trauma center was fairly close. He was in the hospital for a while and recovered fully.

As you can see, scene size-up is an important part of any call. It begins the moment you receive the call and does not end until you clear the scene. Scan for dangers continuously. Determine as soon as possible whether or not you will need help and what kind. Then do not be afraid to call for it.

Chapter Review

The scene size-up is the first step of every call and it begins the instant you receive the call. This is one of the few rules for which there are no exceptions. A proper scene size-up creates a proper foundation for a call.

The individual components of the scene size-up are: scene safety including BSI precautions, observation of the mechanism of injury or nature of illness, and resource determination. Do not rush through them.

You may notice experienced fire service First Responders who seem to respond to each emergency calmly. A calm, observant approach not only allows you to perform an optimal scene size-up, it also allows you to keep your wits about you. Some experienced EMS personnel who feel as if they are starting to rush will stop and count to three. This brief pause, usually not noticed even by crew members, returns the focus to a proper scene size-up and quality patient care.

FIRE COMPANY REVIEW

Page references where answers may be found or supported are provided at the end of each question.

Section 1

1. What are the components of scene size-up? (pp. 151, 152, 155, 163, 166)

2. What are some signs of potential danger at an emergency scene? (p. 154)

3. What is a general rule you can follow if you find a scene to be unsafe? (p. 154)

Section 2

4. What trauma can be expected from the "down-and-under" and "up-and-over" pathways of injury in a car crash? (pp. 156–157)

5. What factors affect the seriousness of the injuries caused by a fall? (p. 160)

Section 3

6. For what emergency situations would you typically need to call for additional resources? (p. 163)

RESOURCES TO LEARN MORE

"Arrival and First Contact" in *EMS Safety: Techniques and Applications*. Federal Emergency Management Agency/U.S. Fire Administration, FA-144. Washington, D.C.: April 1994.

SCENE SIZE-UP

ASSESS PERSONAL SAFETY.
- Take BSI precautions.
- Assure scene safety.

↓

IDENTIFY MECHANISM OF INJURY/NATURE OF ILLNESS.

↓

DETERMINE NECESSARY RESOURCES.
- Determine number of patients.
- Recognize presence of hazardous materials.
- Identify any special rescue needs.
- If appropriate, implement IMS.

9

Patient Assessment

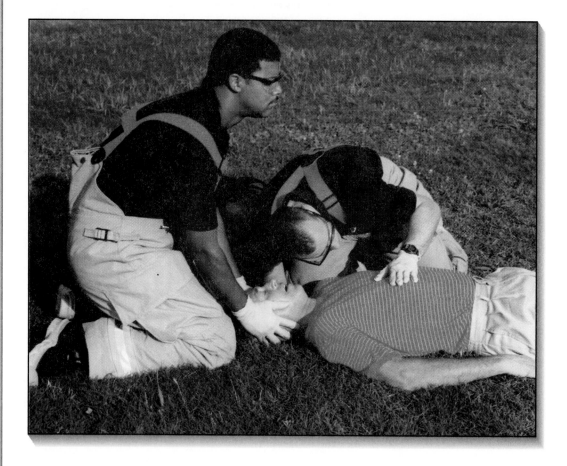

*I*NTRODUCTION   *Fire service First Responders must be able to assess a patient's condition quickly and accurately. This chapter will help you learn how. It presents a step-by-step routine used by many experienced emergency care providers. The routine includes a scene size-up, initial assessment, physical exam, patient history, ongoing assessment, and patient hand-off.*

As you already know from Chapter 8, scene size-up is always the first step in emergency care. When the scene is safe to enter, the next step is to identify immediate threats to the patient's life. This is called the initial assessment. You will perform a more thorough physical exam and get a patient history afterwards. The ongoing assessment follows the physical exam and history. It is an organized way of monitoring the patient's condition while you wait for help. Finally, there is the patient hand-off. It is a summary of the patient's condition, which you report to the EMTs when they take over patient care.

OBJECTIVES

Cognitive, affective, and psychomotor objectives are from the U.S. DOT's "First Responder: National Standard Curriculum." Enrichment objectives, if any, identify supplemental material.

Cognitive

3-1.7 Summarize the reasons for forming a general impression of the patient. (pp. 171–172)

3-1.8 Discuss methods of assessing mental status. (pp. 172–173)

3-1.9 Differentiate between assessing mental status in the adult, child, and infant patient. (pp. 172–173)

3-1.10 Describe methods used for assessing if a patient is breathing. (pp. 173–175)

3-1.11 Differentiate between a patient with adequate and inadequate breathing. (pp. 174–175)

3-1.12 Describe the methods used to assess circulation. (pp. 175–176)

3-1.13 Differentiate between obtaining a pulse in an adult, child, and infant patient. (pp. 175, 183–184)

3-1.14 Discuss the need for assessing the patient for external bleeding. (pp. 175–176)

3-1.15 Explain the reason for prioritizing a patient for care and transport. (p. 177)

3-1.16 Discuss the components of the physical exam. (pp. 177–188)

3-1.17 State the areas of the body that are evaluated during the physical exam. (pp. 178–182)

3-1.18 Explain what additional questioning may be asked during the physical exam. (pp. 188–189)

3-1.19 Explain the components of the SAMPLE history. (pp. 189–191)

3-1.20 Discuss the components of the on-going assessment. (p. 191)

3-1.21 Describe the information included in the First Responder "hand-off" report. (p. 191)

Affective

3-1.24 Explain the importance of forming a general impression of the patient. (pp. 171–172)

3-1.25 Explain the value of an initial assessment. (p. 171)

3-1.26 Explain the value of questioning the patient and family. (pp. 188–189)

3-1.27 Explain the value of the physical exam. (pp. 177–188)

3-1.28 Explain the value of an on-going assessment. (p. 191)

3-1.29 Explain the rationale for the feelings that these patients might be experiencing. (pp. 177–178, 191)

3-1.30 Demonstrate a caring attitude when performing patient assessments. (pp. 177–178, 191)

3-1.31 Place the interests of the patient as the foremost consideration when making any and all patient care decisions during patient assessment. (pp. 177–178, 191)

3-1.32 Communicate with empathy during patient assessment to patients as well as with family members and friends of the patient. (pp. 177–178, 191)

Psychomotor

3-1.34 Demonstrate the techniques for assessing mental status. (pp. 172–173)

3-1.35 Demonstrate the techniques for assessing the airway. (pp. 173–174)

3-1.36 Demonstrate the techniques for assessing if the patient is breathing. (pp. 174–175)

3-1.37 Demonstrate the techniques for assessing if the patient has a pulse. (pp. 175–176, 183–184)

(continued)

3-1.38 Demonstrate the techniques for assessing the patient for external bleeding. (pp. 175–176)

3-1.39 Demonstrate the techniques for assessing the patient's skin color, temperature, condition, and capillary refill (infants and children only). (pp. 184–185)

3-1.40 Demonstrate questioning a patient to obtain a SAMPLE history. (pp. 189–191)

3-1.41 Demonstrate the skills involved in performing the physical exam. (pp. 177–178)

3-1.42 Demonstrate the on-going assessment. (p. 191)

Enrichment

◆ Describe when and how to manually stabilize a patient's head and neck. (pp. 172, 178–179)

◆ Identify the components of an assessment of vital signs. (p. 182)

◆ Describe the methods used to obtain a breathing rate and a pulse rate. (pp. 182–184)

◆ Describe normal breathing rates and pulse rates. (pp. 182–183)

◆ Identify the terms that describe the quality of breathing and the quality of pulse. (pp. 182–184)

◆ Describe the methods used to assess the pupils. (p. 185)

◆ Describe the methods used to assess blood pressure. (pp. 185–188)

◆ Differentiate between a sign and a symptom. (p. 189)

◆ State the importance of accurately reporting and recording the baseline vital signs. (p. 182)

ON SCENE

DISPATCH

My partners and I were returning to the firehouse after extinguishing a brush fire when we were dispatched to 3820 Littler Court for "an unknown problem." The dispatcher told us that the woman caller was extremely upset. However, he believed the woman said her husband was bleeding. I asked if police officers were responding. The dispatcher stated they should be on scene in two minutes.

SCENE SIZE-UP

We approached the scene carefully, as we always do. This call was in a quiet section of town, but you never can tell. We realized that an "unknown problem" could mean anything from a drunk to a cardiac arrest, so we had all our protective equipment ready. We observed a police officer talking to a woman at the door. He turned around and waved at us, signaling that the scene was secure.

The woman was quite upset. She was about 60. She told us that her husband was going to

the bathroom when she heard him shout. Hurrying to the bathroom, she was horrified to see him slumped on the toilet, with drool and blood coming from his mouth. The woman was sure her husband did not fall. We felt we had a medical problem on our hands.

Before we went any farther inside the house, we put on our gloves and eye protection.

INITIAL ASSESSMENT

Entering the bathroom, we observed a white-haired male patient sitting on the toilet, his body slumped against the wall. We identified ourselves to the man, but received no response. One of my partners gently shook the patient while saying, "Sir, are you okay?"

Receiving no response, we immediately decided to place the patient on the floor, where we could more thoroughly assess him. After carefully lowering him to a supine position, I checked his airway. I heard some gurgling, so I suctioned him out, which helped. I noted his respirations were slow and shallow. Meanwhile, my partner found a pulse, which was rapid. There was no external bleeding visible. Our general impression was of a male, unresponsive from unknown causes, who required ventilatory assistance.

INITIAL ASSESSMENT

The **initial assessment** may be the most important part of the patient assessment process. In it you must identify and treat conditions that cause an immediate threat to the patient's life, such as breathing problems or severe bleeding. The initial assessment includes getting a general impression of the patient; assessing responsiveness; assessing the **ABCs** (airway, breathing, and circulation); treating life threats; and updating incoming EMS personnel about the patient's condition.

General Impression

Form a general impression as you approach the patient (Figure 9-1). It should include the patient's

Figure 9-1 Form a general impression as you approach the patient.

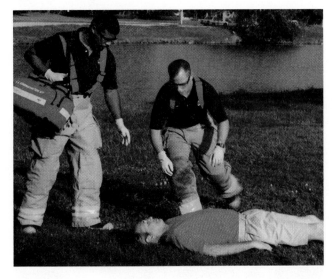

chief complaint and a brief immediate assessment of the environment in which the emergency has taken place.

The **chief complaint** is the reason that EMS was called. It is generally the response to the question "Can you tell me why you called EMS today?" Record the response on your forms in the patient's own words. "I fell down the stairs" or "My chest hurts" are examples of chief complaints. If the patient is unresponsive, get information from the person who called EMS.

The general impression is not designed to be the final word in the patient's condition. Rather, it lets you get started on the right track with patient care. During this phase of the initial assessment, determine if the situation is a trauma (injury) complaint, a medical (illness) complaint, or both. Do this by listening to what the patient or bystanders tell you. Also look around the scene to gain a quick impression of the forces involved in an injury.

If your general impression is that you are facing a patient with a potentially serious injury or illness, you may wish to ask EMS dispatch for advanced life support or additional personnel to assist. (You may have already done this in the scene size-up.)

To complete your general impression, you must also determine the patient's age and sex.

Consider the following general impressions:

◆ A 5-year-old male patient has fallen off the top of a slide at a playground. Your observation of the scene reveals that the fall was from approximately 10 feet (3 m), with the patient landing on soft sand. The boy is sitting upright and crying.

◆ A 32-year-old woman is complaining of abdominal pain. She seems to be experiencing severe pain, as she cannot speak without straining. She is lying on her side with knees drawn up, protecting her abdomen with one hand.

◆ A 55-year-old man is found on the lawn outside his front door, unconscious. No one is quite sure of what happened.

In the first case, the general impression leads you to believe the boy is probably not seriously injured. The second case appears to be a medical patient who is in severe distress. The third case gives you little information on how to proceed. However, the mere fact that the man is unconscious can tell you that immediate action must be taken.

FIRE DRILL

When arriving on the fireground, it is necessary to perform the management function known as "size-up." This rapid, yet deliberate consideration of critical fireground factors leads to the development of an attack plan. Similarly, the general impression can be thought of as a rapid, yet deliberate consideration of critical EMS scene factors leading to the development of a patient-care plan.

Responsiveness

The next part of the initial assessment is determining the patient's **level of responsiveness.** This is especially important for patients having an **altered mental status,** as airway care may be required immediately.

If the patient is confused, be sure to let him or her know who you are. Always make your identity clear as you approach a patient. State your name. Then explain that you are a fire service First Responder who is there to help.

If the mechanism of injury suggests possible spine or head injury, take **spinal precautions** at this time. Hold the patient's head and neck stable and in a neutral position (Figure 9-2). Be sure to keep the patient from moving. Movement in patients with suspected head or spinal-cord injury could cause additional injuries.

To manually stabilize a patient's head and neck, place your hands on either side of the head and then spread your fingers apart. The object is

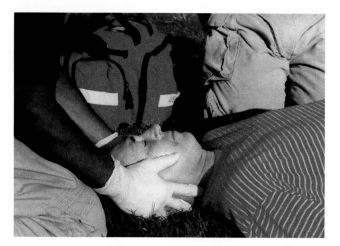

Figure 9-2   If there is possible spine or head injury, immediately stabilize the patient's head and neck.

to prevent movement. If the patient is responsive, explain what you are doing so he or she is not alarmed. Your hand position may reduce the patient's hearing. So, be aware of the anxiety this may cause. (Spine injuries are covered in more detail in Chapter 18.)

There are four levels of responsiveness that are commonly used to classify patients. They are: alert, verbal, painful, and unresponsive. Together, these terms make up the mnemonic **AVPU.** The classifications, when applied to patients, are as follows (Figure 9-3):

A — Alert. A patient who is alert is responsive and oriented. That is, the patient is aware of his surroundings, the approximate time and date, and his name. This is commonly

Figure 9-3   Assess the patient's level of responsiveness.

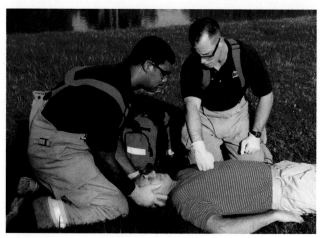

referred to as being "responsive to person, place, and date" (or "oriented × 3"). Each distinction is important. Some patients may appear wide awake but in fact are not aware of the surroundings. A patient's mental status is important to determine because it can indicate injury or illness.

V — Verbal. This patient is disoriented but responds when spoken to. We say that he "responds to verbal stimulus." A patient who answers questions about place and date incorrectly is another example. He may be suffering from a medical condition, such as seizures or diabetes or a traumatic condition, such as shock.

P — Painful. The patient who responds only to a painful stimulus does not answer questions or open his eyes to any verbal commands. A stimulus, such as a pinch or careful but firm rub on the sternum, produces only a flinch or moan. (Remember that you first check responsiveness by observation. If the patient is alert or verbal, there is no need to apply a painful stimulus.)

U — Unresponsive. This patient does not respond to any stimulus. He does not open his eyes, respond verbally, or even flinch when pain is applied. This patient is deeply unconscious, most likely in a critical condition, and in definite need of airway and other supportive care.

PEDIATRIC NOTE

Determining the level of responsiveness is different for infants and children. For these patients, assess the response to the environment. They should recognize their parents, and they usually wish to stay with them. Expect your assessment to cause tears, too. Children who do not recognize their parents or who are indifferent to your assessment and treatment may be very sick.

Airway

The patient's airway status is the most critical aspect of patient care. People who cannot breathe cannot live. Make sure the patient's airway is open and clear (Figure 9-4). The method of

GERIATRIC NOTE

In the elderly there are common diseases and conditions, such as Alzheimer's, that cause changes in responsiveness. In cases such as these, try to find out from the family if there has been a change. You might ask: "In your opinion, is your mother acting in a manner that is normal for her?"

Some elderly patients live alone and have neither the means nor reason to keep track of the date and current events. If an elderly patient does not know the date or another piece of information, try other questions to determine orientation. You might ask about the immediate surroundings, for example, or about what you are doing there.

assessing the patient's airway depends on whether or not the patient is responsive.

◆ *Responsive patient.* When a patient is able to respond to your questions, notice if he can speak clearly. Gurgling or other sounds may indicate that something like teeth, blood, or other matter is in the airway. Also make sure the patient can speak full sentences, instead of short, two-word bursts.

◆ *Unresponsive patient.* The unresponsive patient needs aggressive airway maintenance. Immediately make sure the airway is open. If the patient is ill with no sign of trauma, use the head-tilt/chin-lift maneuver to open the airway. If trauma is suspected, use the jaw-thrust

Figure 9-4 Open the patient's airway and make sure it is clear.

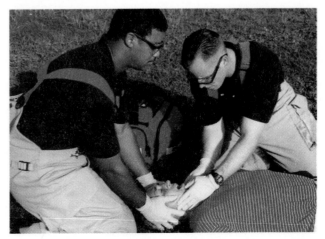

maneuver with great care to avoid tilting the head. Then, inspect the airway for blood, vomit, or secretions. Also look for loose teeth or other foreign matter that could cause an obstruction. Clear the airway using suction or a gloved finger.

Remember that the airway check is not a one-time event. Some patients with serious trauma or unresponsive medical patients who are vomiting will need almost constant suctioning and airway maintenance.

Breathing

After securing an open airway, look, listen, and feel for breathing (Figure 9-5). If there is breathing, determine if respirations are adequate. As you will recall from Chapter 5, breathing is not an all-or-nothing proposition. There will be times when a patient is breathing but not at a sufficient depth or rate to sustain life.

Adequate breathing is characterized by three factors: adequate rise and fall of the chest, ease of breathing (breathing should appear to be effortless), and adequate respiratory rate.

Inadequate breathing may be identified by (Figure 9-6):

◆ Inadequate rise and fall of the chest.

◆ Increased effort of breathing, including use of accessory muscles.

◆ Cyanosis (blue or gray color to the skin, lips, or nail beds).

◆ Mental status changes.

Breathing rates:
   less than 8 in adults
   less than 10 in children
   less than 20 in infants

Inadequate chest wall motion

Cyanosis

Mental status changes

Increased effort to breathe

Gasping and grunting

Slow heart rate accompanied by slow breathing rate

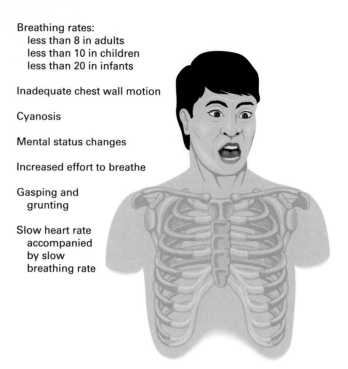

Figure 9-6   Signs of inadequate breathing.

◆ Inadequate respiratory rate (less than 8 per minute in adults, less than 10 in children, and less than 20 in infants).

If the patient is breathing adequately, there may be no need to assist respiration in any way. However, during your assessment, you may determine that your patient would benefit from oxygen therapy. If you are trained and allowed, administer oxygen to these patients (Figure 9-7).

Figure 9-7   Apply oxygen if the patient needs it and if you are permitted to do so.

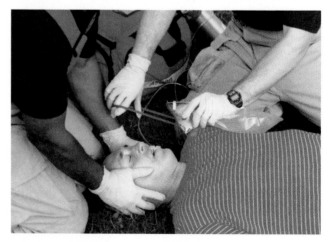

Figure 9-5   Look, listen, and feel for breathing.

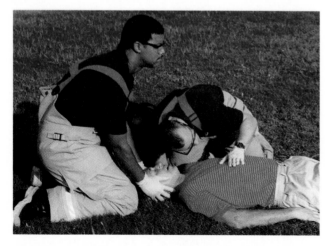

If you determine that the patient's respirations are absent or inadequate, you must begin ventilating immediately. Do not stop until another trained rescuer relieves you or until the patient regains adequate respirations. In most cases, you will continue ventilations until the EMTs arrive.

Circulation

Assessing circulation allows the fire service First Responder to determine if the heart is adequately pumping blood to all parts of the body. It is also used to check for signs of life-threatening external bleeding. To assess circulation (Figure 9-8):

◆ *Responsive patient.* If the patient is a verbally responsive adult, use the radial pulse to assess circulation. Checking the carotid pulse may cause this patient undue anxiety. Always use the brachial pulse point for an infant. Use either the radial or brachial pulse point for the responsive child.

When checking the pulse, note the approximate rate and rhythm. An extremely slow or fast pulse, or one that is irregular, could signify a serious condition. Also note the patient's skin at this time. Check the color and temperature (Figure 9-9). Skin that is pale, cool, and moist may indicate shock.

◆ *Unresponsive patient.* Check the pulse of an unresponsive adult at the carotid artery. Check the pulse of unresponsive children at the carotid or femoral arteries. Remember, the pulse check for all infants is done at the brachial artery. If the pulse is absent, begin CPR.

After checking the patient's pulse, check for serious external bleeding (Figure 9-10). Remember that the initial assessment is designed to

Figure 9-8a   If the patient is responsive, assess the radial pulse.

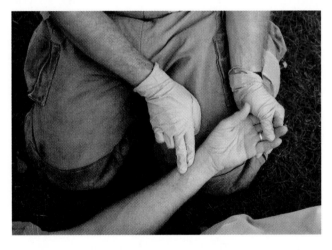

Figure 9-8b   If the patient is unresponsive and there is no radial pulse, assess the carotid pulse.

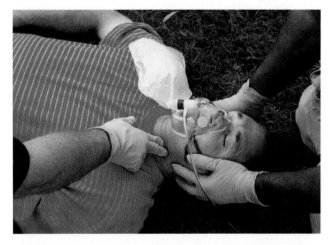

Figure 9-8c   Assessing an infant's brachial pulse.

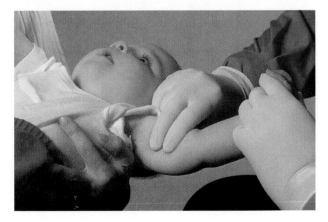

Figure 9-8d   Assessing a child's femoral pulse.

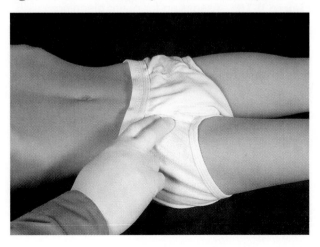

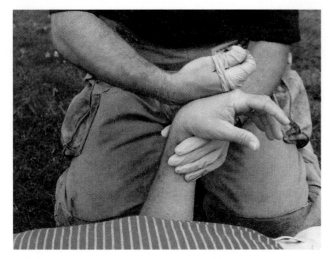

Figure 9-9 Assessing the patient's skin temperature.

identify and treat life-threatening problems. Be alert. Do not let minor wounds sidetrack you or keep you from caring for more serious injuries first.

Scan the patient for serious bleeding. Use your gloved hands to check areas that are hard to see, such as the small of the back and the buttocks. Remember that heavy clothes can absorb large quantities of blood. If serious bleeding is found, use the methods discussed in Chapter 14 to control the blood flow.

EMS Update

At this point, you will know if your patient is barely breathing and requires ventilation (a high priority) or if your patient is stable with a minor complaint. Regardless, the EMS personnel en route to the scene will be interested in an update.

The information you provide will allow them to plan for patient care.

If you have a phone or radio available (Figure 9-11), report the patient's:

◆ Age and sex.
◆ Chief complaint.
◆ Level of responsiveness.
◆ Airway, breathing, and circulatory status.
◆ Any treatment provided.

Also ask the incoming EMS personnel to give you their **ETA** (estimated time of arrival), so you can continue patient care and prepare for their arrival.

The following is an example of a radio report. It might have been given by the fire service First Responders in the "Case Study" that opened this chapter:

> "Dispatcher, we have an approximately 60-year old male who was found unresponsive to painful stimulus. His airway required suctioning, and we are assisting ventilations. The patient's pulse is rapid and weak."

Figure 9-11 The last step of the initial assessment is an EMS update.

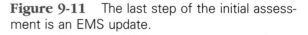

Figure 9-10 Assessing for major bleeding.

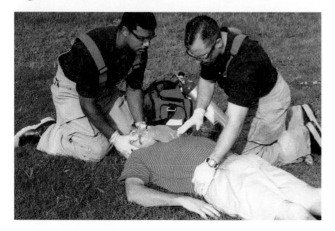

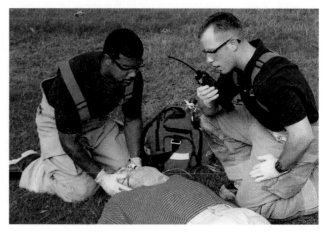

After your report, the dispatcher will acknowledge your transmission and advise you of the ambulance's ETA.

FIRST RESPONDER PHYSICAL EXAM

The initial assessment is designed to help you identify and treat your patient's life-threatening problems. However, not all problems are life threatening. The First Responder physical exam is a survey of the patient's entire body that is meant to reveal all signs of illness or injury.

The physical exam is designed to be more thorough. In some cases, you will have time to perform it. In others, you will only have time for an initial assessment before the EMTs arrive. When time permits and the patient does not need continued life-saving care, begin the physical exam.

The physical exam proceeds in a logical order, usually from head to toe. It will be slightly different for each patient. If a patient falls a considerable height from a ladder, for instance, he could have injuries anywhere on his body. This patient would require a full assessment. In the case of an isolated cut to a finger, a complete, hands-on examination would not be necessary.

Principles of Assessment

Patient assessment is a skill. Like all skills, the more you practice, the better you will be. If you do not practice regularly, the result could be poor performance and missed injuries.

Methods of Examination

The patient assessment process involves the use of your senses. There are three methods used when performing a patient assessment: **inspection** (looking), **auscultation** (listening), and **palpation** (feeling).

◆ *Inspection.* The first method is the easiest. Simply make an overall observation of the patient. Then observe the various parts of the body. What you see is very important throughout a call. As you approach a patient, even before you speak to him, you may observe that he is clutching his fist to his chest and appears to be uncomfortable. This could be your first indication of a heart problem.

◆ *Auscultation.* The most important listening you will do is for the sound of air entering and leaving the lungs. It will help you to determine the status of the patient's breathing. If your EMS system requires auscultation of breath sounds using a stethoscope, practice on your classmates. Become familiar with the sound of normal breathing.

◆ *Palpation.* Palpating, or feeling, with your fingertips is usually done last, because it can cause pain. The actual pressure you apply depends on the area you are palpating and the type of problem you suspect. For example, if you observe a swollen lower leg where the bone is normally near the surface, you only would need to palpate the area gently to determine if tenderness is present. Assessing the abdomen of an obese patient would require more pressure. Palpation will also identify areas where bones are rubbing together, abnormally rigid areas, skin temperature, and sweating.

Inspect and palpate each part of the body before you move on to the next area. You must also auscultate the chest. For example, observe the chest for rise and fall with breathing. Then auscultate for adequacy of breathing and palpate for tenderness or other sensations. After examining the chest, you would move to the abdomen where you would inspect and palpate as appropriate.

When conducting the exam, look for the following signs of injury:

D — Deformities.

O — Open injuries.

T — Tenderness.

S — Swelling.

Look at the first letters of each word listed above. The letters form the mnemonic DOTS. Use it to help you remember the signs you are looking for. Some signs will be obvious, such as a cut in the skin (open injury). Others, such as abdominal tenderness caused by internal injuries, will not be as obvious but are very serious.

As you proceed through the physical exam, be sure to listen to what your patient tells you. Although this seems obvious, a busy emergency scene can be very distracting. Missing what your patient says to you may lead to inappropriate care. Listening also demonstrates a caring attitude.

Finally, remember that your patient will be anxious or scared. It is important to reassure him or her throughout the call. When the EMTs arrive, be sure to introduce your patient to them and relay special concerns or fears the patient may have discussed with you.

FIRE DRILL

Good listening skills have been referred to as providing "psychological oxygen" to a patient. When others feel that you are listening intently to what they say, it reaffirms their self-worth. They think, "Hey, this person really must care about me. He remembered my name the first time." Think about situations where you have just been introduced to someone and, 30 seconds later, he or she spoke your name. How did you feel?

Medical vs. Trauma Patients

An examination of a trauma patient is different from an examination of a medical patient. It has been said that a trauma exam is 80% hands-on and 20% questioning, while a medical exam is 80% questions and 20% hands-on.

For example, once you observe and examine an isolated injury, you do not need too much more information. The physical signs of injury can be observed and palpated. Compare that to a patient who is having a heart problem. Chest pain is something you cannot observe or palpate. Only the patient can feel it. Therefore, in order to provide the appropriate emergency care, you must use questions to encourage the patient to describe the symptoms to you.

The Physical Examination

The balance of this section details the examination of specific areas of the body. Use the DOTS mnemonic to guide your examination.

Examination of the Head

Assess all areas of the head, including the skull, face, and jaw (Figure 9-12). Check the ears and

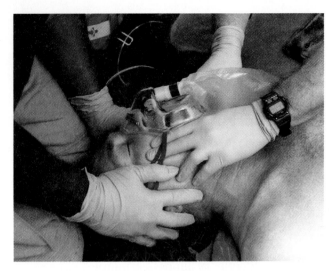

Figure 9-12 Assess all areas of the skull, face, and jaw.

nose for clear or blood-tinged fluid. Also check pupils for size and responsiveness. Note that injuries to the head may be serious. Bleeding can be severe. Many areas are covered by hair and can hide injuries. Use DOTS to guide you:

D — *Deformities.* Examine the skull, face bones, and jaw for signs of deformity (depressions or indentations, for example). Check for deformities such as loose teeth, which can create airway problems.

O — *Open injuries.* Open injuries to the head may bleed profusely. As such, they may have been treated in the initial assessment. Any injury that bleeds into the airway is of particular concern. Also, look in the hair for injuries that may be hidden.

T — *Tenderness.* When you are palpating the head, the patient may complain of pain or tenderness where there is no obvious injury. Make a note of the locations of tenderness.

S — *Swelling.* Swelling frequently accompanies injuries to the head. It may be noted around injuries to the skull and to facial structures such as areas around the eyes, nose, and mouth.

Examination of the Neck

There are large blood vessels and major airway structures in the neck. Injuries can be quite serious. Also, remember that the patient may have a stoma. While you are examining the neck, look to see if the patient is wearing a medical identification tag.

To examine the neck (Figure 9-13):

D — *Deformities.* Gently feel to see that the trachea is not deformed or shifted. Either can indicate a critical condition such as excessive pressure in the chest cavity. Palpate the cervical spine, or the vertebrae in the posterior (back) of the neck.

O — *Open injuries.* Open injuries to the neck may result in serious blood loss. Bandage them immediately. Use an occlusive (airtight) dressing, which prevents air from entering the neck.

T — *Tenderness.* Palpate the soft tissues, trachea, and vertebrae for tenderness.

S — *Swelling.* The neck may accumulate blood, so examine for swelling. Also, note that air may escape from the trachea or other airway structure and cause a popping sound or crackling feeling under the skin (subcutaneous emphysema).

Whenever there is a possibility of spine injury, maintain manual stabilization of the head and neck until the patient can be completely immobilized. If you are equipped, trained, and allowed, apply a rigid cervical collar at this time (Figure 9-14). See Chapter 18 for a detailed discussion.

Examination of the Chest

Any injury to the chest may involve injury to the vital organs or to major blood vessels. Note that you should also include the shoulders in this exam.

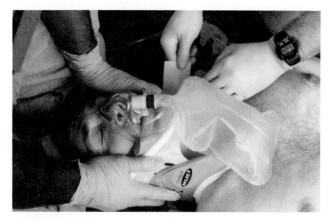

Figure 9-14   Apply a rigid cervical collar, if the patient needs it and if you are permitted to do so.

If you are trained to do so, listen to the chest with a stethoscope. Determine if an adequate amount of air is entering the lungs. When you compare both sides, the sounds you hear should be equal.

D — *Deformities.* Palpate the rib cage for signs of deformity (Figure 9-15). Remember the ribs extend all the way back to the spine. Injuries to the back pose the same grave dangers as those to the front of the chest. Do not move the patient in order to examine the back until appropriate spinal precautions have been taken. Palpate the sternum. If the patient is responsive, ask him or her to take a deep breath. Determine if it causes pain. Also, check that both sides of the chest are expanding equally.

Figure 9-13   Examine both the front and back of the neck.

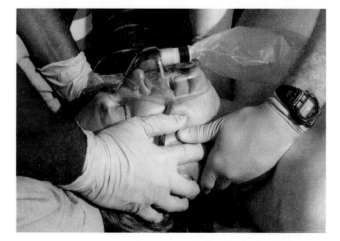

Figure 9-15   Examine the chest.

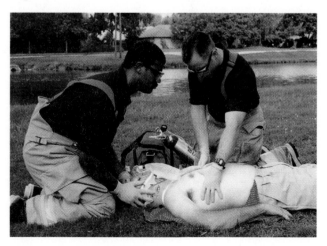

O — *Open injuries.* Open injuries are of particular concern when they occur in the chest. If a wound extends into the chest cavity, air may enter the area around the lungs and cause a serious condition. Bandage open wounds to the chest immediately with an occlusive (airtight) dressing.

T — *Tenderness.* While palpating the chest, ask the patient if he or she feels any pain. Even when no obvious injury is apparent, internal injuries may be present.

S — *Swelling.* Observe for swelling. If there is swelling or any other sign of possible injury, assess for underlying breathing problems.

Examination of the Abdomen

As you will recall from Chapter 4, there are many organs within the abdominal cavity that can be injured. (Though the spine lies to the rear of the abdomen, it is not palpated at this time.) To examine the abdomen (Figure 9-16):

D — *Deformities.* Deformity of the abdomen usually refers to rigidity (hardness) or distention.

O — *Open injuries.* These may include cuts and scrapes (lacerations and abrasions), penetrating wounds (from a knife or gunshot), or protruding organs (eviscerations). These wounds are severe because of the potential for bleeding and infection.

T — *Tenderness.* An important symptom, tenderness can indicate underlying injury. Recall the abdominal quadrants from Chapter 4.

Palpate the quadrant where the patient complains of pain last. If you examine it first, you could increase the pain, making the examination of the other quadrants impossible or inaccurate.

S — *Swelling.* Swelling or discoloration of the skin is another indication of abdominal injury. Be sure to check the flanks (the lateral sides of the hips and buttocks).

Examination of the Back (Posterior)

While it is important to check the patient's back, moving a patient could result in making a neck or spine injury worse. If enough properly trained rescuers are present, you may decide to check the back. If you suspect spine injury and a long backboard is available, you may wish to move it under the patient while the patient is being rolled. However, do so only if you are trained in its use and have enough help to do so safely.

Check the patient's posterior as follows (Figure 9-17):

D — *Deformities.* Check for chest wall deformity, which may indicate broken ribs. Also, look for obvious deformity along the length of the spine.

O — *Open injuries.* Injuries to the posterior chest can cause the same serious conditions that occur to the anterior chest. Look for open or sucking chest wounds (open chest wounds sometimes make a sucking sound with respiration). Observe for scrapes, cuts, and other open injuries. Look for both entry and exit gunshot wounds.

Figure 9-16 Palpate each quadrant of the abdomen.

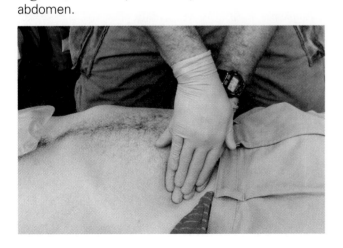

Figure 9-17 Examine the back, keeping the patient's head and neck in alignment at all times.

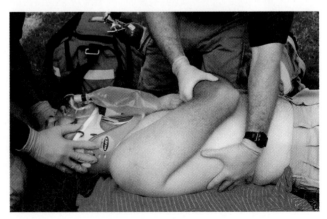

T — *Tenderness*. Tenderness may indicate a broken rib or an abdominal injury. Tenderness along the spine may indicate serious injury to the spinal cord.

S — *Swelling*. Swelling or discoloration of the flanks could indicate bleeding in the abdomen. Swelling anywhere indicates some type of injury.

Examination of the Pelvis

The pelvis is a large, bony structure. As you may recall, it is composed of a left and right ileum, ischium, and the pubis bones. Palpate each of these areas for injury. The pelvis, or hips, may be fractured, which could result in blood loss of two liters or more. This amount of blood loss is life threatening. So, be sure to identify any possibility of pelvic injury during the assessment process. Note also if there has been any loss of bladder or bowel control or bleeding from the rectum, vagina, or urinary opening.

D — *Deformities*. Unlike the bones of the arms and legs, deformities of the pelvis are not always obvious. Palpate the bones to feel for deformity (Figure 9-18).

O — *Open injuries*. These are not as common as in other areas of the body.

T — *Tenderness*. Palpate with less force if the bones of the pelvis are close to the skin. Palpate with more force if the patient is obese with bones under a considerable amount of tissue. Though you may feel awkward assessing the pubis (groin) bone, be sure to check it.

S — *Swelling*. Look for swelling and discoloration around the hips.

Examination of the Extremities

The extremities are common sites of injury. So, do not rush your examination. Be sure to inspect and palpate each one (Figure 9-19). While you are examining the limbs, look to see if the patient is wearing a medical identification tag.

D — *Deformities*. Because bones are close to the surface, deformities may be seen easily in the extremities. Check the entire length of each bone and all joints.

O — *Open injuries*. Look for open injuries, which are quite common in the extremities.

T — *Tenderness*. Just as in other areas of the body, there may be underlying injury without obvious deformity. Palpate each extremity for tenderness.

S — *Swelling*. Since injuries to the extremities are often close to the skin, swelling and discoloration may be evident. Any extremity that is painful, swollen, or deformed may be broken and should be manually stabilized until it can be splinted. Note that swelling also may be the result of a medical problem.

The extremities also should be checked by feeling for a pulse in each extremity (Figure 9-20). The radial pulse in the wrist will tell you if circulation in the entire arm is adequate. There are two pulses in the feet, either of which may be palpated to see if circulation is adequate in the lower extremities. They are the dorsalis pedis (top of the

Figure 9-18 Examine the pelvis by applying gentle pressure.

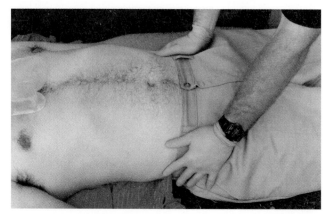

Figure 9-19 Visually inspect and palpate each extremity.

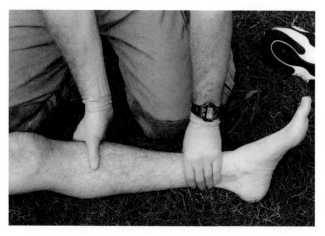

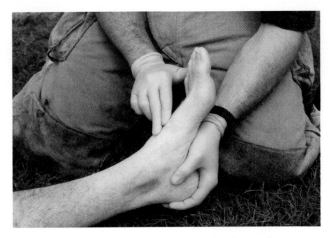

Figure 9-20 Also check the pulse in each extremity.

foot) and the posterior tibial artery (inside surface of the ankle).

The ability to move an extremity, such as wiggling fingers or toes, is an important sign to look for (Figure 9-21). Movement means that impulses from the nervous system can reach these points. If there is no movement, there may be a problem with a nerve. No movement on one side of the body or below a certain point could indicate problems with the central nervous system.

For the same reason, check to see that the patient has sensation in his or her limbs. Gently squeeze one extremity and then the other. As you do, ask questions, such as "Can you feel me touching your fingers?"

Vital Signs

Vital signs include the patient's respiration, pulse, skin, pupils, and blood pressure. You can

Figure 9-21 Check for sensation and the ability to move fingers and toes.

assess and monitor most vital signs by looking, listening, and feeling. However, it is best if you have the proper equipment:

◆ A wristwatch to count seconds.

◆ A pen light to examine pupils.

◆ A stethoscope to listen for adequacy of respiration and to take blood pressure.

◆ A blood-pressure cuff, or sphygmomanometer, to take blood pressure.

◆ A pen and notebook to take notes.

More important than any one vital sign is change in vital signs over time. The vital signs taken by a fire service First Responder are particularly important because they are taken early in the call. EMTs and hospital personnel will refer back to them to see if the patient has improved or gotten worse over time. For example, if you take a pulse and obtain a reading of 90 beats per minute, and later the pulse rises to 120, a serious condition may be developing. Without your early readings, this observation would not be possible. Be sure to record vital signs and the times each is taken.

Respiration

A respiration consists of one inhalation and one exhalation. The normal number of respirations per minute varies with age. For an adult that number is between 12 and 20 times per minute (Table 9-1). To count a patient's respirations (Figure 9-22):

1. **Position your hand.** Place it on the chest or abdomen.

2. **Count** the number of times the chest (or abdomen) rises during a 30-second period. Then multiply that number by 2.

Table 9-1

Normal Respiratory Rates

Patient	Respiratory Rate*
Adult	12-20
Child	15-30
Infant	25-50

*Approximate per minute at rest

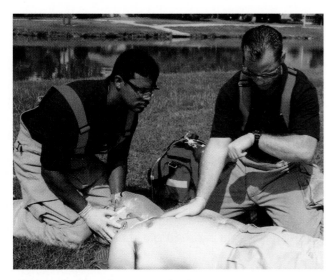

Figure 9-22 Count respirations.

3. Record the result immediately. Never rely on your memory.

The depth of respirations gives a clue to the amount of air that is inhaled. You can gauge depth by placing your hand on the patient's chest and feeling for chest movement. Also, feel the abdomen to see if it is moving instead of the chest.

Normally, the work required by breathing is minimal. Some effort is required to inhale, but almost none is required to exhale. For this reason, normal inspiration takes slightly longer than normal exhalation. When exhaling is prolonged, the patient may have a chronic obstructive pulmonary disorder (COPD) such as emphysema.

Common signs and symptoms of respiratory distress include:

◆ Gasping for air.

◆ Breathing that is unusually fast, slow, deep, or shallow.

◆ Wheezing, gurgling, high-pitched shrill sounds, or other unusual noises.

◆ Unusually moist, flushed skin. Later, the skin may appear pale or bluish as the oxygen level in blood falls.

◆ Difficulty speaking. The patient can say only a few words without catching his or her breath.

◆ Dizziness, anxiety.

◆ Chest pain and tingling in the hands and feet.

Abnormal breathing conditions you should know about include:

◆ Shortness of breath or breathing difficulty.

◆ Use of accessory muscles and retractions (pulling of the skin between the ribs or above the sternum and clavicles).

◆ Abnormally slow breathing.

◆ Abnormally deep, rapid breathing.

If the patient is aware that you are assessing respiration, he may not breathe naturally. This can give a false reading. To get around this, take a pulse with the patient's arm draped over his chest or abdomen. Count the pulse for 15 seconds. Then, without moving the patient's arm or telling the patient what you are about to do, count respirations for the next 30 seconds. Readings are easily obtained by observing and feeling the chest rise and fall with your hand, which is already on the patient's torso.

Pulse

Each time the heart beats, the arteries expand and contract with the blood that rushes into them. The pulse is the pressure wave generated by the heartbeat. It directly reflects the rate, strength, and rhythm of the pumping heart. When taking a pulse, note the following:

◆ *Is the pulse rate slow or fast?* See Table 9-2 for normal rates.

◆ *What is the strength of the pulse?* A normal pulse is full and strong. A "thready" pulse is weak and rapid. A "bounding" pulse is unusually strong.

◆ *What is the rhythm of the pulse?* A normal pulse has regular spaces between each beat. An irregular one is spaced irregularly. You can

Table 9-2

Normal Pulse Rates

Patient	Pulse Rate*
Older than 10 years	60 to 100
2 to 10 years	60 to 140
3 months to 2 years	100 to 190
Newborn to 3 months	85 to 205

*Approximate per minute at rest

describe the pulse of a patient, for instance, as "72, strong, and regular." The rate, strength, and regularity of a pulse tell what the heart is doing at any given time.

The pulse can be felt at any point where an artery crosses over a bone or lies near the skin. The radial artery crosses over the end of one of the forearm bones—the radius. We commonly refer to this site as the wrist, the most common site for assessing a pulse. To take a radial pulse (Figure 9-23):

1. **Position the patient.** Have him or her lie down or sit.

2. **Gently touch the pulse point** on the thumb side of the patient's wrist with the tips of two or three fingers. (Avoid using your thumb. It has a prominent pulse of its own that can be counted by mistake.)

3. **Palpate the pulse.** Count the number of beats you feel for 15 seconds. Then multiply that number by 4 to get the number of beats per minute. If a pulse is irregular, slow, or difficult to obtain, count the beats for 30 seconds and multiply by 2 for a more accurate reading.

4. **Record the pulse** and any other vital sign immediately. Never rely on your memory.

Fire service First Responders may also assess a pulse at the brachial artery (upper arm), carotid artery (neck), femoral artery (groin), dorsalis pedis (top of the foot), and the posterior tibial artery (inside surface of the ankle). If trained, you also

Figure 9-23 Assess the pulse.

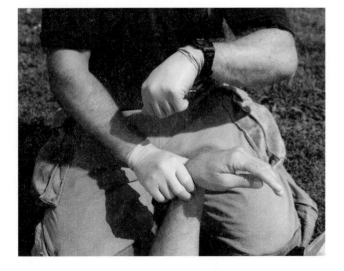

can place a stethoscope on the left anterior chest wall (over the heart).

Check pulses in several areas to determine how well the patient's entire circulatory system is working. For example, the absence of a pulse in a single extremity may indicate a blocked artery. If left untreated, numbness, weakness, and tingling will follow pain. The skin also will gradually turn mottled, blue, and cold.

The Skin

Assessment of skin temperature, color, and condition can tell you more about the patient's circulatory system.

Skin Temperature. Normal body temperature is 98.6°F (37°C). The most common way First Responders take temperature is by touching a patient's skin with the back of the hand. This is called relative skin temperature. It does not measure exact temperature, but you can tell if it is very high or low.

Changes in skin temperature can alert you to certain injuries and illnesses. A patient whose skin temperature is cool, for example, may be suffering from shock, heat exhaustion, or exposure to cold. A high temperature may be the result of fever or heat stroke. Body temperature also can change over a period of time, and it can be different in various parts of the body. A cold arm or leg may indicate circulatory problems, for example. An isolated hot area could indicate a localized infection. Be alert to changes, and record them.

Skin Color. Skin color can tell you a lot about a patient's heart, lungs, and other problems. For example:

◆ *Paleness* may be caused by impaired blood flow, shock, or heart attack. It also may be caused by fright, faintness, or emotional distress.

◆ *Redness* (flushing) may be caused by high blood pressure, alcohol abuse, sunburn, heat stroke, fever, or an infectious disease.

◆ *Blueness* (cyanosis) is always a serious problem. It appears first in the fingertips and around the mouth. Generally, it is caused by reduced levels of oxygen as in shock, heart attack, or poisoning.

◆ *Yellowish color* may be caused by a liver disease.

◆ *Black-and-blue mottling* is the result of blood seeping under the skin. It is usually caused by a blow or severe infection.

If your patient has dark skin, be sure to check for color changes on the lips, nail beds, palms, earlobes, whites of the eyes, inner surface of the lower eyelid, gums, and tongue.

You may also wish to check the patient's nail beds. This is called assessing **capillary refill.** It is one way of checking for shock. Capillary refill is recommended only for children under six years of age. Research has proven that it is not always accurate in adults.

To assess capillary refill in your patient, follow these steps (Figure 9-24):

1. **Squeeze a fingernail or toenail.** When squeezed, the tissue under the nail turns white. (For an infant you can squeeze the palm of a hand or the sole of a foot.)

2. **Release the pressure.**

3. **Measure the time it takes for color to return** to the tissue. Count "one one-thousand, two two-thousand," and so on.

4. **Record the result immediately.**

Two seconds or less is normal. If refill time is greater than two seconds, suspect shock or decreased blood flow to that extremity. Note that when you recheck capillary refill in the ongoing assessment, be sure to do it in the same place. Different parts of the body may have different refill times.

Figure 9-24 Assess capillary refill in children under 6 years of age.

FIRE DRILL

One easy way to remember how to assess any extremity is to use the mnemonic PMS-CR, which stands for **p**ulse, **m**ovement, **s**ensation, and **c**apillary **r**efill.

Skin Condition. Normally, a person's skin is dry to the touch. Wet or moist skin may indicate shock or a diabetes- or heat-related emergency. Abnormally dry skin may be a sign of a heat-related injury or severe dehydration. One method of assessing dehydration is to gently pinch the skin on the forearm. If the skin remains "tented" or peaked, as opposed to returning to its normal flat shape, dehydration may be present.

Pupils

Normally, pupils **constrict** (get smaller) when exposed to light and **dilate** (enlarge) when the level of light is reduced. Both pupils should be the same size unless a prior injury or condition changes this. (See Figure 9-25.)

With these normal responses in mind, assess a patient's pupils. Shine your penlight into one of the patient's eyes and watch for the pupils to constrict in response to the light. Both pupils should get smaller. Do the same with the other eye. Once again, both pupils should react the same way. If you are outdoors in bright light, cover the patient's eyes and observe for dilation of the pupils. Do not expose the patient's eyes to light for more than a few seconds, as this can be very uncomfortable to the patient.

Abnormal findings for pupils include:

◆ Pupils that do not react to light.

◆ Pupils that remain constricted.

◆ Pupils that are unequal.

Blood Pressure

Fire service First Responders may be taught to assess blood pressure. Be sure to follow all local protocols.

Blood pressure is the amount of pressure the surging blood exerts against the arterial walls. It is an important index of the efficiency of the whole circulatory system. In part, it tells how well the organs and tissues are getting the oxygen they

Normal pupils

Constricted pupils

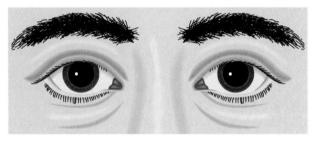

Dilated pupils

Unequal pupils

Figure 9-25 Check pupils for size, reactivity, and equality.

need. The blood-pressure cuff, or **sphygmomanometer,** is the instrument used to measure blood pressure.

The result of a contraction of the heart, which forces blood through the arteries, is called **systolic pressure.** The result of the relaxation of the heart between contractions is called **diastolic pressure.** With most diseases or injuries, these two pressures rise or fall together.

Blood pressure normally varies with the age, sex, and medical history of the patient. (See normal ranges in Table 9-3.) The usual guide for systolic pressure in the adult male is 100 plus the patient's age, up to 150 mmHg. Normal diastolic pressure in the male is 65 mmHg to 90 mmHg. Both the systolic and diastolic pressures are about 10 mmHg lower in the adult female. Blood pressure is reported as systolic over diastolic (for example, 120/80).

Measuring Blood Pressure. There are two methods of obtaining blood pressure with a blood pressure cuff. One is by auscultation, or by listening for the systolic and diastolic sounds through a stethoscope. The second method is by palpation, or by feeling for the return of the pulse as the cuff is deflated. To assess blood pressure by auscultation, follow the steps described below (Figure 9-26):

1. **Choose the proper size blood pressure cuff.** It must be able to encircle the arm so that the Velcro on opposite ends meets and fastens securely. The cuff's bladder should cover half the circumference of the arm. If it covers less, it will not compress the blood vessels properly. If it covers more, it will suppress the pulse too quickly.

2. **Place the cuff on the patient's arm.** It should fit snugly with the lower edge at least an inch (25 mm) above the **antecubital space** (the hollow, or front, of the elbow). Center the bladder over the brachial artery. The cuff should not be too tight. You should be able to place one finger easily under its bottom edge. Some cuffs have markers for overlap placement, but they are not always in the correct location.

3. **Palpate the radial pulse, while you rapidly inflate the cuff.** Continue until the pulse can no longer be felt. Make a mental note of the reading and, without stopping, continue to inflate the cuff to 30 mm above the level where the pulse disappeared.

4. **Apply the stethoscope.** Place the diaphragm of the stethoscope over the brachial artery just above the antecubital space. The diaphragm may be held with the thumb.

5. **Deflate the cuff** at approximately 2 mm per second (faster if skill permits). Watch the mercury column or needle indicator drop.

TABLE 9-3

Normal Blood Pressure Ranges

Patient	Systolic	Diastolic
Adults	Age + 100 (up to 150 mmHg)	60 to 90 mmHg
Infants and children	Approximately 80 + (age × 2)	Approximately 2/3 systolic

6. **Record the systolic pressure,** when you hear two or more consecutive beats (clear tapping sounds of increasing intensity). Continue releasing air from the bulb.

7. **Record the diastolic pressure** at the point where you hear the last sound. Continue to deflate slowly for at least 10 mm. Remember that slow pulses require slower-than-normal rates of deflation. With children and some adults, you may hear sounds all the way to zero. In such cases, record the pressure when the sound changes from clear tapping to soft, muffled tapping.

When it is too noisy for you to hear well enough to measure by auscultation, palpate the blood pressure (Figure 9-27). First, inflate the cuff rapidly. As you do so, palpate the patient's radial pulse. Make a mental note of the level at which you can no longer feel the pulse. Without stopping, continue to inflate the cuff another 30 mmHg. Then slowly deflate it. Note the pressure at which the radial pulse returns. This is the systolic pressure. Record it as a palpated systolic pressure (for example, 120/P).

Take several blood pressure readings during the time the patient is in your care. Watch for changes, as this may indicate changes in the patient's condition. Carefully record the blood pressure when you measure it, including the time it was taken.

It is not unusual for a patient's blood pressure to vary between the first reading on scene and the reading at the hospital emergency department. Record the pressure accurately so that the receiving physician can tell how much it has changed. Also, be sure to record the limb on which it was taken, the patient's position if not supine, and the size of the cuff if not standard.

Figure 9-26 Taking blood pressure by auscultation.

Figure 9-27 Taking blood pressure by palpation.

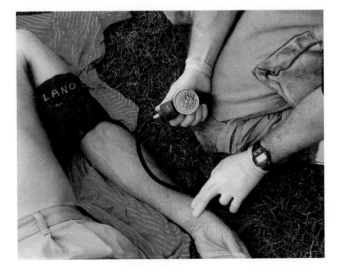

Blood pressure may be normal even if the patient is seriously injured. You will learn in Chapter 14 that by the time the blood pressure drops, the patient already is in serious condition.

Standards for Adults. Blood pressures vary greatly from person to person. In general, systolic blood pressures above 180 and below 90 usually indicate problems. Diastolic blood pressures above 90 and below 60 also indicate problems. In general, blood pressure changes occur late in an emergency. Always consider the blood pressure reading as only one element of the patient's picture. Be sure to look at other vital signs, too.

Standards for Children. It is often difficult to take a child's blood pressure. Carrying a variety of cuff sizes is not always possible. But you must have a correctly fitting cuff if you are to get an accurate reading. Always try to get a complete set of vital signs. However, do not waste time attempting multiple blood pressure readings on a critical child if your first tries were unsuccessful. Note that a blood pressure reading is not recommended on children under three years of age.

Average blood pressure in children may be determined by the formula: 80 plus two times the age in years. For a nine-year-old child, for example, we could expect an average blood pressure of 98, or 80 + (2 × 9). This formula works until the child is about 12 years of age. From that point on, adult blood pressure values apply.

In children, adequate airway management is vital. Do not place taking a blood pressure reading over assessment and treatment of life-threats. A child's blood pressure may not begin to drop until well over 40% of blood volume is lost. If the mechanism of injury suggests it, treat for shock regardless of vital signs.

Variable Factors. Heart failure, trauma, and most types of shock can decrease blood pressure. Factors that may increase blood pressure include conditions and substances that constrict blood vessels, such as:

◆ Cold environment.

◆ High altitude.

◆ Physical and emotional stress.

◆ Pain.

◆ Full bladder.

◆ Upper arm lower than heart level.

◆ Caffeine (coffee, tea, cola, and some analgesic drugs).

◆ Decongestants.

With the many possible variables that affect blood pressure, it is essential for you to recognize the mistakes that can occur in taking a reading. The most critical of possible errors are:

◆ Rescuer does not hear accurately due to noise, head cold, or distractions.

◆ Stethoscope ear pieces are improperly placed.

◆ Improper conditions exist, such as a cuff not at heart level or a patient not sitting or lying down.

◆ Systolic pressure is not palpated at the highest level.

◆ Cuff is the wrong size, either too wide or too narrow.

◆ Bladder is too wide.

◆ Cuff is deflated too fast.

COMPANY OFFICER'S NOTE

At times, it may be difficult to obtain a blood pressure on a patient. This does not mean you are a poorly trained First Responder. Emphasize to your personnel that it is okay and normal to sometimes need help in obtaining a blood pressure reading. Encourage them to ask another member of the squad to take a reading if needed.

SECTION 3

PATIENT HISTORY

The **patient history** is an important part of a thorough patient assessment. It involves gathering facts that you would not be able to gather otherwise. For example, the answer to a simple question such as "What happened?" can provide a good amount of information on the patient's condition and the events leading up to it. Document the facts given to you, and report them to the EMS personnel who take over patient care.

If the patient is unresponsive, gather facts by observing the scene, by looking for medical identification tags, and by questioning family members and bystanders (Figure 9-28).

Remember the differences between a trauma patient and a medical patient. In trauma patients, you will most likely perform a physical exam first. For a medical patient, you may take a history first.

The *SAMPLE* History

One way to remember the questions you need to ask is by using the mnemonic **SAMPLE.** Each letter identifies an important area of questioning:

S — Signs and symptoms.

A — Allergies.

M — Medications.

P — Pertinent medical history.

L — Last oral intake.

E — Events.

Signs and Symptoms

Note a patient's signs and symptoms (Figure 9-29). A **sign** is something you can observe directly. That is, you can see, feel, or hear it. Examples include deformities (see), skin temperature (feel), and wheezing (hear). A **symptom** cannot be observed by anyone but the patient. In order for you to be aware of a symptom, the patient must describe it. Examples include pain, tenderness, or difficulty breathing.

Figure 9-28   Gather facts from the patient or from family and bystanders.

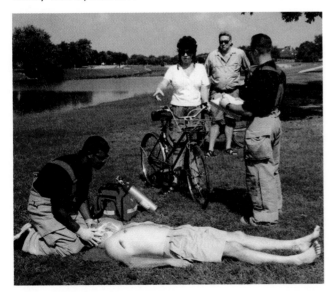

Figure 9-29a   A sign is something one can observe—a deformed wrist, for example.

Figure 9-29b   A symptom is something that the patient feels and describes, such as stomach pain.

A good starting point to determine signs and symptoms is to ask the patient an open-ended question, such as "Why did you call today?" or "Describe how you feel." This technique allows patients to answer without restriction. They may even give you important information you may not have thought to ask about.

Responses to questions like "Do you have chest pain?" are restricted to yes-or-no answers. They can cause you to miss important information. Such questions might be better phrased as

"What do you feel in your chest?" or "Tell me what you feel in your chest."

You will find signs and symptoms—as well as emergency care—for specific conditions in later chapters. Remember that you are not required to diagnose any medical condition. A First Responder's emergency care of a patient is based on assessment findings only.

Allergies

Determine if your patient is allergic to anything. This includes allergies to medications, foods, or substances in the environment. An allergic reaction can be very serious, in fact it can be life-threatening. Awareness of an allergy can help to determine possible causes of the patient's condition. Determining if a patient is allergic to a medication also will help prevent other medical personnel from administering it.

Medications

Identify all medications the patient is currently taking or has recently taken. This information may help to identify a medical condition. For example, a patient who takes insulin has diabetes. Other specific medications are used for seizures, cardiac conditions, and respiratory problems. Many EMS rescuers carry a pocket guide that lists common prescription medications and what they are used for.

Pertinent Medical History

Most patients have had some type of medical condition in their lifetimes. Some of these may be pertinent to the emergency care you provide to the patient. What is pertinent depends on the type of emergency. For example, if a patient has shortness of breath, the fact that he has a history of heart problems is pertinent. The fact that he had foot surgery many years ago is not. However, if the patient's present emergency involves dropping a bowling ball on that foot, then the surgery is pertinent.

Patients and family members who are in the middle of a medical crisis may not know what to tell you. Some people say very little. Others may give you enough information to write a short novel. It is your job to guide them. Sometimes you just need to ask more than one question to guide a patient to an answer. For example, you

may begin by asking a patient to tell you about any medical problems he may have. Your patient may answer "none," even though he has diabetes. This is because he may have understood his physician to say only that he has "a little sugar problem," which is controlled by diet or pills. So your next question might be, "Do you see a doctor for anything?" or "Have you ever been admitted to the hospital?" It may cause the patient to disclose the information you need.

Most patients do not withhold answers or answer incorrectly on purpose. Remember, they may be scared, confused, or both. Taking time to help reduce their anxiety may help you obtain accurate medical information.

FIRE DRILL

When performing a search in a smoked-filled structure, it is important to do so methodically. For example, "rescue right" tells the team the search will occur from right to left throughout the structure. Similarly, it is useful to use a methodical approach when determining a patient's medical history. One useful method is to begin at the head and proceed to the abdomen. For example, you might ask these questions:

◆ "Have you ever had a stroke or seizure?" (head, or brain)

◆ "Have you ever had any breathing problems such as asthma, emphysema, or bronchitis?" (lungs)

◆ "Have you ever had a heart attack or problems with your heart?" (heart)

◆ "Have you ever had an ulcer?" (stomach)

◆ "Have you ever had hepatitis?" (liver)

◆ "Have you ever had diabetes?" (pancreas)

Last Oral Intake

You must find out the time of your patient's last oral intake. It may be pertinent to the patient who is unresponsive or confused, as well as to the patient needing immediate surgery. Do not ask "When was your last meal?" People may not consider a snack or several drinks a "meal." Instead ask: "When was the last time you had anything to eat or drink?" This may include anything from a glass of water to a large meal.

Events

Questions such as "What were you doing when this happened?" may help determine the events leading up to the emergency. This information might not be as clear-cut as it seems. Say a patient falls from a ladder. You arrive to find her complaining of a possible broken arm. The patient is obviously a trauma patient, right? The answer is "yes," if the patient slipped on a broken rung of the ladder. However, if she fell as a result of getting dizzy or losing consciousness, then she also may be a medical patient.

If a person is driving, has a heart attack, and passes out behind the wheel, he will surely crash. Upon your arrival at the scene, you could assume that the patient has suffered serious trauma and is unresponsive from that trauma. But it will be your assessment of the events that will help you determine what really happened. A passenger might tell you, for example, that she saw the driver clutch his chest before the crash, or a medical information tag may alert you to a heart condition.

As you can see, the SAMPLE history is an important part of the patient assessment process where information vital to patient care is obtained.

SECTION 4   ONGOING ASSESSMENT

Some patients are stable, others are not. So, your patient assessment process must be ongoing. Continually reassess your patient until he or she is turned over to the EMTs. Complete the following every 5 minutes for an unstable patient and every 15 minutes for a stable one:

- Reassess level of responsiveness (AVPU).
- Reassess and correct any airway problems.
- Reassess breathing for rate and quality. Ventilate if necessary.
- Reassess pulse for rate and quality.
- Reassess skin temperature, color, and condition.
- Reassess blood pressure, if you are allowed.
- Repeat any portions of the physical exam that might be necessary.

- Reassess your interventions (treatment) to see if they are effective.
- Continue to calm and reassure the patient.

Remember that all the patients you encounter as a fire service First Responder are in crisis. They would not have called you if they were not. They may be uncomfortable, confused, and possibly afraid that they will die. It is important for you to have a professional, calm, and caring attitude. Try to address the patient's concerns. For example, if you can protect the patient's modesty, do so. Do not leave the patient alone. If the patient feels cold, even if you are not, turn up the heat or provide another blanket.

The kindness and compassion you show will help to calm the patient. It also will be remembered for a long time to come.

SECTION 5   HAND-OFF REPORT

When the EMTs arrive, you must be prepared to tell them appropriate information about your patient and the care you have given. This is called your hand-off report. The hand-off report contains:

- Age and sex of the patient.
- Chief complaint.
- Level of responsiveness (AVPU).
- Airway and breathing status.
- Circulation status.
- Physical exam findings.
- SAMPLE history.
- Treatment, interventions, and the patient's response to them.

The hand-off report is designed to give the transporting EMTs an up-to-the-minute account of the condition of the patient, what treatments have been performed, and other information that you feel is important. If you must complete a written report, many agencies require that you provide one copy to the EMTs at the scene.

At the beginning of this chapter, you read that fire service First Responders were on scene with an unresponsive male patient. To see how chapter skills apply to this emergency, read the following. It describes how the call was completed.

PATIENT HISTORY

Since the patient was unresponsive, we asked his wife if he had any problems prior to going to the bathroom. She did not know of any. We also learned that he had no allergies. He had taken his insulin in the morning (for diabetes), but had been working in the yard since breakfast. His wife added that he had a similar episode last year.

PHYSICAL EXAMINATION

Although it seemed like we had a medical problem, you never can be too sure, especially with a patient who cannot talk. So I continued to assist ventilations and monitored the airway, while one of my partners did a quick head-to-toe exam, including vital signs, to check for any hidden injuries. There were none.

ONGOING ASSESSMENT

The patient required ventilation the entire time we were at the scene. I suctioned him one more time, when I felt his secretions were building up. We continued to monitor the patient's mental status, and spoke to him by name just in case he could hear. His wife was quite upset, so we also tried to reassure her and keep her calm.

PATIENT HAND-OFF

When the paramedics arrived, one of them performed an initial assessment. "Good job," he said after assessing the patient's airway. I gave them the hand-off report:

"This is Rick Townsend. He is 62, and his wife found him unconscious in the bathroom. He responds only to a painful stimulus. We had to suction him and assist ventilations. The physical exam did not turn up anything, but his wife says he has diabetes. He took his insulin, had a small meal, then worked all morning in the yard. His wife says this has happened once before. His vitals are pulse 110 and weak, respirations 10 and shallow, blood pressure 110/68. Pupils are equal and reactive. Skin is cool and moist."

The paramedics thanked us and took over care. We then headed back to the firehouse. A few days later, I saw one of the paramedics. He told me that after Mr. Townsend received glucose through an IV, he began to come around. By the time he and his partner had left the hospital, Mr. Townsend was sitting up and talking.

Patient assessment is performed on all patients you come in contact with. However, the procedure will vary depending on whether your patient is suffering from a medical problem or trauma and whether the patient has minor or serious injuries. In either case, it is important to be thorough and to work in a logical order. Remember: Perfect practice is the only way to become proficient at this critical skill.

Chapter Review

FIREFIGHTER FOCUS

Even after you have memorized the steps of the assessment process, you still may be amazed at the ease in which experienced fire service First Responders and EMTs perform that complicated task. If you were to ask, many of them would tell you that patient assessment is an art.

You will see throughout the remaining chapters of this text that there are countless injuries and illnesses from which patients can suffer. In every condition the patient assessment is designed to find current signs and symptoms and histories of past illnesses, all of which are important to patient care.

One experienced First Responder said it like this: "Patient assessment is like a puzzle. It is important to find all the pieces—the examination, the history—and put them together so that they make sense." A proper patient assessment helps you to make sense of the patient's condition and provide quality patient care.

FIRE COMPANY REVIEW

Page references where answers may be found or supported are provided at the end of each question.

Section 1

1. What are the components of the initial assessment? (p. 171)

2. Why would you apply manual stabilization to your patient's head and neck? (p. 172)

3. What are the components of the EMS update? (p. 176)

Section 2

4. What are the three basic methods of performing a physical exam? (p. 177)

5. What mnemonic helps you to recall the conditions to look for in the physical exam? What does each letter in the mnemonic stand for? (p. 177)

6. In what patients is capillary refill used? (p. 185)

Section 3

7. What mnemonic can help you to recall the parts of a patient history? What does each letter in the mnemonic stand for? (p. 189)

Section 4

8. What are the components of ongoing assessment? (p. 191)

Section 5

9. What information should be included in the patient hand-off report? (p. 191)

RESOURCES TO LEARN MORE

Bates, B., et al. *A Guide to Physical Examination and History Taking/A Guide to Clinical Thinking,* Sixth Edition. Philadelphia: Lippincott-Raven, 1995.

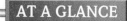

PATIENT ASSESSMENT PLAN

INITIAL ASSESSMENT
- Form a general impression.
- Assess responsiveness.
- Assess ABCs—airway, breathing, and circulation.
- Update EMS.

PHYSICAL EXAMINATION
- Include DOTS of head, neck, chest, abdomen, pelvis, extremities.
- Assess vital signs, if allowed.

PATIENT HISTORY
- Include signs and symptoms, allergies, medications, pertinent medical history, last oral intake, events.

ONGOING ASSESSMENT
- Repeat the initial assessment.
- Repeat the physical exam, including vital signs.
- Reassess treatment and interventions.
- Calm and reassure the patient.

HAND-OFF REPORT
- Include patient's sex and age, chief complaint, level of responsiveness, status of ABCs, physical exam findings, SAMPLE history, treatment/interventions/response.

Cardiac and Respiratory Emergencies

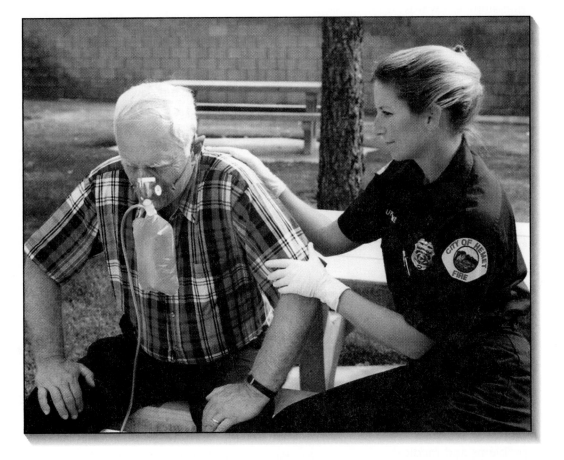

*I*NTRODUCTION   *Heart disease is a killer. In the United States alone it causes nearly one million deaths, 350,000 of which occur suddenly and without warning. Illnesses that affect the respiratory system are also very common. Emphysema and chronic bronchitis alone affect more than 23 million people in the U.S. Asthma affects nearly 10 million. Both types of problems—respiratory and cardiac—can be life threatening. When they occur, patients can benefit from immediate care.*

Cognitive, affective, and psychomotor objectives are from the U.S. DOT's "First Responder: National Standard Curriculum." Enrichment objectives, if any, identify supplemental material.

Cognitive

No objectives are identified.

Affective

5-1.16 Attend to the feelings of the patient and/or family when dealing with the patient with a specific medical complaint. (pp. 198, 201–205)

5-1.21 Demonstrate a caring attitude towards patients with a specific medical complaint who request emergency medical services. (pp. 198, 201–205)

5-1.22 Place the interests of the patient with a specific medical complaint as the foremost consideration when making any and all patient care decisions. (pp. 198, 201–205)

5-1.23 Communicate with empathy to patients with a specific medical complaint, as well as with family members and friends of the patient. (pp. 198, 201–205)

Psychomotor

No objectives are identified.

Enrichment

◆ State the signs and symptoms of the patient having a cardiac emergency. (pp. 197–198)

◆ Describe the emergency care of the patient experiencing chest pain or discomfort. (p. 201)

◆ List signs of adequate breathing. (p. 201)

◆ State the signs and symptoms of a patient in respiratory distress. (pp. 201–202)

◆ Describe the emergency care of the patient in respiratory distress. (pp. 202–203)

ON SCENE

DISPATCH

I was working at the chemical plant during my regular 4-to-12 shift. I also was assigned to the emergency response team (ERT), which is responsible for hazmat problems and medical emergencies. At about 6:30 p.m. I heard our team paged to respond. A man was having chest pain.

SCENE SIZE-UP

I approached the scene and looked around. Other than a few concerned coworkers, everything was quiet. It appeared that there was only one patient. I recognized the man, Harry

Nowack, because I used to work with him. He told me that his chest hurt.

INITIAL ASSESSMENT

Harry was alert. His airway was clear. He could speak in full sentences, but his breathing was labored. He had no obvious external bleeding. Harry told me that he didn't fall or have any injuries. "My chest just hurts." I couldn't help but notice Harry's color. He looked ashen. It was definitely not normal.

I radioed my findings to the plant office, and told them to notify the 9-1-1 ambulance. I informed them that my general impression was of an alert 55-year-old man who had chest pain, poor color, and labored breathing. The ETA of the ambulance was 10 minutes.

What do you think this patient's problem may be? What should be done to assess and treat his condition? Consider this patient as you read Chapter 10.

CARDIAC EMERGENCIES

Cardiac emergencies can occur from abnormal heart rhythm patterns. They also occur when there is an interruption of oxygen to some part of the heart muscle. The reduction of oxygen causes chest pain or discomfort, one of the most common symptoms of a cardiac emergency.

Coronary artery disease affects the inner lining of the arteries, which supply the heart with blood. People who have it usually suffer from arteriosclerosis, a condition that causes the walls of the arteries to become thick and hard. With this disease, the opening of the coronary artery is narrowed (Figure 10-1). This restricts the amount of oxygen-carrying blood that can reach and nourish the heart. The rough artery surfaces then cause a build up of debris, further narrowing the artery. The more the artery narrows, the less oxygen gets to the heart. At some point, the patient may have chest pain. When the artery is blocked, the patient may suffer a heart attack that results in death of the heart muscle.

Researchers have identified a number of risk factors for heart attack. Obviously, some risks cannot be controlled. With awareness and determination, however, a person can lessen the risks. The major risk factors are (Figure 10-2):

◆ Physical inactivity, a sedentary lifestyle.
◆ Cigarette smoking.
◆ Obesity.
◆ High serum cholesterol and triglycerides.
◆ Diabetes.
◆ Male gender.
◆ Age (incidence increases over 30 years of age).
◆ Hypertension (blood pressure above 140/90).

◆ Family history of coronary heart disease under age 60.
◆ Oral contraceptive use in women over 40.

There are many possible causes of a cardiac emergency (Table 10-1). However, you do not have to identify them. A First Responder's assessment and treatment of a patient with chest pain is the same, no matter what the cause.

Patient Assessment

Signs and Symptoms

Signs and symptoms of a cardiac emergency include the following (Figure 10-3):

Figure 10-1 Fatty deposits build up in arteries, depriving the heart muscles of blood and oxygen.

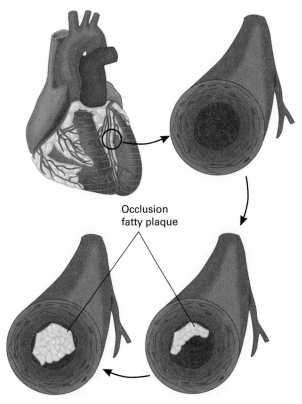

Occlusion
fatty plaque

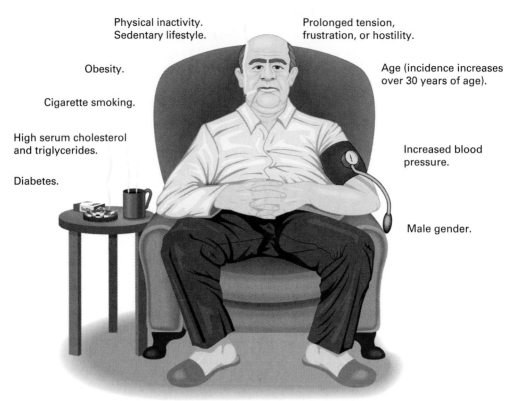

Physical inactivity. Sedentary lifestyle.

Obesity.

Cigarette smoking.

High serum cholesterol and triglycerides.

Diabetes.

Prolonged tension, frustration, or hostility.

Age (incidence increases over 30 years of age).

Increased blood pressure.

Male gender.

Family history of premature coronary heart disease (usually under age 60).

Figure 10-2   Cardiac risk factors.

◆ Chest pain, or discomfort described as heaviness or squeezing. May radiate to the arms, shoulder, neck, or jaw (Figure 10-4).

◆ Difficulty breathing, shortness of breath.

◆ Unusual pulse (rapid, weak, slow, or irregular).

◆ Indigestion, nausea, vomiting.

◆ Sweating.

◆ Skin and mucous membranes may be pale, gray, or cyanotic.

◆ A feeling of impending doom.

◆ History of heart problems, or previous similar experience.

Note that signs and symptoms differ among individuals. Some people may present with only one symptom, such as breathing difficulty. Women, especially, may not experience any chest pain at all. Men, however, more than likely will. A few patients may report that they "just don't feel well." Be suspicious. It may be that your only clue to a cardiac emergency is the patient's lifestyle (smoking, obesity, lack of exercise) and family history.

GERIATRIC NOTE

Many older patients do not have chest pain with a heart attack. This is known as a "silent" heart attack. These patients may exhibit respiratory distress, general feelings of weakness, and other symptoms, but no pain. A "silent" heart attack also is common in patients who have diabetes.

Remember, patients with chest pain will be very anxious. They may feel as if they are going to die. Be compassionate and reassuring. Make sure they understand that everything possible is being done. Advise them and their families that more help is on the way.

Keep in mind that patients frequently deny a heart problem. This causes a significant delay in getting the medical care they desperately need. In fact, research has shown that many patients do not call for help until at least two hours after they develop symptoms. So be sure that the EMS sys-

Table 10-1

Some Common Cardiac Conditions

Condition	Description	Commonly Prescribed Medications
Angina pectoris	Patient experiences chest pain or discomfort due to narrowed coronary arteries, which cannot provide enough blood flow to the heart, especially during exertion when the blood is needed.	Nitroglycerin tablets or spray (under the tongue)
Myocardial infarction	Also known as heart attack, it results from a total occlusion of a coronary artery. This may be due to narrowing or a clot. Heart muscle supplied by the artery dies due to the lack of blood supply.	Clot-busting drugs (thrombolytics) may be used in the hospital only. A variety of cardiac medications may be administered to the patient after a heart attack, including aspirin.
Congestive heart failure	Occasionally occurring after a heart attack, the heart cannot pump as efficiently causing fluid to accumulate in the lungs and other parts of the body. Swelling in the ankles may be noted.	Diuretics ("water pills") such as furosemide (Lasix) are used to prevent fluid accumulation.
Heart-rhythm disturbances	The most notable heart rhythm problem is ventricular fibrillation ("V-fib"), which is treated by shocking with a defibrillator.	A wide variety of medications may be administered including: digoxin (Lanoxin) and verapamil (Calan)

tem has been activated as soon as you suspect a cardiac emergency.

Another reason for activating EMS immediately is **sudden death,** a condition caused by sudden heart rhythm disturbance. It usually occurs within two hours of the onset of symptoms. Though not all cardiac patients will experience sudden death, it is important for you to activate the EMS system and update incoming EMS units as soon as possible. Request advanced care, if possible. Monitor the patient constantly, because he or she may become unstable rapidly.

SIGNS AND SYMPTOMS OF A CARDIAC EMERGENCY

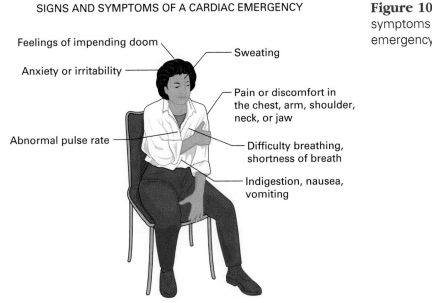

Figure 10-3 Signs and symptoms of a cardiac emergency.

Feelings of impending doom

Anxiety or irritability

Abnormal pulse rate

Sweating

Pain or discomfort in the chest, arm, shoulder, neck, or jaw

Difficulty breathing, shortness of breath

Indigestion, nausea, vomiting

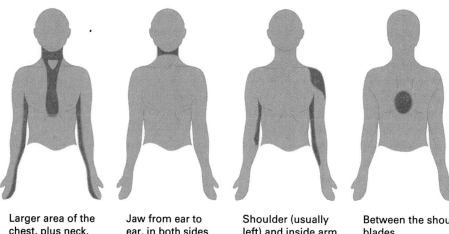

Just under sternum, midchest, or the entire upper chest.

Midchest, neck, and jaw.

Midchest and on left more frequently, the shoulder and inside arm.

Upper abdomen, often mistaken for indigestion.

Larger area of the chest, plus neck, jaw, and inside arms.

Jaw from ear to ear, in both sides of upper neck, and in lower center neck.

Shoulder (usually left) and inside arm to the waist, plus opposite arm, inside to the elbow.

Between the shoulder blades.

Figure 10-4 Pain or discomfort can occur in any one location or any combination of locations.

One more reason for activating EMS quickly is that there are many new treatments available. Special clot-busting drugs, for example, may open clogged arteries and prevent death of heart muscle. These treatments are time dependent. That is, they must be performed within the first few hours after the onset of symptoms.

Chest Pain and Cardiac Patients

Continue to monitor the patient's airway and breathing during and after the initial assessment. Patients who experience chest pain may need airway maintenance and ventilation. Also be

FIRE DRILL

Every time you respond to a fire, you bring certain equipment with you. It would be dangerous and inefficient not to have the proper tools when they are needed. A required "tool" on a call for chest pain is the AED. If your EMS system allows you to operate one, remember always to have it available and ready to go. Take it with you when you exit your truck. If you approach the patient without it, too many valuable minutes may be lost when you go back to your truck to get it.

alert for changes in the patient's mental status. These patients can lose consciousness rapidly. If the patient becomes unresponsive, be prepared to perform CPR and, if you are trained, to use an AED.

When you take a SAMPLE history, you also may want to use the **OPQRRRST** mnemonic to help you get a good description of the pain. Each letter identifies an important area of questioning:

O — *Onset.* When did the pain begin?

P — *Provocation.* Did anything cause or start the pain (exercise, an activity)?

Q — *Quality.* What is the pain like (crushing, stabbing, etc.)?

R — *Region.* Where is the pain?

R — *Radiation.* Does the pain begin in one place and then seem to travel somewhere else?

R — *Relief.* Does anything relieve the pain?

S — *Severity.* On a scale of 1 to 10, with 10 the worst, how bad is the pain?

T — *Time.* How long have you had the pain?

When you perform a physical exam, palpate the chest for DOTS. Make a note if your touch causes pain. If you are trained to do so, listen to the lungs to determine if air is moving in and out of both sides equally.

◆ **Emergency Care**

To provide emergency care, first be sure you have taken BSI precautions. Then proceed as follows:

1. **Have the patient stop all activity.**

2. **Position the patient.** Place the responsive patient in a position of comfort. This is usually a semi-reclining or sitting position.

3. **Monitor breathing.** Make sure that the airway is open. Administer high-flow oxygen by way of a nonrebreather mask. (Follow local protocol.) If needed, provide artificial ventilation or CPR.

4. **Loosen any tight clothing.**

5. **Maintain body temperature** as close to normal as possible.

6. **Comfort and reassure the patient.** Your goal is to lower the patient's anxiety level. (High anxiety produces high levels of adrenaline, which in turn produce a faster heart rate.

A faster heart rate causes the heart to use more oxygen, which can increase chest pain.)

7. **Activate the EMS system** immediately, if it was not done previously.

Patients may tell you that they have had heart surgery. They also may have a pacemaker or implanted defibrillator. (An implanted defibrillator delivers shocks to a patient but at much less power than an AED.) Treat these patients in the same way as described above. Note that a malfunctioning pacemaker may cause a slow heart rhythm. If this occurs, monitor the patient carefully. Provide oxygen and be prepared to administer CPR.

RESPIRATORY EMERGENCIES

Without oxygen, cells such as those in the brain and heart can die within minutes. A variety of diseases and injuries can affect the body's ability to get enough oxygen. The need for rapid treatment in those cases is essential.

Patient Assessment

As you learned in Chapter 5, adequate breathing occurs at a normal rate. For adults, that is 12 to 20 breaths per minute. For children, it is 15 to 30 breaths per minute. For infants, it is 25 to 50 breaths per minute. Adequate breathing is regular in rhythm and free of unusual sounds, such as wheezing or whistling. The chest should expand adequately and equally with each breath. The depth of the breaths should be adequate, too. In addition, breathing should be effortless. That is, it should be accomplished without the use of accessory muscles in the neck, shoulders, or abdomen.

Respiratory distress is shortness of breath or a feeling of air hunger with labored breathing. It is one of the most common medical complaints. Two circumstances may cause it—either air cannot pass easily into the lungs or air cannot pass easily out of them. Signs and symptoms include (Figure 10-5):

◆ Inability to speak in full sentences without pausing to breathe.

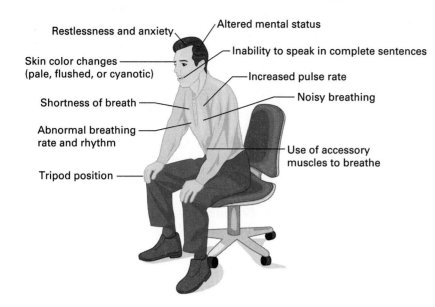

Restlessness and anxiety

Altered mental status

Inability to speak in complete sentences

Skin color changes
(pale, flushed, or cyanotic)

Increased pulse rate

Shortness of breath

Noisy breathing

Abnormal breathing
rate and rhythm

Use of accessory
muscles to breathe

Tripod position

Figure 10-5 Signs and symptoms of respiratory distress.

- Noisy breathing.
- Use of accessory muscles to breathe. That includes neck muscles, muscles between the ribs, and abdominal muscles. Note that this is especially common in pediatric patients.
- **Tripod position.** In this position, the patient is sitting upright, leaning forward, fighting to breathe (Figure 10-6).
- Abnormal breathing rate and rhythm.
- Increased pulse rate.
- Skin color changes (pale, flushed, or cyanotic).
- Altered mental status.

Figure 10-6 Patients with emphysema or chronic bronchitis often lean forward as they breathe.

In patients with respiratory distress, breathing generally is rapid and shallow. Patients may feel short of breath whether they are breathing rapidly or slowly. Remember, a certain amount of shortness of breath is normal following exercise, fatigue, coughing, or with the production of excess sputum.

FIRE DRILL

Patients with respiratory distress are very fearful. Relate this to the feeling of the warning bell on your SCBA going off when you are performing an interior fire attack. The next time you put on your SCBA mask and cover the end to check the seal, observe what it is like to breathe in and not get any air. This is what a severe respiratory distress patient feels with every breath.

There are a variety of conditions that can lead to respiratory distress (Table 10-2). Remember that you do not need to diagnose them.

◆ **Emergency Care**

Respiratory distress may be a symptom of an injury or an illness. It is not a disease in itself.

Table 10-2

Some Common Respiratory Conditions

Condition	Description	Commonly Prescribed Medications
Emphysema	Alveoli (air sacs in the lungs) lose elasticity, fill with trapped air, and stop working.	Home oxygen systems, inhalers, or "puffers." Also theophylline (Theo-Dur).
Chronic bronchitis	Inflammation, edema, and mucus are found in the bronchial tree. Patients will describe a productive cough that has persisted for some time.	Home oxygen systems, inhalers, or "puffers."
Asthma	Airways spasm and increase mucus production causing reduction in airflow to the lungs.	Inhalers or "puffers" such as albuterol (Ventolin).
Pulmonary edema	Fluid in the lungs (usually caused by congestive heart failure).	Diuretics ("water pills") such as furosemide (Lasix).

Whatever the cause, treatment is the same. First, take BSI precautions. Then:

1. **Determine if breathing is adequate.** If it is, monitor it throughout the call. If it is not, provide artificial ventilation immediately.

2. **Position the patient.** Place the responsive patient with adequate breathing in a position of comfort. If the patient has difficulty breathing, the sitting position almost always is preferred.

3. **Administer oxygen,** if you are allowed to do so. Use a nonrebreather mask at 12–15 liters per minute.

4. **Comfort and reassure the patient** (Figure 10-7). There are few feelings as terrifying as not being able to breathe. The patient may become agitated or even angry. Do not take it personally.

5. **Activate the EMS system immediately,** if it was not done previously.

Note: You may have heard that patients with emphysema and chronic bronchitis, which are chronic obstructive pulmonary diseases, should not get oxygen. In some of these patients, supplemental oxygen can cause a decrease in the body's drive to breathe. However, this occurs only under prolonged administration, which is not common in the field. This should not prevent the administration of oxygen to any patient. When oxygen is provided to any patient, monitor him or her carefully. If respiratory efforts decrease, assist ventilations and advise incoming EMS units.

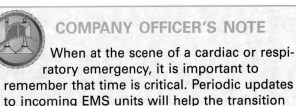

COMPANY OFFICER'S NOTE

When at the scene of a cardiac or respiratory emergency, it is important to remember that time is critical. Periodic updates to incoming EMS units will help the transition of care and reduce on-scene time.

Hyperventilation

Hyperventilation is a condition characterized by breathing too fast. It is normal for most people occasionally, such as when they are surprised. It remains normal if the breathing rate quickly returns to normal.

Hyperventilation syndrome is an abnormal state in which rapid breathing persists. It is a common disorder usually associated with anxiety. As a patient becomes more anxious, he or she breathes more rapidly. This in turn makes the patient more anxious, and so on, creating a vicious cycle.

The syndrome is characterized by rapid, deep, or abnormal breathing. The lungs overinflate, and the patient blows off too much carbon

Figure 10-7 Reassure patients with respiratory difficulty to help reduce stress and fear.

dioxide. In prolonged cases, the patient may pass out. It typically occurs in young, anxious patients, most of whom are not aware that they are breathing too fast.

Signs and symptoms include:

◆ Air hunger, or "gulping" air.

◆ Deep, sighing, rapid breathing with rapid pulse.

◆ Sensation of choking.

◆ Dryness or bitterness of the mouth.

◆ Tightness or a "lump" in the throat.

◆ Marked anxiety escalating to panic and a feeling of impending doom.

◆ Dizziness, lightheadedness, fainting.

◆ Giddiness or unusual behavior.

◆ Drawing up the hands at the wrist and knuckles with the fingers flexed.

◆ Blurring of vision.

◆ Numbness or tingling of the hands and feet or around the mouth.

◆ Pounding of the heart with stabbing pains in the chest.

◆ Fatigue, great tiredness, or weakness.

◆ A feeling of being in a dream.

Not every patient who is breathing rapidly or deeply is hyperventilating. Several serious conditions may be the cause, including heart disease, diabetes, asthma, shock, and trauma. It also may have a medical origin, such as aspirin overdose. If you are certain that no life-threats exist, then try to calm your patient. Be reassuring, and listen carefully to his or her concerns. Try to talk the patient into breathing slowly. If the patient does not respond immediately to your efforts, then administer oxygen. It will not make hyperventilation worse.

You may have heard that putting a paper bag over a patient's mouth is a cure for hyperventilation. This treatment is dangerous, especially if an underlying medical condition exists. As noted above, calming is a powerful benefit to the hyperventilating patient. Follow local protocol.

ON SCENE FOLLOW-UP

At the beginning of this chapter, you read that a First Responder was caring for a patient with chest pain, labored breathing, and poor skin color. To see how chapter skills apply to this emergency, read the following. It describes how the call was completed.

PHYSICAL EXAMINATION

Two other members of the ERT arrived to help. They performed a head-to-toe exam, while I talked to Harry about his history.

PATIENT HISTORY

Harry told me that he was working at his station when he started getting severe pain in his chest. He had never felt anything like it before, and I had never seen Harry that scared. I reassured him and asked a few questions, using the OPQRRRST mnemonic.

Harry was working when the pain came on. He told me it was crushing, like someone sitting right on the center of his chest. He held his fist there to show me. It didn't radiate. Nothing, he said, helped to relieve the pain. On a 1-to-10 scale, he said the pain was an "8." It started about 10 minutes ago.

Harry hadn't seen a doctor in years. He took no medications and had no medical problems or allergies. He ate a big spaghetti dinner during his break.

When I heard Harry's vital signs—pulse 92 and irregular, respirations 18 and adequate, blood pressure 146/96, with skin that was cool, gray, and moist, I began to wish our plant had an AED.

ONGOING ASSESSMENT

After helping to make Harry comfortable, I continued to talk and reassure him. We rechecked his ABCs, which remained okay. The chest pain did not diminish. A second assessment of vital signs revealed a pulse of 96 and irregular, and respirations 20 and still labored. Unfortunately, we didn't have oxygen to give him.

PATIENT HAND-OFF

The paramedics arrived a short time later. We immediately called their attention to Harry's chief complaints, chest pain, labored breathing, and poor color to let them know how serious I thought it was. They agreed. Then we filled them in on the rest:

"This is Harry Nowack, 55 years old. His chest pain started while he was working. It is in the center of his chest and is crushing. It is an 8 on a scale of 10, and it doesn't radiate. He has no medical problems that he knows of, but he hasn't been to a doctor in years. He has no meds or allergies. He ate a big spaghetti dinner a short time ago. His pulse is 96 and irregular, respirations 20 and slightly labored, blood pressure 146/96."

The paramedics thanked us, and I helped them put the oxygen on Harry. He wanted me to come with him, and the paramedics didn't mind some help. I made sure that someone called Harry's wife.

Well, it turned out that Harry had a major heart attack. The word spread around the plant the next day. The way he looked was just like they described in the books. I always take chest pain seriously. I'm glad we did with Harry.

> Problems with cardiac and respiratory systems are frequently the reasons why First Responders are summoned. Quickly recognizing these problems as potentially life-threatening and making sure the proper medical help is on the way can save your patient's life. Also keep in mind that these emergencies require compassionate emotional care as well as management of the patient's physical condition.

Chapter Review

FIREFIGHTER FOCUS

The condition of a patient with respiratory distress—or a patient with chest pain—can deteriorate rapidly. So make sure EMS has been notified and be prepared to provide basic life support as soon as it is needed. Also, remember that artificial ventilation is not only for patients in respiratory arrest. Be prepared to provide ventilations if your patient is breathing inadequately, too. You may want to review Chapters 5, 6, and 7 at this time.

FIRE COMPANY REVIEW

Page references where answers may be found or supported are provided at the end of each question.

Section 1

1. What are the risk factors for heart disease? (p. 197)
2. What are the signs and symptoms of a cardiac emergency? (pp. 197–198)

3. What is the emergency care for a chest-pain emergency? (p. 201)

Section 2

4. What are four signs and symptoms of respiratory distress in a patient? (pp. 201–202)
5. What is the emergency medical care for a patient with respiratory distress? (pp. 202–203)

RESOURCES TO LEARN MORE

"Dyspnea" in Harwood-Nuss, A.L., et al. *The Clinical Practice of Emergency Medicine,* Second Edition. Philadelphia: Lippincott-Raven, 1995.

Graves, J.R., et al. *RapidZap: Automated Defibrillation.* Upper Saddle River, NJ: Brady/Prentice-Hall, 1989.

Textbook of Advanced Life Support. Dallas: American Heart Association, 1993.

Weigel, A., et al. *Automated Defibrillation.* Upper Saddle River, NJ: Brady/Prentice-Hall, 1988.

AT A GLANCE

Refer to Chapter 5, page 112; Chapter 6, pages 135–136; and Chapter 7, page 150.

Other Common Medical Complaints

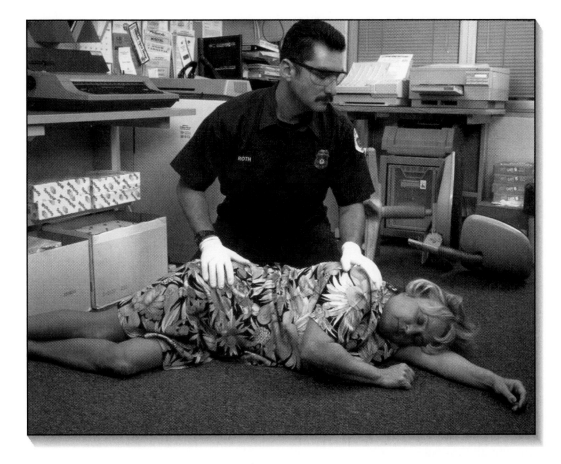

*I*NTRODUCTION   *A* medical complaint *is any chief complaint that is not caused by trauma. There will be many such calls in your career. They may involve abdominal pain, altered mental status, or even complaints such as "I don't feel well." As with any patient, your responsibility in the emergency care of these patients is to follow your patient assessment plan from scene size-up to patient hand-off.*

Cognitive, affective, and psychomotor objectives are from the U.S. DOT's "First Responder: National Standard Curriculum." Enrichment objectives, if any, identify supplemental material.

Cognitive

5-1.1 Identify the patient who presents with a general medical complaint. (p. 209)

5-1.2 Explain the steps in providing emergency medical care to a patient with a general medical complaint. (p. 209)

5-1.3 Identify the patient who presents with a specific medical complaint of altered mental status. (p. 210)

5-1.4 Explain the steps in providing emergency medical care to a patient with an altered mental status. (p. 210)

5-1.5 Identify the patient who presents with a specific medical complaint of seizures. (pp. 219–220)

5-1.6 Explain the steps in providing emergency medical care to a patient with seizures. (pp. 221–222)

Affective

5-1.15 Attend to the feelings of the patient and/or family when dealing with the patient with a general medical complaint. (p. 209)

5-1.16 Attend to the feelings of the patient and/or family when dealing with the patient with a specific medical complaint. (pp. 210, 218–219, 222)

5-1.18 Demonstrate a caring attitude towards patients with a general medical complaint who request emergency medical services. (p. 209)

5-1.19 Place the interests of the patient with a general medical complaint as the foremost consideration when making any and all patient care decisions. (p. 209)

5-1.20 Communicate with empathy to patients with a general medical complaint, as well as with family members and friends of the patient. (p. 209)

5-1.21 Demonstrate a caring attitude towards patients with a specific medical complaint who request emergency medical services. (pp. 210, 218–219, 222)

5-1.22 Place the interests of the patient with a specific medical complaint as the foremost con-

sideration when making any and all patient care decisions. (pp. 210, 218–219, 222)

5-1.23 Communicate with empathy to patients with a specific medical complaint, as well as with family members and friends of the patient. (pp. 210, 218–219, 222)

Psychomotor

5-1.27 Demonstrate the steps in providing emergency medical care to a patient with a general medical complaint. (p. 209)

5-1.28 Demonstrate the steps in providing emergency medical care to a patient with an altered mental status. (p. 210)

5-1.29 Demonstrate the steps in providing emergency medical care to a patient with seizures. (pp. 219–220)

Enrichment

◆ Establish the relationship between airway management and the patient with altered mental status. (p. 210)

◆ Describe the assessment and emergency medical care of the patient with a diabetic emergency. (pp. 210–214)

◆ List various ways that poisons enter the body. (pp. 214–217)

◆ List signs and symptoms of poisoning. (pp. 214–217)

◆ Describe the assessment and emergency medical care of a patient with suspected poisoning. (pp. 214–217)

◆ Describe the assessment and emergency care of the patient with an altered mental status and a loss of speech, sensory, or motor function. (pp. 218–219)

◆ Recognize the common signs and symptoms of a generalized seizure. (p. 220)

◆ Explain the assessment and emergency care of a seizing patient. (pp. 221–222)

◆ Identify the signs and symptoms of abdominal pain or distress. (pp. 222–223)

◆ Describe the emergency medical care of a patient with abdominal pain or distress. (pp. 222–223)

DISPATCH

My volunteer fire service EMS squad was dispatched for an unconscious man at 1243 Martin Luther King Jr. Avenue. I realized that the location was right on my way to the fire station, so I responded to the scene in my own car.

SCENE SIZE-UP

I always carry protective gloves in my car, so after parking my vehicle, I put them on. I then approached the scene especially carefully, since I had no radio contact with dispatch. When I saw a woman looking out the window waiting for EMS, I identified myself as a fire service First Responder. She took me to her husband.

INITIAL ASSESSMENT

I identified myself to him. He responded with rambling speech that didn't make any sense. I saw that he was sweaty and pale. His airway was clear, and his breathing appeared to be strong and at a reasonable rate. His pulse was strong but seemed a bit rapid. Based on his color and mental status, my general impression was that of a patient with a medical condition who seemed like he could worsen at any minute. With the wife's permission, I used the phone to update EMS, reporting that the patient is awake but confused. I also requested an ALS unit.

> *Consider this scenario as you read Chapter 11. What else may be done to assess and care for this patient?*

GENERAL MEDICAL COMPLAINTS

As a fire service First Responder, you will be called to the scenes of patients with *specific* medical complaints such as "my chest hurts" or "I can't breathe." Every once in a while, however, your medical patient will have a *general* complaint such as "I feel weak" or "I don't feel well."

Handle patients with a general medical complaint the same as you would any other. After your scene size-up, complete an initial assessment and treat any life-threatening conditions you observe. Perform a physical exam as needed, and be especially thorough gathering the patient's history. It could provide important clues to the underlying problem.

If you are unable to determine a more specific complaint or unable to obtain a pertinent medical history, do the following:

1. **Monitor the airway, breathing, and circulation.** Be sure there is a patent airway with adequate breathing and circulation.

2. **Allow the patient to get in a position of comfort,** if he is responsive and there are no suspected spine injuries.

3. **Perform an ongoing assessment.** Continue to do so until the incoming EMTs take over patient care. Be sure to report any changes in the patient's condition.

These patients may be just as frightened and worried as patients with more specific problems. Consider their feelings as you assess and care for them. Be gentle and empathetic. If the family is present, they may be very concerned and ask you to tell them "what's wrong." Be truthful and kind. For example, you might tell them that though you do not know exactly what the problem is, you are doing all that is possible. Also, reassure them that you have arranged for the patient to be transported to a hospital for further assessment and care.

Altered Mental Status

A change in a patient's normal level of responsiveness and understanding is called an **altered mental status.** It can occur quickly or slowly. It can range from disoriented to combative to unresponsive. There are many medical reasons for it. A few examples are:

◆ Hypoxia (decreased levels of oxygen in the blood).

◆ Hypoglycemia (low blood sugar).

◆ Stroke (loss of blood flow to part of the brain).

◆ Seizures.

◆ Fever, infections.

◆ Poisoning, including drug and alcohol poisoning.

◆ Head injury.

◆ Psychiatric conditions.

COMPANY OFFICER'S NOTE

The term "altered mental status" refers to a wide range of medical and/or psychological conditions. It is not always necessary to know the reason the condition exists, especially since patient care is so similar for the conditions that cause altered mental status. Provide airway support and oxygen. Calming the patient may help prevent further injury.

As a First Responder, you do not need to figure out why your patient has an altered mental status. Your job is to recognize it as soon as possible and to support the patient appropriately. So after assuring scene safety, proceed with patient assessment. Gather an accurate patient history as soon as appropriate. A patient with an altered mental status may deteriorate rapidly. If you wait too long, the history could be lost to the EMTs and hospital staff who take over care. A history of diabetes or seizures, for example, may be important to the EMTs and hospital personnel when they try to determine the cause of the altered mental status and subsequently treat the patient.

◆ *Emergency Care*

Emergency care of a patient with altered mental status is as follows:

1. **Assess and monitor the patient's airway and breathing closely.** These patients may not be able to protect their own airways. It is up to you to be aware of this danger. If the patient is unresponsive, secure the airway with an adjunct. Suction as needed.

2. **Position the patient.** If there is no reason to suspect head or spine injury, place the patient in a recovery position. Continue to monitor the patient's breathing closely.

3. **Administer high-flow oxygen.** One of the most serious and most common causes of altered mental status is hypoxia. If the patient is breathing adequately, apply high-flow oxygen via nonrebreather mask. If the patient is not breathing adequately, assist with a bag-valve-mask device attached to an oxygen source. If you cannot provide oxygen, be prepared to assist ventilations.

A patient with an altered mental status may be aware of her condition. This can be very frightening. If the patient has had a seizure, she could lose control of the bowels and bladder, which adds to embarrassment and anxiety. A caring attitude on your part, as well as helping the patient maintain some privacy, will help. Note that the patient's condition may be very upsetting to the family, too. Take time, if possible, to make sure they understand that you are caring for the patient and that an ambulance is on the way.

Hypoglycemia and Hyperglycemia

The human body needs both oxygen and sugar to produce the energy that sustains it. When blood sugar is too low or too high, the body reacts. The most common reaction is altered mental status.

Hypoglycemia

Low blood sugar, or hypoglycemia, is the result of two conditions. One is too much insulin, a drug used by many people with diabetes. The other is too little sugar, such as occurs when a patient with diabetes does not eat properly. (See Figure 11-1.)

People with diabetes are not the only ones who can suffer from low blood sugar. Alcoholics, people who have ingested certain poisons, and people who are ill also may suffer from it. Some common causes of low blood sugar are:

◆ Skipped meals, particularly for a patient with diabetes.

◆ Vomiting, especially with illness.

◆ Strenuous exercise.

◆ Physical stress from extreme heat or cold.

◆ Accidental overdose of insulin.

A common cause of hypoglycemia is the accidental overdose of insulin by a patient with diabetes. After a time, diabetes can inflict a degree of blindness in patients. This can make it very hard for them to give themselves the proper amount of

insulin. The result is an insulin overdose and hypoglycemia.

Signs and symptoms of hypoglycemia may include any of the following (Figure 11-2):

◆ Rapid onset of altered mental status, even unresponsiveness.

◆ Intoxicated appearance, staggering, slurred speech.

◆ Rapid pulse rate.

◆ Cool, clammy skin.

◆ Hunger.

◆ Headache.

◆ Seizures.

Hyperglycemia

By definition, patients with diabetes have elevated blood sugar, or hyperglycemia. This condition is basically one of too little insulin and too much

Figure 11-1 Diabetes has long been recognized as a serious disorder.

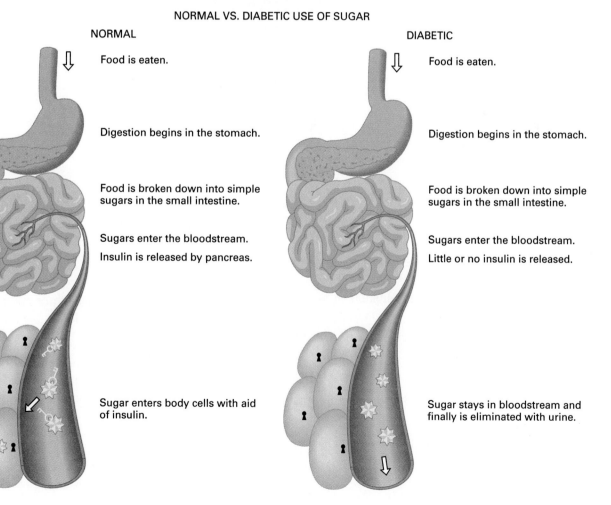

NORMAL VS. DIABETIC USE OF SUGAR

NORMAL

Food is eaten.

Digestion begins in the stomach.

Food is broken down into simple sugars in the small intestine.

Sugars enter the bloodstream.
Insulin is released by pancreas.

Sugar enters body cells with aid of insulin.

DIABETIC

Food is eaten.

Digestion begins in the stomach.

Food is broken down into simple sugars in the small intestine.

Sugars enter the bloodstream.
Little or no insulin is released.

Sugar stays in bloodstream and finally is eliminated with urine.

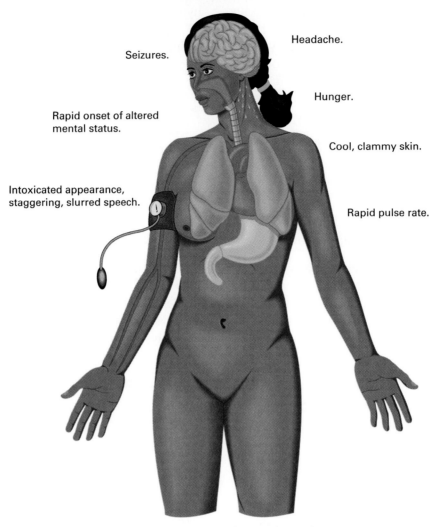

Seizures.

Headache.

Rapid onset of altered mental status.

Hunger.

Cool, clammy skin.

Intoxicated appearance, staggering, slurred speech.

Rapid pulse rate.

Figure 11-2   Signs and symptoms of hypoglycemia.

blood sugar, or the inability of the body to properly use the insulin it does produce. Certain conditions may worsen hyperglycemia in diabetics, including:

◆ Infection, such as a respiratory infection.

◆ Failure of the patient to take insulin or to take a sufficient amount.

◆ Eating too much food that contains or produces sugar.

◆ Increased or prolonged stress.

Signs and symptoms of hyperglycemia may include (Figure 11-3):

◆ Sweet, fruity, or acetone-like breath.

◆ Flushed, dry, warm skin.

◆ Hunger and thirst.

◆ Rapid, weak pulse.

◆ Altered mental status.

◆ Intoxicated appearance, staggering, slurred speech.

◆ Frequent urination.

◆ Reports that the patient has not taken the prescribed diabetes medications.

The onset of severe hyperglycemia is gradual. In most cases, it develops over a period of 12 to 48 hours. At first, the patient experiences excessive hunger, thirst, and urination. The patient appears extremely ill and becomes sicker and weaker as the condition progresses. If left untreated, the patient may die. With treatment, improvement is gradual, occurring 6 to 12 hours after insulin and intravenous fluid are administered.

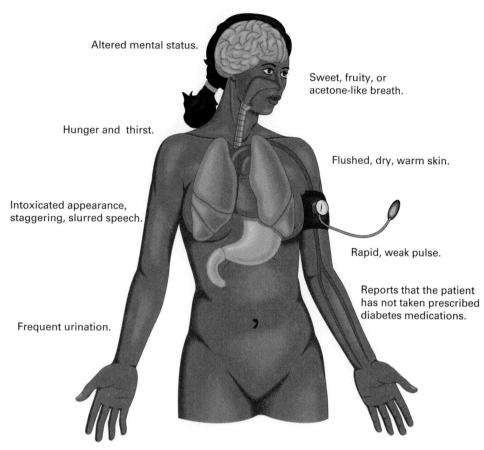

Altered mental status.

Sweet, fruity, or acetone-like breath.

Hunger and thirst.

Flushed, dry, warm skin.

Intoxicated appearance, staggering, slurred speech.

Rapid, weak pulse.

Reports that the patient has not taken prescribed diabetes medications.

Frequent urination.

Figure 11-3   Signs and symptoms of hyperglycemia.

Patient Assessment

When you gather the SAMPLE history, try to find out about the onset of the emergency. Be sure to ask, "Do you have diabetes?" If the patient says yes, ask: "Have you eaten today? Did you take your insulin or diabetic oral medication?" Also ask about any current illness, stress, and problems with medications.

Look for a medical identification tag during the physical exam. If the police are present, ask them to check the patient's wallet, too. If the patient is at home, check the refrigerator for insulin or other medications. Also check around the house for needles and syringes. Special needle containers are often present in the house.

◆ Emergency Care

Proceed with emergency care as you would for any patient with altered mental status. However, if you suspect hypoglycemia or hyperglycemia, alert the incoming EMS crew immediately. While

COMPANY OFFICER'S NOTE
Never assume that a patient who is acting in an unusual or even abusive manner is intoxicated. Your patient may be suffering from a diabetic condition, head injury, or shock. Assuming a patient with one of these conditions is simply drunk could cause you to withhold necessary and appropriate care. Even if your patient is intoxicated, remember that chronic drinkers are much more prone to diabetes than others. Keep an open mind.

waiting for them, monitor the airway closely. Note that this patient may suddenly have a seizure. Be prepared. (See "Seizures" later in this chapter.)

Remember, patients with diabetes can suffer from either hypoglycemia or hyperglycemia. When in doubt, give sugar. You will not harm a hyperglycemic patient with sugar, and you may save the life of a patient with hypoglycemia by this emergency treatment.

Your EMS system may permit you to help a patient take some sugar. To do so, the patient must be awake and able to control his or her own airway. Follow local protocols. The patient may benefit from one of the following:

◆ Dissolve some sugar in a glass of water.

◆ Pour a drink that is naturally rich in sugar such as orange juice.

◆ Squeeze a commercially prepared glucose paste onto a tongue depressor (Figure 11-4). Note that a tube of glucose should only be used for a single patient and then discarded.

Never give patients who cannot control their own airways anything to eat or drink. They could aspirate the substance into their lungs. This can have grave results, including eventual death. If you are in doubt, call for medical direction. If sugar or glucose is administered, be sure to tell the EMTs who take over care. Also report any changes in mental status that occurred while the patient was in your care.

Poisoning

A poison is a substance that can impair health if left untreated. Poisoning can occur in a variety of settings. However, many occur in the relative safety of the home (Figure 11-5). Poisons may be inhaled, ingested, injected, or absorbed. So when-

Figure 11-4 You may be allowed to help a patient self-administer oral glucose.

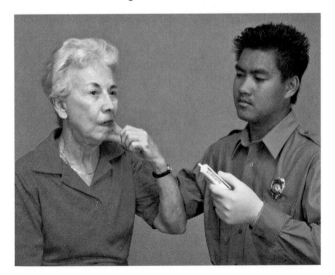

Figure 11-5 Poisoning is a leading cause of accidental death among children.

ever you suspect a poisoning, and when it is safe to do so, try to answer these questions: What substance is involved? How much is involved? When did the poisoning occur? What has the patient done to relieve symptoms?

Ingested Poisons

An ingested poison is one that is introduced into the digestive tract by way of the mouth. Every year in the U.S. there are over eight million reported ingested poisonings. Drugs such as aspirin and alcohol are among the top offenders.

Patient Assessment. Be alert for clues when you size up the scene and after. A clue to a poisoning may be an overturned or empty pill box, scattered pills, chemical containers, household cleaners, empty alcohol bottles, or overturned plants.

Keep in mind as you do your initial assessment that patients who are poisoned often vomit. So, be alert to airway obstruction and breathing difficulty. They can lead to hypoxia and death.

In addition to altered mental status, signs and symptoms of poisoning may include:

◆ History of ingesting poisons.

◆ Burns around the mouth.

◆ Odd breath odors.

◆ Nausea, vomiting.

◆ Abdominal pain.

◆ Diarrhea.

Notice that the symptoms follow a path of ingestion. Start at the mouth. Look for chemical burns around the mouth or a chemical odor on the breath (Figure 11-6). Then notice if there is or has been any nausea and vomiting. Finally, the patient may complain of abdominal cramps and diarrhea.

Often poisons will affect the central nervous system. You may see dilated or constricted pupils, or you may hear the patient complain of double vision. There may be excessive saliva or foaming at the mouth. There may be excessive tearing or sweating. Finally, the patient may become unresponsive or have seizures.

◆ *Emergency Care.* Your top priority is the patient's airway. With that in mind, proceed as you would for any patient with altered mental status. To help limit the damage a poison can cause, make sure the EMS system is activated. If there will be a delay, call the poison control center in your area or medical oversight for instructions. Follow local protocols.

You may be instructed to give the patient activated charcoal (Figure 11-7). This is a finely ground charcoal that is very absorbent. It binds with the poisons in the stomach and then passes through the body harmlessly. It may be effective in reducing poisons for up to four hours after ingestion. Use it only by order of poison control or according to your local protocols. Never use it with a patient who is not awake and cannot control his or her own airway.

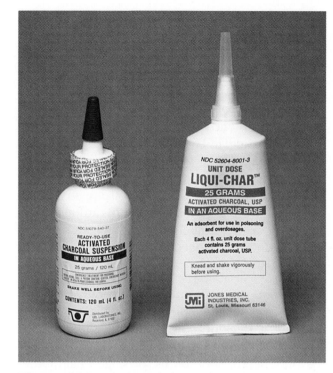

Figure 11-7 Activated charcoal.

Most activated charcoal is premixed with water. If it is dry, then you must mix two tablespoons of it in a glass of water to make a slurry. Be careful. It stains most clothing easily.

You may be ordered to induce vomiting in the patient. This is usually done with syrup of ipecac. Ipecac can have side effects. Use it only on the direct order of either poison control or medical oversight. In the field, it is never given with activated charcoal unless under direct medical orders. Follow all local protocols. Remember that you should never induce vomiting if the patient:

◆ Is unresponsive.

◆ Cannot maintain an airway.

◆ Has ingested an acid, a corrosive such as lye, or a petroleum product such as gasoline or furniture polish.

◆ Has a medical condition that could be complicated by vomiting, such as heart attack, seizures, and pregnancy.

If the patient swallows an acid, corrosive, or petroleum product, you may have to dilute it. Use either several glasses of water or milk. Whatever the situation, always follow local protocols. If that means calling poison control, do so and follow their instructions exactly.

Figure 11-6 Burns or stains around the mouth may indicate poisoning.

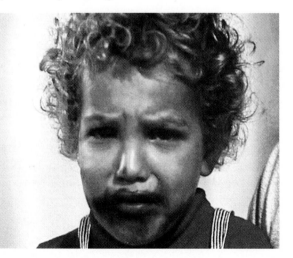

Inhaled Poisons

An **inhaled poison** is a poison that is breathed in. A common source of poisonous gas is fire, which can produce carbon monoxide and cyanide. However, fire is not the only source of poison gas. Large amounts of carbon dioxide can come from sewage treatment plants or industrial sites. Even the chlorine gas in swimming pools can be lethal. Do not become a victim yourself. Remember, many poison gases are colorless, odorless, and tasteless. You may not know you are in danger until it is too late. Look out for hazardous materials. Pay constant attention to the nature of the incident and the dangers it might contain. Protect yourself! And keep others away from the scene.

Patient Assessment. It is imperative for you to give special attention to this patient's airway. Once in a safe location, open the airway. Then inspect the mouth and nose. Be careful to note the presence of soot, burns, or singed hair. Other signs and symptoms include:

- History of inhaling poisons.
- Breathing difficulty.
- Chest pain.
- Cough, hoarseness, burning sensation in the throat.
- Cyanosis (bluish discoloration of skin and mucous membranes).
- Dizziness, headache.
- Seizures, unresponsiveness (advanced stages).

Carbon monoxide is a poison gas that is especially lethal. Kerosene heaters, hot water heaters, and car exhaust fumes are some of the most common sources. Be particularly alert to carbon monoxide poisoning if several members of a household have the same signs and symptoms. Suspect it if they say they are only sick when they are in a certain location. Be alert if the family pet seems sick as well.

Signs and symptoms of carbon monoxide poisoning include:

- Throbbing headache and agitation.
- Nausea, vomiting.
- Confusion, poor judgment.
- Diminished vision, blindness.
- Breathing difficulty with rapid pulse.
- Dizziness, fainting, unresponsiveness.

- Seizures.
- Paleness.
- Cherry red color to skin (very late sign).

♦ ***Emergency Care.*** The first rule of EMS is safety. You absolutely must protect yourself. Do not enter the scene of a poisonous gas. Call dispatch for specialized rescue teams who will have the appropriate safety equipment, including a self-contained breathing apparatus. When it is safe to do so, quickly remove the patient from the source of the poison. Then proceed as you would for any patient with altered mental status. Verify that an ambulance is en route. Consider helicopter evacuation, if it would be quicker.

Note that all patients who are exposed to carbon monoxide need medical care, even those who seem to recover.

Absorbed Poisons

An absorbed poison is one that enters the body upon contact with the skin. Examples of natural sources include poison ivy, sumac, and oak (Figure 11-8), which may cause reactions in certain individuals. Other sources are corrosives, insecticides, herbicides, and cleaning products.

Patient Assessment. Some absorbed poisons cause harm only to the point of contact. Others may cause life-threatening reactions. In general, signs and symptoms include:

- History of exposures.
- Liquid or powder on the skin.
- Burns.
- Itching, irritation.
- Redness, rash, blisters (Figure 11-9).

Once a poison is identified, advise the incoming EMS units. If hazardous materials are suspected, follow local protocols. Note that an oil-based poison can spread easily from person to person. Protect yourself. Impervious gloves are essential. Also consider a gown, mask, and eye protection.

♦ ***Emergency Care.*** For emergency care of a patient who has been poisoned, contact poison control or medical direction. General guidelines for emergency care are as follows:

1. **Remove the clothing** that came in contact with the poison.

Figure 11-8a   Poison ivy.

Figure 11-8b   Poison sumac.

Figure 11-8c   Poison oak.

2. **Remove the poison.** With a dry cloth, blot the poison from the skin. If the poison is a dry powder, gently brush it off.

3. **Flood the area** with copious amounts of water. A shower or garden hose is ideal for this purpose. Continue until other EMS units arrive. Note that you may need to use alcohol

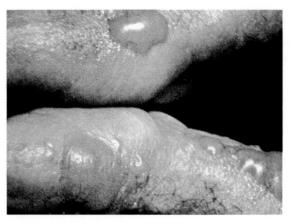

Figure 11-9   Blisters from poisonous plant contact.

or vegetable oil with some poisons. Follow instructions from poison control or medical direction. A "material safety data sheet" also may be available (see Chapter 24); if so, obtain a copy and transport it with the patient.

4. **Monitor vital signs continually.** Be alert for sudden changes. Seizures and shock are common.

The eyes are especially vulnerable to absorbed poisons. If ordered to do so, flood the eyes with copious amounts of water. If only one eye is affected, be sure to avoid running contaminated water into the other eye. Advise incoming EMS units of the patient's condition. Follow local protocol.

Injected Poisons

An injected poison is one that enters the body by way of an object that pierces the skin. For example, an illegal drug may enter the body by way of a hypodermic needle. An overdose may be the result. Other causes of injected poisons include the bites and stings of insects, spiders, snakes, and marine animals. The venom of these creatures can cause serious allergic reactions, even death.

First Responder assessment and emergency care of the patient with an injected poison is the same as for any patient with a medical complaint: assess and monitor the patient's ABCs, position the patient properly and, if you are allowed, administer high-flow oxygen. For more information on the overdose patient, see Chapter 13.

Stroke

A patient may suffer a **cerebrovascular accident (CVA)** or stroke, when an area of the brain is deprived of blood. This can occur when a *thrombus* (blood clot) blocks an artery, when an *embolus* (matter) lodges in an artery, or in the event of an *aneurysm* (when an artery bursts). (See Figures 11-10 and 11-11.)

The National Stroke Association refers to a stroke as a "brain attack." Strokes are the third leading cause of death in the U.S. and the leading cause of adult disability. They are more common in people over the age of 65 but can affect anyone. People who are at risk for heart attack also may be in danger of having a stroke. They include patients with high blood pressure or diabetes, and patients who smoke tobacco.

Patient Assessment

The signs and symptoms of stroke are the result of the location and amount of brain damage.

Figure 11-10a   A cerebrovascular accident (CVA), or stroke, from cerebral hemorrhage.

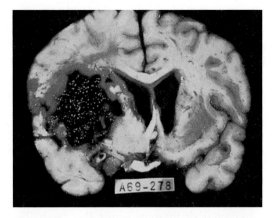

Figure 11-10b   Brain damaged by stroke.

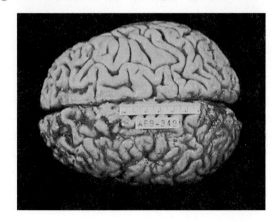

Signs and symptoms may be mild, or they may be life threatening. Sometimes they are temporary. Temporary ones indicate a "mini-stroke." Called transient ischemic attacks (TIAs), these mini-strokes are warning signs of an impending larger stroke.

Signs and symptoms of stroke include (Figure 11-12):

◆ *Inability to communicate.* The patient may either fail to speak or fail to understand what is spoken.

◆ *Impairment in one part of the body.* For example, loss of muscle control on one side of the face or loss of movement on one entire side of the body.

◆ *Altered mental status.* This can range from a change in personality to seizures and unresponsiveness.

About 50% of stroke patients have an elevated blood pressure during a stroke. The combination of an elevated blood pressure, slow pulse, and rapid or irregular breathing is a sign of a major stroke. Be prepared if the patient should convulse suddenly.

◆ *Emergency Care*

Proceed as you would for any patient with altered mental status. However, note that as pressure increases in the skull from swelling tissues and bleeding, the patient's breathing will be affected. Be especially alert to the airway of a patient who has difficulty speaking or slurred speech. Never give a suspected stroke patient anything to eat or drink. Administer oxygen, and be prepared to provide artificial ventilation.

Following your initial assessment, try to gather a history from the patient, family, and bystanders. Be sure to find out if there is a medical history of stroke, high blood pressure, diabetes, or heart disease. When you examine the patient, be sure to handle any paralyzed limbs carefully. They may have no feeling, so you could injure them without being aware of it. The patient also may unintentionally cause them to strike an object.

The loss of a mental or motor function is a frightening reality for stroke patients. Try to remain calm and never express surprise about abnormal physical findings. Instead, maintain a professional attitude. Reassure the patient. Do not make any statements about long-term disability.

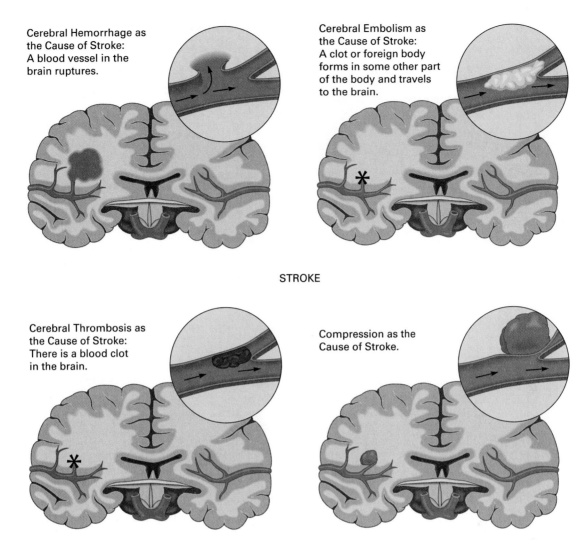

Cerebral Hemorrhage as the Cause of Stroke: A blood vessel in the brain ruptures.

Cerebral Embolism as the Cause of Stroke: A clot or foreign body forms in some other part of the body and travels to the brain.

STROKE

Cerebral Thrombosis as the Cause of Stroke: There is a blood clot in the brain.

Compression as the Cause of Stroke.

Figure 11-11   Causes of stroke.

Continue to talk to the patient even if he or she cannot speak. These patients often can hear very well. Explain to them what it is you are doing. Do not talk down to them or treat them like children.

Seizures

There are many causes of seizures. Sometimes the cause is unknown. All of the conditions described in this chapter can lead to them. Common causes include:

◆ Chronic medical conditions.

◆ Epilepsy.

◆ Hypoglycemia.

◆ Poisoning, including alcohol and drug poisoning.

◆ Stroke.

◆ Fever (most common in children).

◆ Infection.

◆ Head injury or brain tumors.

◆ Hypoxia (decreased levels of oxygen in the blood).

◆ Complications of pregnancy.

A seizure is the result of a nervous system malfunction. It may last five minutes or it may be prolonged. Its symptoms can range from a twitch of a limb to whole body muscle contractions. Most patients become unresponsive. Many vomit during the seizure. Typically, patients are tired and sleep afterwards. Seizures are rarely life threatening, but they do indicate a very serious condition.

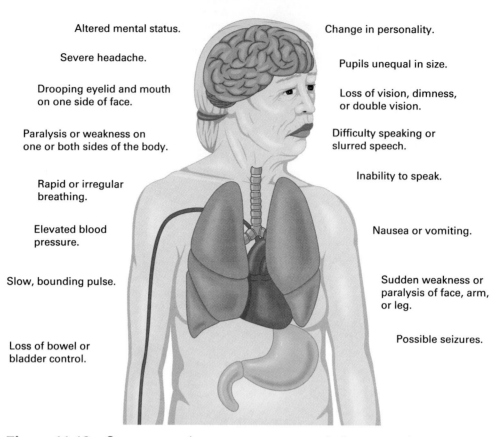

Altered mental status.

Severe headache.

Drooping eyelid and mouth on one side of face.

Paralysis or weakness on one or both sides of the body.

Rapid or irregular breathing.

Elevated blood pressure.

Slow, bounding pulse.

Loss of bowel or bladder control.

Change in personality.

Pupils unequal in size.

Loss of vision, dimness, or double vision.

Difficulty speaking or slurred speech.

Inability to speak.

Nausea or vomiting.

Sudden weakness or paralysis of face, arm, or leg.

Possible seizures.

Figure 11-12   One or more signs or symptoms may indicate a stroke.

Patient Assessment

You will probably be called most often to a *grand mal,* or generalized seizure. There are four phases to this type of seizure. They are called in order (Figure 11-13):

◆ Aura phase.

◆ Tonic phase.

◆ Clonic phase.

◆ Postictal phase.

In the *aura phase,* the patient may become aware that a seizure is coming on. The aura is often described as an unusual smell, taste, or a flash of light. It lasts a split second. Not all patient's experience this phase.

In the *tonic phase,* the patient becomes unresponsive and collapses to the ground. Then all of the muscles of the body contract. This often forces a scream out of the patient. It also can force out sputum, which can look like

foam. During this phase, the patient may stop breathing.

In the *clonic phase,* the patient's muscles alternate between contracting and relaxing. The patient may become incontinent of urine (unable to retain it). Because the patient may bite the tongue and cheek, there may be blood in the mouth.

In the *postictal phase,* the patient gradually regains responsiveness. At first, the patient is confused and even combative. Gradually, over several minutes, the patient becomes aware of his or her surroundings.

Note that a continuous seizure, or two or more seizures without a period of responsiveness, is called *status epilepticus.* This is a true medical emergency, which can be fatal. Complications include aspiration, hypoxia, hyperthermia (fever), and heart problems. If you suspect this type of seizure, advise responding EMS units and request ALS intervention. Transportation to a medical facility must not be delayed.

1. AURA PHASE. Often described as unusual smell or flash of light that lasts a split second.

2. TONIC PHASE. 15 to 20 seconds of unresponsiveness followed by 5 to 15 seconds of extreme muscle rigidity.

3. CLONIC PHASE. 1 to 5 minutes of seizures.

4. POSTICTAL PHASE. 5 to 30 minutes to several hours of deep sleep with gradual recovery.

Figure 11-13 Stages of a generalized seizure.

◆ *Emergency Care*

During your scene size-up, ask yourself if the patient was injured when he or she fell to the ground. Pay careful attention to the potential for spine or head injury. When you arrive on scene, if the patient is still seizing, then:

1. **Stay calm.** Just wait. The seizure usually is over in a few minutes.

2. **Prevent any further injury** by moving objects away from the patient (Figure 11-14). If they cannot be moved, then put something between them and the patient. If needed, drag the patient a few feet away from the danger.

3. **Place padding under the patient's head.** A coat or blanket will do. Remove the patient's eyeglasses. Do not force anything into the patient's mouth. Do not try to restrain the patient.

4. **If you suspect status epilepticus, do your best to prevent aspiration.** Position the patient on his or her side or suction if possible. Assist ventilations with a bag-valve-mask attached to 100% oxygen. Notify the incoming EMS unit immediately.

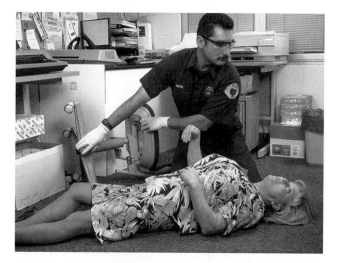

Figure 11-14 Move objects away from the seizing patient.

Most often you will arrive on scene when a seizure has passed or when the patient is in the last phase of seizing. When the seizure stops, assess and monitor the patient's airway and breathing closely. If there is no reason to suspect head or spine injury, place the patient in the recovery position (Figure 11-15). If you are allowed, administer high-flow oxygen. If you suspect the patient was injured during a fall, use a jaw-thrust to open the airway. Offer comfort and reassurance as the patient recovers. Remember that he or she will have muscle soreness as well as fatigue.

While you wait for the EMTs to arrive on scene, consider the patient's feelings. Often he or

Figure 11-15 When the seizure stops, position the patient to allow drainage of saliva and vomit.

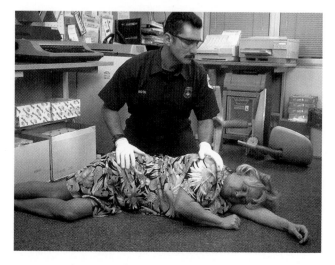

she is embarrassed. Consider asking onlookers to move away to provide some privacy. Place a sheet or towel over the patient's body, if there was incontinence.

When you give your hand-off report to the EMTs, be sure to include a description of the seizure, including details such as which body part was affected. It may be important in determining its cause.

Abdominal Pain and Distress

The abdominal cavity contains many different organs and blood vessels. Complaints of pain or discomfort could actually be caused by a number of problems or conditions. Pain that is the result of a problem within the abdominal cavity may be located directly over a problem organ, or it may be in a totally different part of the body. This is called *referred pain*.

Abdominal problems may cause referred pain to the shoulders and chest. Sometimes pain begins in the anterior abdomen and radiates to the back. Abdominal emergencies require aggressive care to prevent shock and to save lives. All abdominal pain should be taken seriously. Do not spend time trying to determine its cause. Rather, complete a thorough patient assessment and history.

Patient Assessment

Severe abdominal pain should be considered an emergency. Any abdominal pain that is persistent or is significant enough for the patient or family to call for assistance should be considered an emergency. Signs and symptoms include:

- Abdominal pain, local or diffuse.
- Colicky pain (cramps that occur in waves).
- Abdominal tenderness, local or diffuse.
- Anxiety, reluctance to move.
- Loss of appetite, nausea, vomiting.
- Fever.
- Rigid, tense, or distended abdomen.
- Signs of shock.
- Vomiting blood, bright red or like coffee grounds.
- Blood in the stool, bright red or tarry black in color.

Patients with abdominal distress or pain will appear to be very ill. A patient with acute

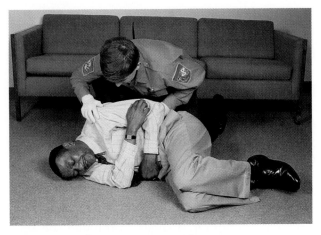

Figure 11-16 Guarding position.

abdominal distress often gets in a guarding position (on the side with knees drawn up toward the abdomen). (See Figure 11-16.) This position reduces tension on the muscles of the abdomen, which in turn helps to reduce pain.

In assessing a patient with acute abdominal distress, the initial assessment is the first priority. Even after assuring the patient's ABCs, stay alert for signs of shock, which include a rapid thready pulse, restlessness, cold clammy skin, and falling blood pressure. Shock is common with internal bleeding and continued vomiting and diarrhea.

As with all medical patients, gather a good patient history. It may identify clues to the patient's condition, such as prior similar problems and factors that may have caused the pain.

During the physical exam, find out if movement causes pain. Check to see if the abdomen is distended, and ask the patient to confirm your observation. Note if the patient can relax the abdominal wall when asked to do so. Palpate the abdomen gently to determine if it is rigid or soft. If you know one area is causing pain, examine that area last.

Do not spend too much time on assessment before making sure the EMS system has been activated. Too much palpation can worsen pain. It also can aggravate the medical condition that caused it.

◆ Emergency Care

The goals of emergency care for acute abdominal distress are to prevent any possible life-threatening complications, to make the patient comfortable, and to arrange transport as quickly as possible. In addition:

1. **Maintain an open airway.** Be alert for vomiting and possible aspiration. If the patient is nauseated, position the patient on his or her left side if it does not cause too much pain.

2. **Administer oxygen** by way of a nonrebreather mask at 10 to 15 liters per minute, if you are trained and allowed to do so.

3. **Be alert for signs of shock.** If vital signs and other observations point to it, position the patient on his or her back with legs elevated. If there are no signs of shock, then allow the patient to get into a position of comfort.

Protect the patient from any rough handling. Never give anything by mouth. Do not allow the patient to take any medications. Medications could mask symptoms and complicate the physician's diagnosis and treatment.

ON SCENE FOLLOW-UP

At the beginning of this chapter, you read that a fire service First Responder is providing emergency care to an "unconscious man." To see how chapter information applies to this emergency, read the following. It describes how the call was completed.

INITIAL ASSESSMENT (CONTINUED)

I didn't have any oxygen with me, so I couldn't administer any. I'm not sure he would have understood to keep the mask on anyway.

PATIENT HISTORY

I continued trying to talk to the patient, but if he could understand me, he didn't show it. With patients like this, I always talk to them even though they don't respond. Sometimes they can understand but just not speak.

His wife told me that he was 59 years old, and he was experiencing flu-like symptoms for several days. There was no history of cardiac problems, but he is a diabetic. I asked about his diet and medication for the past several days and found that it has been somewhat erratic. He hadn't eaten properly and had been trying to adjust his insulin accordingly.

PHYSICAL EXAMINATION

It seemed like I found the cause, but you never can be too sure. I asked his wife about recent falls or injury, and there were none. Since the patient couldn't tell if something hurt, I looked for deformity, swelling, bruising, or other signs of injury. The exam revealed none.

ONGOING ASSESSMENT

When I heard the sirens of the squad coming down the street, I was glad. When the other firefighters arrived, I advised them of what I had done. They confirmed my request for paramedics in case the patient had a diabetic problem.

PATIENT HANDOFF

The paramedic unit arrived shortly thereafter. By this time, we had a full set of vitals and oxygen applied to the patient. We reported how the patient was found, and the results of the history and physical exam. The patient's level of responsiveness had changed in that he had become slightly agitated in addition to his apparent confusion.

The medics agreed that it could be a diabetic condition. They did a glucose check right at the house, found the man's blood sugar dangerously low, and administered sugar. You could see improvement almost immediately. Finally, they packaged the patient and transported him to the hospital.

> *Whether or not you know the cause of a medical emergency, your job is the same. Assess, care for, and monitor the airway until the EMTs arrive to take over. Be prepared to provide basic life support if needed. And try to get a complete patient history from the patient, family, or bystanders.*

Chapter Review

Medical complaints will be the most common type of emergency for most First Responders. The percentage of the population that is elderly is increasing dramatically. This means even more calls for medical problems in the future.

This chapter covered a wide range of medical problems—from stroke to diabetes to poisoning to generalized complaints. Remember that it is never necessary for you to diagnose a patient's medical problem. Sometimes with these patients, diagnosis in the hospital is difficult, even with the tests and procedures available to physicians there.

Though it is not practical for you to determine the cause of a patient's medical condition in the field, there are many things you can do. In the initial assessment, for example, you will treat all life-threatening problems. Taking an accurate history is very important to a medical patient, too. It will provide clues to the patient's condition, which will benefit you, the EMTs, and hospital personnel.

Finally, never jump to conclusions. In this chapter, for example, you learned that under certain conditions a patient who has diabetes may appear to be drunk. Consider all patients who are exhibiting an altered mental status or unusual behavior as having a medical problem. Never assume that the patient is drunk, drugged, or mentally ill.

FIRE COMPANY REVIEW

Page references where answers may be found or supported are provided at the end of each question.

Section 1

1. How would you treat a patient with a general medical complaint? (p. 209)

Section 2

2. Why must you monitor the airway and breathing of all patients with altered mental status? (p. 210)

3. What are three possible indicators of diabetes in a patient? (p. 213)

4. What are various ways a poison can enter the body? (p. 214)

5. Why do inhaled poisons pose a risk for First Responders? (p. 216)

6. In addition to altered mental status, what are the signs and symptoms of an ingested poison? An inhaled poison? An absorbed poison? (pp. 214, 216)

7. What are the characteristic signs of a stroke? (p. 218)

8. Why must you be careful with the limbs of a stroke patient? (p. 218)

9. What can you do for a patient who is seizing? (pp. 221–222)

10. What can you do for a seizure patient when the seizure has stopped? (pp. 221–222)

11. What signs and symptoms are related to abdominal pain and abdominal distress? (pp. 222–223)

12. What are the goals of emergency care of a patient with acute abdominal distress? (p. 223)

RESOURCES TO LEARN MORE

Goldfranks, L.R., et al., eds. *Toxicologic Emergencies,* Fifth Edition. Norwalk, CT: Appleton-Lange, 1994.

Rapid Identification and Treatment of Acute Stroke. NIH Publications No. 97-4239. August 1997.

Tintinalli, J.E., et al. *Emergency Medicine: A Comprehensive Study Guide,* Fourth Edition. New York: McGraw-Hill, 1995.

AT A GLANCE

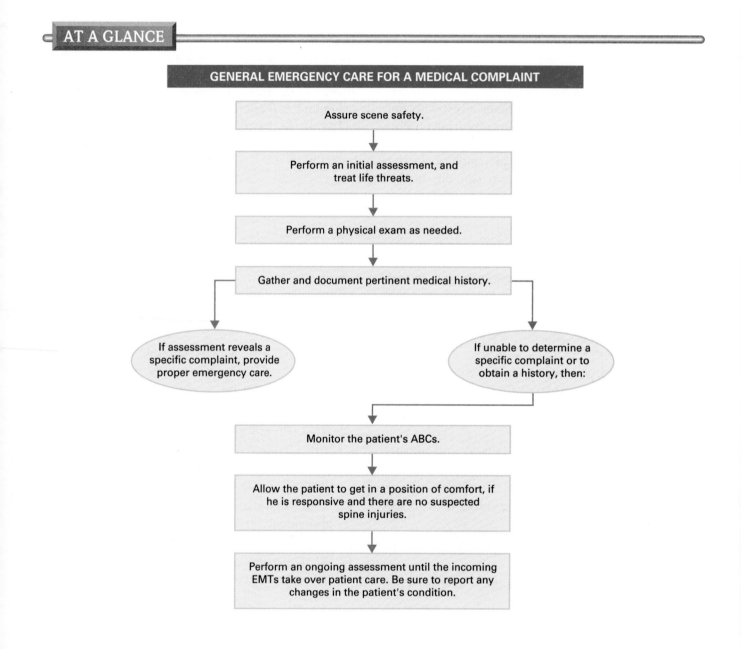

GENERAL EMERGENCY CARE FOR A MEDICAL COMPLAINT

Assure scene safety.

Perform an initial assessment, and treat life threats.

Perform a physical exam as needed.

Gather and document pertinent medical history.

If assessment reveals a specific complaint, provide proper emergency care.

If unable to determine a specific complaint or to obtain a history, then:

Monitor the patient's ABCs.

Allow the patient to get in a position of comfort, if he is responsive and there are no suspected spine injuries.

Perform an ongoing assessment until the incoming EMTs take over patient care. Be sure to report any changes in the patient's condition.

Environmental Emergencies

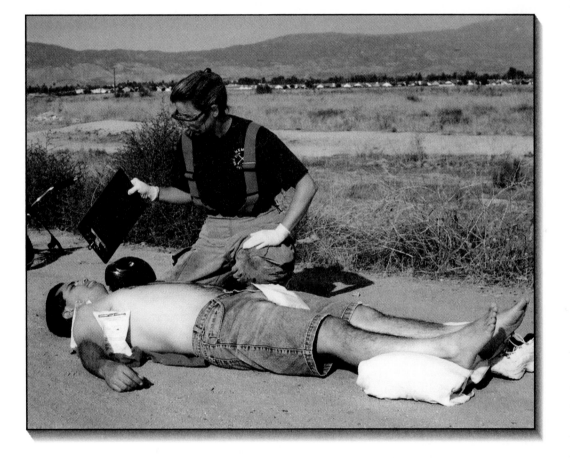

*I*NTRODUCTION   *A heat- or cold-related emergency can happen to anyone, anywhere, and anytime. Firefighters, letter carriers, farmers, police officers, and countless others who work or play outdoors are at risk. Among the people who work and play outdoors, the elderly and very young are especially at risk. While some environmental emergencies—heat cramps or frostbite, for example—are not usually life threatening, others such as heatstroke and hypothermia may prove fatal.*

Cognitive, affective, and psychomotor objectives are from the U.S. DOT's "First Responder: National Standard Curriculum." Enrichment objectives, if any, identify supplemental material.

Cognitive

5-1.7 Identify the patient who presents with a specific medical complaint of exposure to cold. (pp. 230–231, 232)

5-1.8 Explain the steps in providing emergency medical care to a patient with an exposure to cold. (pp. 231–232, 232–234)

5-1.9 Identify the patient who presents with a specific medical complaint of exposure to heat. (pp. 234, 236)

5-1.10 Explain the steps in providing emergency medical care to a patient with an exposure to heat. (p. 237)

Affective

5-1.16 Attend to the feelings of the patient and/or family when dealing with the patient with a specific medical complaint. (p. 232)

5-1.21 Demonstrate a caring attitude towards patients with a specific medical complaint who request emergency medical services. (p. 232)

5-1.22 Place the interests of the patient with a specific medical complaint as the foremost consideration when making any and all patient care decisions. (p. 232)

5-1.23 Communicate with empathy to patients with a specific medical complaint, as well as with family members and friends of the patient. (p. 232)

Psychomotor

5-1.30 Demonstrate the steps in providing emergency medical care to a patient with an exposure to cold. (pp. 231–232, 232–234)

5-1.31 Demonstrate the steps in providing emergency medical care to a patient with an exposure to heat. (pp. 234, 236–237)

Enrichment

◆ Describe the various ways that the body creates and loses heat. (pp. 229–230)

 ON SCENE

DISPATCH

It was a hot, very humid day with the temperature around 90 degrees. Our volunteer fire department had been called for a mutual aid assist to our neighboring department to the south. A large brush fire was threatening several structures, and they requested we respond two units. We donned our wildland firefighting gear and began driving south. It would take us approximately 45 minutes to reach their command post.

SCENE SIZE-UP

We had arrived at the staging area, when we were approached by a woman who identified herself as the staging officer. With concern in her voice, she asked if any of our personnel were certified as First Responders. I replied that we all were. She looked relieved, and asked if we could immediately follow her to the rehabilitation area. She said that one of their new volunteers was complaining of dizziness.

Grabbing our jump kit, we got out of our vehicles and followed her on foot. After a minute or so, we reached the rehab area. There we found a group of concerned firefighters surrounding a man, who appeared to be around 55 years old. He was sitting on the ground, and I noted he was wearing bunker pants and a sweat-soaked T-shirt. Off to his side, I observed a dirty, sweat-soaked bunker coat. The man was conscious, but his head was hanging down. Even so, I could tell his face was very red.

INITIAL ASSESSMENT

My partner positioned himself next to the patient and asked, "How are you doing?"

"I'm OK," he answered, "Feeling a little bit dizzy. That's all. Just give me a minute or two, and I'll be ready to go."

Meanwhile, I found out that the firefighter's name was Bob Moore, and that he had been actively fighting the fire for almost two hours without a rest. His airway, breathing, and circulation all appeared normal. I checked his pulse, and noted that it was strong and rapid. Also, his skin was cool and sweaty.

> *Consider this patient as you read Chapter 12. Does he need emergency care?*

BODY TEMPERATURE

Heat and cold can produce a number of emergencies. To respond to them appropriately, you need a basic understanding of how people adjust to heat and cold.

The body produces and conserves heat mainly through the process of metabolism, including the digestion of food. In cold, the body holds onto its heat by constricting blood vessels near its surface. The body can produce more heat, if needed, by shivering and by producing certain hormones such as epinephrine.

In general, the body loses heat in these different ways (Figure 12-1):

Figure 12-1 Mechanisms of heat loss.

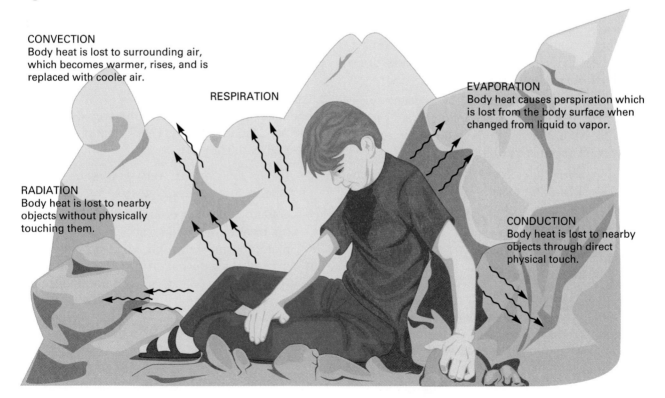

CONVECTION
Body heat is lost to surrounding air, which becomes warmer, rises, and is replaced with cooler air.

RESPIRATION

EVAPORATION
Body heat causes perspiration which is lost from the body surface when changed from liquid to vapor.

RADIATION
Body heat is lost to nearby objects without physically touching them.

CONDUCTION
Body heat is lost to nearby objects through direct physical touch.

WIND SPEED (MPH)	WHAT THE THERMOMETER READS (degrees °F.)											
	50	40	30	20	10	0	−10	−20	−30	−40	−50	−60
	WHAT IT EQUALS IN ITS EFFECT ON EXPOSED FLESH											
CALM	50	40	30	20	10	0	−10	−20	−30	−40	−50	−60
5	48	37	27	16	6	−5	−15	−26	−36	−47	−57	−68
10	40	28	16	4	−9	−21	−33	−46	−58	−70	−83	−95
15	36	22	9	−5	−18	−36	−45	−58	−72	−85	−99	−112
20	32	18	4	−10	−25	−39	−53	−67	−82	−96	−110	−121
25	30	16	0	−15	−29	−44	−59	−74	−88	−104	−118	−133
30	28	13	−2	−18	−33	−48	−63	−79	−94	−109	−125	−140
35	27	11	−4	−20	−35	−49	−67	−82	−98	−113	−129	−145
40	26	10	−6	−21	−37	−53	−69	−85	−100	−116	−132	−148

Source: U.S. Army

Little danger if properly clothed | Danger of freezing exposed flesh | Great danger of freezing exposed flesh

Figure 12-2 Wind-chill index.

◆ *Convection.* This occurs when moving air passes over the body and carries heat away. (See the "wind chill index," Figure 12-2.)

◆ *Conduction.* This occurs when direct contact with an object carries heat away. For example, a swimmer is in direct contact with water. If it is cooler than the body, the water will take away the swimmer's body heat. It can do so 25 times faster than air.

◆ *Radiation.* This method involves the transfer of heat to an object without physical contact. Most loss is from the head and neck, areas rich in blood and blood vessels.

◆ *Evaporation.* The process by which sweat changes to vapor has a cooling effect on the body. Note that it stops when the relative humidity of the air reaches 75%.

◆ *Respiration.* This occurs when a person inhales cold air and exhales air that was warmed inside the body.

FIRE DRILL

When a building is on fire, firefighters must be aware of the way heat can be "lost" to adjacent exposures. Of the five types of heat loss mentioned above, which types apply to firefighting?

COLD-RELATED EMERGENCIES

Exposure to cold can cause two kinds of emergencies. One is a generalized cold-related emergency, or **generalized hypothermia.** It involves an overall reduction of body temperature, which can be deadly. The other kind of emergency is called a **local cold injury,** or damage to body tissues in a specific (local) part of the body.

Generalized Hypothermia

Exposure to extreme cold for a short time or moderate cold for a long time can cause hypothermia. There are several risk factors you should know about. They are:

◆ *Age of the patient.* Very young or very old patients are especially at risk.

◆ *Medical condition of the patient.* Any underlying problem—such as shock, head or spine injury, burns, infection, diabetes, and hypoglycemia—can weaken the body's responses to heat and cold. So can certain medications.

◆ *Drugs, alcohol, and poisons.* These also impede the body's ability to maintain body temperature.

Did you know that many outdoor clubs and organizations discourage or even ban the use of alcohol? The momentary flush of warmth felt after a drink actually increases heat loss. Coupled with impaired judgment, a walk in the woods can quickly turn into a deadly excursion.

PEDIATRIC NOTE

The anatomy of infants puts them at risk for hypothermia. The head is large in proportion to the body. The body surface is large compared to body mass. The result is that infants lose more heat more rapidly than adults do. Infants also have an immature nervous system, which means they cannot shiver well enough to warm themselves when needed.

GERIATRIC NOTE

The elderly, too, are especially at risk. If they are on a fixed income, for example, they may not be able to afford to heat their homes properly. Sudden illness or injury can limit their ability to escape the cold. Impaired judgment due to a medication, or limited mobility due to a medical condition, also contributes to their risk.

Patient Assessment

During scene size-up, notice the location of your patient. Ask yourself these questions: Does the environment suggest the possibility of hypothermia? How long has the patient been exposed to those conditions? If scene size-up suggests the possibility of a cold-related emergency, put your hand on the patient's abdomen during the physical exam. If it is cool or cold, treat for hypothermia.

Note that hypothermia is a progressive condition (Figure 12-3). At first patients will shiver. When shivering stops, they may appear to be clumsy, confused, and forgetful. They may even appear to be intoxicated. Often witnesses will say that patients had mood swings, one moment calm and the next animated or even combative.

The hypothermic patient's level of responsiveness will decrease, too. They become less communicative and more difficult to rouse. They may display poor judgment and do things like remove clothing while still out in the cold. There may

be muscle stiffness, a rigid posture, and loss of sensation. The most ominous sign of a life-threatening condition is unresponsiveness. These patients are unstable and need immediate transport if they are to survive. Signs and symptoms of hypothermia are summarized in Figure 12-4.

If the environment is cold, remember that your sense of touch may be less than it should be. Consider putting your fingers in your armpits or your groin before taking a pulse. Note that when assessing circulation, you may find no pulses in the patient's limbs. When the body is in severe hypothermia, it is a metabolic icebox. It does not need normal circulation to sustain life, because everything is slowed down.

Because hypothermic patients may have extremely slow pulse rates, assess the patient's carotid pulse for 30-45 seconds before starting CPR. If your patient is breathless and pulseless, begin CPR. The rule of thumb in EMS is: "you're not dead until you're warm and dead." That means, you cannot consider resuscitation a failure until the heart has been given a chance to restart at a near-normal temperature. So even a patient who is stiff or rigid and has no apparent pulse should be given CPR and immediate transport to a hospital.

Remember to handle the patient gently. Any rough handling of a severely hypothermic patient can induce ventricular fibrillation or sudden cardiac death.

◆ Emergency Care

Provide the following emergency care for all patients suspected of having generalized hypothermia (Figure 12-5):

1. **Remove the patient from the cold environment.** Move to a shelter, away from the cold wind or water. If the patient is on the ground, get him off or put a blanket between him and the ground.

2. **Handle the patient very gently.** Rough handling can make the patient's condition worse and even cause further harm.

3. **Administer high-flow oxygen,** if you are allowed to do so. If possible, it should be warm and humidified.

4. **Use techniques of passive rewarming,** if the patient is unresponsive or does not respond appropriately. These techniques are: remove any cold, wet, or restrictive clothing;

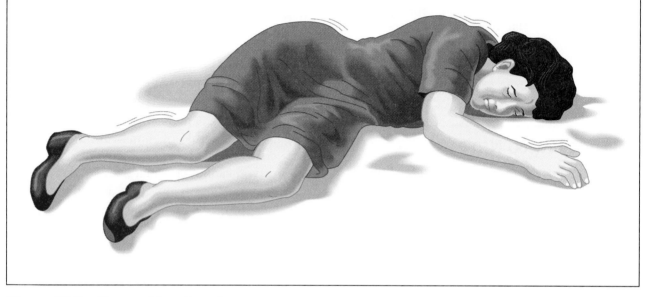

STAGES OF HYPOTHERMIA (Cold-Related Injury)

Stage 1: **Shivering** is a response by the body to generate heat. It does not occur below a body temperature of 90°F.

Stage 2: **Apathy and decreased muscle function**. First fine motor function is affected, then gross motor functions.

Stage 3: **Decreased level of responsiveness** is accompanied by a glassy stare and possible freezing of the extremities.

Stage 4: **Decreased vital signs**, including slow pulse and slow respiration rate.

Stage 5: **Death**.

Figure 12-3   Stages of hypothermia.

apply blankets; turn the heat up in your location, if possible.

5. **Comfort, calm, and reassure the patient.** Tell him that everything that can be done will be done. Communicate with empathy.

Keep in mind that in all aspects of emergency care, you must be extremely gentle with the patient to avoid further injury. In addition, do not massage the patient's extremities. Do not allow the patient to walk or exert himself. Do not allow the patient to eat or drink stimulants.

Local Cold Injuries
Patient Assessment

Frostbite, or local cold injury, is the freezing or near freezing of a body part. The toes, fingers, face, nose, and ears are most at risk (Figure 12-6).

Frostbitten areas are usually easy to identify. With early or superficial frostbite, light skin will redden. Dark skin will turn pale. When the skin is depressed gently, it will blanch and then return to its normal color. The patient often will complain of loss of feeling and sensation in the injured area.

In the later stages of frostbite (called "late or deep cold injury"), the skin may appear to be pale and waxy. Upon palpation, the skin may feel hard, like wood. Blisters or local swelling may also be seen. In the most severe cases, tissue as deep as the muscles and bones may be frozen. As deep injuries begin to thaw, the affected skin may show a purple-blue or blotchy, spotted, or mottled color.

◆ Emergency Care

Note that if you suspect hypothermia, treat it before you care for a frostbitten extremity. Emergency care for a local cold injury is as follows:

1. **Remove the patient from the cold environment.** Do not allow the patient to walk on a frostbitten limb.

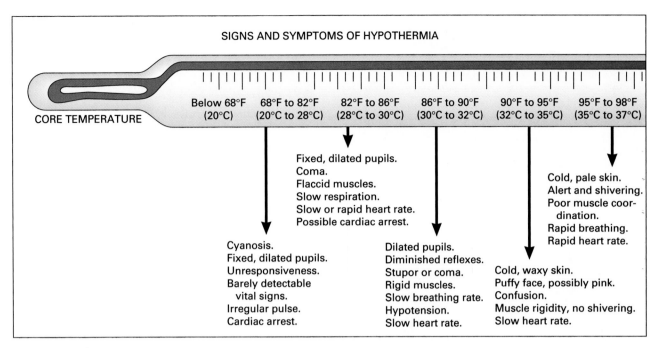

SIGNS AND SYMPTOMS OF HYPOTHERMIA

CORE TEMPERATURE

| Below 68°F (20°C) | 68°F to 82°F (20°C to 28°C) | 82°F to 86°F (28°C to 30°C) | 86°F to 90°F (30°C to 32°C) | 90°F to 95°F (32°C to 35°C) | 95°F to 98°F (35°C to 37°C) |

Fixed, dilated pupils.
Coma.
Flaccid muscles.
Slow respiration.
Slow or rapid heart rate.
Possible cardiac arrest.

Cold, pale skin.
Alert and shivering.
Poor muscle coor-
   dination.
Rapid breathing.
Rapid heart rate.

Cyanosis.
Fixed, dilated pupils.
Unresponsiveness.
Barely detectable
   vital signs.
Irregular pulse.
Cardiac arrest.

Dilated pupils.
Diminished reflexes.
Stupor or coma.
Rigid muscles.
Slow breathing rate.
Hypotension.
Slow heart rate.

Cold, waxy skin.
Puffy face, possibly pink.
Confusion.
Muscle rigidity, no shivering.
Slow heart rate.

Figure 12-4   Signs and symptoms of hypothermia.

2. Administer oxygen, if appropriate, and if you are allowed to do so.

3. Remove all wet clothing.

4. Protect the frostbite area from further injury. If the injury is to an extremity, manually stabilize it. If the injury is superficial, cover it with a blanket. If it is late and deep, cover it with a dry cloth or dressings. Do not rub or massage the area. Ice crystals under the skin could damage the fragile capillaries and tissues, making the injury worse.

5. Comfort, calm, and reassure the patient. Tell him everything that can be done will be done.

6. Monitor the patient for signs of hypothermia.

Rewarming

If transport will be delayed, consider rewarming the affected area. Check with medical direction or follow local protocol for directions. *Never rewarm an area with late or deep frostbite.* Never rewarm an area if there is a chance that it may refreeze. The injury from the second freezing would be much worse than the original one.

First, carefully remove all rings, bracelets, and so on. Then, warm the entire frostbite area in warm water. Be sure to pick a container that permits the entire area to be immersed (Figure 12-7). The ideal water temperature for rewarming is 108°F (42°C). The frozen body part will quickly cool the water, so monitor water temperature closely. Add warm water as necessary to keep the bath as close to 108°F (42°C) as possible. Continue until the body part softens and color and sensation begin to return. (This usually takes 20 or 30 minutes.)

Figure 12-5   Protect a patient with generalized hypothermia from further injury from the cold.

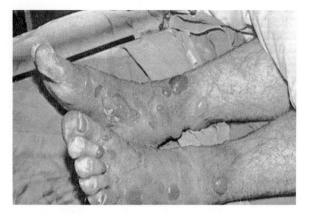

Figure 12-6a   Frostbite, or local cold injury.

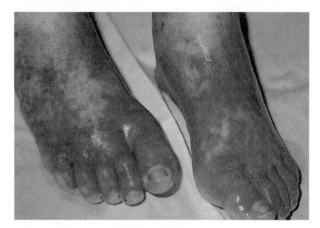

Figure 12-6b   Late or deep frostbite.

Once the affected part is thawed, apply dry sterile dressings to the area. If the hands or feet have been frozen, place sterile dressings between the fingers or toes. As thawing occurs and blood flow returns to the affected part, the patient may experience severe pain. Be prepared to comfort and reassure the patient.

Figure 12-7   Rewarming a local cold injury.

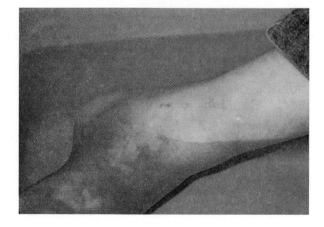

HEAT-RELATED EMERGENCIES

When a person cannot lose excessive heat, he or she may experience a heat-related emergency. The three most common heat-related emergencies you are likely to encounter are "heat cramps," "heat exhaustion," and "heat stroke." Heat stroke, the most serious, is life threatening.

Contributing Factors

Factors that contribute to the risk of a heat-related emergency include the following:

◆ *Heat and humidity.* High air temperature can reduce the body's ability to lose heat by radiation. High humidity can reduce its ability to lose heat by way of evaporation. (See "Heat and Humidity Risk Scale," Figure 12-8.)

◆ *Exercise and strenuous activity.* Each can cause a person to lose more than one liter of perspiration (fluid and essential salts) per hour.

◆ *Age of the patient.* Very young and very old patients may be unable to respond to overheating effectively.

◆ *Medical condition of the patient.* Any number of conditions, such as heart or lung disease, diabetes, dehydration (fluid loss), obesity, fever, and fatigue can inhibit heat loss.

◆ *Certain drugs and medications.* Alcohol, cocaine, barbiturates, hallucinogens, certain prescription drugs taken by psychiatric patients, and others can affect heat loss in many ways, including through side effects such as dehydration (fluid loss).

Patient Assessment

The general signs and symptoms of a heat-related emergency include the following:

◆ Muscle cramps.

◆ Weakness, exhaustion.

◆ Dizziness, faintness.

◆ Rapid pulse rate that is strong at first, but becomes weak as damage progresses.

◆ Headache.

◆ Seizures.

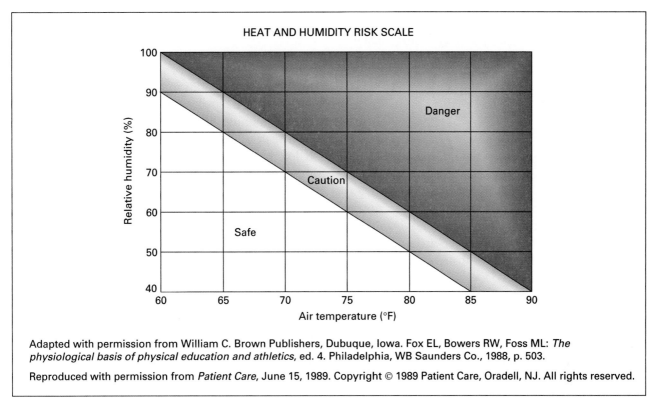

HEAT AND HUMIDITY RISK SCALE

Figure 12-8   Heat and humidity risk scale.

Figure 12-9   Heat cramps are the most common but least serious heat-related emergency.

SIGNS AND SYMPTOMS OF HEAT CRAMPS

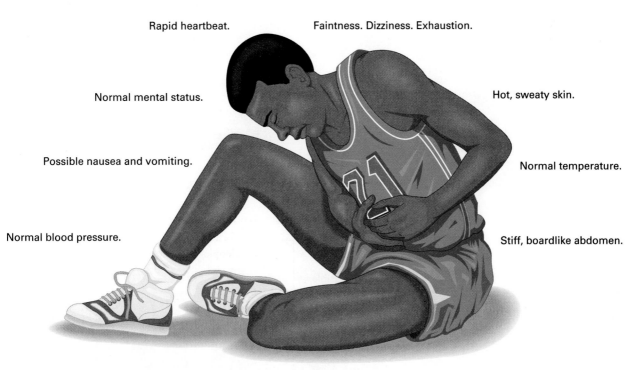

Severe muscular cramps and pain, especially of the arms, fingers, legs, calves, and abdomen.

◆ Loss of appetite, nausea, vomiting.

◆ Altered mental status, possibly unresponsiveness.

◆ Skin may be moist, pale, and normal-to-cool in temperature ("heat cramps" or "heat exhaustion"). Or it may be hot and dry or hot and moist ("heat stroke").

Heat cramps involve acute spasms of the muscles of the legs, arms, or abdomen (Figure 12-9). This may be the result of losing too much salt during profuse sweating. Heat cramps usually follow hard work in a hot environment. Hard work in a hot, humid environment also can cause loss of fluids through sweat. This can result in a mild state of shock, or "heat exhaustion."

In some cases, if the patient does not stop work, move to a cool environment, and replace lost fluid, his condition will get worse. The result can be "heat stroke," which is very serious and life threatening. It occurs when the body becomes overheated and, in many patients, sweating stops. If left untreated, brain cells begin to die, causing permanent disability or death. (See Figures 12-10 and 12-11.)

Feel the abdomen to check the body temperature of a patient with a heat-related emergency. Remember that the chief characteristic of heat stroke is hot skin.

Figure 12-10 Heat stroke is a life-threatening emergency.

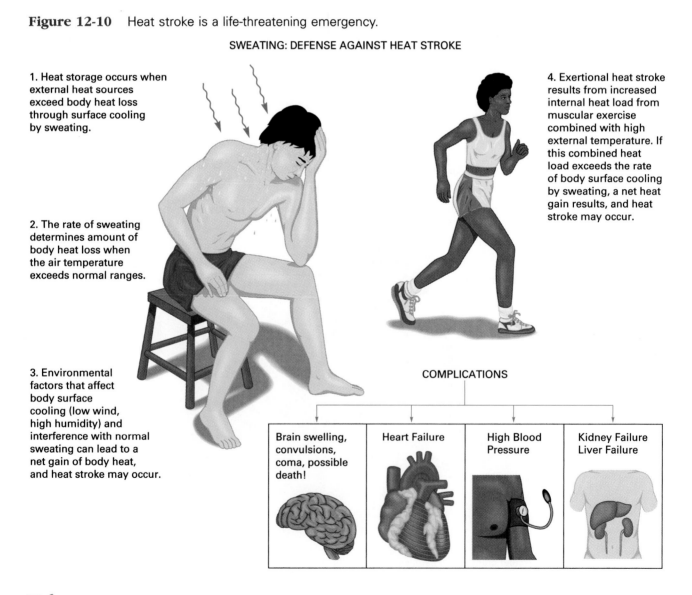

SWEATING: DEFENSE AGAINST HEAT STROKE

1. Heat storage occurs when external heat sources exceed body heat loss through surface cooling by sweating.

2. The rate of sweating determines amount of body heat loss when the air temperature exceeds normal ranges.

3. Environmental factors that affect body surface cooling (low wind, high humidity) and interference with normal sweating can lead to a net gain of body heat, and heat stroke may occur.

4. Exertional heat stroke results from increased internal heat load from muscular exercise combined with high external temperature. If this combined heat load exceeds the rate of body surface cooling by sweating, a net heat gain results, and heat stroke may occur.

COMPLICATIONS

Brain swelling, convulsions, coma, possible death!

Heart Failure

High Blood Pressure

Kidney Failure Liver Failure

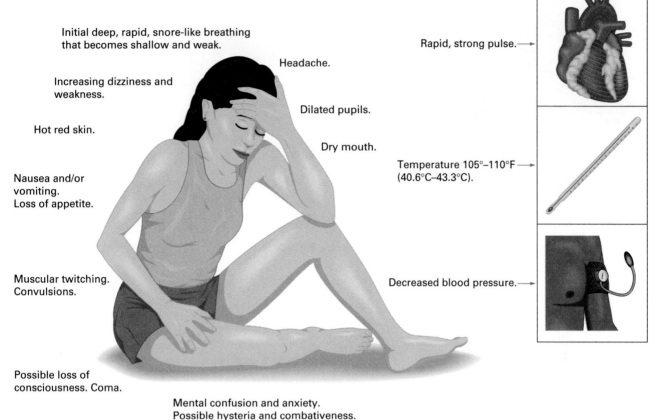

Initial deep, rapid, snore-like breathing that becomes shallow and weak.

Headache.

Increasing dizziness and weakness.

Dilated pupils.

Hot red skin.

Dry mouth.

Nausea and/or vomiting.
Loss of appetite.

Rapid, strong pulse.

Temperature 105°–110°F (40.6°C–43.3°C).

Muscular twitching.
Convulsions.

Decreased blood pressure.

Possible loss of consciousness. Coma.

Mental confusion and anxiety.
Possible hysteria and combativeness.
Delirium.

Figure 12-11   Signs and symptoms of heat stroke.

◆ EMERGENCY CARE

For a patient with moist, pale, and normal-to-cool skin temperature, provide emergency care as follows (Figure 12-12):

1. **Remove the patient from the hot environment.** Place him in a cool one if possible. If the source of heat is the sun, place the patient in the shade.

2. **Administer oxygen,** if you are allowed to do so.

3. **Cool the patient.** Loosen or remove clothing. Then fan the surface of his body while applying a light mist of water. Be careful not to cool the patient so fast that he becomes chilled.

4. **Position the patient.** Place him in a supine position with legs elevated 8 to 12 inches (20 to 30 cm).

5. **Monitor the patient.** Take vital signs frequently. Advise incoming units or other EMS personnel on scene if the patient develops signs of shock.

If the patient is responsive and not nauseated, encourage him to drink about one-half glass of cool water every 15 minutes or so. Follow local protocol or consult medical direction.

A patient with hot and moist skin or hot and dry skin must be removed from the hot environment. Administer oxygen if possible. If the patient's breathing becomes shallow, assist it with a bag-valve-mask. Cool the patient with hot skin as follows (Figure 12-13):

1. **Loosen or remove clothing.**

2. **Apply cold packs** to neck, armpits, and groin.

3. **Keep the skin wet.** Apply water with wet towels or a sponge.

4. **Fan aggressively.** An effective way is to direct an electric fan over the patient's body while you wet the skin.

5. **Monitor the patient** as appropriate during your ongoing assessment.

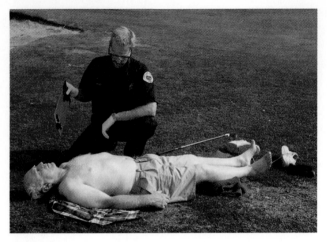

Figure 12-12 Cooling a patient with normal-to-cool skin temperature.

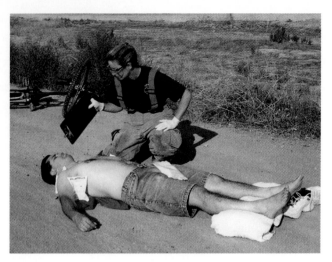

Figure 12-13 Cooling a patient with hot skin temperature.

ON SCENE FOLLOW-UP

At the beginning of this chapter, you read that First Responders were on scene with a firefighter, who was complaining of dizziness. To see how chapter skills apply to this emergency, read the following. It describes how the call was completed.

PHYSICAL EXAMINATION

I performed a complete physical exam on Firefighter Moore. I noted that his skin was cool and sweaty, and I confirmed that his pulse was a bit high—132/minute. Bob also told me that he was thirsty, and was wondering if he could have a drink.

PATIENT HISTORY

Bob told us that he was healthy. He denied taking any medications, but did tell us that he was allergic to penicillin. He also said he had not eaten anything for about eight hours. His engine company captain told us that Bob had been fighting this fire since early that morning. Being new in the department, he had not yet received his protective clothing for wildland firefighting because of a problem with the order. He had been wearing his structural firefighting gear instead.

ONGOING ASSESSMENT

We took a cool, moist towel and placed it on the back of Bob's neck. We also helped him out of his bunker pants, and allowed him to drink another glass of water. Rechecking his vital signs, we noted that his pulse was down to 104. He also told us that "the dizziness was gone." He asked if he could go back and continue fighting the fire. His captain told him that he was going to have an ambulance take him to the hospital to get him checked out first.

PATIENT HAND-OFF

When the paramedics arrived, I told them:

"This is Firefighter Bob Moore. He had been fighting a brushfire for about two hours with no rest when he began to feel dizzy. He was wearing his regular structural firefighting gear, not the lighter wildland firefighting clothing. He has been alert and oriented, with cool and clammy skin. His pulse was 132, but is now down to 104. His BP and breathing have been normal. We allowed him to drink two 16-ounce glasses of water, which he was able to keep down. We removed his heavy bunker pants and also placed a cool, moist towel around his neck to help cool him off. He says he is now "ready to go back and continue fighting the fire."

Later that week, I ran into Firefighter Moore's captain. He said that Bob had suffered from heat exhaustion. They placed an IV in the ambulance and gave him some fluids. He was released from the hospital emergency room two hours later.

Heat- and cold-related emergencies often occur in isolated areas to such people as campers, hikers, skiers, and, yes, firefighters! They also occur in our own neighborhoods. Always consider the environmental conditions as soon as you get your call. Early recognition and the appropriate emergency care can save a life.

Chapter Review

Conditions resulting from extremes in heat and cold are common in most parts of the country. Even though many people are aware of the effects of these temperatures, they still can fall victim. Very young and old patients are at increased risk for heat- and cold-related emergencies. Because their temperature control systems are not working optimally, it would not take bitter cold or extreme heat to cause an emergency. Patients who are exposed to moderate temperatures for long periods of time also may be stricken. Patients with wet clothes in the cold, and patients in a hot environment who do not maintain their fluid levels, will be overcome quickly. Patients may even suffer from these conditions indoors.

The most important concept in this chapter is that the fire service First Responder can provide immediate life-saving help. Remove patients from extremes in temperature and begin to treat them. Be sure you do not become stricken by the same conditions yourself.

FIRE COMPANY REVIEW

Page references where answers may be found or supported are provided at the end of each question.

Section 1

1. What are the five major mechanisms of heat loss? (pp. 229–230)

Section 2

2. What factors contribute to the possibility that a patient may be at risk for hypothermia? (p. 230)

3. What are the signs and symptoms of hypothermia? (pp. 231–232)

4. How can you minimize the injury your patient experiences from hypothermia? (pp. 231–232)

5. If you find a hypothermic patient breathless and pulseless, and with signs of death, should you begin CPR? Explain your answer. (pp. 231–232)

6. How can you identify a local cold injury? (p. 232)

7. What is the basic emergency care of a local cold injury? (pp. 232–233)

Section 3

8. What factors contribute to the possibility that a patient may be at risk for a heat-related emergency? (p. 234)

9. What are the signs and symptoms of a heat-related emergency? (pp. 234, 236)

10. What is the appropriate emergency care for a patient with a heat-related emergency? (p. 237)

RESOURCES TO LEARN MORE

Auerbach, P.S., ed. *Management of Wilderness and Environmental Emergencies,* Third Edition. St. Louis: C.V. Mosby, 1996.

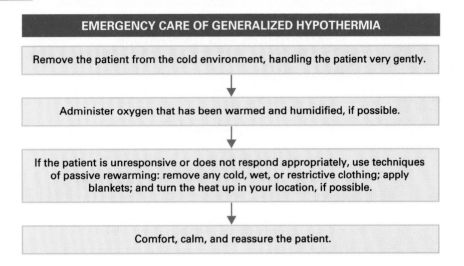

EMERGENCY CARE OF GENERALIZED HYPOTHERMIA

Remove the patient from the cold environment, handling the patient very gently.

↓

Administer oxygen that has been warmed and humidified, if possible.

↓

If the patient is unresponsive or does not respond appropriately, use techniques of passive rewarming: remove any cold, wet, or restrictive clothing; apply blankets; and turn the heat up in your location, if possible.

↓

Comfort, calm, and reassure the patient.

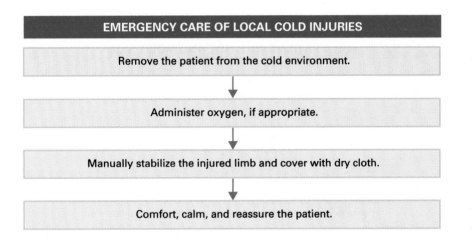

EMERGENCY CARE OF LOCAL COLD INJURIES

Remove the patient from the cold environment.

↓

Administer oxygen, if appropriate.

↓

Manually stabilize the injured limb and cover with dry cloth.

↓

Comfort, calm, and reassure the patient.

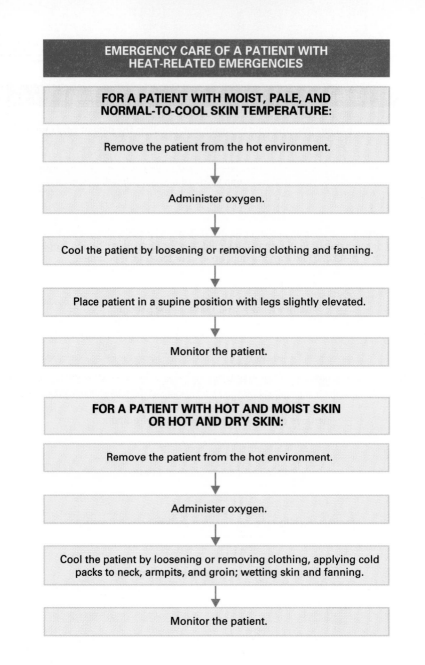

EMERGENCY CARE OF A PATIENT WITH HEAT-RELATED EMERGENCIES

FOR A PATIENT WITH MOIST, PALE, AND NORMAL-TO-COOL SKIN TEMPERATURE:

Remove the patient from the hot environment.

Administer oxygen.

Cool the patient by loosening or removing clothing and fanning.

Place patient in a supine position with legs slightly elevated.

Monitor the patient.

FOR A PATIENT WITH HOT AND MOIST SKIN OR HOT AND DRY SKIN:

Remove the patient from the hot environment.

Administer oxygen.

Cool the patient by loosening or removing clothing, applying cold packs to neck, armpits, and groin; wetting skin and fanning.

Monitor the patient.

Psychological Emergencies and Crisis Intervention

*I*NTRODUCTION   *When providing emergency care for injury or illness, you can actually see the effects of that injury or illness. Often times, you also may see the benefit your treatment has on your patient. Emergency care for behavioral emergencies is different. You cannot easily see the comfort that your words or your presence provides to someone who is emotionally ill or injured. However, your kind words and compassionate actions may heal or even save more lives than any physical care you may provide.*

Cognitive, affective, and psychomotor objectives are from the U.S. DOT's "First Responder: National Standard Curriculum." Enrichment objectives, if any, identify supplemental material.

Cognitive

5-1.11 Identify the patient who presents with a specific medical complaint of behavioral change. (pp. 245–248)

5-1.12 Explain the steps in providing emergency medical care to a patient with a behavioral change. (pp. 246–249)

5-1.13 Identify the patient who presents with a specific complaint of a psychological crisis. (pp. 245–246)

5-1.14 Explain the steps in providing emergency medical care to a patient with a psychological crisis. (pp. 245–249)

Affective

5-1.17 Explain the rationale for modifying your behavior toward the patient with a behavioral emergency. (pp. 245–246)

5-1.24 Demonstrate a caring attitude towards patients with a behavioral problem who request emergency medical services. (p. 246)

5-1.25 Place the interests of the patient with a behavioral problem as the foremost consideration when making any and all patient care decisions. (p. 247)

5-1.26 Communicate with empathy to patients with a behavioral problem, as well as with family members and friends of the patient. (pp. 246–247)

Psychomotor

5-1.32 Demonstrate the steps in providing emergency medical care to a patient with a behavioral change. (pp. 246–249)

5-1.33 Demonstrate the steps in providing emergency medical care to a patient with a psychological crisis. (pp. 246–249)

Enrichment

◆ Discuss the guidelines for restraining patients with a behavioral emergency. (pp. 247–249)

◆ List the legal considerations involved in providing emergency care to a patient with a behavioral emergency. (p. 249)

◆ Identify the signs and symptoms of a patient with a drug or alcohol emergency. (pp. 249–251)

◆ Describe management of a patient with a drug or alcohol emergency. (pp. 251–252)

◆ Discuss the four general stages of rape trauma syndrome. (p. 253)

◆ Describe the proper management of a rape scene. (p. 253)

 *ON SCENE*

DISPATCH

I was driving home after a training exercise with our volunteer fire department, when I saw a naked man running along the road. I noticed that he stopped every few hundred feet or so to throw punches in the air.

SCENE SIZE-UP

I realized this person might be dangerous. So I dialed 9-1-1 using the cell phone in my car. I remained a safe distance from the person until police and the fire department arrived on scene.

INITIAL ASSESSMENT

The man had to be restrained by the police. Afterwards, I offered assistance to the fire service First Responders, who had been careful not to restrict the man's breathing. We covered him with a blanket to keep him warm. All the while, he was screaming at the police. So I knew he had adequate breathing. There also were no indications of external bleeding.

My general impression was that of a patient with an altered mental status, possibly from a psychiatric emergency, alcohol, or drugs. We called for an ambulance.

> *Consider this patient as you read Chapter 13. What else might be done to assess or provide emergency care to this patient?*

1 BEHAVIORAL EMERGENCIES

Behavior is the manner in which a person acts or performs. A **behavioral emergency** is a situation in which a patient exhibits "abnormal" behavior, or behavior that is unacceptable or intolerable to the patient, family, or community. Such an emergency may be due to extremes of emotion, a psychological condition such as a mental illness, or even a physical condition such as lack of oxygen or low blood sugar.

A number of factors can cause a change in a patient's behavior. They include:

- Situational stresses, such as the death of spouse or a child.
- Illness or injury, including head trauma, lack of oxygen, inadequate blood flow to the brain, low blood sugar in a person with diabetes, or excessive heat or cold.
- Mind-altering substances, such as alcohol, depressants, stimulants, psychedelics, and narcotics.
- Psychiatric problems, such as phobias (irrational fears of specific things), depression, paranoia, or schizophrenia.
- Psychological crises, such as panic and bizarre thinking.

Patients with behavioral emergencies may act in unusual and unexpected ways. They can pose a danger to themselves through suicide or self-inflicted injuries. They also can pose a danger to others through violence or actions they are not able to understand.

Patient Assessment

Consider the need for law enforcement during your scene size-up and throughout a call. If you suspect that the patient is a threat to himself or others, arrange for backup law enforcement at the scene. Remember, your own safety is of the utmost importance.

The following guidelines may help you determine if your patient is likely to become violent:

- During scene size-up, look around. Locate the patient before approaching (Figure 13-1). Check to see if there are any weapons or items that could be used as weapons such as a knife or blunt object. If there are, assume that he may use them. Overturned furniture or other signs of chaos also can indicate violent behavior.
- If family members, friends, or bystanders are at the scene, ask them if the patient has a history of being aggressive or combative. Also find out if the patient has been violent or has threatened violence at the scene.
- Expect violence if the patient is standing or sitting in a way that threatens anyone (includ-

Figure 13-1   Locate the patient before approaching. Check to see if there are any weapons on scene.

ing himself). Clenched fists, even when he is holding them at his side, may be a sign.

◆ Listen to the patient. Expect violence if he is yelling, cursing, arguing, or verbally threatening to hurt himself or others.

◆ Signs of possible violence in a patient include moving toward you, carrying a heavy or threatening object, making quick or irregular movements, and muscle tension.

◆ Plan an escape route. Never allow yourself to get cornered by your patient or bystanders.

Keep the following basic principles in mind whenever you are on the scene of a behavioral emergency:

◆ Identify yourself. Let the patient know you are there to help.

◆ Inform the patient of exactly what you are doing. Uncertainty will make the patient more anxious and fearful.

◆ Ask questions in a calm, reassuring voice. Speak directly to the patient. Stay polite, use good manners, and show respect. Make no unsupported assumptions.

◆ Without being judgmental, allow the patient to tell you what happened.

◆ Show you are listening by rephrasing or repeating part of what is said. Ask questions to show you are paying attention. Also use gestures such as a nod of the head or verbal responses such as "I see" or "Go on."

◆ Acknowledge the patient's feelings. Use phrases like, "I can see that you are very depressed" or "I am not surprised that you feel frightened."

◆ Assess the patient's mental status by asking specific questions. Try to determine whether or not the patient is oriented to time, person, and place. Watch the patient's appearance, level of activity, and speech patterns.

◆ EMERGENCY CARE

While caring for a patient who has a behavioral emergency, be sure to comfort, calm, and reassure the patient as you proceed. And never leave the patient alone. All such patients are escape risks, and violence is a real possibility. Once you have responded to a behavioral emergency, the patient's safety is legally your responsibility until

someone with more training arrives on scene. Even if the patient pleads to be alone for just a few minutes, do not do it. Firmly explain that you could get in trouble if you did.

If you suspect the patient may have overdosed, provide emergency medical care as described in Chapter 11 for a poisoning patient. Give any medications or drugs you find on scene to the transporting EMS personnel.

Methods to Calm Patients

The situations presented by behaviorally disturbed patients are often difficult. However, a number of techniques can help:

◆ Acknowledge that the patient seems upset, and restate that you are there to help.

◆ Inform the patient of exactly who you are and what you are going to do to help (Figure 13-2).

◆ Ask questions in a calm, reassuring voice. Speak directly to the patient.

Figure 13-2 Explain who you are and that you are trying to help.

- Maintain a comfortable distance between you and the patient. Many patients are threatened by physical contact. Unwanted touching could set off a violent response. After you have established some rapport with the patient, get his or her permission before moving in any closer.

- Encourage the patient to tell you what the trouble might be. Ask the patient to explain the problem.

- Never assume that it is impossible to communicate with the patient until you have tried, even if others insist it cannot be done.

- Do not make any quick movements. Act quietly and slowly. Let the patient see that you are not going to make any sudden moves.

- Respond honestly to the patient's questions. Instead of saying, for example, "You have nothing to worry about," say something like "Even with all your problems, you seem to have lots of people around who really care about you."

- Never threaten, challenge, belittle, or argue with disturbed patients. Remember that the patient is ill. His or her comments are not about you personally.

- Always tell the truth. Never lie to a patient.

- Do not "play along" with a patient's visual or auditory disturbances. Instead, reassure the patient that they are temporary and can clear up with treatment.

- Involve the patient's family members or friends when you can. Some patients are calmed and reassured by their presence. However, others may be upset or embarrassed. Let the patient decide.

- Be prepared to stay on scene for a long time.

- Avoid unnecessary physical contact. Enlist the help of law enforcement if you are unable to maintain control on your own.

- Maintain good eye contact with the patient. It communicates your control and confidence. Also, the patient's eyes can reflect his or her emotions. They may tell you if the patient is terrified, confused, struggling, or in pain. The eyes can telegraph intentions, too. If a patient is about to reach for a weapon or make a dash, the eyes may alert you.

- If it is safe to do so, get on the same eye level as the patient. When the rescuer towers over a patient, the patient may feel intimidated.

Just as with any other patient, place the interests of the patient with a behavioral problem first in all patient care decisions. Communicate with empathy with the patient, as well as with his or her family and friends.

COMPANY OFFICER'S NOTE

Make sure your personnel know and understand the difference between "listening with empathy" and "listening with sympathy." Empathy refers to listening with an intent to understand. It is listening with your eyes and heart, as well as your ears. Listening with empathy means you do not necessarily agree but you do understand the patient's feelings. Sympathetic listening, on the other hand, is a form of agreement. It may cause patients to believe that what they are doing is correct.

Restraining Patients

Avoid restraining patients unless one poses a danger to himself (or herself) or others. Before restraining a patient, involve your chain of command, seek medical direction, and follow local protocols. For an example of how to restrain a patient, see Figure 13-3.

Remember, never inflict pain or use unnecessary force on a patient. Use only as much force as needed for restraint. What is reasonable force? That depends on the amount of force needed to keep a patient from injuring himself or someone else. It depends on:

- *Size and strength of the patient.* What may seem reasonable force on a 275-pound athlete may not be reasonable on a 150-pound homemaker.

- *Type of abnormal behavior.* You would not expect to use the same kind of force against a frightened patient who is huddling quietly in a corner as you would against an angry patient who is loudly threatening to kill you.

- *Mental state of the patient.* It may be reasonable to use more force on a patient who is

Assisting in Restraining a Patient

Figure 13-3a If asked to assist other EMS providers, stay beyond the range of the patient's arms and legs.

Figure 13-3b If restraints are needed, work in conjunction with an adequate number of other EMS providers.

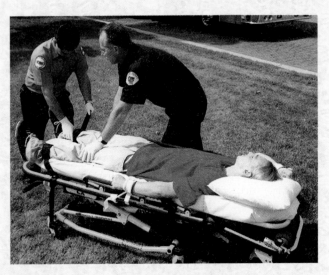

Figure 13-3c You also may be asked to assist the EMTs when they apply ankle and wrist restraints.

loud and threatening than on a patient who is quiet and subdued.

◆ *Method of restraint.* Soft leather or cloth straps are called "humane restraints." They are generally considered reasonable. Metal cuffs are not.

Law enforcement personnel should be involved when you need to restrain a patient, when you need to give care without consent, and when there is any threat of violence. The police can help protect you from injury, and they can serve as credible witnesses if a legal case arises.

The best way to protect yourself against false accusations by a patient is to carefully and completely document everything that occurs during the call. Document your actions and the behavior that warranted the restraint. In most areas, anything you document during the call is legally admissible evidence. Anything that is not documented is considered hearsay (not legally admissible).

Another source of protection is witnesses, preferably throughout the entire course of treatment. It is common for emotionally disturbed patients to accuse medical personnel of sexual misconduct. To protect yourself against such allegations:

◆ Involve other EMS responders who can testify that there was no misconduct.

◆ Include EMS responders who are the same gender as the patient.

◆ Involve third-party witnesses whenever possible.

Legal Considerations

Your legal problems are greatly reduced if an emotionally disturbed patient consents to care. However, such patients commonly refuse treatment, especially intoxicated or overdose patients. They may even threaten you or others. Unless the patient is considered mentally incompetent, legally he or she must provide consent before you can treat. Remember, the patient—not concerned family members—must consent to care.

Generally, you may provide care against a patient's will only if the patient threatens to hurt himself or others and only if you can demonstrate reason to believe that the patient's threats are real. A good rule of thumb to follow is to consult with medical direction and involve law enforcement.

DRUG AND ALCOHOL EMERGENCIES

Drug abuse is the self-administration of one or more drugs in a way that is not in accord with approved medical or social practice. An **overdose** is an emergency that involves poisoning by drugs or alcohol. The term **withdrawal** refers to the effects on the body that occur after a period of abstinence from the drugs or alcohol to which the body has become accustomed. Note that withdrawal—especially from alcohol—can be as serious as an overdose emergency.

Most drug overdoses involve drug abuse by long-time drug users. However, a drug overdose also can be the result of miscalculation, confusion, use of more than one drug, or a suicide attempt.

Various drugs can cause changes in respiration, heart rate, blood pressure, and central nervous system function. (See Figure 13-4 for examples of illegal drugs.) In addition, several major medical problems can result from a drug or alcohol overdose or from sudden withdrawal (Figure 13-5). Among them are respiratory problems, seizures, and cardiac arrest.

Patient Assessment

Signs and symptoms that indicate a life-threatening emergency include (Figure 13-6):

◆ Unresponsiveness.

◆ Breathing difficulties or inability to maintain an open airway.

Figure 13-4 Examples of illegal drugs. *(Source: Michael A. Gallitelli)*

Effects of Alcohol and Drug Abuse

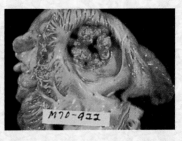

Figure 13-5a Fungal-damaged heart related to drug injections.

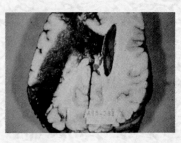

Figure 13-5b Bullet wound to the brain, alcohol-related.

Figure 13-5c Chronic gastric ulcer from alcohol abuse.

Figure 13-5d Alcoholic cirrhosis of the liver.

Figure 13-5e Enlarged weak heart, alcohol-related.

Figure 13-5f Ruptured vein in esophagus, alcohol-induced.

◆ Abnormal or irregular pulse.

◆ Fever.

◆ Vomiting with an altered mental status or without a gag reflex.

◆ Seizures.

If these signs and symptoms are present, your patient is a high priority for transport. Report to dispatch immediately. Additional signs and symptoms will vary widely. They may include:

◆ Altered mental status.

◆ Extremely low or high blood pressure.

◆ Sweating, tremors, and hallucinations (with alcohol withdrawal).

◆ Digestive problems, including abdominal pain and bleeding.

◆ Visual disturbances, slurred speech, uncoordinated muscle movement.

◆ Disinterested behavior, loss of memory.

◆ Combativeness.

◆ Paranoia.

If your patient is unresponsive and you suspect a drug or alcohol emergency, after your initial assessment proceed with the following:

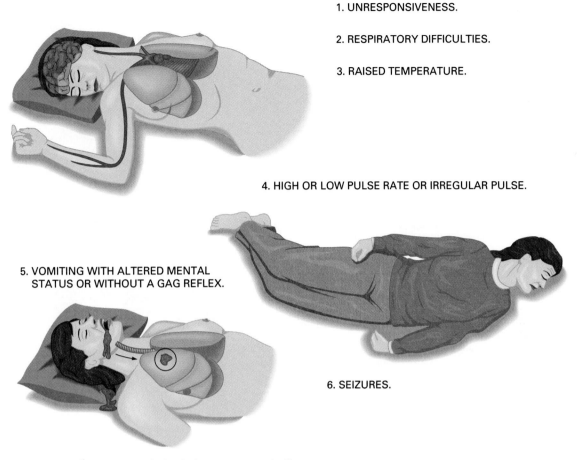

1. UNRESPONSIVENESS.

2. RESPIRATORY DIFFICULTIES.

3. RAISED TEMPERATURE.

4. HIGH OR LOW PULSE RATE OR IRREGULAR PULSE.

5. VOMITING WITH ALTERED MENTAL STATUS OR WITHOUT A GAG REFLEX.

6. SEIZURES.

Figure 13-6   Drug and alcohol emergency indicators.

◆ Check the patient's mouth for partially dissolved pills or tablets. If you find any, remove them so they cannot block the patient's airway or be swallowed.

◆ Smell the patient's breath for traces of alcohol. Do not confuse the smell of alcohol with a musky, fruity, or acetone odor (Figure 13-7). Those three can indicate an emergency related to diabetes.

◆ Ask the patient's friends or family members what they know about the incident.

◆ Continue to examine the scene for evidence of drugs or alcohol use, such as vials, bottles, and so on.

Because signs and symptoms vary so widely and are so similar to many medical conditions, the most reliable indications of a drug- or alcohol-related emergency are likely to come from the scene and the patient history.

◆ EMERGENCY CARE

Your immediate goals are to protect your own safety, maintain the patient's airway, and manage life-threatening conditions. If you believe the patient has overdosed, follow emergency care directions offered below and in Chapter 11 for poisoning.

After assuring scene safety and taking BSI precautions, follow these steps:

1. **Establish and maintain an open airway.** Remove anything from the patient's throat or mouth that might obstruct the airway, including loose false teeth, blood, or mucus. In case of vomiting, turn the patient's head to the side for drainage (unless trauma is suspected).

2. **Monitor the patient's mental status and vital signs frequently.** Overdose patients can be alert one minute and unresponsive the

CAUTION: Do not immediately decide that a patient with apparent alcohol on the breath is drunk. The signs may indicate an illness or injury such as epilepsy, diabetes, or head injury.

SIGNS OF INTOXICATION
• Odor of alcohol on the breath.
• Swaying and unsteadiness.
• Slurred speech.
• Nausea and vomiting.
• Flushed face.
• Drowsiness.
• Violent, destructive, or erratic behavior.
• Self-injury, usually without realizing it.

EFFECTS
• Alcohol is a depressant. It affects judgment, vision, reaction time, and coordination.
• When taken with other depressants, the result can be greater than the combined effects of the two drugs.
• In very large quantities, alcohol can paralyze the respiratory center of the brain and cause death.

MANAGEMENT
• Give the same attention as you would to any patient with an illness or injury.
• Monitor the patient's vital signs constantly. Provide life support when necessary.
• Position the patient to avoid aspiration of vomit.
• Protect the patient from hurting him-herself.

Figure 13-7 Alcohol emergencies.

next. Be prepared to provide basic life support if needed.

3. **Maintain the patient's body temperature.** If the patient is cold, cover him or her with blankets. If the patient is abnormally hot, sponge with tepid water.

4. **Take measures to prevent shock.** Shock can result from vomiting, profuse sweating, or inadequate fluid intake. Be alert for allergic reactions, too.

5. **Care for any behavioral problem.** Follow the guidelines given earlier in this chapter for managing behavioral emergencies.

6. **Support the patient.** Comfort, calm, and reassure him or her while waiting for additional EMS personnel to arrive.

If the patient is responsive, try to get him to sit or lie down. Do not restrain a patient unless he poses a risk to safety—his, yours, or that of others. If you suspect trauma in an unresponsive patient, begin emergency care by immediately stabilizing the patient's head and neck. If there is vomiting, roll the patient as a unit to facilitate drainage.

SECTION

3

RAPE AND SEXUAL ASSAULT

Rape is one of the most devastating crises that can occur in a person's life. It involves both emotional and physical trauma. Legally, **rape** is defined as sexual intercourse that is performed without consent and by compulsion through force, threat, or fraud. **Sexual assault** is defined as any touch that the victim did not initiate or agree to and that is imposed by coercion, threat, deception, or threats of physical violence.

Such crimes often are committed by someone the victim knows, such as a relative, friend, classmate, date, neighbor, or a friend of the parents.

Rape Trauma Syndrome

An intensely personal experience under forced or terrifying circumstances can destroy a person's inner defenses. Most rape victims go into acute emotional shock during or shortly after the attack. Common physical reactions to rape include:

- Struggling and screaming to avoid penetration.
- Physical and psychological paralysis.
- Pain and shock from penetration or physical abuse.
- Choking, gagging, nausea, vomiting.
- Urinating.
- Hyperventilating.
- Dazed state, unresponsiveness.

Following rape, most patients experience a great deal of disorganization in their lives. This emotional trauma follows a pattern described as **rape trauma syndrome.** It involves four general stages:

- Acute (impact) reaction, which takes effect immediately after the rape and continues for several days.
- Outward adjustment, which lasts for weeks or months after the rape.
- Depression, which is recurring for days and months after the rape.
- Acceptance and resolution, which can take months or years.

Rape is a difficult and complex problem. It involves physical and emotional trauma, as well as significant legal and criminal issues. Supporting the patient is of critical importance, especially during the acute reaction stage. So when you care for such a patient, remember that his or her coping system has already been stressed to the limit by the attack.

Note that too often the seriousness of rape is equated with physical damage alone. This is a mistake. Even if there are no external visible injuries, the rape victim will suffer profound emotional trauma.

Managing the Rape Scene

Keep the following considerations in mind:

- Rape is a crime. As such, you should always consider your own safety first. Also, protect any evidence you find at the scene.
- Your immediate reaction to the patient is important. Do not impose your own feelings.

Instead, try to find out the patient's emotional state.

- Action can minimize the helplessness a patient may be feeling. Tell him or her what can and should be done immediately.
- The patient might be comforted by a rescuer of his or her own gender (Figure 13-8).
- Perform patient assessment and emergency care as you would for any patient. Treat all life threats. Check for trauma, especially around the thighs, lower abdomen, and buttocks. If vaginal or rectal bleeding is significant, give appropriate care. That is, control bleeding with direct pressure over a bulky dressing or sanitary pad. If the patient is alert, he or she may prefer doing this. Do not remove undergarments unless necessary and do not pack the anus or vagina.
- Do not clean the patient. The patient should avoid showering or bathing, brushing teeth, gargling, douching, or urinating. Cleaning could destroy important evidence.
- Follow local protocols on the preservation of evidence. If possible, seek out training by law enforcement on how to handle issues related to preventing contamination of evidence.

Note that documentation of the call should include the patient's chief complaint, information about the incident that relates to injuries and your care, plus your objective observations and physical findings. Your notes may be used later as evidence in court.

Figure 13-8 It may be best for a First Responder of the same sex to assist the rape patient.

At the beginning of this chapter, you read that a male patient was having a behavioral emergency. To see how chapter skills apply, read the following. It describes how the call was completed.

PHYSICAL EXAMINATION

I spoke to the patient. He asked if I was a Golden Glove. I identified myself by stating "No, I'm not. My name is Khalid Reaves, and I'm a fire service First Responder with the Augusta Volunteer Fire Department." I asked his name. He told me "Andre."

He was beginning to calm down a little. He denied any injuries. Knowing that head injuries or other conditions could affect mental status, I did a head-to-toe exam. It had to be quick, since the patient was still quite agitated. Pulse was 80 and bounding. His respirations were 22 and deep.

PATIENT HISTORY

When the patient was quieter, I tried to gather a history. He was talking in a very confused manner. One minute he said he was a witch. The next he was cawing like a crow. He didn't have any medical identification tags on him. Since he had no clothes, he certainly didn't have a wallet I could check.

One of the officers said he thought he knew the patient's name from a prior call. The dispatcher checked the police computer and found that the man had done this several times before.

He was a frequent patient at the psychiatric center in the next county.

ONGOING ASSESSMENT

While the dispatcher's information answered some questions, I still felt that I should monitor the patient in case there was an underlying medical problem. However, the patient soon became agitated again, so I couldn't recheck vitals. There was really no other change that I could see.

PATIENT HAND-OFF

I told the EMTs what I saw and why I requested the police and our fire department to respond. I told them about the possible psychiatric history, the vitals, and that there were no apparent physical injuries. The EMTs asked the police to ride along with them to the hospital.

I later found out that the local hospital transferred the patient back to the psychiatric center. It seemed that whenever he stopped taking his medications, incidents like this would occur.

Your patient's well being is your responsibility as a fire service First Responder. However, your well being is important, too. Call for law enforcement and for additional EMS resources whenever they are needed at the scene. Remember, too, that a patient with a behavioral emergency may have a medical problem such as diabetes or overdose. Monitor these patients carefully.

Chapter Review

A psychological crisis can be difficult for any EMS professional to deal with. Medical and trauma emergencies have specific sets of signs and symptoms. Psychological emergencies do not. This can make a psychological emergency awkward for both the patient and the fire service First Responder.

Always assure your personal safety first. Not all patients with psychological emergencies will want to harm you, but you must be cautious. The best way to care for this patient is to use good "people skills." Be empathetic. Listen. Use body language. Usually, if you convey the message that you care about the patient and his or her problems, you have the best chance for successful patient care.

Finally, always keep in mind that there are many medical causes for unusual behavior. Never assume the problem is psychological or alcohol-related until medical conditions such as diabetes have been ruled out.

FIRE COMPANY REVIEW

Page references where answers may be found or supported are provided at the end of each question.

Section 1

1. How can you recognize a patient with a behavioral emergency? (pp. 245–246)

2. What are the signs of potential violence in a patient with a behavioral emergency? (pp. 245–246)

3. What is the general emergency care of a patient with a behavioral emergency? (p. 246)

4. What are some techniques that can help calm a patient with a behavioral emergency? List at least five. (pp. 247–248)

5. What is the definition of "reasonable force"? What factors would it depend on? (pp. 247–248)

6. Under what conditions would an EMS responder consider using restraints on a patient? (p. 247)

Section 2

7. What are six signs and symptoms that indicate a life-threatening emergency in a drug or alcohol overdose patient? (pp. 249–250)

8. What are the general guidelines for First Responder care of a patient with an alcohol- or drug-related emergency? (pp. 251–252)

Section 3

9. What is the basic management of a scene in which a rape has occurred? (p. 253)

RESOURCES TO LEARN MORE

Hafen, B.Q., and K.J. Frandsen. *Psychological Emergencies and Crisis Intervention.* Upper Saddle River, NJ: Brady/Prentice-Hall, 1985.

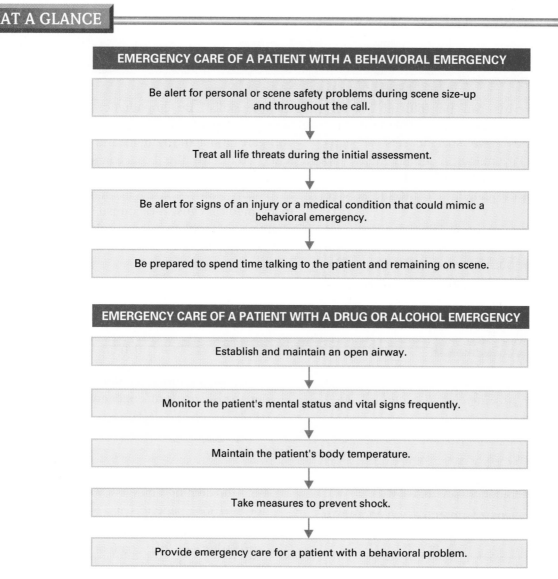

EMERGENCY CARE OF A PATIENT WITH A BEHAVIORAL EMERGENCY

Be alert for personal or scene safety problems during scene size-up and throughout the call.

Treat all life threats during the initial assessment.

Be alert for signs of an injury or a medical condition that could mimic a behavioral emergency.

Be prepared to spend time talking to the patient and remaining on scene.

EMERGENCY CARE OF A PATIENT WITH A DRUG OR ALCOHOL EMERGENCY

Establish and maintain an open airway.

Monitor the patient's mental status and vital signs frequently.

Maintain the patient's body temperature.

Take measures to prevent shock.

Provide emergency care for a patient with a behavioral problem.

Bleeding and Shock

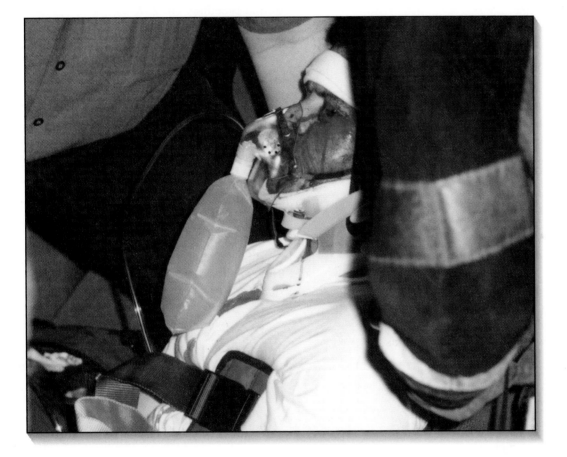

*I*NTRODUCTION   *Any unchecked bleeding can create a life-threatening situation for your patient. It does not really matter if it is visible (external) or invisible (internal) bleeding. Before the patient dies, however, the phenomena known as "shock" occurs. What is shock? It is defined as a lack of tissue* **perfusion.** *This inability of the body to deliver oxygenated blood to the cells, or* **hypoperfusion,** *has characteristics that are clearly identifiable by the well-trained fire service First Responder. Early recognition of the signs and symptoms of shock can allow you to deliver quick, life-saving treatment to your patient.*

Cognitive, affective, and psychomotor objectives are from the U.S. DOT's "First Responder: National Standard Curriculum." Enrichment objectives, if any, identify supplemental material.

Cognitive

5-2.1 Differentiate between arterial, venous, and capillary bleeding. (p. 260)

5-2.2 State the emergency medical care for external bleeding. (pp. 260, 262–265)

5-2.3 Establish the relationship between body substance isolation and bleeding. (p. 259)

5-2.4 List the signs of internal bleeding. (pp. 265–266)

5-2.5 List the steps in the emergency medical care of the patient with signs and symptoms of internal bleeding. (p. 266)

Affective

5-2.15 Explain the rationale for body substance isolation when dealing with bleeding and soft tissue injuries. (p. 259)

5-2.16 Attend to the feelings of the patient with a soft tissue injury or bleeding. (pp. 262–263)

5-2.17 Demonstrate a caring attitude towards patients with a soft tissue injury or bleeding who request emergency medical services. (pp. 262–263)

5-2.18 Place the interests of the patient with a soft tissue injury or bleeding as the foremost consideration when making any and all patient care decisions. (pp. 260, 262–263)

5-2.19 Communicate with empathy to patients with a soft tissue injury or bleeding, as well as with family members and friends of the patient. (p. 263)

Psychomotor

5-2.20 Demonstrate direct pressure as a method of emergency medical care for external bleeding. (p. 262)

5-2.21 Demonstrate the use of diffuse pressure as a method of emergency medical care for external bleeding. (p. 262)

5-2.22 Demonstrate the use of pressure points as a method of emergency medical care for external bleeding. (p. 262)

5-2.23 Demonstrate the care of the patient exhibiting signs and symptoms of internal bleeding. (p. 266)

Enrichment

◆ Discuss the use of splints and tourniquets as methods of bleeding control. (pp. 263–264)

◆ Describe emergency care of a patient with a nosebleed. (pp. 264–265)

◆ List the causes of shock. (p. 266)

◆ Describe the compensatory, decompensated, and irreversible stages of shock. (pp. 266–268)

◆ Describe the emergency care of a patient in shock. (pp. 268–269, 271)

ON SCENE

DISPATCH

While training outside the fire station, the engine company was dispatched for a fall injury. Since we had been practicing hose evolutions, most of us were already in our turnout gear. En route, we were notified by dispatch that a 41-year-old male had fallen from his roof while removing holiday lights.

SCENE SIZE-UP

Before arriving on scene, all responders had taken BSI precautions. Upon arrival, we encountered a woman kneeling beside a person who was supine in the front yard. The woman got up and approached us. She identified herself as the patient's spouse and stated that her husband, Dan Williams, had fallen from the roof but was conscious. We noted that the fall was about 18 to 20 feet (6 m to 7 m) onto a grassy lawn.

INITIAL ASSESSMENT

My general impression revealed a conscious male patient, whose eyes were open and who was talking coherently. I observed his breathing, which appeared to be a little too fast. There were no obvious signs of external bleeding.

As I introduced myself to the patient, another member of our engine company provided manual stabilization of the patient's head and neck. I assessed a radial pulse, which was rapid. I also noted that the patient's skin was cool and clammy despite the 80°F (27°C) air temperature. I then asked one of my partners to administer oxygen via a nonrebreather mask at a flow rate of 15 liters per minute.

PHYSICAL EXAMINATION

Dan complained that his stomach hurt. When I asked what happened, he stated he was removing holiday lights from his roof when he slipped and fell face first onto the ground. He said he hadn't lost consciousness. He stated he was thirsty and asked if we could get him a drink of water. I informed him that we could not allow him to drink anything at this time. Dan then became upset with us, and wondered aloud why we didn't "leave him alone."

I explained to Dan that we were there to help him, and asked if I might remove his shirt to further assess his condition. He gave us permission, and I began assessing his chest and abdomen. When I gently palpated the left upper abdominal quadrant, I noticed it was unusually hard—more so than the right. Dan also winced in pain while I did this, stating that his stomach was "really sore." At this point, I asked my partner to contact the ambulance and determine their ETA.

Dan's baseline vital signs were pulse 124, blood pressure 130/70, and respirations 28.

PATIENT HISTORY

Dan was 41 years old, and he told us he had no medical conditions. I asked if he was taking any medications or was allergic to anything. He stated he was taking ibuprofen for an arthritic knee, and was not allergic to anything. I told Dan I was going to give him a number to remember: the number 51. I asked if he would repeat it and he said "the number is 51." I also explained to Dan why we were doing this.

> As you read the following chapter, consider these questions: Is this patient bleeding? Is this patient in shock? How would you treat him?

BLEEDING

Body Substance Isolation (BSI)

Always take BSI precautions to protect against diseases transmitted by way of blood and body fluids. This is especially true when the patient is bleeding. You should:

◆ Keep a barrier between you and the patient's blood and body fluids. Wear the appropriate protective equipment, such as gloves and eyewear.

◆ Never touch your nose, mouth, or eyes or handle food while providing emergency care.

◆ Keep all of the patient's open wounds—*and all of yours*—covered with dressings or sterile bandages.

◆ Wash your hands properly as soon as possible after treating a patient.

If any equipment, turnout gear, helmets, and so on have been exposed to a patient's blood or body fluids, they must be cleaned as soon as possible. Follow your local disinfecting protocol.

How the Body Responds to Blood Loss

Blood is the fluid component of the circulatory system (Figure 14-1). One of its critical functions is the transport of oxygen to cells. This is called **perfusion.** After unloading its cargo of oxygen, blood then transports the waste product of

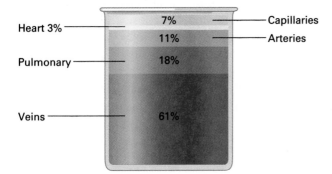

DISTRIBUTION OF BLOOD IN THE BODY

Heart 3%

Capillaries — 7%

Arteries — 11%

Pulmonary — 18%

Veins — 61%

Figure 14-1 Blood is part of the circulatory system.

FIRE DRILL

Just as turnout gear and a self-contained breathing apparatus (SCBA) are necessary when entering a burning building, protective equipment, such as gloves and eyewear, are necessary when treating patients.

cells—carbon dioxide—back to the lungs. This process can be impaired by as little as one liter (1000 cc) of sudden blood loss in an adult, and as little as one-half liter (500 cc) in a child. Left unchecked, death may result. (See Figure 14-2.)

External Bleeding

External bleeding occurs from three types of blood vessels: arteries, veins, and capillaries. Bleeding from any of these vessels can be life-threatening if not properly treated. Each type of bleeding has its own unique characteristics (Figure 14-3):

◆ *Arterial bleeding.* Bright red blood spurting from an open wound usually indicates a damaged or severed artery. The blood is a bright red color because it is rich in oxygen. Since arteries are "high pressure" vessels, this type of bleeding can be difficult to control.

◆ *Venous bleeding.* This is usually a darker red color than arterial blood. This type of bleeding is characterized by a slow, yet steady flow. The blood is a darker red color because it has little, if any, oxygen. The flow is steady,

because veins are "low pressure" vessels. Venous bleeding is usually easier to control than arterial bleeding.

◆ *Capillary bleeding.* This is the most common type of bleeding encountered with injured patients. It is characterized by oozing, dark red blood. Because capillaries are tiny "low pressure" vessels, this type of bleeding is usually easily controlled.

FIRE DRILL

One way to picture the differences among arteries, veins, and capillaries is to think of your pumper and the different types of hoses it carries.

Let's assume you have a 1.75-inch (45 mm) pre-connect line used for a quick attack on fires requiring an offensive mode. Since that line is plumbed from the discharge side of the pump, it will be under a great deal of pressure. What would happen if it was punctured or if it developed a tear? The escaping water would be under a great deal of pressure—similar to an arterial bleed.

Another type of hose typically carried on all pumpers is a supply line bringing water from the hydrant to the pumper. Whether it is a 3-inch (77 mm) hose or larger, the pressure required to supply water to a pumper from a hydrant is typically lower than the pressure in an attack line. If a puncture, or tear, were to develop in this line, we would still see water escaping but not as forcefully as in our attack line. This would be similar to venous bleeding.

Finally, imagine the pump on your fire engine is engaged, but with the engine at idle. If the attack line were to develop a puncture or tear, we may not see any water escaping. If we did notice the tear, the water might well be "oozing" out of the line. This would be similar to capillary bleeding.

◆ **Emergency Care**

All external bleeding must be controlled as soon as possible during the initial assessment. Although airway and breathing take precedence, in most instances bleeding control will occur at the same time that airway and breathing are being managed.

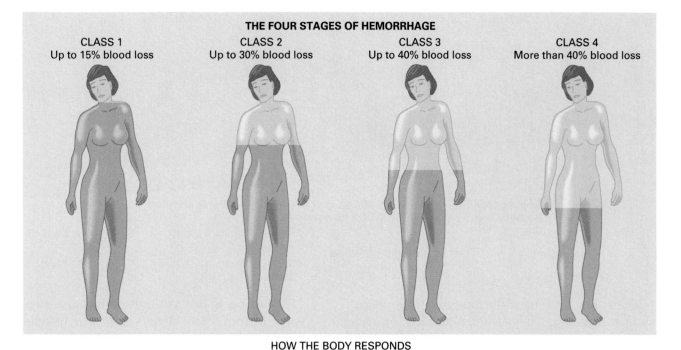

THE FOUR STAGES OF HEMORRHAGE

CLASS 1
Up to 15% blood loss

CLASS 2
Up to 30% blood loss

CLASS 3
Up to 40% blood loss

CLASS 4
More than 40% blood loss

HOW THE BODY RESPONDS

The body compensates for blood loss by constricting blood vessels (vasoconstriction) in an effort to maintain blood pressure and delivery of oxygen to all organs of the body.

EFFECT ON PATIENT

• Patient remains alert.
• Blood pressure stays within normal limits.
• Pulse stays within normal limits or increases slightly; pulse quality remains strong.
• Respiratory rate and depth, skin color and temperature all remain normal.

*The average adult has 5 liters (1 liter = approximately 1 quart) of circulating blood; 15% is 750 ml (or about 3 cups). With internal bleeding 750 ml will occupy enough space in a limb to cause swelling and pain. With bleeding into the body cavities, however, the blood will spread throughout the cavity, causing little, if any initial discomfort.

• Vasoconstriction continues to maintain adequate blood pressure, but with some difficulty now.
• Blood flow is shunted to vital organs, with decreased flow to intestines, kidneys, and skin.

EFFECT ON PATIENT

• Patient may become confused and restless.
• Skin turns pale, cool, and dry because of shunting of blood to vital organs.
• Diastolic pressure may rise or fall. It's more likely to rise (because of vasoconstriction) or stay the same in otherwise healthy patients with no underlying cardio-vascular problems.
• Pulse pressure (difference between systolic and diastolic pressures) narrows.
• Sympathetic responses also cause rapid heart rate (over 100 beats per minute). Pulse quality weakens.
• Respiratory rate increases because of sympathetic stimulation.
• Delayed capillary refill.

• Compensatory mechanisms become overtaxed. Vaso-constriction, for example, can no longer sustain blood pressure, which now begins to fall.
• Cardiac output and tissue perfusion continue to decrease, becoming potentially life threatening. (Even at this stage, however, the patient can still recover with prompt treatment.)

EFFECT ON PATIENT

• Patient becomes more confused, restless, and anxious.
• Classic signs of shock appear—rapid heart rate, decreased blood pressure, rapid respiration and cool, clammy extremities.

• Compensatory vasoconstriction now becomes a complicating factor in itself, further impairing tissue perfusion and cellular oxygenation.

EFFECT ON PATIENT

• Patient becomes lethargic, drowsy, or stuporous.
• Signs of shock become more pronounced. Blood pressure continues to fall.
• Lack of blood flow to the brain and other vital organs ultimately leads to organ failure and death.

Figure 14-2 Four stages of hemorrhage.

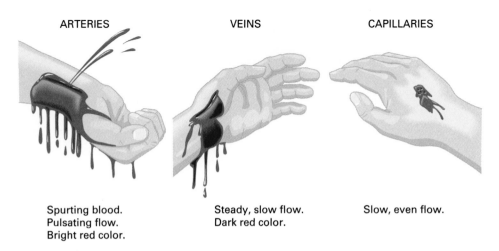

ARTERIES	VEINS	CAPILLARIES
Spurting blood. Pulsating flow. Bright red color.	Steady, slow flow. Dark red color.	Slow, even flow.

Figure 14-3 Types of external bleeding.

There are four basic methods of controlling external bleeding. They can be recalled by the mnemonic **DEPS:**

D — Direct pressure.
E — Elevation.
P — Pressure points.
S — Splinting.

For a patient with external bleeding, take BSI precautions and follow these steps (Figure 14-4):

1. **Apply direct pressure to the wound.** Direct pressure will stop blood from flowing into the damaged vessel, helping it to clot. Do this by placing a gloved hand over the wound until a sterile dressing can be applied. If the wound is small, apply pressure directly over it with your fingertips. If the wound is large, the flat part of your hand may be necessary to control the bleeding. Large gaping wounds may require sterile gauze and direct hand pressure if fingertip pressure fails to control bleeding.

   One effective way to maintain direct pressure over a wound on an extremity is through the use of an air splint. Apply it directly on top of the wound (Figure 14-5).

2. **Elevate the bleeding extremity.** Do this at the same time you are applying direct pressure. Ideally, the limb should be elevated above the level of the heart. However, do not elevate it if you suspect a bone or joint injury.

3. **Assess bleeding.** After applying direct pressure and using elevation, assess the wound. If the bleeding has soaked through the dressing, apply another dressing on top of it.

4. **If bleeding is not yet controlled, use pressure points.** (See Figure 14-6.) For bleeding in the forearm, find the brachial pulse point, which overlies the upper arm. Using the flat surface of your fingers, compress the artery against the bone, slowing the flow of blood below that point.

   For bleeding in the leg, find the femoral pulse point. Then compress the artery against the pelvis using the heel of your hand.

FIRE DRILL

Your patients do not care how much you know until they know how much you care! Therefore, administer a large dose of reassurance as often as possible.

NOTE: In some EMS systems, bleeding control procedures are slightly different. After you find that direct pressure and elevation do not work to stop bleeding, in those systems you are required to remove the first dressing to assess the bleeding point. If it is still bleeding or if there is more than one bleeding point, then you are to apply more pressure directly to the point or points. If this still does not control bleeding, you are to use pressure points as described above. Removing the first dressing to assess the bleeding is a controversial technique. Be sure to follow your own local protocols.

Throughout emergency care, provide emotional support to your patient. The sight of an arm

Methods of Bleeding Control

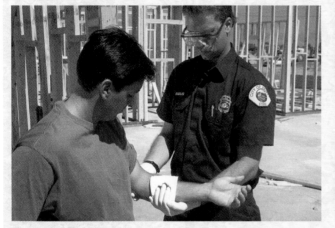

Figure 14-4a Apply direct pressure.

Figure 14-4b Elevate an extremity.

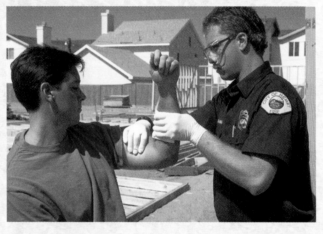

Figure 14-4c Use pressure points.

or leg soaked in blood can be quite unnerving to anyone, young or old. Reassure your patient. Be honest, and communicate with empathy. Remember, staying calm throughout emergency care is a must.

Spinting

Splinting, or immobilization, is another way to control bleeding in an extremity. It works because movement promotes blood flow, which can disrupt the clotting process. So if a limb does not move, less blood reaches it and clotting is enhanced. Any splint will work. However, using an air splint can immobilize the limb as well as maintain direct pressure, which can free you up

for other duties. NOTE: It is probable that a combination of direct pressure, elevation, pressure points, and splinting will be used to control external bleeding. (See Chapter 17 for information on how to apply a splint.)

Some EMS systems do not permit First Responders to splint, or immobilize, an injured limb. In many of those systems, manual stabilization of the limb is recommended instead. Always follow local protocol.

Tourniquets

A tourniquet should be used to control life-threatening bleeding when all other methods have failed. Because it can stop all blood flow to

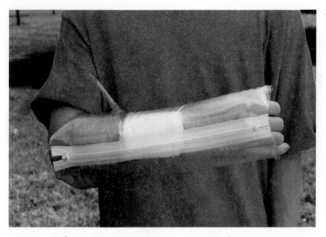

Figure 14-5 One way to maintain direct pressure is by use of an air splint.

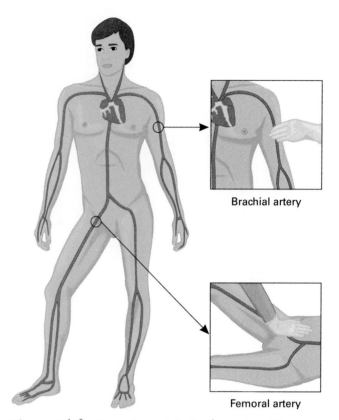

Brachial artery

Femoral artery

Figure 14-6 Pressure points in the extremities.

an extremity, use a tourniquet only as a last resort. A tourniquet can cause permanent damage to nerves, muscles, and blood vessels, and it can result in the loss of the affected limb. Always seek medical direction before using a tourniquet. Follow all local protocols. To apply a tourniquet:

1. **Select a bandage.** It should be 4 inches (100 mm) wide and six to eight layers deep.
2. **Wrap it around the extremity twice** at a point above but as close to the wound as possible.
3. **Tie a knot in the bandage material.** Then place a stick or rod on top of it. Tie the ends of the bandage again in a square knot over the stick.
4. **Twist the stick until the bleeding stops.** Then secure the stick or rod in place.
5. **Note the time.**
6. **Notify the EMTs** who take over patient care that you have applied a tourniquet.

In some cases an inflated blood pressure cuff may be used as a tourniquet until bleeding stops. If you choose to do this, you need to monitor the cuff continuously to make sure pressure is maintained.

When using any type of tourniquet, take the following precautions:

◆ Always use a wide bandage, and secure it tightly. Never use a wire, belt, or any other material that could cut the skin or underlying soft tissues.

◆ Once applied, never loosen or remove a tourniquet unless you are directed to do so by medical direction.

◆ Never apply a tourniquet directly over a joint.

◆ Always make sure the tourniquet is in open view. If it is covered by bandages or clothing, it may be overlooked, which can result in permanent tissue damage.

Nosebleeds

Causes of nosebleeds can range from injury, to disease, to the environment. Often more annoying than serious, nosebleeds also can be life-threatening. For example, nosebleeds associated with a fracture can be severe enough to compromise the airway and cause shock (hypoperfusion).

In most cases a nosebleed may be treated as follows (Figure 14-7):

Method of Nosebleed Control

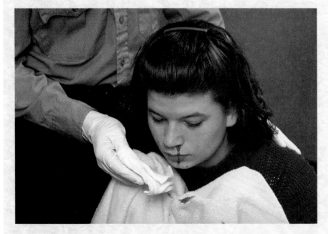

Figure 14-7a Keep the patient quiet and leaning forward in a sitting position.

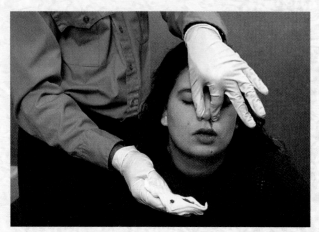

Figure 14-7b Apply pressure by pinching the nostrils. Also apply cold compresses if needed.

1. **Keep the patient calm.**

2. **Position the patient.** Have the patient sit and lean forward in order to prevent aspiration of blood into the lungs. If that is not possible, the patient should lie down with head and shoulders elevated. Be sure to maintain an open airway.

3. **Have the patient pinch the nostrils together,** if you do not suspect a broken nose.

4. **Slow the flow of blood.** Apply ice packs or cold compresses to the nose and face. Instruct the patient to avoid blowing his or her nose for several hours. It could dislodge the clot and restart bleeding.

5. **Activate the EMS system,** if bleeding continues and cannot be stopped.

If a fractured skull is suspected, do not try to stop a nosebleed. To do so might increase pressure on the brain. Instead, cover the nasal opening loosely with a dry, sterile dressing. Do not apply pressure. Treat the patient for a skull fracture as outlined in Chapter 18.

Internal Bleeding

External bleeding is usually easy to spot. Internal bleeding is much more difficult to detect. This

GERIATRIC NOTE

A nosebleed in an elderly patient is often caused by severe hypertension. So, during scene size-up, look around the area. If you spot any medications, ask the patient what they are for. If there is reason to suspect hypertension, activate the EMS system. Hypertension could pose more of a problem than the nosebleed itself.

"invisible" blood loss can quickly result in death. So always maintain a high degree of suspicion whenever scrapes, bruises, swelling, deformity, or impact marks are present. Also assume internal bleeding with any penetrating wounds to the skull, chest, or abdomen.

The signs and symptoms of internal bleeding include:

- Pale, cool, clammy skin.
- Increased respiratory and pulse rates.
- Thirst.
- Changes in mental status, such as confusion, disorientation, and anxiety or agitation.
- Nausea and sometimes vomiting.
- Vomitus with blood.

- Dark, tarry stools.
- Discolored, tender, swollen, or hard tissue.
- Tender, rigid, or distended abdomen.
- Weakness, faintness, or dizziness.

◆ Emergency Care

When you suspect your patient is bleeding internally, immediately determine the ETA of the transporting unit. This patient needs rapid transport to a hospital. In the meantime, follow these steps:

1. **Establish and maintain an open airway.** If allowed, apply oxygen by way of a nonrebreather mask at 15 liters per minute. Ventilate, if necessary.

2. **Control external bleeding.** Since you cannot control internal bleeding, it becomes even more critical to control any external bleeding you observe.

3. **Keep your patient warm,** but take care to not overheat. Remember that a patient who loses a large amount of blood cannot conserve body heat effectively.

4. **Treat for shock.**

SECTION 2 SHOCK

To the lay person, shock can mean an unpleasant surprise or an electrical stimulus. For First Responders, however, shock (hypoperfusion) has a very precise definition. It is the inadequate delivery of oxygen-rich blood to the body's tissues. When the body's cells do not receive enough oxygen, they die. As the cells die, body tissue begins to suffer and die. As tissue begins to die, organs—such as the brain, heart, and kidneys—begin to fail. Death of the person eventually occurs if shock is not corrected. In fact, shock has been called "a momentary pause on the road to death."

Yet, shock—if recognized—proves a valuable ally to the health-care professional. Its early recognition may be the only sign of internal bleeding.

Causes of Shock

Shock, or poor tissue perfusion, can best be understood if you know what it takes for proper tissue oxygenation to occur. That is:

- *A properly functioning pump.* Heart attack, coronary artery disease, and cardiac tamponade (fluid leaking into the sac around the heart) all can contribute to a malfunction in the pumping action of the heart.

- *Enough oxygen-rich blood for the heart to pump.* Blood volume can decrease as a result of internal and external bleeding, fluid loss during prolonged illness or burns, and dehydration. (See Figure 14-8.)

- *Intact blood vessels through which blood can be pumped throughout the body.* When these "pipes," or vessels, are damaged, their ability to deliver oxygen-rich blood to cells is disrupted. Blood vessel failure also may be caused by extreme vessel dilation, which may occur with severe allergic reactions or spinal-cord injury.

What causes shock? An interruption of any of these three mechanisms—the pump (heart), fluid (blood), or pipes (arteries, veins, or capillaries).

🔔 FIRE DRILL

On the fireground, in order to "put the wet stuff on the red stuff," it is critical that the same three mechanisms be present. They are a functioning pump (centrifugal single or dual stage pump), a fluid to be pumped (water), and pipes or vessels (fire hose) to deliver the water to the fire. If any one of these three mechanisms is not working properly, we will not be able to extinguish the fire:

- Pump—if it fails, we cannot deliver water (blood) to the nozzle (cells).

- Fluid—if we run out of water (blood), no fluid will be available at the nozzle (cells).

- Pipes or vessels—if our hoseline breaks (arterial bleed), the amount of water (blood) delivered to the nozzle (cells) may be insufficient for extinguishment (proper oxygenation).

. . . And "shock" to the building—and its residents—will result!

Watch for shock in all trauma patients. They can lose fluids not only externally through hemorrhage, vomiting, or burns, but also internally through crush injuries and organ punctures.

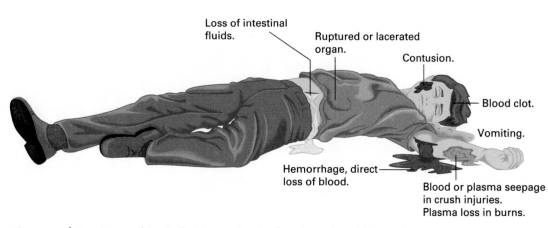

Loss of intestinal fluids.

Ruptured or lacerated organ.

Contusion.

Blood clot.

Vomiting.

Hemorrhage, direct loss of blood.

Blood or plasma seepage in crush injuries. Plasma loss in burns.

Figure 14-8   Loss of body fluids can be both external and internal.

Stages of Shock

Regardless of which component is malfunctioning —pump, blood, or blood vessels, shock (hypoperfusion) has three distinct phases. They are called *compensatory shock, decompensated shock,* and *irreversible shock* (Figures 14-9 and 14-10).

Compensatory Shock

In this first stage of shock, the body uses defense mechanisms to maintain normal function. They are so effective that an adult can lose up to 15% of total blood volume with little change in vital signs. If the condition does not get worse, the body will overcome it. Although subtle, the signs and symptoms of shock at this stage can be recognized:

◆ *Pale skin.* The vessels have constricted, forcing blood away from the skin into larger vessels lying further beneath the skin surface.

◆ *Slightly rapid heart rate.* With less blood, each blood cell must now make more trips through the body in order to deliver an adequate supply of oxygen to the cells.

◆ *Restlessness or anxiety.* The brain may not be receiving as much oxygen as it would like.

◆ *Blood pressure in the normal range.* With vessel constriction and increased heart rate, blood pressure will remain the same.

Decompensated Shock

At this stage, the body can no longer make up for reduced tissue perfusion of oxygenated blood.

> ### FIRE DRILL
> Waiting for blood pressure to drop in a bleeding patient is like waiting for the roof to collapse in a structure fire. If you wait, you probably won't be able to save the patient or the building!

Signs and symptoms are quite recognizable and include:

◆ *Extreme thirst,* which reflects the fluid deficit within the body.

◆ *Rapid, weak pulse.* It is rapid because the heart is beating faster than before in order to circulate what little fluid is in the vessels. It is weak because the vessels are now very narrow, and the volume moving through them is decreased.

◆ *Decreased blood pressure.* With little volume left and the vessels constricted as tight as they can be, pressure will begin to fall.

◆ *Cool and moist skin that is pale, gray, or mottled.* The body has now diverted all blood flow from the extremities and surface of the skin to the major organs.

◆ *Noticeable changes in the patient's mental status.* When oxygen delivery is compromised, the brain is the first organ to react.

When you perform an initial assessment, get an objective measurement of the patient's level of

Developing Shock

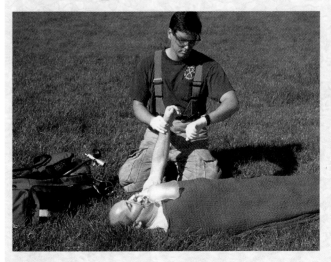

Figure 14-9a Compensatory shock: slight increase in pulse.

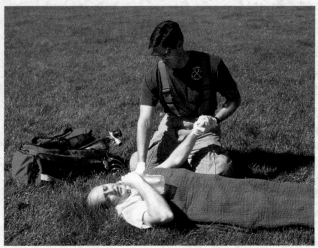

Figure 14-9b Compensatory shock: restlessness or anxiety.

Figure 14-9c Decompensated shock: rapid, weak pulse.

responsiveness. For example, in addition to asking name, place, day, and date, have the patient repeat a certain number. You might say: "I'm giving you a number to repeat. It is the number 51. Can you say that number?" Throughout the assessment repeatedly ask "What is that number?" Note any changes in the patient's responses.

Irreversible Shock

Even with treatment, irreversible shock leads to death. At this stage, the blood vessels lose their ability to constrict. This "relaxation" causes blood to pool away from vital organs. This leads to a dangerously low blood pressure, so low that entire organ systems begin to die.

◆ Emergency Care

It is vital for you to learn to see what is not obvious, but life threatening. Since trauma is a major cause of shock in patients, assume that any trauma patient is in compensatory shock until proven otherwise.

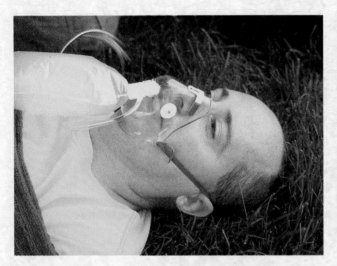

Figure 14-9d Decompensated shock: skin color changes and sweating.

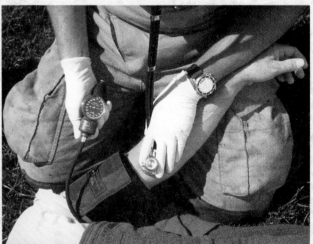

Figure 14-9e Decompensated shock: decreasing blood pressure.

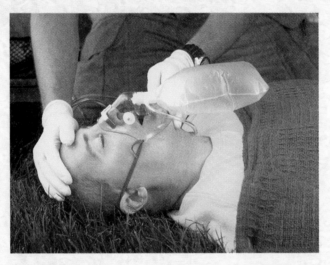

Figure 14-9f Decompensated shock: unresponsiveness.

Time is critical. Patients in shock (hypoperfusion) need to be transported quickly to an appropriate facility, such as a trauma center. In fact, the most important care you can provide is to identify the serious trauma patient and prepare him or her for transport as quickly as possible.

To provide emergency care, take BSI precautions and follow these steps (Figures 14-11 and 14-12):

1. **Maintain an open airway.** If breathing is adequate, administer high-flow oxygen with a nonrebreather mask at 15 liters per minute. Provide artificial ventilation, if necessary.

2. **Prevent further blood loss.** Since you cannot control internal bleeding, make sure you control all external bleeding. Use the DEPS principle.

3. **Elevate the lower extremities** about 8 to 12 inches (200 mm to 300 mm), if they are not injured. Also do not elevate if there are serious injuries to the head, neck, spine, chest, abdomen, or pelvis.

Skin around mouth
may be grayish

Lips may be blue

Tongue may be blue

Nail beds may
be blue

Mucous membranes
of mouth may be blue
or have a pale,
grayish, waxy pallor.

Figure 14-10   Signs of shock in a dark-skinned patient.

Figure 14-11   Administer oxygen to a patient in
shock, if you are allowed to do so.

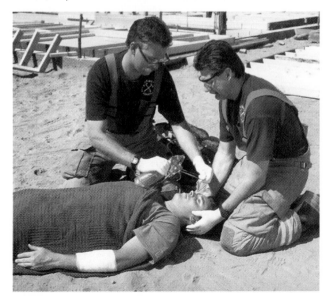

Figure 14-12   Elevate the lower extremities 8 to
12 inches (200 mm to 300 mm), if appropriate.

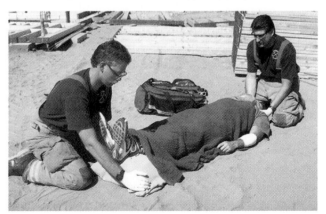

4. Keep the patient warm. But do not overheat. Use a blanket over and under the patient to help maintain a normal body temperature.

5. Do not give the patient anything to eat or drink. If the patient is experiencing nausea, eating or drinking may result in vomiting. The last thing you need is to create an airway problem!

6. Provide emergency care for specific injuries. But do so only if it will not delay transport.

Report your initial assessment findings to the transport crew as soon as possible. This will allow them to choose the appropriate response. They also may be able to initiate other EMS system resources available in your community (such as air transport, trauma center notification, and so on).

ON SCENE FOLLOW-UP

At the beginning of this chapter, you read that fire service First Responders were providing emergency care to a patient who fell from a roof while removing holiday lights. To see how chapter information applies to this emergency, read the following. It describes how the call was completed.

ONGOING ASSESSMENT

I performed another assessment of Dan's vital signs. They were breathing 28, pulse 144, and blood pressure 128/70. I asked Dan if he would repeat the number I gave him and he said "Sure. The number is, ah, 51 . . . I think, right?"

PATIENT HAND-OFF

At this time, the ambulance crew had arrived. I gave the following report to Paramedic Bill Post:

"This is Dan Williams, a 41-year-old male who fell about 18 to 20 feet (6 m to 7 m) from this roof onto the lawn where he is lying. Upon arrival, Dan was alert and oriented. He responded appropriately to our questions, and stated that he fell face first. His wife confirms this. He denies any loss of consciousness. His chief complaint is abdominal pain. Our physical exam reveals left

upper quadrant tenderness, with some rigidity. Dan states he has no other medical conditions, is taking ibuprofen for arthritis of his knee, and has no allergies.

Baseline vital signs: BP was 130/70, pulse 120, breathing 28. He denies any shortness of breath. He would like something to drink. When we told him no, he became agitated. I also gave him the number 51 to repeat back to me, which he could do 10 minutes ago. However, he wasn't sure what the number was, when I asked him just as you arrived on scene. Also, his latest vitals are BP 128/70, pulse 144, breathing 28. We have taken spinal precautions and placed him on oxygen at 15 liters per minute with a nonrebreather mask. How else can we help you?"

If in the initial assessment you observe significant external bleeding or any signs of internal bleeding in your patient, treat the patient for shock immediately. However, remember that when the mechanism of injury suggests possible internal bleeding—even if you observe no signs or symptoms, treat the patient for shock until proven otherwise. Do not wait for signs of shock to develop.

Chapter Review

Severe bleeding is controlled during the initial assessment. Only airway and breathing care would take priority. Be assured, however, that even severe bleeding can be controlled by the simple methods described in this chapter: direct pressure, elevation, pressure points, and splinting. Experienced EMS personnel will tell you that these methods work. Use them appropriately and confidently.

FIRE COMPANY REVIEW

Page references where answers may be found or supported are provided at the end of each question.

Section 1

1. How would you describe the differences between arterial, venous, and capillary bleeding? (p. 262)

2. What four methods of controlling external bleeding are available for the fire service First Responder? (pp. 262, 264–266)

3. What are at least four signs and symptoms that the fire service First Responder may en-counter in a patient suffering from internal bleeding? (pp. 267–268)

4. What is the emergency medical care of a patient with internal bleeding? (p. 268)

Section 2

5. What is the definition of the term "shock"? (p. 268)

6. What are the signs and symptoms of a patient suffering from shock? (pp. 269–270)

7. How should you provide emergency care for a patient in shock? (pp. 270–271, 273)

RESOURCES TO LEARN MORE

"Management of Shock" in Moore, E.E., et al., eds. *Trauma,* Second Edition. Norwalk, CT: Appleton-Lange, 1991.

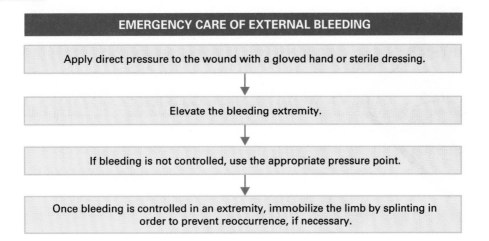

EMERGENCY CARE OF EXTERNAL BLEEDING

Apply direct pressure to the wound with a gloved hand or sterile dressing.

Elevate the bleeding extremity.

If bleeding is not controlled, use the appropriate pressure point.

Once bleeding is controlled in an extremity, immobilize the limb by splinting in order to prevent reoccurrence, if necessary.

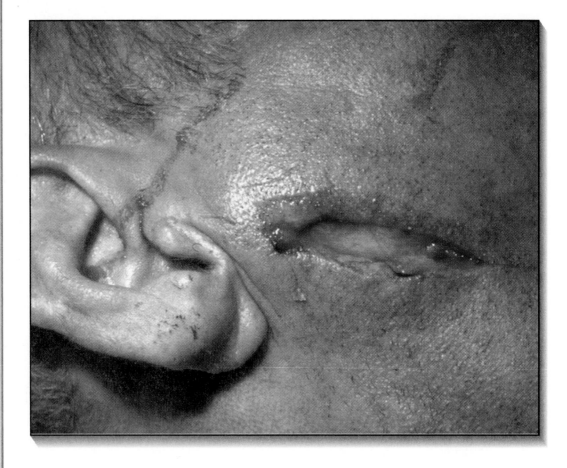

I **NTRODUCTION** *Injuries to a patient's skin, muscles, blood vessels, and internal organs are among the most common you will care for. Some will be minor cuts, scrapes, and bruises. Others may be life threatening. Regardless of the severity, your priorities will include controlling bleeding, preventing shock and further injury, and reducing the risk of infection.*

Cognitive, affective, and psychomotor objectives are from the U.S. DOT's "First Responder: National Standard Curriculum." Enrichment objectives, if any, identify supplemental material.

Cognitive

5-2.6 Establish the relationship between body substance isolation (BSI) and soft tissue injuries. (pp. 277, 280)

5-2.7 State the types of open soft tissue injuries. (pp. 277–278, 280)

5-2.8 Describe the emergency medical care of the patient with a soft tissue injury. (pp. 277, 280)

5-2.9 Discuss the emergency medical care considerations for a patient with a penetrating chest injury. (pp. 281–282)

5-2.10 State the emergency medical care considerations for a patient with an open wound to the abdomen. (pp. 283–284)

5-2.11 Describe the emergency medical care for an impaled object. (pp. 282–283)

5-2.12 State the emergency medical care for an amputation. (pp. 284–285)

5-2.14 List the functions of dressing and bandaging. (pp. 288–290)

Affective

5-2.15 Explain the rationale for body substance isolation when dealing with bleeding and soft tissue injuries. (pp. 277, 280)

5-2.16 Attend to the feelings of the patient with a soft tissue injury or bleeding. (p. 282)

5-2.17 Demonstrate a caring attitude towards patients with a soft tissue injury or bleeding who request emergency medical services. (p. 282)

5-2.18 Place the interests of the patient with a soft tissue injury or bleeding as the foremost consideration when making any and all patient care decisions. (p. 282)

5-2.19 Communicate with empathy to patients with a soft tissue injury or bleeding, as well as with family members and friends of the patient. (p. 282)

Psychomotor

5-2.24 Demonstrate the steps in the emergency medical care of open soft tissue injuries. (p. 280)

5-2.25 Demonstrate the steps in the emergency medical care of a patient with an open chest wound. (pp. 281–282)

5-2.26 Demonstrate the steps in the emergency medical care of a patient with open abdominal wounds. (pp. 283–284)

5-2.27 Demonstrate the steps in the emergency medical care of a patient with an impaled object. (pp. 282–283)

5-2.28 Demonstrate the steps in the emergency medical care of a patient with an amputation. (pp. 284–285)

5-2.29 Demonstrate the steps in the emergency medical care of an amputated part. (p. 285)

Enrichment

◆ Describe the emergency medical care for a patient with a large open neck wound. (p. 283)

◆ Describe the emergency medical care for a patient with an object impaled in the eye. (pp. 282–283)

◆ State the general principles of dressing and bandaging soft-tissue injuries. (p. 290)

DISPATCH

I work as a quality improvement coordinator at a chemical plant. I am also a member of my company's Emergency Response Team (ERT) and the volunteer fire chief of the small town where I live. While at the chemical plant one afternoon, my pager went off for an emergency. Other members of the ERT and I proceeded quickly to the scene.

SCENE SIZE-UP

As our team approached the scene, a security guard directed us to the injury site. She advised us that there were no dangers. A worker had a piece of glass impaled in one hand. We put on protective gloves and glasses and approached the scene.

INITIAL ASSESSMENT

The patient was sitting near a workbench. We immediately became aware of the fact that her mental status, airway, and breathing were okay. This was because of the language she was using as she complained about her injury. I noticed a broken glass beaker in the lab sink. We identified ourselves.

I asked the patient to explain what happened while I glanced at the patient's wound. She told us that she had been cleaning glassware in the lab sink when a beaker broke. I noted the absence of protective gloves on her hands. We were surprised to see very little bleeding from the wound. While calming the patient down, she gave us permission to assess her injury.

After a minute, we reported the patient's status and the nature of the injury to the incoming EMTs, who were 10 minutes away.

> *Consider this patient as you read Chapter 15. What may be done to treat her condition?*

SOFT-TISSUE INJURIES

Soft-tissue injuries are injuries to the skin, muscles, nerves, and blood vessels. They are often dramatic, but they are rarely life threatening. They can be serious, however, if they lead to airway or breathing problems, uncontrolled bleeding, or shock.

A soft-tissue injury is commonly referred to as a **wound.** Wounds may be classified as open or closed, single or multiple. They also are classified by location (head wounds or chest wounds, for example).

In general, emergency medical care focuses on controlling bleeding, preventing further injury, and reducing the risk of infection. Unless it is a life-threat, a soft-tissue injury is usually cared for after the initial assessment.

Closed Wounds

In a closed wound, soft tissues beneath the skin are damaged (Figure 15-1). The skin itself is not broken. Closed wounds generally are caused by **blunt trauma.** This may be due to anything from being struck by a vehicle to dropping a block on a foot. Always suspect internal injuries with blunt trauma. For example, blunt trauma to the torso or head may cause significant organ damage and internal bleeding. When there is blunt trauma to the extremities, broken bones are possible.

A **contusion,** or bruise, is a closed wound that is characterized by swelling and pain at the injury site. If small blood vessels have been broken, the patient will have **ecchymosis** (black and blue discoloration). If there is a larger collection of blood under the skin, a **hematoma** is evident as a lump with bluish discoloration.

A clamping or crushing injury usually involves a finger or limb stuck in an area smaller than

Closed Wounds

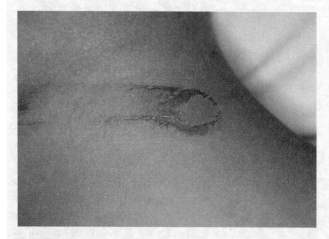

Figure 15-1a Contusions. *(Source: Charles Stewart & Associates)*

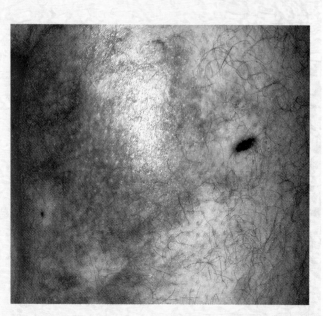

Figure 15-1b Hematoma. *(Source: Charles Stewart & Associates)*

COMPANY OFFICER'S NOTE

Blunt trauma also may be thought of as "invisible" trauma. Although there may be no apparent injury, a life-threatening situation may be developing internally. Therefore, always assume that all patients suffering from blunt trauma are seriously injured until proven otherwise.

itself. This type of injury occurs when a patient reaches into a space and is stuck there. The body part can swell rapidly, making the condition worse. Another type of clamping injury occurs when an object, such as a ring, strangles a body part that was previously injured.

◆ Emergency Care

Small contusions generally do not need treatment. For larger contusions, cold compresses can help to relieve pain and reduce swelling. Note that large areas of discoloration of the skin can indicate serious internal bleeding. A contusion the size of a fist, for example, could mean significant blood loss. If the patient has a large contusion, or if the mechanism of injury suggests blunt trauma, treat the patient for internal bleeding. Be sure to assess for broken bones, especially when swelling or deformity is present.

For a clamping injury, a lubricant such as soap may help to free the body part. If not, then it may need to be removed at the hospital. Apply a cold pack to help reduce swelling. Since circulation to the body part may be reduced, do not cool it for longer than 15 to 30 minutes at a time. Elevate the injured part above the level of the heart. Keep it well supported.

If you are ever in doubt about the seriousness of a closed wound, treat the patient for internal bleeding.

NOTE: Always take BSI precautions when there is any possibility that you will come in contact with a patient's blood or body fluids. When caring for a patient with closed wounds, at a minimum wear protective gloves.

Open Wounds

When skin breaks as a result of a blow, the wound is referred to as an open wound. (See

Figure 15-2.) Open wounds place the patient at risk for contamination, which can lead to infection. An open wound also may be the first indicator of a deeper, more serious injury such as a broken bone.

Abrasions

An **abrasion** is an open wound caused by scraping, rubbing, or shearing away of the epidermis (outermost layer of skin). Even though an abrasion is considered a superficial injury, it is often very painful because of exposed nerve ends. In most cases, there is capillary bleeding from an abrasion, though there may be no bleeding at all.

Small abrasions usually are not life threatening. Large ones, however, may be cause for concern. For example, a motorcycle rider who is thrown and slides across the pavement will sustain head-to-toe abrasions ("road rash"). Bleeding in such a case may not be serious, but contamination, infection, and the potential for underlying injuries may be.

COMPANY OFFICER'S NOTE

One useful way to estimate the size of an abrasion is to look at the patient's palm as 1% of his or her total body surface area. Using the palm as a guide, give an estimated percentage of body surface area abraded when you report to other medical professionals.

Lacerations

A **laceration** is a break of varying depth in the skin. It may occur in isolation. It also may occur with other types of soft-tissue injuries (Figure 15-3). Lacerations are caused by forceful impact with a sharp object. Bleeding may be severe, especially if an artery is involved.

The edges of a laceration may be regular or irregular. Regular lacerations are usually caused by a knife or razor, and are called "incisions." They may heal better because their edges are

Figure 15-2 Open wounds.

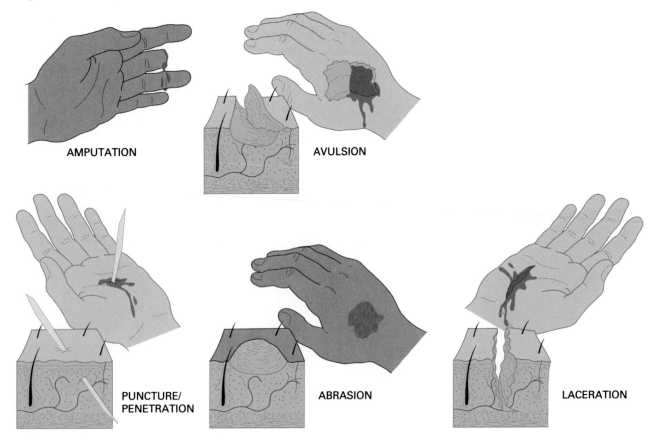

AMPUTATION

AVULSION

PUNCTURE/PENETRATION

ABRASION

LACERATION

Abrasions and Lacerations

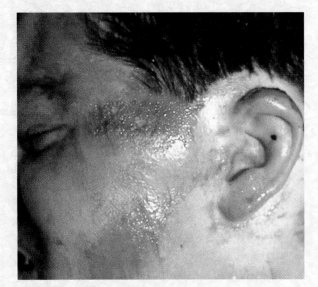

Figure 15-3a Abrasions.

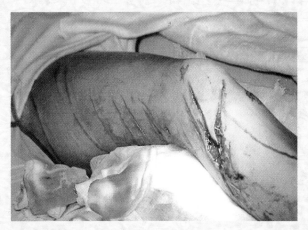

Figure 15-3b Lacerations.

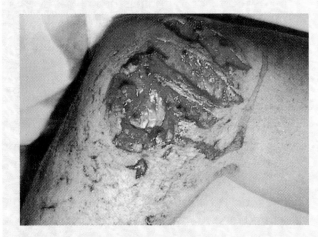

Figure 15-3c Deep abrasions and lacerations.

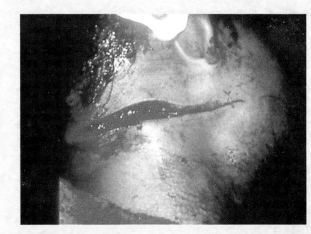

Figure 15-3d Laceration.

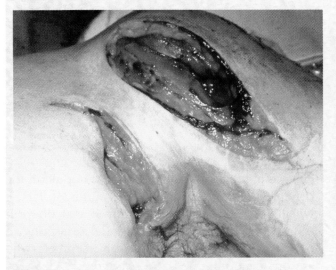

Figure 15-3e Lacerations (stab wounds). *(Source: Shout Picture Library)*

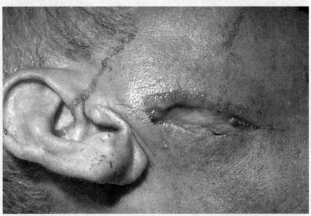

Figure 15-3f Laceration. *(Source: Shout Picture Library)*

smoother. Irregular lacerations are commonly caused by a blunt object, serrated knife, or broken glass. The edges of the wound are jagged, and healing is usually prolonged unless the wound is properly closed by a physician.

Penetration/Puncture Wounds

A penetration/puncture wound is usually the result of a sharp, pointed object being pushed or driven into soft tissues. This type of injury may have both an entry wound and an exit wound.

The entry wound may be small, and there may be little or no external bleeding (Figure 15-4). However, such injuries may be deep, damaging, and cause severe internal bleeding. A gunshot injury is a good example (Figure 15-5). The entry wound in many cases is smaller than the exit wound. If the patient was shot at close range, the entry wound may be surrounded by powder burns. The larger exit wound generally bleeds more profusely.

The overall severity of a penetration or puncture wound depends on the following factors:

◆ Location of the injury.

◆ Size of the penetrating object.

◆ Forces involved in creating the injury.

It can be difficult to determine the extent of these injuries based only on external signs. So treat penetration or puncture wounds with great caution, especially when they occur on the head, neck, trunk, or proximal extremities.

Figure 15-4 Puncture wound to the foot.

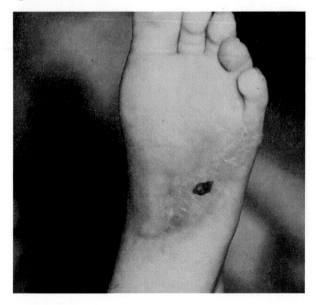

◆ Emergency Care

Since it is likely that you will be exposed to your patient's blood and body fluids, take all appropriate BSI precautions. Wear protective gloves. Also protect your face and eyes, and wear a disposable gown as necessary. After patient care, wash your hands, even if you wore gloves. Handwashing is still the single most important thing you can do to prevent the spread of infection.

To care for a patient with an open soft-tissue injury, follow the guidelines outlined below.

1. **Assess and treat all life threats.** Maintain a patent airway and adequate breathing. Administer oxygen by way of a nonrebreather mask. If the patient's breathing is inadequate, provide ventilations with supplemental oxygen, if allowed.

2. **Expose the entire injury site.** Cut away clothing, if needed. Note: To help preserve evidence, try to avoid cutting through holes that were made by bullets or knives. If possible, cut around the holes. Then clear the area of blood and debris with sterile gauze or the cleanest material available. Remember to look for additional wounds or injuries, especially in the case of a gunshot wound. (Often multiple gunshot wounds are involved.)

3. **Control bleeding,** and treat for shock. Begin with direct pressure and elevation. If bleeding still is not controlled, then use a pressure point.

4. **Prevent further contamination.** Keep the wound as clean as possible. If there are loose particles of foreign matter around the wound, wipe them away from the wound, never toward it. Never try to pick embedded particles or debris out of the wound.

5. **Dress and bandage the wound.** Apply a dry sterile dressing. Then secure it with a bandage. Check distal pulses both before and after applying the bandage to make sure it is not cutting off circulation.

As you care for a patient's soft-tissue injuries (especially when there are open, bleeding ones) be careful what you say and do. Do not alarm the patient by your reaction to the wounds. In fact, it will help the patient significantly if you do your best to be comforting, calming, and reassuring. If possible, keep the patient's family members who are on scene informed and reassured.

Gunshot Wounds

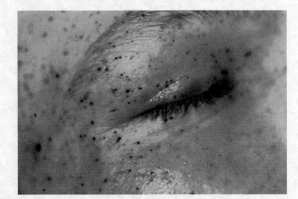

Figure 15-5a Powder burns from a gunshot.

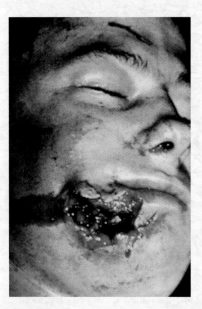

Figure 15-5b Gunshot wound to the chin.

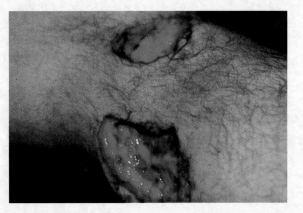

Figure 15-5c Entrance and exit gunshot wounds.

Special Considerations

Some open wounds need special consideration. Basic emergency care is the same as for all other open wounds, but there are certain exceptions, which are described below. In all of these cases, administer oxygen by way of a nonrebreather mask or, if breathing is inadequate, provide ventilations with supplemental oxygen.

Chest Injuries

A penetrating chest wound can prevent a patient from breathing adequately. In this case, apply an **occlusive dressing.** (This is a special type of dressing used to form an air-tight seal.) Secure the dressing with tape on three sides (Figure 15-6). Leave one side untaped to allow air to escape as the patient exhales. This will help to prevent a condition called *tension pneumothorax,* a severe build-up of air within the chest cavity that compresses the lungs and heart toward the uninjured side of the chest.

Let the patient assume a position of comfort, if you do not suspect spine injury. Generally, the patient will favor the position that allows for the greatest chest expansion. This is usually sitting

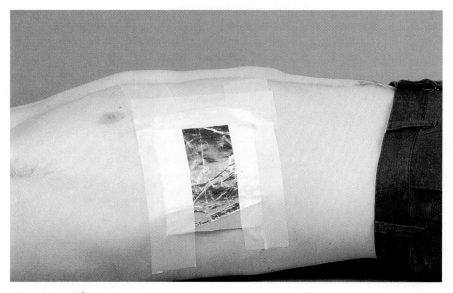

Figure 15-6 Occlusive dressing taped on three sides for a penetrating chest wound.

upright. Assume spine injury if there is any significant mechanism of injury to the chest, including a gunshot wound.

Impaled Objects

An **impaled object** is an object that is embedded in an open wound. It should never be removed in the field unless it is through the patient's cheek or it interferes with airway management or CPR. To provide emergency care to a patient with an impaled object, follow these guidelines:

1. **Manually stabilize the object to prevent any motion.** Motion could cause further damage and bleeding.

2. **Expose the wound area.** Remove clothing from around it, but remember to take care not to move the object at all.

3. **Control bleeding.** Apply direct pressure to the edges of the wound. Avoid putting pressure directly on the impaled object.

4. **Stabilize the object with bulky dressings** (Figure 15-7). Surround the entire object. Pack the dressings around it, and tape it securely in place. A ring or doughnut pad may be used (Figure 15-8).

Object Impaled in the Cheek. When an object is impaled in a patient's cheek, bleeding can interfere with breathing. If this is the case, then remove the object (Figure 15-9). First, while maintaining an open airway, feel inside the patient's

mouth. Find out if the object has penetrated completely. Then remove the object in the direction in which it entered. Control bleeding from the outside of the cheek, and dress the wound. If the object penetrated the cheek completely, pack sterile gauze between the cheek wall and the teeth, taking care not to occlude the airway. Continue to monitor the airway and suction when necessary.

You may encounter resistance when you try to remove the object from the cheek. If so, maintain an open airway and suction as needed. Stabilize the object while you wait for the EMTs to arrive on scene. Position the patient on his or her side for drainage.

Object Impaled in the Eye. Only a physician should remove objects impaled in the eye. You must protect the patient from further injury until

Figure 15-7 Bulky dressings can help to stabilize an impaled object.

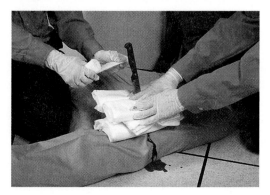

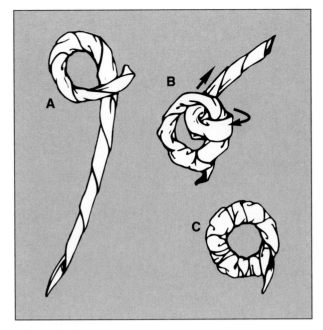

Figure 15-8 An alternative is to use a doughnut-type ring pad to stabilize an impaled object.

he or she can reach a doctor. To do so, stabilize the object in place (Figure 15-10).

Begin by stabilizing the patient's head with sandbags or large pads. Keep the patient supine. Then encircle the eye with a gauze dressing or soft sterile cloth. Do not apply pressure. You can then cut a hole in a single bulky dressing and slip it over the impaled object. Place a metal shield, a crushed cup, or a cone over the object and the eye. The sides of the shield or cup should not touch the object at all. Hold the cup and the dressing in place with a self-adhering bandage and a roller bandage that covers both eyes.

Figure 15-9 An impaled object in the cheek may be removed.

After covering the patient's eyes, do not leave him alone. He could panic. Keep in hand contact so that he knows someone is there.

Large Open Neck Wounds

Severe bleeding from a wound involving a major blood vessel of the neck is a serious emergency. In addition to the possible loss of a great deal of blood, there is danger of air being sucked into a neck vein and carried to the heart. This can be lethal. Also, suspect spine injury with any injury to the neck.

For emergency care, control of bleeding and prevention of an **air embolism** (air bubble) are your major goals. Follow these steps:

1. **Place a gloved hand over the wound immediately.**

2. **Apply an occlusive dressing.** Make sure it extends beyond the wound on all sides to prevent it from being sucked in. Tape the dressing on all four sides. Then cover it with a regular dressing.

3. **Apply direct pressure to control bleeding.** Compress the carotid artery only if it is severed.

4. **Apply a pressure dressing,** after bleeding is controlled. Do not restrict air flow or compress major blood vessels. Do not apply a dressing that circles the neck.

FIRE DRILL

Allowing an air embolus to enter the body's "pump"—or heart—will have a result similar to allowing air to enter the centrifugal pump of your fire truck. When air enters the fire truck pump, you can lose the ability to pump water as well as damage your pump. Similarly, when you allow air to enter the heart, you can lose the ability to pump adequate amounts of blood to the lungs. This can kill a person—quickly.

Eviscerations

An **evisceration** occurs when internal organs protrude from an open wound. This most commonly occurs with abdominal wounds.

Impaled Object Injury

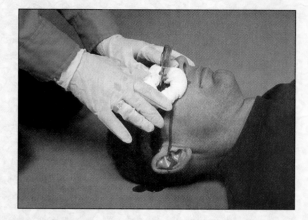

Figure 15-10a Place padding around the object.

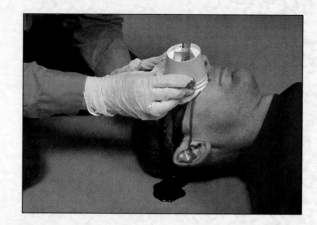

Figure 15-10b Stabilize the object with a shield or cup.

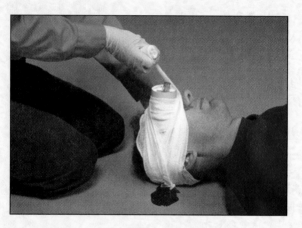

Figure 15-10c Secure the cup in place.

When you care for a patient with an evisceration, never try to replace protruding organs and never touch them. You could cause further damage and contaminate both the organs and the cavity from which they protrude.

Cover exposed organs with a thick, sterile dressing moistened with sterile water or saline. The dressing, preferably sterile gauze, should be large enough to cover all protruding organs. Never use absorbent materials such as toilet tissue or paper towels, which can shred and cling to the organs. Loosely cover the moistened dressing with an occlusive one. (See Figure 15-11.)

Maintain the temperature of the wound area. Cover the occlusive dressing with more layers of bulky dressings such as a particle-free bath blanket or towel. These dressings may be held loosely in place with a bandage or clean sheet.

Amputations

In an **amputation,** a body part has been completely severed from the body (Figure 15-12). This can be the result of ripping or tearing forces, often from an industrial accident or motor-vehicle crash. Massive bleeding may occur. In many cases, however, the elasticity of the severed arteries

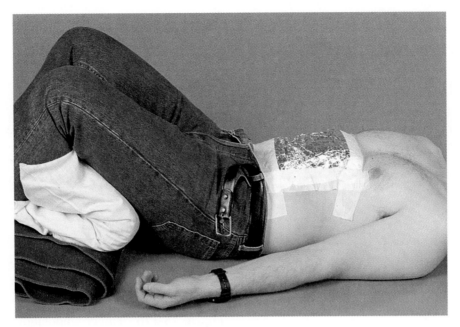

Figure 15-11 Abdominal evisceration with a thick, moist, sterile dressing and an occlusive covering.

helps them to contract, which partially controls bleeding. Care for an amputation in the same way you care for any other open injury.

With amputations, you also must care for the amputated part. First, provide emergency care to the patient. Do not spend time looking for the amputated body part. If possible, have other fire service First Responders or support personnel search for it. Once the body part is found, follow these guidelines (Figure 15-13):

1. **Wrap the amputated part.** Use sterile gauze that is moistened with sterile saline.

2. **Bag the part.** Place it in a plastic bag, labeled with the patient's name, date, and time. Never immerse the part in water.

3. **Keep the amputated part cool.** Place the bagged part in a larger bag or container of ice and water. Do not use ice alone. Never use dry ice. Never place the part directly on ice. Mark the container with the patient's name, date, and body part.

4. **Give the packed part to arriving EMS personnel,** so they can transport it with the patient to the hospital.

Note: If the amputation is partial, never complete it. Care for the injury as you would any other. Make sure that the partially amputated part is not twisted or constricted.

Avulsions

An **avulsion** is a flap of skin or soft tissue that has been torn loose or pulled off completely. (See Figure 15-14.) Healing generally is prolonged and scarring may be extensive.

Avulsions are commonly the result of accidents with industrial or home machinery and motor vehicles. They frequently involve the fingers, toes, hands, feet, forearms, legs, ears, and nose. The seriousness of the injury depends on how well blood can circulate to the avulsed skin. If it is still attached and the flap is folded back, circulation may be compromised severely. If this is the case, emergency care includes making sure the flap is lying flat and aligned in its normal position. Bandage loosely.

Bites

Bite wounds can be quite serious, whether from animals or humans (Figure 15-15). Even when they look minor, soft tissues may be badly lacerated and the threat of infection is usually high. Medical intervention may be required, especially when a tetanus shot or antibiotics are needed. Note that human bites are known to be even more dangerously infectious than most animal bites.

First Responder treatment is the same as for any patient with an open wound. However,

Amputations

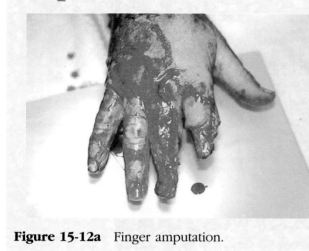

Figure 15-12a Finger amputation.

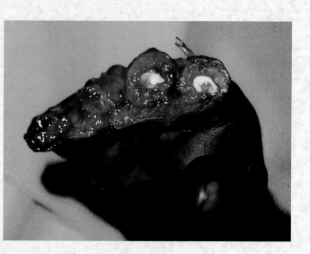

Figure 15-12b Finger amputation.

during emergency care, wash a bite wound with plenty of warm, soapy water. Check it for any teeth fragments, too.

Do not kill the animal that bit your patient unless absolutely necessary to stop an attack. If you do kill the animal, call an animal control officer and request that the corpse be examined for **rabies,** a viral infection that affects the nervous system. If you do not kill the animal, try to trap it in some kind of enclosure so that it can be exam-

ined for rabies. Take care not to injure the animal's head. Remember to protect yourself from any danger.

If the animal is not present, find out where it can be located. If getting an address is not possible, obtain a description of it, where it was encountered, and whether or not the attack might have been provoked. Follow local protocols on reporting requirements.

Figure 15-13 Emergency care for an amputated part.

(1) Wrap completely in saline-moistened sterile dressings.

(2) Place in plastic bag and seal shut.

(3) Place sealed bag on top of a cold pack or another sealed bag of ice. Do not allow the tissue to freeze.

Avulsions

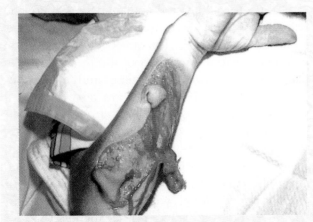

Figure 15-14a Avulsion to the forearm.

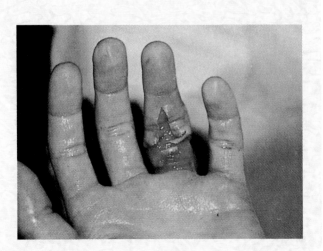

Figure 15-14b Ring avulsion.

Bites

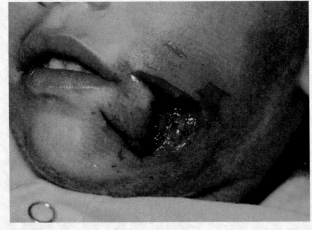

Figure 15-15a Horse bite.

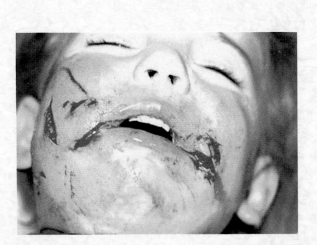

Figure 15-15b Dog bite.

SECTION 2 — DRESSING AND BANDAGING WOUNDS

The basic purposes of dressing and bandaging are to control bleeding, prevent further contamination and damage to the wound, keep the wound dry, and immobilize the wound site. Proper wound care also enhances healing. It adds to the comfort of the patient, and it promotes more rapid recovery. Improper wound care can cause infection, severe discomfort and, in rare cases, result in the loss of a limb.

Dressings

A **dressing** is a covering for a wound (Figure 15-16). It should be **sterile** (free of all microorganisms and spores). Ideally, a dressing is layered and consists of coarse mesh gauze. It also should be absorbent and large enough to protect the entire wound from contamination. In an emergency, you can use clean handkerchiefs, towels, sheets, cloth, or sanitary napkins as dressings. Never use elastic bandages, which have a tourniquet effect. Never use toilet tissue, paper towels, or other materials that can shred and cling to a wound.

Types of dressings include:

◆ *Gauze pads.* These are usually individually wrapped and sealed to prevent contamination. Take care not to touch the portion of gauze that is to make contact with the wound. Unless otherwise specified, all gauze dressings should be covered with open triangular, cravat, or roller bandages.

◆ *Bandage compress.* This is a gauze pad that is attached to the middle of a strip of bandaging material. The pad can be applied directly to an open wound with virtually no exposure to the air or your fingers. The strips of bandage can be folded back and used to tie it in place. When necessary, the sterile pad may be extended to twice its normal size by continued unfolding.

◆ *Occlusive dressing.* Made of plastic wrap, aluminum foil, or other material, this dressing is used to form an air-tight, moisture-proof seal over a wound.

◆ *Petroleum gauze.* This is sterile gauze saturated with petroleum jelly to prevent it from sticking to a wound.

Large, thick, layered, bulky pads (some with water-proof surfaces) are also available. They come in several sizes for quick application to an extremity or to a large area of the trunk. They are used to help control bleeding and to stabilize impaled objects. These pads are also referred to as *bulky dressings, trauma dressings, multi-trauma dressings, trauma packs, general purpose dressings, burn pads,* or *ABD dressings.* If a commercial bulky dressing is not available, you can improvise one with a sanitary napkin. If purchased in individual wrappers, sanitary napkins have the added advantage of cleanliness.

Bandages

A **bandage** does not make contact with a wound. It is used to hold a dressing in place, create pressure to help control bleeding, or provide support for an injured body part. Properly applied, a bandage promotes healing. It also helps the patient to remain comfortable during transport.

Bandages should be applied firmly and fastened securely. They should not be so tight as to stop circulation. They should not be so loose as to let dressings slip. If a bandage becomes unfastened, the wound could bleed or become infected.

Before bandaging, remove the patient's jewelry and other potentially restricting materials such as tape. In case of swelling, these items can restrict circulation. Loosen bandages if the skin around them becomes pale or cyanotic, if pain develops, or if the skin distally is cold, tingly, or numb.

If the pain or discomfort caused by a bandage disappears after several hours, severe damage may have already occurred. Permanent muscle paralysis may result. Please note that improper bandaging can be defined in a court of law as negligence.

Types of bandages include:

◆ *Triangular bandage.* This is used to support injured limbs, to secure splints, to form slings, and to make improvised tourniquets. It also can be used to bandage the forehead or scalp (Figure 15-17). The standard triangular bandage is made from a piece of unbleached cotton about 40 inches square, which is folded

Types of Dressings

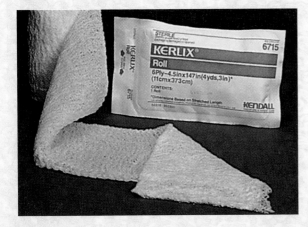

Figure 15-16a Nonelastic, self-adhering dressing and roller bandage.

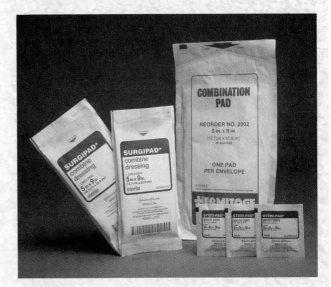

Figure 15-16b Sterile gauze pads.

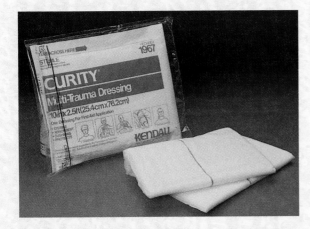

Figure 15-16c Multi-trauma dressing.

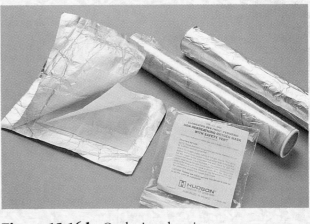

Figure 15-16d Occlusive dressings.

diagonally and cut along the fold. It can be handled and applied easily. If applied correctly, it usually remains secure. In an emergency, one can be improvised from a clean handkerchief or clean piece of shirt.

◆ *Cravat.* This is a triangular bandage that has been folded (Figure 15-18). For a wide cravat, make a one-inch fold along the base of the triangle. Bring the point to the center of the folded base, placing the point under the fold. For a medium cravat, fold lengthwise along a line midway between the base and the new top of the bandage. For a narrow cravat, fold-

ing is repeated. To complete the procedure, the ends of the bandage are tied securely.

◆ *Roller bandage.* The self-adhering, form-fitting, nonelastic roller bandage is the most popular and easy to use. Overlapping wraps cling together and can be cut and tied or taped in place. (See Figure 15-19 for how to apply a roller bandage.) Elastic roller bandages should not be used because an unintentional tourniquet effect may result.

To apply a **pressure dressing** to a bleeding wound, first cover the wound with a sterile bulky

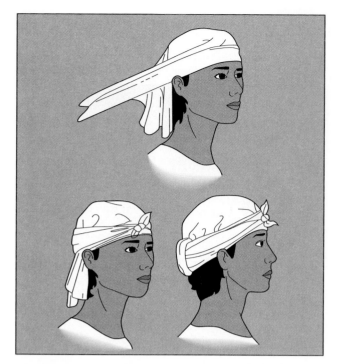

Figure 15-17 Triangular bandage for the forehead or scalp.

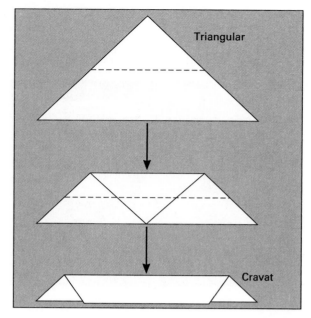

Figure 15-18 Triangular and cravat bandages.

dressing. Apply hand pressure over the wound until bleeding stops. Then apply a firm roller bandage, preferably the self-adhering type. The pressure dressing also may be used to hold some manual pressure while you use a pressure point to stop bleeding.

Principles of Application

There are no hard-and-fast rules for dressing and bandaging wounds. Often, adaptability and creativity are far more important. In dressing and bandaging, use materials you have on hand and meet the general conditions listed below (Figure 15-20):

◆ Material used for dressings should be sterile. If sterile items are not available, use a cloth that is as clean as possible.

◆ Make sure that the dressing is opened carefully. Avoid contaminating it before it reaches the wound surface.

◆ A dressing should adequately cover the entire wound.

◆ If the wound bleeds through the dressing, place additional dressings on top of the first one.

◆ Do not bandage a dressing in place until bleeding has stopped. The exception is a pressure dressing, which is meant to stop bleeding.

◆ All edges of a dressing should be covered by the bandage. There also should be no loose ends of cloth or tape.

◆ Do not bandage a wound too loosely. Bandages should not slip or shift or allow the dressings to slip or shift.

◆ Bandage wounds snugly, but not too tightly. Be sure to ask the patient how the bandage feels. Be careful not to interfere with circulation.

◆ If you are bandaging a small wound on an extremity, cover a larger area with the bandage. This will help avoid creating a pressure point, and it will distribute pressure more evenly.

◆ Always place the body part to be bandaged in the position in which it is to remain. You can bandage across a joint, but do not try bending a joint after the bandage has been applied to it.

◆ Tape bandages in place or tie them by using a square knot (Figure 15-21).

◆ Leave fingers and toes exposed when arms and legs are bandaged, so that you can check for problems with circulation.

◆ Keep the bandage neat in appearance. You will be perceived as more professional, which can result in easier patient management.

Applying a Roller Bandage

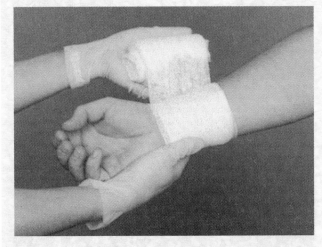

Figure 15-19a Secure the bandage with several overlapping wraps.

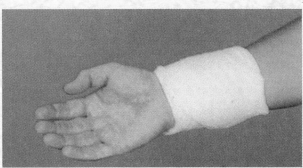

Figure 15-19b When the bandage covers an area larger than the wound, secure it with tape or tie it in place.

Bandaging

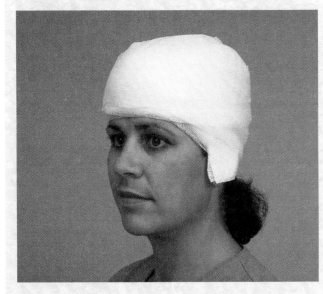

Figure 15-20a Head or ear bandage.

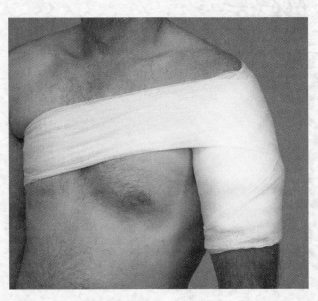

Figure 15-20b Shoulder bandage.

(continued)

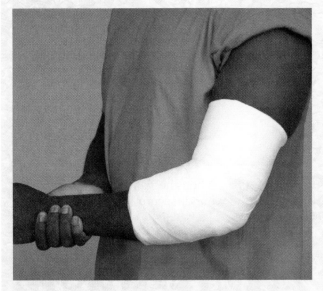

Figure 15-20c Elbow bandage.

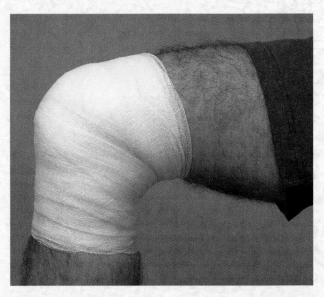

Figure 15-20d Knee bandage.

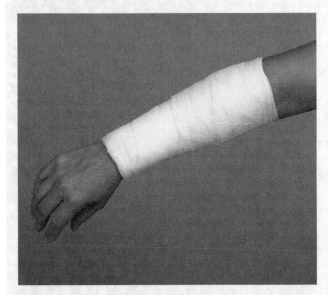

Figure 15-20e Lower arm bandage.

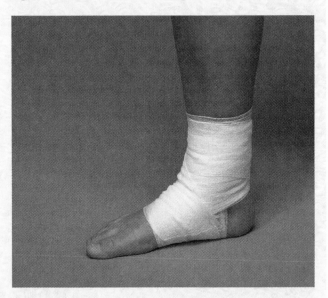

Figure 15-20f Foot or ankle bandage.

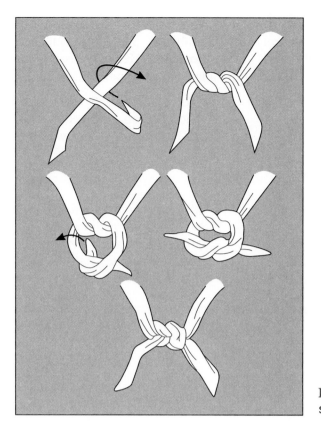

Figure 15-21 Tying a square knot.

ON SCENE FOLLOW-UP

At the beginning of this chapter, you read that fire service First Responders were on scene with a female patient who has an impaled object in one hand. To see how chapter skills apply to this emergency, read the following. It describes how the call was completed.

PHYSICAL EXAMINATION

A sliver of glass about five inches long had penetrated the palm of the patient's hand. About two inches of the glass shard protruded from each side. The patient's name was Sylvia Ortega. Sylvia hadn't passed out or fallen after the injury. She had sensation in her fingers distal to the injury, and there was no numbness or tingling. I didn't ask her to move her fingers, because I didn't want to take the chance of causing further problems. Sylvia's pulse was 88, strong, and regular. Her respirations were 18 and adequate. Her skin was cool and dry.

A member of our team stabilized the impaled object while I obtained a patient history.

PATIENT HISTORY

Though Sylvia was still upset that she had been so careless by not following standing operating procedures, we were able to find out that she had no allergies to medications or the environment. She was in good health and took no medications. Sylvia told us she never missed a day of work. She had a doughnut and coffee about an hour before the injury. She confirmed that she had placed her hand where it wasn't supposed to be. She denied any loss of consciousness before or after the object entered her hand. We also asked Sylvia and her boss to identify the substance the glass beaker had originally held.

ONGOING ASSESSMENT

Sylvia remained alert, calming down after a bit. Her airway and breathing were adequate. There was no bleeding from her wound through the bandage. We made sure the bandage wasn't too tight and that the object was being held securely. She had sensation in her fingers. Her pulse and respirations were unchanged.

PATIENT HAND-OFF

When the EMTs arrived, we told them what we had:

"This is Sylvia Ortega. She is 34 years old and has a five-inch shard of glass impaled in her left hand. She never lost consciousness and hasn't fallen or suffered any other injury. We applied some bulky dressings and then bandaged around the object so it wouldn't move. She has no allergies or medications, and she tells us that she has no medical problems. She ate a doughnut and coffee an hour ago. Her pulse is 88, strong, and regular. Her respirations are 18 and adequate."

The EMTs took Sylvia to the hospital, where a hand surgeon removed the glass from her hand. Fortunately, there was no permanent damage.

> *Whatever your trauma patient's injuries may be, remember your priorities will always be control of bleeding, preventing further injury, and reducing the risk of infection.*

Chapter Review

In your experience as a fire service First Responder, you will come across many patients with soft-tissue injuries. Some injuries will be minor, while others will be large and gaping. Keep your evaluation of these wounds in perspective with the patient's overall condition. If the patient has a laceration that is bleeding profusely, then it is your first priority. Remember, treat all life-threatening conditions first.

Also, dress and bandage a wound to keep it clean. Your patient will be very aware of these injuries. So offer reassurance as you provide care.

FIRE COMPANY REVIEW

Page references where answers may be found or supported are provided at the end of each question.

Section 1

1. How are soft-tissue injuries classified? Name the different types. (p. 276)

2. What is the recommended emergency care of closed wounds? (p. 277)

3. In general, how would you care for a patient's open wounds? (p. 280)

4. What special considerations are made in caring for open chest injuries? Impaled objects? Eviscerations? Amputated body parts? (pp. 281–285)

Section 2

5. What are the functions of dressing and bandaging? (p. 288)

6. What are five basic types of dressings? What are the special characteristics of each? (p. 288)

7. What are the general principles of dressing and bandaging? (p. 290)

RESOURCES TO LEARN MORE

Moore, E.E., et al., eds. *Trauma,* Second Edition. Norwalk, CT: Appleton-Lange, 1991.

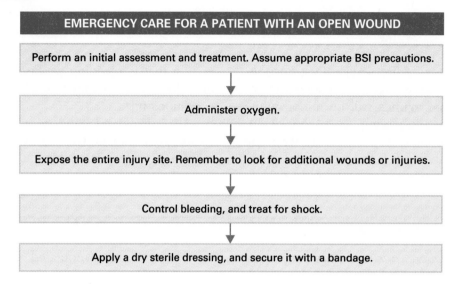

EMERGENCY CARE FOR A PATIENT WITH AN OPEN WOUND

Perform an initial assessment and treatment. Assume appropriate BSI precautions.

↓

Administer oxygen.

↓

Expose the entire injury site. Remember to look for additional wounds or injuries.

↓

Control bleeding, and treat for shock.

↓

Apply a dry sterile dressing, and secure it with a bandage.

Burn Emergencies

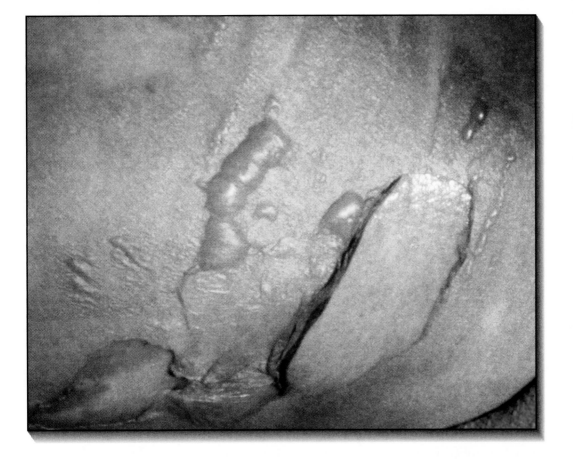

*I*NTRODUCTION   *Burns are a leading cause of accidental death in the U.S. More than two million burn accidents occur each year. Of those people who are burned, many die and many more need long-term rehabilitation. In this chapter, you will learn how to assess burns and provide emergency medical care. You also will learn about common causes of burns.*

Cognitive, affective, and psychomotor objectives are from the U.S. DOT's "First Responder: National Standard Curriculum." Enrichment objectives, if any, identify supplemental material.

Cognitive

5-2.6 Establish the relationship between body substance isolation (BSI) and soft tissue injuries. (p. 304)

5-2.13 Describe the emergency medical care for burns. (pp. 304–306)

Affective

5-2.17 Demonstrate a caring attitude towards patients with a soft tissue injury or bleeding who request emergency medical services. (p. 306)

5-2.19 Communicate with empathy to patients with a soft tissue injury or bleeding, as well as with family members and friends of the patient. (p. 306)

Psychomotor

No objectives are identified.

Enrichment

◆ List the classifications of burns. (p. 300)

◆ Define the characteristics of superficial burns, partial-thickness burns, and full-thickness burns. (pp. 300, 302)

◆ Establish the relationship between airway management and patients with burns. (pp. 299, 306, 312)

◆ Describe the emergency medical care for a patient with inhalation injuries. (p. 306)

◆ Describe the emergency medical care for a patient with chemical burns. (p. 307)

◆ Describe the emergency medical care for a patient with electrical burns. (pp. 308–309)

◆ ON SCENE

DISPATCH

It was a Saturday afternoon. I was finishing mowing the lawn when our volunteer fire department was dispatched to a house fire. I grabbed my gear, put it in my car, and responded to the scene.

As I was approaching the address, I noticed thick, dark smoke rising into the sky. I parked my car and hurried to the command post. Our fire chief, who was the incident commander, told me that there had been a woman found wandering outside the burning house by the first-due engine company. She had suffered burns to her hands and arms. As one of the only certified First Responder's currently on-scene, I was assigned to treat the patient.

SCENE SIZE-UP

The patient had been moved to a neighbor's front porch. I was joined by one of our department's other First Responders. Taking BSI precautions, we approached the patient.

INITIAL ASSESSMENT

Our general impression was that of an elderly female who appeared to be alert and talking with her neighbor. Seeing us approach, she asked, "Have you seen Fluffy, my cat?"

Our initial assessment revealed a patent airway, respirations that were adequate and of good quality, and no visible bleeding. I noted that there appeared to be soot on the patient's face. We elected to place the patient on a nonrebreather mask at 15 liters per minute. We also noted that both arms were covered by wet kitchen towels placed there by the neighbor.

> *What else should be done to assess the patient's condition? What is the proper emergency medical care? Consider this patient as you read Chapter 16.*

The skin is the largest organ of the body. Its outermost layer is the epidermis, which contains cells that give the skin its color. The dermis, or second layer, contains a vast network of blood vessels. The deepest layers of the skin contain hair follicles, sweat and oil glands, and sensory nerves. Just below the skin is a layer of fat called subcutaneous tissue. (See Chapter 4 for a cross-section illustration of the skin.)

The function of the skin includes protecting the deep tissues from injury, drying out, and invasion by bacteria and other foreign bodies. It helps to regulate body temperature, and aids in getting rid of water and various salts. It also acts as the receptor organ for touch, pain, heat, and cold. When the skin is damaged by burns, some or all of its functions may be compromised or destroyed.

Patient Assessment

Always make sure the scene of a burn accident is safe to enter. If the emergency involves noxious fumes, chemical spills, or electricity, call for specialized personnel to secure the scene before entering. Never try to rescue people trapped by fire unless you are properly equipped to do so.

Most burn patients who die in the prehospital setting die from a blocked airway, inhaled toxins, or other trauma, and not from the burn itself. As with all patients, perform an initial assessment. After you have assessed the ABCs and cared for all life-threats, determine the severity of your patient's burns. (See Table 16-1.) Severity of a burn depends on many factors, including:

◆ Depth of the burn.

◆ Extent of body surface burned.

Table 16-1

Determining Severity of Burns

	Adults	Infants and Children
CRITICAL BURNS	Full-thickness burns involving the hands, feet, face, or genitalia. Burns associated with respiratory injury. Full-thickness burns covering more than 10% of body surface area. Partial-thickness burns covering more than 30% of body surface area. Burns complicated by a painful, swollen, deformed extremity. Burns that encircle any body part; e.g., arm, leg, or chest.	Any full-thickness or partial-thickness burn greater than 20%, or burns involving hands, feet, face, airway, or genitalia.
MODERATE BURNS	Full-thickness burns of 2% to 10% of the body surface area excluding hands, feet, face, genitalia, and upper airway. Partial-thickness burns of 15% to 30% of body surface area. Superficial burns of greater than 50% body surface area.	Partial-thickness burns of 10% to 20% of body surface area.
MINOR BURNS	Full-thickness burns of less than 2% body surface area. Partial-thickness burns less than 15% body surface area.	Partial-thickness burns less than 10% of body surface area.

◆ Which part of the body was burned.

◆ Other complicating factors.

Depth of Burns

Burns typically are classified by depth (Figure 16-1). A **superficial burn** involves only the first layer of skin. A **partial-thickness burn** involves the epidermis and the dermis. In a **full-thickness burn,** the burn extends through all layers of skin and may involve subcutaneous tissue, muscles, organs, and bone. Note that burns are seldom only one depth. They usually involve a combination.

You can recognize superficial, partial-thickness, and full-thickness burns as follows:

◆ *Superficial burns* (Figure 16-2). A superficial burn is caused by flash, flame, scald, or the sun. It is the most common of all burns and is considered minor. The patient's skin surface will be dry, and there may be some swelling. Though the skin is red and painful, the burn involves only the epidermis. A superficial burn heals in 2 to 5 days with no scarring, though peeling of the burned skin may occur. Some temporary discoloration may result.

◆ *Partial-thickness burns* (Figure 16-3). This type of burn usually results from contact with hot liquids or solids, flash or flame contact with clothing, direct flame from fire, contact with chemicals, or the sun. The skin appears moist and mottled, ranging in color from white to red. The burn area is blistered and intensely painful. It usually requires 5 to 21 days to heal. If infection occurs, healing time can take longer.

◆ *Full-thickness burns* (Figure 16-4). This burn results from contact with hot liquids or solids, flame, chemicals, or electricity. The skin is dry and leathery and may be a mix of colors from white to dark brown to charcoal. Often charred blood vessels are visible. While this burn can be very painful, the patient may feel little if nerve endings have been destroyed. Small full-thickness burns require weeks to heal. Large ones, which may need skin grafting and other specialized burn care, can take

Figure 16-1   Classification of burns by depth.

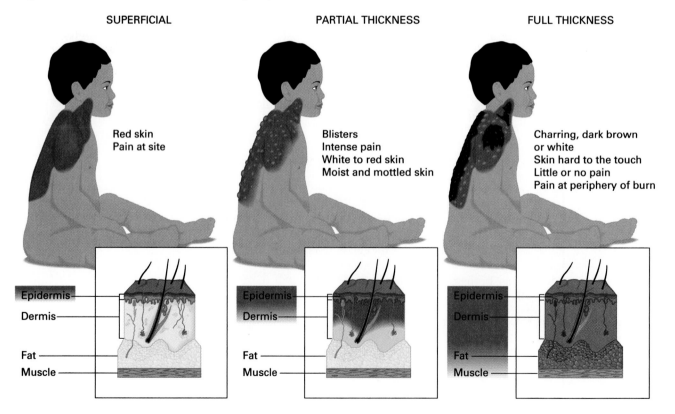

SUPERFICIAL

Red skin
Pain at site

PARTIAL THICKNESS

Blisters
Intense pain
White to red skin
Moist and mottled skin

FULL THICKNESS

Charring, dark brown or white
Skin hard to the touch
Little or no pain
Pain at periphery of burn

Epidermis
Dermis
Fat
Muscle

Epidermis
Dermis
Fat
Muscle

Epidermis
Dermis
Fat
Muscle

Superficial Burns

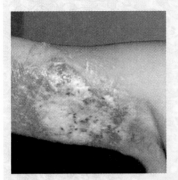

Figure 16-2a

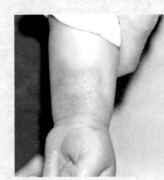

Figure 16-2b

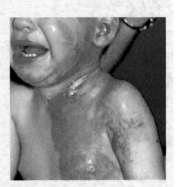

Figure 16-2c

Partial-Thickness Burns

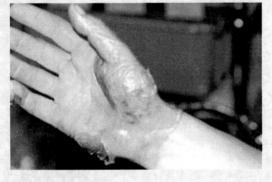

Figure 16-3a

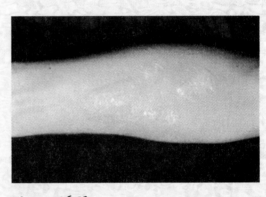

Figure 16-3b

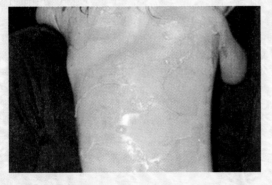

Figure 16-3c

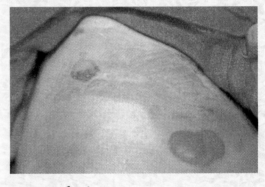

Figure 16-3d

Full-Thickness Burns

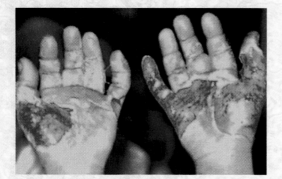

Figure 16-4a

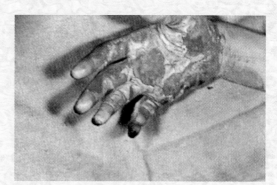

Figure 16-4b

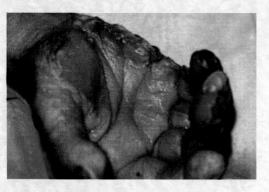

Figure 16-4c

Figure 16-4d

months or years to heal. These burns often result in scarring. Even with optimal care, full-thickness burns can prove fatal.

Extent of Body Surface Burned

The **rule of nines** is a standardized way to estimate the amount of burned **body surface area (BSA).** The head and neck region is considered to be 9% of the total body surface area. The posterior trunk is 18%, and the anterior trunk is 18%. Each upper extremity is 9%. Each lower extremity is 18%. In an infant, the head is considered to be 18% BSA and each lower extremity is 14% BSA. External genitalia are estimated as 1% BSA in all patients. (See Figure 16-5.)

An alternative method of estimating BSA is called the **palmar surface method.** With this method, use the palm of the patient's hand—approximately 1% BSA—to estimate of the size of

a burn. For example, if a burn area is equal to "7 palms," the burn would be estimated as 7% BSA.

You will find it useful to use the rule of nines to estimate the BSA of larger burn injuries and the palmar surface method for smaller ones. Follow local protocols. However, do not spend time trying to determine a burn's exact percent of BSA. Slight differences in percentages will not affect proper fire service First Responder care.

FIRE DRILL

Being overly concerned about correctly estimating BSA is the same as being overly concerned about correctly estimating the percentage of involvement of a burning structure. Close is good enough!

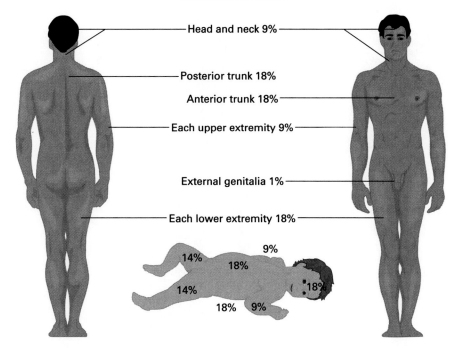

Head and neck 9%

Posterior trunk 18%

Anterior trunk 18%

Each upper extremity 9%

External genitalia 1%

Each lower extremity 18%

9%

14%

18%

14%

18%

18%

9%

Figure 16-5 Rule of nines.

Location of Burns

Burns to certain areas of the body are more critical than others (Figure 16-6). For example, burns to the face can acutely compromise breathing or cause injury to the eyes. Burns that encircle the chest can limit its ability to expand, which can result in inadequate breathing. Burns that encircle a body part—such as a joint, arm, or leg—are considered critical because of the possibility of blood-vessel and nerve damage. Other serious complications can occur as a result of scarring and the healing process; for example, burns to the hands or feet can cause a loss of function, and burns to the genital area may result in loss or impairment of genitourinary function.

For patients with burns to any of the areas mentioned above, arrange for transport to a hospital or burn center immediately.

Complicating Factors

Patients with other injuries or chronic diseases, such as heart disease or diabetes, will always react more severely to burns, even minor ones. So, try to determine the patient's medical history early in the course of care.

The age of a patient may be a complicating factor. Children under the age of 5 years and adults over the age of 55 tolerate burns poorly. In an elderly patient, a burn covering only 20% of body surface can be fatal. Because the elderly and very young generally have thin skin, they can sustain much deeper burns. Fluid loss from a burn also can affect the elderly and young more critically. Even a small fluid loss can result in serious problems. Infection may be an additional problem, because the immune system is immature in children and usually compromised in the elderly patient.

PEDIATRIC NOTE

Please note that burns may be the result of child abuse. Look for burn patterns that indicate a child might have been dipped in scalding water. Cigarettes also are used to burn children as a form of abuse. Follow local protocols for reporting your observations.

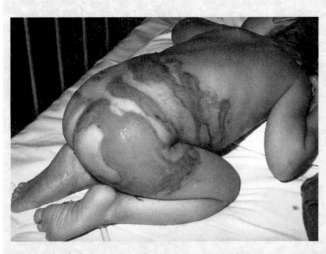

Figure 16-6a Scalds burns encircling the torso of a child. *(Source: Shout Picture Library)*

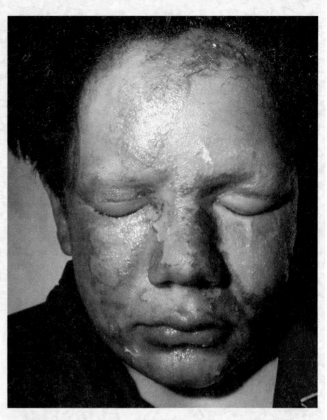

Figure 16-6b Burns to the face from an exploded gas cylinder. *(Source: Shout Picture Library)*

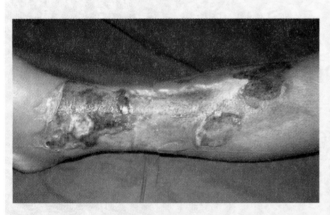

Figure 16-6c Petroleum burns to the leg. *(Source: Shout Picture Library)*

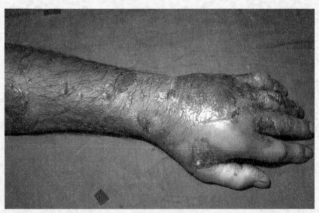

Figure 16-6d Cooking oil burns to the arm and hand. *(Source: Shout Picture Library)*

◆ *Emergency Care*

Once the patient has been removed from the source of the burn, provide emergency care. Take BSI precautions and follow these steps (Figure 16-7):

1. **Stop the burning process.** If flames or smoldering clothing are present, extinguish them with water or saline. If a semi-solid material like grease, wax, or tar is burning on the patient's skin, cool it with water or saline, but

Emergency Care of Burn Injuries

Figure 16-7a Stop the burning process.

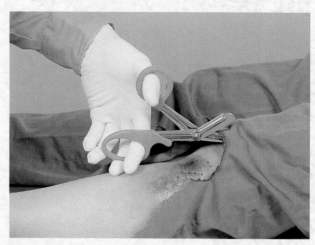

Figure 16-7b Remove smoldering clothing and jewelry.

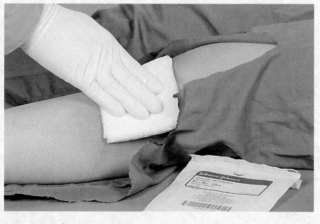

Figure 16-7c After life-threats have been treated and a physical exam completed, cover burns with dry sterile dressings.

do not attempt to remove the material. Remove any smoldering clothing and jewelry, but if you meet resistance, avoid the area by cutting around it. Flush away chemicals with water for 20 minutes or more. Note: if you use water or saline on a large body surface area, stay alert to avoid hypothermia in your patient.

2. **Perform an initial assessment**. Treat all life-threats. Administer oxygen by nonrebreather mask, if allowed. If your patient's breathing is inadequate, provide ventilations with supplemental oxygen.

3. **Determine the severity of the patient's burns.** During the physical exam, take into account the depth of burns, extent of BSA involved, location of burns, and complicating factors. Do not forget to look for other possible injuries.

4. **Cover the burns.** Use dry sterile dressings or a disposable sterile burn sheet. Do not use grease or fat, ointment, lotion, antiseptic, or ice on the burns. Do not break any blisters. If a burn involves an eye, be sure to apply dressings to both eyes (Figure 16-8).

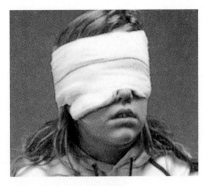

Figure 16-8 Apply dressings to both eyes, even if only one is burned.

5. **Keep the patient warm,** and treat other injuries as needed.

Proper care of burns must start as soon as possible after the injury. Loss of body fluids, pain contributing to shock, swelling, and infection may quickly follow a burn injury. Be especially alert to any sign of breathing difficulty. If burns were caused by an electric source, monitor the patient closely for cardiac arrest. Be prepared to administer CPR and, if you are trained and equipped, to apply an automated external defibrillator (AED). Remember that the patient's status can change suddenly, so monitor vital signs continually.

As always, do your best to calm and reassure the patient. He or she may be in a great deal of pain. The patient and family members also may be afraid of permanent scarring and disfigurement. Tell them that you are doing what is necessary to prevent further injury and contamination. Let them know that additional EMS personnel are on the way.

Also, note that patients and family members may have tried to treat the burns before you arrived on scene. Find out what they did. One reason why burns are so often critically damaging or even fatal is that some individuals are poorly informed about methods of care. Include this information in your patient hand-off report.

SPECIAL TYPES OF BURN INJURIES

Inhalation Injuries

Greater than half of all fire-related deaths are caused by smoke inhalation. About 80% of those who die in residential fires do so only because they have inhaled heated air or smoke and other toxic gases. Suspect inhalation injury in any patient who was burned in a fire, especially if the patient was confined in an enclosed space.

The severity of an inhalation injury is determined by the following factors:

◆ Products of combustion (what was burned).

◆ Degree of combustion (how completely it was burned).

◆ Duration of exposure (how long the patient was exposed to the smoke or gas).

◆ Whether or not the patient was in a confined space.

Most upper airway damage from heat consists of scorched mucous membranes and swelling, which can block the airway. Specific signs and symptoms include (Figure 16-9):

◆ Singed nasal hairs.

◆ Burns to the face (Figure 16-10).

◆ Burned specks of carbon in the sputum.

◆ Sooty or smoky smell on the breath.

◆ Respiratory distress.

◆ Noisy breathing.

◆ Hoarseness, cough, difficulty speaking.

◆ Restricted chest movement.

◆ Cyanosis.

If any of these signs and symptoms are present, administer high-concentration, humidified oxygen if available. Note that this type of injury may appear to be mild at first and then become more severe. Monitor the patient's airway and breathing closely. Be prepared to assist ventilations if necessary.

COMPANY OFFICER'S NOTE

Suspect inhalation injury in any patient who has been burned at a structure fire. Be especially wary if that patient was burned in other parts of the body. Also, take care not to become distracted by a visible burn. Do not forget the patient's ABCs—airway, breathing, and circulation.

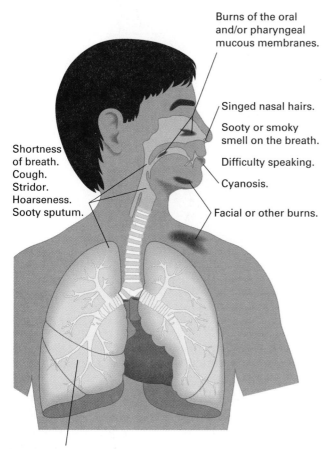

Burns of the oral and/or pharyngeal mucous membranes.

Singed nasal hairs.

Sooty or smoky smell on the breath.

Difficulty speaking.

Cyanosis.

Facial or other burns.

Shortness of breath. Cough. Stridor. Hoarseness. Sooty sputum.

Respiratory distress. Noisy breathing. Restricted chest movement.

Figure 16-9 Signs and symptoms of inhalation burns.

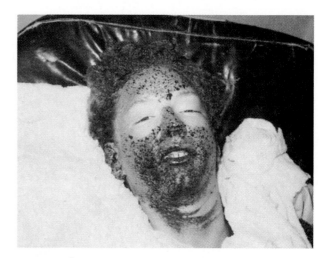

Figure 16-10 Burns to the face.

Chemical Burns

It is very difficult to assess the severity of chemical burns in the field. The general guideline is to treat all chemical burns as critical. Speed is essential. The faster you stop the burning process and initiate care, the less severe the burn will be. Follow these guidelines:

◆ Remember scene safety, and make sure that it is safe to approach the patient. If not, wait for the hazmat team to secure the scene. When you can approach your patient, wear the appropriate protective gear to avoid contamination.

◆ Immediately begin to flush the patient's burns vigorously with water. Do not waste time trying to find an antidote. If the patient is at home, use the shower or a garden hose. Irrigate the area continuously under a steady stream for at least 20 minutes.

◆ If chemical burns affect the eyes, flush them with water (Figure 16-11). Use a faucet or a hose running at low pressure. If necessary, use a pan, bucket, cup, or bottle. Have the patient remove any contact lenses.

◆ Minimize further contamination by making sure the water runs away from the injury and not toward any uninjured parts.

Note that you should brush off dry chemicals, such as lime powder, before flushing with water (Figure 16-12). Be sure you do not contaminate yourself. Also, ask for a copy of the material safety data sheets (MSDS) and other reference materials for specific care of a chemical exposure. Often these are available at the work site. Give them to the EMTs when they arrive. Poison control or medical direction will also have information about specific chemical agents.

 COMPANY OFFICER'S NOTE

Remember to irrigate where chemicals can "hide." For example, run water over the folds of the buttocks, under the breasts, over the scalp, and in the arm and leg "pits." Chemicals also can pool in shoes. It may be appropriate to immerse or flush the patient in copious amounts of water in order to decontaminate him or her.

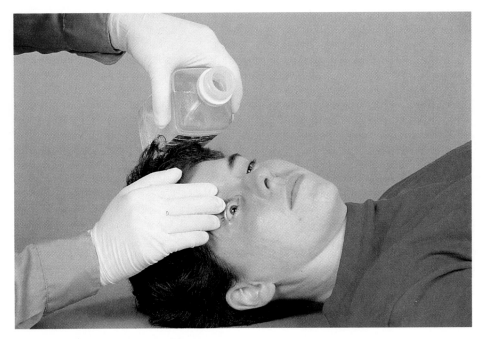

Figure 16-11 Flushing chemical burns to the eye.

Electrical Burns

Electric burns may occur in a variety of household and industrial settings. If you approach an emergency scene involving electrical hazards, make a visual sweep for power cords. Pull the plug before you approach or touch the patient. Remember that a power tool does not have to be "on" to present a shock hazard. In an industrial setting, make sure all power sources are locked out and tagged out before approaching the patient. In general, you should never try to remove a patient from an electrical source unless you are trained and equipped

Figure 16-12 Lime powder should be brushed off the skin before flushing.

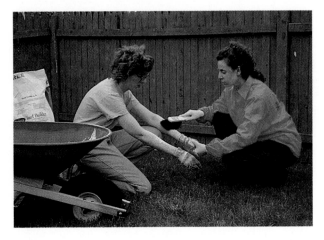

to do so. Never touch a patient still in contact with an electrical source.

In any incident involving a car crash into a power pole, look for downed power lines. Sometimes they are hidden from sight by grass or bushes. So look at the next pole down the line, and count the number of power lines at the top crossarm. There should be the same number of lines at the top of the damaged pole. If the number is not the same, then proceed as follows:

◆ If you suspect that lines are down or the power pole has been weakened, notify all rescue personnel of the possible danger. Then notify the power company and request an emergency crew.

◆ If the soles of your feet tingle when you enter the area, go no further. You are entering an energized zone.

◆ Assume that a downed power line is live until the power company crew tells you otherwise. Remember that vehicles, guard rails, metal fences, and so on conduct electricity.

◆ If the patient's vehicle is in contact with a downed power line, tell the patient to stay inside the car. Maintain a safe distance. Never have a patient try to jump clear unless there is immediate danger of fire or explosion. Do not touch the vehicle and the ground at the same time. If you do, the current can kill you.

Never try to remove a power line. Personnel from the electrical company must do it. They have the training and the proper equipment to handle the line safely.

Signs and symptoms of electric shock may include: altered mental status; obvious severe burns; weak, irregular, or absent pulse; shallow, irregular, or absent breathing; multiple fractures due to intense muscle contractions.

Care for a patient with electrical burns the same way you would care for any patient with burns. However, note that an electric shock can throw a patient a significant distance, so stabilize the patient's head and neck during assessment and treatment. Be sure to look for both entry and exit burns (Figure 16-13). Note also that patients who experience an electric shock may experience cardiac problems, including cardiac arrest. Be prepared to provide basic life support. (See examples of electrical burns, Figure 16-14.)

Lightning Injuries

Thousands of electrical injuries occur each year in the U.S. About 25% of them are lightning injuries.

Figure 16-13   Look for an entry burn and an exit burn.

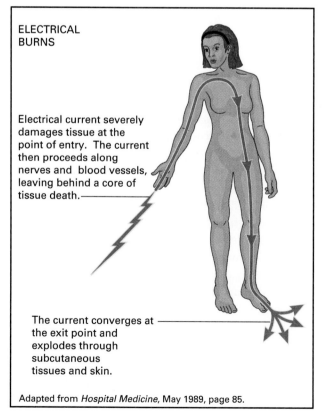

ELECTRICAL BURNS

Electrical current severely damages tissue at the point of entry. The current then proceeds along nerves and blood vessels, leaving behind a core of tissue death.

The current converges at the exit point and explodes through subcutaneous tissues and skin.

Adapted from *Hospital Medicine*, May 1989, page 85.

A lightning bolt can pack more than a trillion watts of electricity and up to 100 million volts. Much of the electrical energy from lightning flows around, not through, a strike victim. A patient who has been struck by lightning does not hold a charge, so it is safe to approach him or her.

People are struck by lightning most often in open fields, under trees, on or near water, near tractors and heavy equipment, on golf courses, and at telephones. A person may be struck directly by lightning or lightning may "splash" off a nearby object. Whole groups of people can be affected by a ground strike in which lightning hits the ground and electricity ripples outward.

Most victims of lightning are knocked down or thrown, so assume spine injury. Also assume that a victim of a lightning strike has multiple injuries.

◆ *Nervous system.* In many instances of lightning strike, the patient becomes unresponsive. Few actually remember being struck. Some patients suffer partial paralysis. Occasionally paralysis of the respiratory system causes death.

◆ *Senses.* Some patients experience a loss of sight, hearing, and the ability to speak. Rupture of one or both tympanic membranes (eardrums) occurs in 50% of patients who have been struck by lightning.

◆ *Skin.* A lightning burn may be red, mottled, blue, white, swollen, or blistered. The skin also may appear to be feathery, patchy, or in a scattered pattern resembling flowers. This is called "ferning," which usually fades away in a few hours.

◆ *Heart.* The lightning strike itself can disrupt the heart's rhythm and lead to full cardiac arrest.

◆ *Vascular system.* Within seconds following a strike, the patient may become unresponsive, appear pale and mottled, have cool arms and legs, and lose pulses. If the injury is moderate, the conditions may correct themselves quickly. In case of severe injury, blood may coagulate and tissues in the arms and legs may die, leading to amputation. Kidney failure may result.

Care for lightning burns as you would any other type of burn. Provide manual stabilization of the patient's head and neck, and be prepared to provide basic life support. Such measures should continue even if the patient appears to be lifeless. Victims of lightning have been resuscitated as long as 30 minutes after a strike without any lasting damage.

Electrical Burns

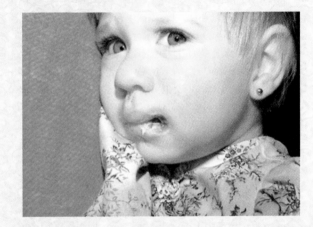

Figure 16-14a Electrical burn caused by chewing on an electrical cord.

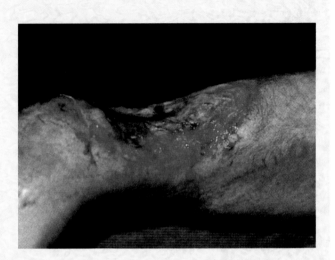

Figure 16-14b Full-thickness electrical burns.

ON SCENE FOLLOW-UP

At the beginning of this chapter, you read that fire service First Responders were on the scene of a structure fire where an elderly female patient had suffered burns to both arms. To see how chapter skills apply to this emergency, read the following. It describes how the call was completed.

PHYSICAL EXAMINATION

We asked our patient her name and she told us it was Adelia O'Connell. My partner began monitoring Mrs. O'Connell's breathing, and because of the soot on her face, placed her on a nonrebreather mask and administered oxygen at 15 liters per minute.

Meanwhile, I conducted a head-to-toe exam. Gently lifting the towels from both arms, I noted that the lateral surface of each arm was red and beginning to form blisters. I estimated the burn surface to be about 3% to 5% on each arm. I also observed that the burns were not circumferential. There appeared to be no other injuries. I replaced the towels with dry sterile dressings. After I took a set of vital signs, which were

within normal ranges, Mrs. O'Connell once again asked about her cat. We assured her that other firefighters were doing everything they could to locate it.

PATIENT HISTORY

We asked our patient what happened. She said that she had been outside hanging clothes on the line, when she noticed smoke blowing out her kitchen window. Remembering that she had been cooking cabbage and potatoes, she rushed to the kitchen but could not enter because of fire and smoke. She then ran outside to ask a neighbor to call the fire department. While doing so, she watched her cat walk into the house. Mrs. O'Connell went back into the house to get her pet but was unable to do so due to the heat. She said she inhaled quite a bit of smoke and burned her arms while protecting her face from the smoke. She told us she had no medical problems and was not allergic to anything.

ONGOING ASSESSMENT

We took the patient's vital signs every five minutes or so until the EMTs arrived. There were no changes noted. We continued administering oxygen, and closely watched our patient's breathing, continually asking her if she was experiencing any tightness in her chest or throat.

PATIENT HAND-OFF

My hand-off report to the EMTs was as follows:

"This is Adelia O'Connell. She is 60 years old. She suffered what appears to be partial-thickness burns on both of her forearms. I would estimate that the total burn area is no more than 5%. She was burned when she ran into her burning house to find her cat. I am concerned that she may have suffered an inhalation injury, as we noted soot around her mouth and on her face. She has been alert throughout. Vital signs are: respirations of 20 and of good quality; pulse is 98 and strong; BP is 138/84. Her skin is warm and dry. She is not complaining of any difficulty breathing at this time. Both arms were wrapped in wet kitchen towels when we arrived. We replaced these with dry, sterile dressings. We also placed her on a nonre-breather mask and have been administering humidified oxygen at 15 liters per minute."

After the EMTs took over, we returned to the command post, where we were assigned to assist with salvage and overhaul. The patient's cat was never found.

Treat all burn injuries in basically the same way. Stop the burning process, remove smoldering clothing, and dress the wounds. However, remember that after the initial assessment and treatment of life-threats, you must perform a thorough physical exam to find and treat other injuries the patient may have. Gather a good patient history. Then perform an ongoing assessment until the EMTs arrive to take over patient care.

Chapter Review

FIREFIGHTER FOCUS

Burns can be painful, disfiguring injuries. Burns to the airway can cause swelling, which leads to obstruction, which leads to inadequate breathing or respiratory arrest. Monitor the patient very carefully for these conditions. Remember that the care you provide as a fire service First Responder will be vital to the patient's survival.

Safety also is a primary concern at the scene of burn injuries. Remember that the source of a burn, such as flames or chemicals, and smoke- or vapor-filled environments are dangerous. Never enter a hazardous scene unless you are properly trained and equipped to do so. If you are properly trained and equipped, remove the patient from the hazardous environment and begin emergency medical care. Remember, do not become part of the problem. Instead, become part of the solution.

FIRE COMPANY REVIEW

Page references where answers may be found or supported are provided at the end of each question.

Section 1

1. When you estimate how severe a patient's burn may be, what four factors should you consider? (pp. 299–300)

2. What is a superficial burn? Briefly describe its characteristics. (p. 300)

3. What is a partial-thickness burn? Briefly describe its characteristics. (p. 300)

4. What is a full-thickness burn? Briefly describe its characteristics. (pp. 300, 302)

5. What is the "rule of nines"? Describe how it is used. (p. 302)

6. What is the "palmar surface method"? (p. 302)

7. Are burns to certain areas of the body more critical than others? Explain. (p. 303)

8. Do age and chronic illness have anything to do with how severe a burn may be? Explain. (p. 303)

Section 2

9. What are the general guidelines for emergency medical care of burns? (pp. 304–306)

10. What special care would you provide if your patient had inhalation burns? (p. 306)

11. How would you stop the burning process if a dry chemical such as lime powder is involved? (p. 307)

12. What special precautions related to scene safety should you take if you are called to the scene of an electrical hazard? (pp. 308–309)

RESOURCES TO LEARN MORE

Ball, R.A. "Hot Stuff: Assessing and Treating Burns." *JEMS,* February 1993.

Bourne, M.K. "Fire and Smoke: Managing Skin and Inhalation Injuries." *JEMS,* September 1989.

"Electrical/Lightning Injuries," *Emergency Medical Update,* June 1992.

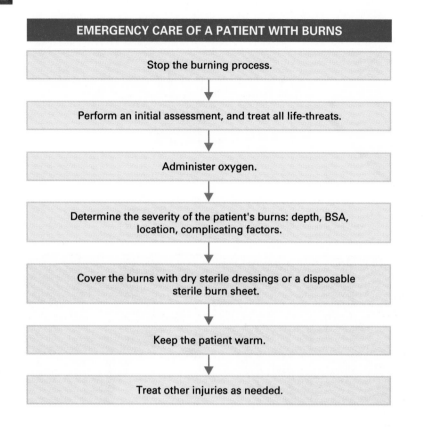

EMERGENCY CARE OF A PATIENT WITH BURNS

Stop the burning process.

Perform an initial assessment, and treat all life-threats.

Administer oxygen.

Determine the severity of the patient's burns: depth, BSA, location, complicating factors.

Cover the burns with dry sterile dressings or a disposable sterile burn sheet.

Keep the patient warm.

Treat other injuries as needed.

CHAPTER 17

Musculoskeletal Injuries

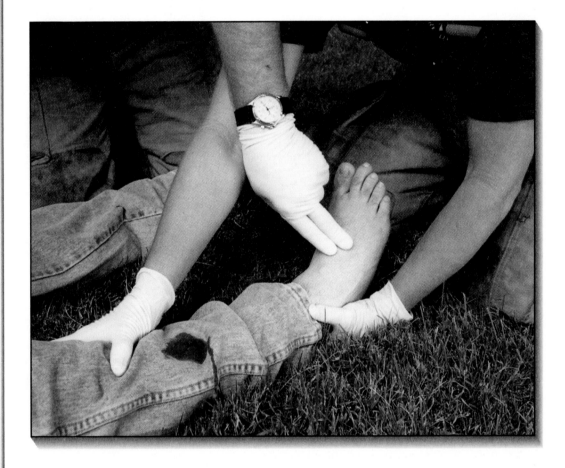

*I*NTRODUCTION *As a fire service First Responder, you will encounter many patients with injuries to muscles, joints, and bones. The injuries will range from mild, such as a pulled muscle or twisted ankle, to life-threatening femur or pelvic fractures. Regardless of whether the injury is mild or severe, your ability to assess your patient and provide the appropriate emergency care can help prevent permanent disability and disfigurement.*

Cognitive, affective, and psychomotor objectives are from the U.S. DOT's "First Responder: National Standard Curriculum." Enrichment objectives, if any, identify supplemental material.

Cognitive

5-3.1 Describe the function of the musculoskeletal system. (p. 316)

5-3.2 Differentiate between an open and a closed painful, swollen, deformed extremity. (pp. 316–317)

5-3.3 List the emergency medical care for a patient with a painful, swollen, deformed extremity. (pp. 318–319)

Affective

5-3.9 Explain the rationale for the feeling patients who have need for immobilization of the painful, swollen, deformed extremity. (p. 323)

5-3.10 Demonstrate a caring attitude towards patients with a musculoskeletal injury who request emergency medical services. (pp. 320–321, 323)

5-3.11 Place the interests of the patient with a musculoskeletal injury as the foremost consideration when making any and all patient care decisions. (p. 321)

5-3.12 Communicate with empathy to patients with a musculoskeletal injury, as well as with family members and friends of the patient. (pp. 320–321, 323)

Psychomotor

5-3.13 Demonstrate the emergency medical care of a patient with a painful, swollen, deformed extremity. (pp. 318–319)

Enrichment

◆ State the reasons for splinting. (p. 320)
◆ List the general rules of splinting. (p. 321)
◆ List the complications of improper splinting. (p. 321)
◆ Describe several different types of splints. (p. 320)
◆ Describe splinting of the upper extremities. (pp. 321, 323–324)
◆ Describe splinting of the lower extremities. (pp. 324–328)

ON SCENE

DISPATCH

Dinner was done, and we were watching what promised to be an exciting game on Monday Night Football when the tone went off for our station. "R2, M2, respond to a fall injury at Walsh's Department Store. This is a 55-year-old female who slipped and fell at the checkout counter. The manager will meet you at the front entrance."

SCENE SIZE-UP

As we pulled into the plaza parking lot nearest Walsh's, we could see a woman waving us to come closer. We stopped our vehicle in front of her. She identified herself as store manager Maggie McGrail.

"Hi, guys," she said. "I think she might have hurt her hip. She's right inside the front door."

We pulled our units into a safe position, one that would not block the flow of traffic through the shopping plaza. As we exited our vehicle and gathered our equipment, a police officer arrived and entered the store.

INITIAL ASSESSMENT

As we followed Maggie into the store, I noticed the police officer kneeling beside a woman who was supine on the floor. She had a jacket underneath her head as a pillow and appeared to be carrying on an animated conversation with the police officer. Looking up as we approached, she smiled and said, "Oh my, four handsome young firefighters to my rescue. If I could get up, I'd ask each and every one of you to dance!"

Smiling, I introduced myself, simultaneously noting that her right leg appeared to be turned slightly outward.

"Ma'am, do you hurt anywhere?" I asked.

"Well, yes I do," she replied. "My right hip is really paining me. You know, I had a hip replacement last year. Why, I was just standing at the counter, flirting with the gentleman in front of me when I felt a 'pop' and fell forward against the counter. Thank goodness, these nice folks helped me to the floor."

She appeared to be alert and very cooperative. Gently taking her wrist, I could tell her pulse was strong and of a normal rate. Her skin was warm and dry. Judging by the way she was speaking, her breathing appeared to be fine. I noted no bleeding. She denied having any pain other than to her right hip.

Consider this patient as you read Chapter 17. What may be done to assess and treat her condition?

INJURIES TO BONES AND JOINTS

The musculoskeletal system is made up of more than 200 bones and over 600 muscles. Together they give the body shape, protect internal organs, and provide for movement. Any time bones and muscles are injured, one of those functions is either temporarily or permanently impaired. (You may wish to turn to Chapter 4 to review system components at this time.)

Bones and muscles may be injured in four basic ways:

- A bone is broken *(fracture)*.
- A muscle or a muscle and tendon are overextended *(strain)*.
- A joint and ligament are injured *(sprain)*.
- A bone is moved out of its normal position in a joint and remains that way *(dislocation)*.

Mechanisms of Musculoskeletal Injury

As you conduct your scene size-up, consider the mechanism of injury. A mechanism of musculoskeletal injury may involve direct, indirect, or twisting forces (Figure 17-1). They can give you a good idea of how extensive an injury may be.

With a direct force, an injury occurs at the point of impact. For example, imagine that a patient is in a car crash. When she is thrust forward, one of her knees strikes the dashboard. The resulting injury to the knee is caused by that direct force, or direct blow.

With an indirect force, the energy of a blow travels along a path away from the point of impact. For example, think of a patient who falls onto her outstretched hand. The force of the blow can travel from her hand and wrist up through her arm and shoulder. The injuries caused by the indirect force could include broken arm bones and even a broken clavicle. So, look beyond the injury caused by a direct force when you examine a trauma patient. Additional injuries may be involved.

With a twisting force, one part of a limb remains stationary while the rest of it twists. An example would be the case of a firefighter jumping off a fire truck and getting her foot caught in a rut on the ground. When she falls, the body would pull the leg one way, while the trapped foot would hold it firmly in its original position. That could twist the limb, causing any of its bones or joints to break. Again, suspect injuries beyond the most obvious one when you examine your patient.

Patient Assessment

A musculoskeletal injury is classified as either closed or open (Figure 17-2). In a closed injury, the skin is not broken at the injury site. It remains intact. In an open injury, the skin is broken, perhaps by protruding bone ends. Signs and symptoms of musculoskeletal injury include (Figure 17-3):

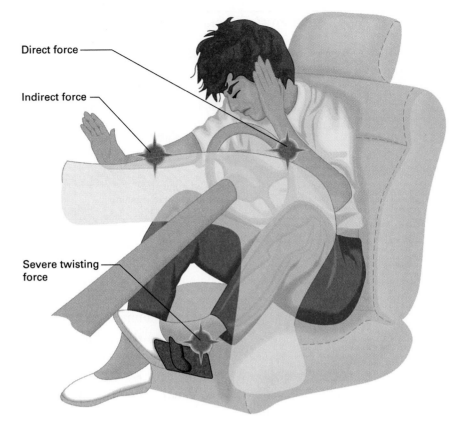

Direct force

Indirect force

Severe twisting force

Figure 17-1 Different types of force can cause different types of injuries.

Figure 17-2 Closed injury vs. open injury.

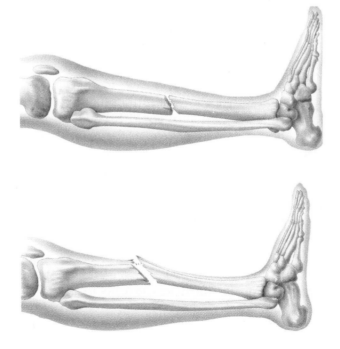

◆ Deformity, or **angulation.** When compared to the uninjured limb, the injured one is a different size or has a different shape (Figure 17-4).

◆ Pain and tenderness.

◆ Grating, or **crepitus.** This is the sound or feeling of broken bones grinding against each other.

◆ Swelling.

◆ Bruising, or discoloration.

◆ Exposed bone ends (Figure 17-5).

◆ Joint locked in position.

An injury that causes pain, swelling, or deformity in an extremity may be the result of a fracture, sprain, strain, or dislocation. Because these injuries look so much alike in the field, you do not need to figure out which is which. Instead, always treat a painful, swollen, or deformed extremity as if it involved a broken bone.

During your physical exam of the patient, be sure to assess an injured extremity for distal pulses, movement, and sensation. Also, remember that a patient with a musculoskeletal injury may

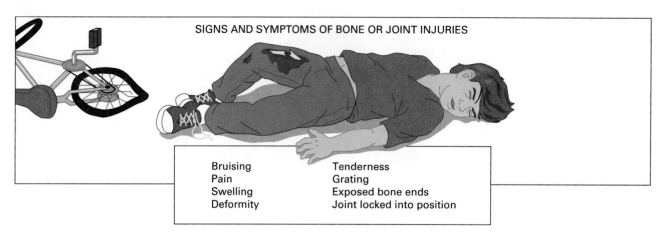

SIGNS AND SYMPTOMS OF BONE OR JOINT INJURIES

Bruising	Tenderness
Pain	Grating
Swelling	Exposed bone ends
Deformity	Joint locked into position

Figure 17-3   Signs and symptoms of bone or joint injuries.

be in a great deal of pain. Be careful not to move the injured limb or jar the body. Be gentle and reassuring to the patient and his or her family.

◆ **Emergency Care**

As a fire service First Responder, you must not be distracted by gruesome-looking injuries, especially when treating a patient with multiple trauma. Simply, your priority is "life before limb." Remain focused on treating the life-threats you identify in the initial assessment. Once that is done, you can turn to limb-threatening injuries.

Generally, after taking BSI precautions, emergency care proceeds in the following manner:

1. **Identify and treat life-threats.** Maintain manual stabilization of the patient's spine, if indicated. Treat for shock. Administer oxygen, if it is available.

2. **Stabilize the injured extremity** after you have completed a physical exam. Hold it man-

Figure 17-4   Deformity, or angulation, of an injured limb. (*Source: Shout Picture Library*)

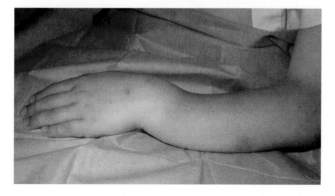

ually above and below the injury site. Maintain manual stabilization until the limb is completely immobilized in a splint.

3. **Expose the injury site.** To avoid jarring the limb, you may need to cut away clothing. Remove jewelry, too, if possible.

4. **Treat any open wounds.** Control bleeding. If necessary, apply pressure to the appropriate pressure point. Be careful to avoid applying any pressure to broken bone ends. Then dress any open wounds with sterile dressings.

5. **Allow the patient to rest in a position of comfort** while you wait for the arrival of the EMTs. Apply a cold pack to the injured area to help reduce pain and swelling. You also may wish to pad under the patient's injured limb to prevent discomfort. Continue to assess for pulse, movement, and sensation below the injury site. Record any changes.

Manual stabilization of the injured limb is maintained to help prevent further injury. Without it, a simple closed injury could become an open one. Another important reason for manual stabilization is to reduce pain.

While you are stabilizing an injured limb, do not intentionally (or unintentionally, if you can help it) replace any protruding bones. Also, do not apply **manual traction** (pull the limb) in an attempt to straighten it or realign the bones, except when you are authorized to do so. Only specially trained medical personnel should attempt traction in the field. Be sure to follow all local protocols.

Open Musculoskeletal Injuries

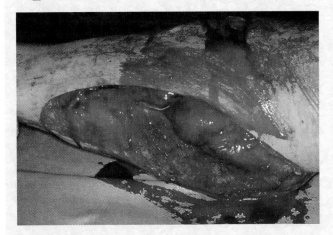

Figure 17-5a Open injury. *(Source: Charles Stewart & Associates)*

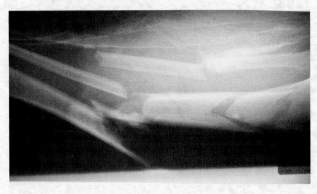

Figure 17-5b X-ray of limb in Figure 17-4a, showing broken bones both above and below the surface. *(Source: Charles Stewart & Associates)*

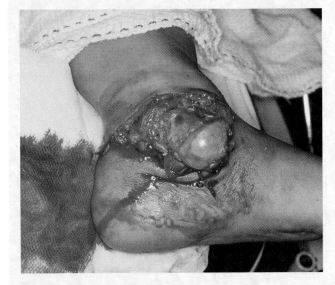

Figure 17-5c Open injury. *(Source: Charles Stewart & Associates)*

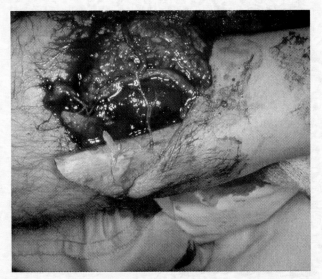

Figure 17-5d Open injury. *(Source: Charles Stewart & Associates)*

Maintain manual stabilization of an injured extremity until it is completely immobilized with a splint. Even if you find that you have to stay in an uncomfortable position for some time, maintain stabilization. The patient's best interests must be your foremost consideration. If you are trained and allowed to do so, splint the injured extremity after you have performed the steps described above.

SECTION

2 SPLINTING MUSCULOSKELETAL INJURIES

Any device used to immobilize a body part is called a **splint.** A splint may be soft or rigid. It can be commercially manufactured or it can be improvised from virtually any object that can immobilize the limb. (See Figure 17-6.)

Figure 17-6 Examples of splints.

There are five good reasons for splinting a musculoskeletal injury:

◆ To prevent motion of bone fragments or dislocated joints.

◆ To minimize damage to surrounding tissues, nerves, blood vessels, and the injured bone itself.

◆ To help control bleeding and swelling.

◆ To help prevent shock.

◆ To reduce pain and suffering.

Note that fire service First Responders may not be allowed to immobilize musculoskeletal injuries in your EMS system. Make sure you follow all local protocols.

Types of Splints

Some common types of splints are rigid splints, traction splints, circumferential splints, improvised splints, and the sling and swathe. All are designed to accomplish the same task. They must immobilize an injured extremity.

Rigid Splints

Rigid, padded boards are the most common type of splint. They may be made of wood, aluminum, wire, plastic, cardboard, or compressed fibers. Some are shaped specifically for arms or legs. Others are pliable enough to be molded to fit any appendage. Some come with washable pads. Others must be padded before being applied.

A rigid splint must be applied in line with the bone. Then it must be anchored to the limb with cravats that are secured with square knots (or straps or Velcro closures). Remember, never place a cravat across the injury site. It could cause further injury and pain.

Traction Splints

A traction splint is a mechanical device that provides a counter-pull to alleviate pain, reduce blood loss, and minimize further injury. It does not realign broken bones. Several types of traction splints are available, and application procedures vary according to manufacturer. As a fire service First Responder, you should use a traction splint only if you are specifically trained and allowed to do so. Follow local protocols.

Circumferential Splints

This type of splint completely surrounds, or envelopes, the injured limb. An example is an air splint. It can be inflated, either mechanically or manually, until it forms a semi-rigid sleeve around the injured limb. An advantage of an air splint is that the compression it provides helps to reduce swelling.

Improvised Splints

An improvised splint can be made from a cardboard box, cane or walking stick, ironing board, rolled-up magazine, umbrella, broom handle, catcher's shin guard, or any similar object. It must be long enough to extend past the joints and prevent movement on both sides of the injury. It also should be as wide as the thickest part of the injured part.

A "self-splint" also may be effective. In fact, in some cases, a patient will not permit any other type of splint to be applied. In a "self-splint," the injured limb is secured against the patient's body with a cravat or roller bandage. Voids between the limb and body are then padded with bulky dressings or similar material as appropriate.

Sling and Swathe

An injured limb can be supported by a sling, while a swathe keeps the limb protected and immobile against the body. When applying a sling, be sure to keep the knot off the back of the patient's neck. It can be very uncomfortable there.

General Rules of Splinting

As you assess and provide emergency care for an injured extremity (Figure 17-7), keep these general rules of splinting in mind:

◆ Be sure you have taken BSI precautions before splinting.

◆ Do not release manual stabilization of an injured limb until it is properly and completely immobilized.

◆ Never intentionally replace protruding bones or push them back below the skin.

◆ You cannot assess what you cannot see. So cut away all clothing around the injury site before applying a splint. Also remove all jewelry from the injury site and below it. Bag the jewelry and give it to the patient, a family member, or the police.

◆ Control bleeding and dress all open wounds before applying a splint.

◆ If a long bone is injured, immobilize it and the joints above and below it.

◆ If a joint is injured, immobilize it and the bones above and below it.

◆ If a limb is severely deformed by the injury, or if the limb has no pulse or is cyanotic below the injury site, align it with gentle manual traction (pulling). If there is pain or grating (crepitus), stop pulling immediately. NOTE: Apply manual traction only if you are specifically trained and allowed to do so. Follow all local protocols.

COMPANY OFFICER'S NOTE

If there is no pulse distal to the fracture and your transport time to the hospital is short, do not bother to try to re-align the injury. Skeletal muscle can be deprived of oxygen for over 60 minutes before irreversible damage occurs. Spare your patients the pain and immobilize the injury as described above.

◆ Pad a splint before applying it to help keep the patient as comfortable as possible.

◆ Before and after applying a splint, assess pulses, movement, and sensation below the injury site. Reassess every 15 minutes after applying a splint, and record your findings.

For all the obvious benefits splints provide, they also can cause complications if they are applied incorrectly. Improper splinting can:

◆ Compress nerves, tissues, and blood vessels under the splint, which can aggravate the injury and cause further damage.

◆ Move displaced or broken bones, causing even further injury to nerves, tissues, and blood vessels.

◆ Reduce blood flow below the injury site, risking the life of the limb.

◆ Delay transport of a patient who has a life-threatening problem.

Remember that patients who have a painful, swollen, deformed extremity may be in considerable pain. They also may be concerned about regaining full use of the limb. So, as you provide emergency care, consider their feelings. Be gentle and reassuring.

Splinting the Upper Extremities

Clavicle

Often an injury to a shoulder will result in a fracture of a clavicle. When a clavicle is broken, the patient's shoulder may appear to have "dropped." The clavicle itself may look crooked and deformed. The best way to splint it is to apply a sling and swathe (Figure 17-8). Assess distal pulses, movement, and sensation before and after splinting.

Shoulder

A dislocated shoulder is a common injury. Patients with one often have had the same injury many times before. The dislocated shoulder will appear to be deformed. You also may see a "hollow" in the upper arm below the clavicle. The patient frequently complains of severe pain and may refuse to let anyone touch the arm.

Attempt to apply a sling and swathe to the arm. Padding the void between the body and the arm may be helpful. Use a small pillow, towels, or even trauma dressings for padding.

In a shoulder dislocation, there is a danger of injuring nerves and arteries. So a great deal of care must be taken when applying the sling and

General Rules of Splinting

Figure 17-7a Stabilize the limb, and assess pulse, movement, and sensation below the injury site.

Figure 17-7b Cut away clothing to expose the injury.

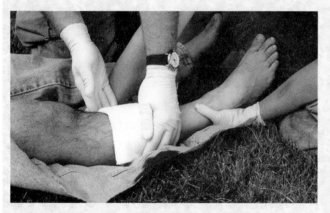

Figure 17-7c Place a sterile dressing over the wound. If bleeding control is required, use a pressure point.

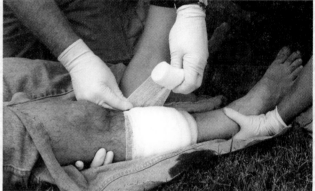

Figure 17-7d After bleeding has stopped, bandage the wound.

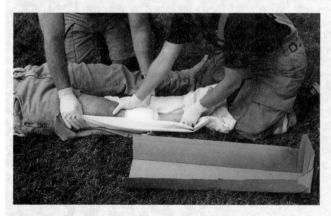

Figure 17-7e Pad the splint.

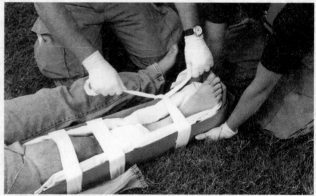

Figure 17-7f Secure the limb to the splint, and then reassess pulse, movement, and sensation.

Applying a Sling and Swathe

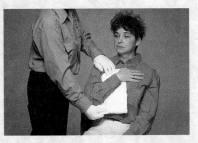

Figure 17-8a Place a pad between the arm and chest.

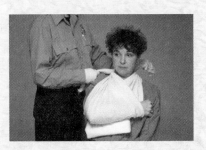

Figure 17-8b Support the injured arm with a sling.

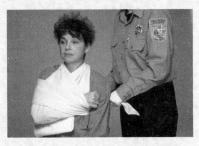

Figure 17-8c Immobilize the arm with a swathe.

swathe. Be sure to assess distal pulses, movement, and sensation before and after splinting.

Shoulder and Humerus

The humerus may break at midshaft or at the shoulder. It is thick and fairly strong. So if it is injured, suspect other injuries nearby.

Manually stabilize the arm as soon as possible. Apply a rigid splint to the outside of the arm, and pad the voids. Then apply a sling and swathe (Figure 17-9). Assess distal pulses, movement, and sensation before and after splinting. (See alternative, Figure 17-10.)

Elbow

The elbow should be splinted in the position in which it was found. Do not attempt to straighten it. If the arm is bent at the elbow, splint the injury with a sling and swathe (Figure 17-11). However, if the deformity is severe, you may elect to use a large, flat pillow or even a blanket wrapped around the limb and secured to the chest with a strap.

If the elbow is straight, then the entire arm should be splinted from the armpit to the fingertips (Figure 17-12). Be sure to assess distal pulses, movement, and sensation before and after splinting.

Figure 17-9 Fixation or rigid splint with a sling and swathe.

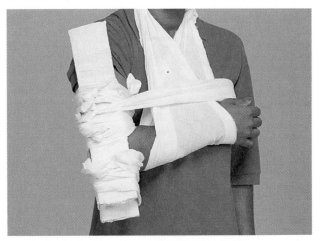

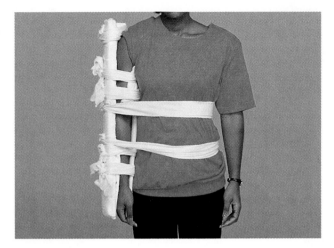

Figure 17-10   Fixed splint for humerus injury.

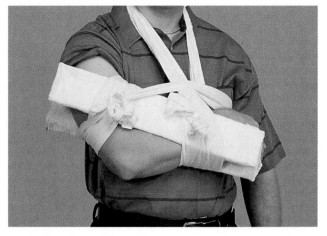

Figure 17-11   Injured elbow immobilized in a bent position.

Forearm and Wrist

Forearm and wrist injuries are very common. They must be supported from the elbow to the fingertips. First splint the injured area with a short arm board. Then a sling and swathe should be applied (Figure 17-13). If the injury is a closed one, a circumferential splint may be used instead. Be sure the splint extends from the elbow to beyond the hand (Figure 17-14). Assess distal pulses, movement, and sensation before and after splinting.

Hands and Fingers

If just one finger is injured, then it may be taped to the uninjured finger beside it. This is called "buddy taping." You may also use a tongue depressor as a splint (Figure 17-15). If there is

more than one finger involved, or if the hand injury is the result of a fight, then the entire hand needs to be immobilized.

A hand must be splinted in the position of function. The easiest way to do it is to place a four-inch roll of bandage, a rolled hand towel, or a small ball inside the palm of the injured hand. Then wrap the entire hand and place it on an arm board to immobilize the wrist (Figure 17-16). Assess distal pulses, movement, and sensation before and after splinting.

Splinting the Lower Extremities
Pelvis

Pelvis injuries can be life threatening, because a large amount of blood can be lost into the lower

Figure 17-12   Injured elbow immobilized in a straight position.

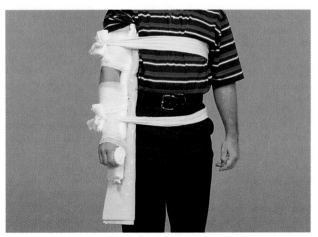

Figure 17-13   Immobilization of an injury to the forearm, wrist, or hand.

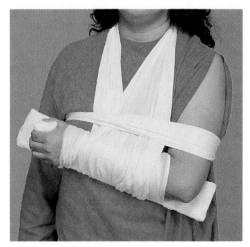

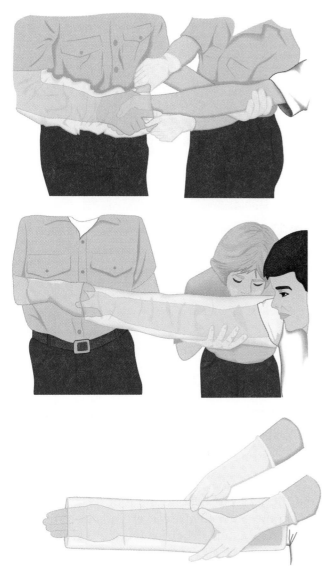

Figure 17-14 Applying an air splint.

abdomen quickly. So suspect shock with any pelvis injury.

The patient must be placed on a long backboard. Pad between the legs, and consider putting a blanket on each side of the patient's hips. Then secure the patient's whole body to the backboard. Keep the patient warm. If you suspect shock, the foot end of the backboard may be elevated slightly if it does not compromise the splinting. Assess distal pulses, movement, and sensation before and after splinting.

Hip

The hip is actually the proximal end of the femur, where the femur fits into the pelvis. Fractures of the hip are common to severe frontal car crashes.

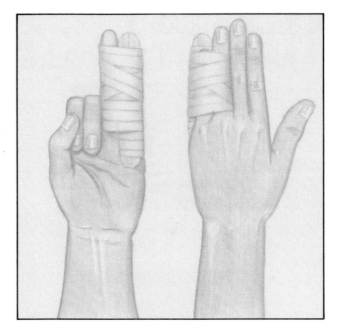

Figure 17-15 A tongue depressor used as a splint.

They also are common in the elderly as the result of a fall. With a broken hip, the leg on the injured side may be shorter than the other leg and rotated. The patient will complain of pain when the leg is moved or when the hips are gently compressed.

Immobilize this patient's whole body on a long backboard. Assess distal pulses, movement, and sensation before and after immobilization.

Femur

It takes a great deal of force to break a femur, such as in sky diving and skiing accidents. The

Figure 17-16 Cardboard splint of the forearm, wrist, or hand.

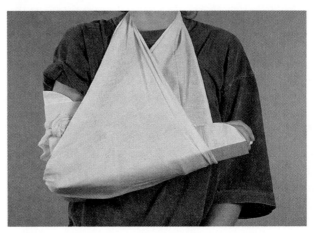

Applying a Traction Splint

Figure 17-17a Assess pulse, movement, and sensation below the injury site.

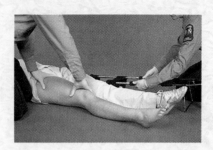

Figure 17-17b Manually stabilize the limb.

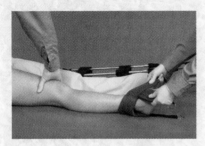

Figure 17-17c Apply the ankle hitch.

Figure 17-17d Apply and maintain manual traction. Position the splint.

result of a break is usually a marked deformity of the thigh, as well as a great deal of pain and swelling. Emergency care consists of immobilizing the bone ends to prevent further injury.

The preferred method of immobilization is a traction splint (Figure 17-17). Remember, only use a traction splint if you are specially trained and allowed to do so.

Alternative care involves using two long boards to create a fixation splint. The inner board must extend from the groin to below the bottom of the foot. The outer board must extend from the armpit to below the bottom of the foot (Figure 17-18). Pad the voids, then secure the boards to the patient with cravats at the shoulders, hips, knees, and ankles. Assess distal pulses, movement, and sensation before and after splinting.

Knee

There are many types of knee injuries. Emergency treatment is basically the same. If you find the injured leg in a straight position, use two padded long boards to splint it in the position found. Place the first on the inner thigh so it extends

from the groin to beyond the foot. Place the second on the outer thigh so it extends from the hip to beyond the foot. Then secure the boards to the patient with cravats.

If you find the knee in a bent position, immobilize it in the position found. The bones above and below it should be splinted with two padded

Figure 17-18 High femur fracture immobilized in a fixation splint.

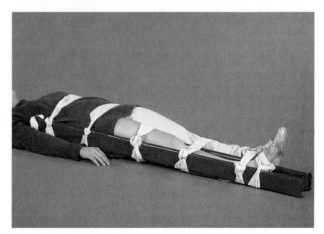

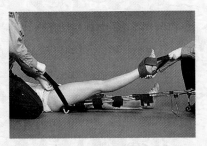

Figure 17-17e Attach the ischial strap.

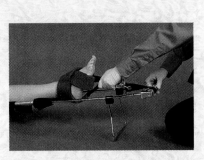

Figure 17-17f Fasten the splint to the ankle hitch. Apply mechanical traction.

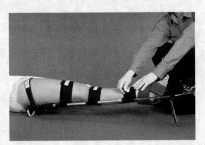

Figure 17-17g Fasten leg support straps in place.

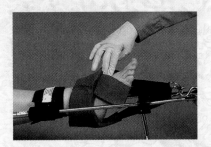

Figure 17-17h Reassess pulse, movement, and sensation below the injury site.

short boards. (See Figure 17-19.) Assess distal pulses, movement, and sensation before and after splinting.

Tibia and Fibula

Open fractures of the tibia are common because only thin layers of skin protect it. Usually fractures

Figure 17-19 Splinted knee.

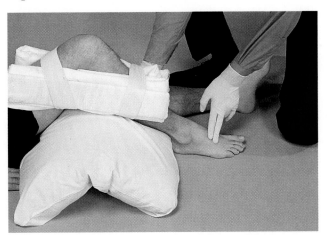

of the fibula are not so readily apparent, since it is not a weight-bearing bone. (Remember the football players who play in spite of a "broken leg?" In their cases, it is the fibula that is fractured, not the tibia.) Whichever one of the two bones is injured, the procedure for splinting remains the same.

Use two padded long boards (Figure 17-20). Place the first on the inner thigh so it extends from the groin to below the foot. Place the second on the outer thigh so it extends from the hip bone to below the foot. Then secure the boards to the patient with cravats.

An alternative method for a closed injury to the tibia or fibula is to use a circumferential splint. Make sure it extends beyond the knee and covers the entire foot (Figure 17-21). Assess distal pulses, movement, and sensation before and after splinting.

Ankle and Foot

The foot is commonly injured by heavy objects falling onto it or by twisting forces during a fall. The ankle bears so much weight, it does not take

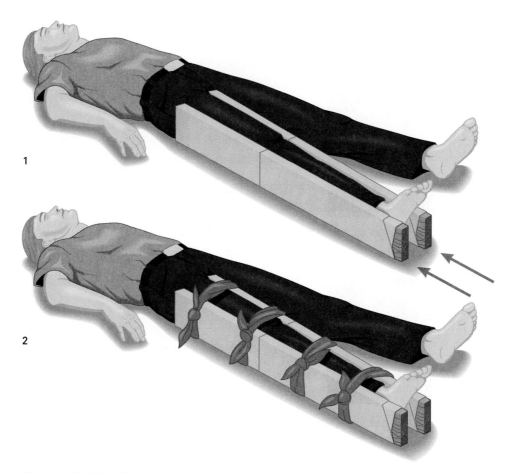

1

2

Figure 17-20 Fixation splint of the tibia/fibula using padded boards.

much movement in the wrong direction to make it unstable. No matter if the injury is to the ankle or foot, it is splinted in the same way.

Circumferential splints work well in these cases. However, the easiest splint may be a pillow. Simply wrap the pillow, or a blanket, around the foot. Then secure it with cravats at the toes and the shin. The more cravats applied, the better (Figure 17-22). Be sure to assess distal pulses, movement, and sensation before and after splinting.

Figure 17-21 Air splint of the lower leg.

Figure 17-22 Blanket-roll splint of the ankle and foot.

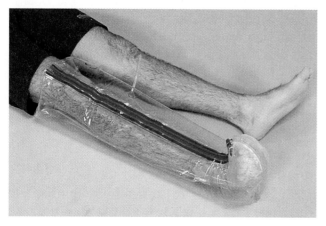

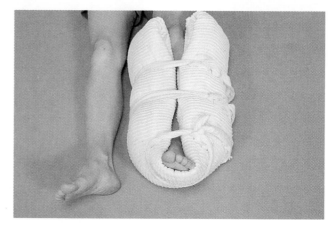

> *At the beginning of this chapter, you read that fire service First Responders were caring for a patient who had fallen and was complaining of hip pain. To see how chapter skills apply to this emergency, read the following. It describes how the call was completed.*

PHYSICAL EXAMINATION

A quick physical exam revealed no other deformities or injuries. I asked the patient to point to where she hurt the most. "It hurts on the right side, and in the back," she told me. Once again, I asked if she was having any other pain. "No," she said.

PATIENT HISTORY

I asked the patient what she thought might be wrong. "I'm sure I dislocated my hip," she said. "It's exactly what happened the last time. I turned and 'pop', out it went."

I also asked if she was taking any medicine or had any allergies. She told me she was on Premarin and was allergic to sulfa drugs.

ONGOING ASSESSMENT

We assessed her vital signs. They were pulse of 90, BP 142/94, and respirations 24. I determined that she was able to wriggle her toes on the affected side. Also, her toes were pink, warm, and dry.

PATIENT HAND-OFF

When the EMTs arrived on-scene, I gave a quick hand-off report:

"This is Tamara Meadors. She is 55, and is complaining of pain in the right hip area. She believes she may have dislocated her right hip again. This hip was surgically repaired last year. She is conscious and alert, with quite a sense of humor. She states she was standing and turned to speak to someone in front of her and felt a 'pop.' She is able to wriggle her toes, and her skin is pink, warm, and dry distally. Her right leg appears to be rotated outward. Here's a list of her latest vitals. She denies any other pain, and states she is taking Premarin and she's allergic to sulfa drugs."

The ambulance crew took over patient care. We assisted in placing Ms. Meadors on a backboard with pillows supporting her right leg in a slightly bent position. We later found out she had dislocated her hip. She underwent surgery that night and was convalescing well at home.

> *Musculoskeletal injuries usually are painful and obvious. Even so, you should always assess for and treat life-threatening problems first. Remember, life before limb!*

Chapter Review

Patients may be aware of only one obvious injury, but it is your responsibility to perform a thorough patient assessment. Do not let an injury that is gruesome but relatively minor make you miss life-threats such as an open chest wound or shock. As you assess a patient with musculoskeletal injuries, keep the following tips in mind:

- The mechanism of injury can alert you to hidden injuries.
- Broken bones can be serious. They bleed, and they cause shock. If there is a possibility that your patient has more than one broken bone, the potential for shock and other hidden injuries is very high.

- Examine the extremities last. More important areas—the head, neck, chest, and abdomen—must be examined first because they contain the vital organs.

In the first 10 minutes, it is important to make accurate decisions based on your assessment findings. Treat injuries and conditions that are a threat to life first.

Page references where answers may be found or supported are provided at the end of each question.

Section 1

1. What is the function of the musculoskeletal system? (p. 316)

2. What is the difference between an open and a closed painful, swollen, deformed extremity? (pp. 316–317)

3. Is the emergency care you give to a patient with a fracture any different from the care you give to a patient with a strain, a sprain, or a dislocation? Why or why not? (p. 317)

4. What are the general emergency care guidelines for a painful, swollen, or deformed extremity? (pp. 318–319)

Section 2

5. What are the reasons for splinting a painful, swollen, or deformed extremity? (p. 320)

6. What are five general rules of splinting? (p. 321)

7. What are some of the possible complications of improper splinting? (p. 321)

Simon, R.R., and S.J. Koenigsknecht. *Emergency Orthopedics: The Extremities,* Third Edition. Norwalk, CT: Appleton-Lange, 1995.

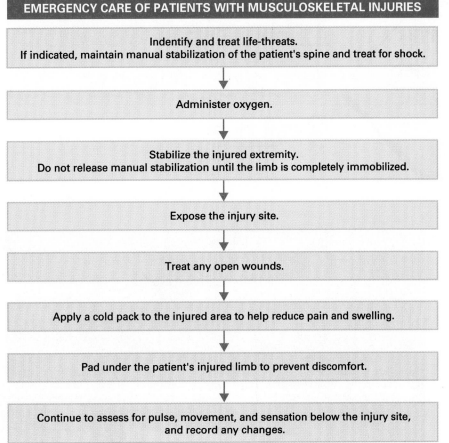

EMERGENCY CARE OF PATIENTS WITH MUSCULOSKELETAL INJURIES

Indentify and treat life-threats.
If indicated, maintain manual stabilization of the patient's spine and treat for shock.

Administer oxygen.

Stabilize the injured extremity.
Do not release manual stabilization until the limb is completely immobilized.

Expose the injury site.

Treat any open wounds.

Apply a cold pack to the injured area to help reduce pain and swelling.

Pad under the patient's injured limb to prevent discomfort.

Continue to assess for pulse, movement, and sensation below the injury site, and record any changes.

Injuries to the Head, Neck, and Spine

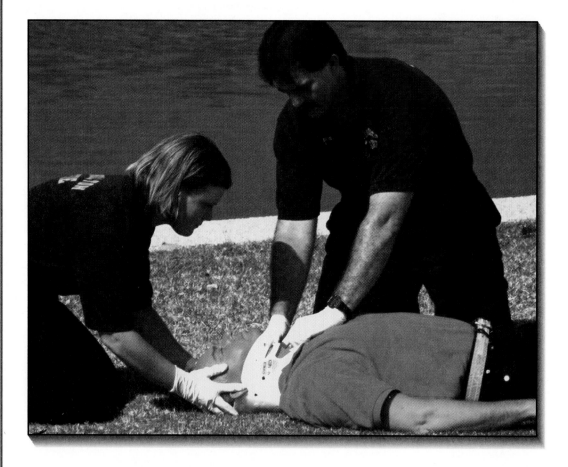

I **NTRODUCTION** *The head, neck, and spine are important structures housing vital organs and tissue. Injuries can damage the brain, structures of the airway, and the spinal cord. These are among the most serious emergencies. They run a high risk of causing life-long complications and death. Your role as the first medically trained rescuer on scene can be critical.*

Cognitive, affective, and psychomotor objectives are from the U.S. DOT's "First Responder: National Standard Curriculum." Enrichment objectives, if any, identify supplemental material.

Cognitive

5-3.4 Relate mechanism of injury to potential injuries of the head and spine. (pp. 335, 336, 337, 339, 340, 343)

5-3.5 State the signs and symptoms of a potential spine injury. (p. 340)

5-3.6 Describe the method of determining if a responsive patient may have a spine injury. (p. 340)

5-3.7 List the signs and symptoms of injury to the head. (pp. 334–335)

5-3.8 Describe the emergency medical care for injuries to the head. (pp. 335–336)

Affective

5-3.10 Demonstrate a caring attitude towards patients with a musculoskeletal injury who request emergency medical services. (p. 340)

5-3.11 Place the interests of the patient with a musculoskeletal injury as the foremost consideration when making any and all patient care decisions. (p. 340)

5-3.12 Communicate with empathy to patients with a musculoskeletal injury, as well as with family members and friends of the patient. (p. 340)

Psychomotor

5-3.14 Demonstrate opening the airway in a patient with suspected spinal cord injury. (pp. 340, 343)

5-3.15 Demonstrate evaluating a responsive patient with a suspected spinal cord injury. (p. 340)

5-3.16 Demonstrate stabilizing the cervical spine. (p. 343)

Enrichment

◆ Describe the implications of not properly caring for potential spine injuries. (pp. 332, 355)

◆ Discuss sizing and using a cervical spine immobilization device. (pp. 343–345)

◆ Describe how to log roll a patient with a suspected spine injury. (pp. 346–347)

◆ Describe how to secure a patient to a long backboard. (p. 347)

◆ Describe when and how to perform a rapid extrication. (pp. 348–351)

◆ Discuss the circumstances in which a helmet should remain on a patient and when it should be removed. (p. 351)

◆ Explain the preferred method of removing a helmet. (pp. 351–353)

 ON SCENE

DISPATCH

Our engine company crew was preparing to go off duty when a call came in. It was 0750 hours. My partner and I jumped in our rescue truck and responded to the intersection of Loantaka and Miller Roads for a reported motor-vehicle collision.

SCENE SIZE-UP

Although dispatch did not give us reason to anticipate an unsafe scene, we approached with our usual caution. We arrived before the police did, and parked our vehicle in a position that afforded us the most protection from the flow of traffic.

Two vehicles were blocking the intersection. Neither appeared to have suffered much damage. As I approached the first vehicle, I noted that no one was inside. There was a man in the driver's seat of the second vehicle. He appeared to be conscious. I asked my partner to find the occupants of the empty car, while I performed an assessment on the man in the second vehicle.

INITIAL ASSESSMENT

As I approached the second vehicle, I noticed the windshield was "spidered" in front of the driver. I introduced myself, and asked the man if he was okay. He said his neck hurt, and I noticed some swelling above his left eye.

At this time, my partner returned and told me that there had been no occupants in the first vehicle, as it had run out of gas and the driver was standing on the curb when his car was struck. I asked him to stabilize my patient's head and neck, while I began the assessment.

The patient was alert, oriented, and speaking coherently when he gave me permission to treat. He denied being knocked out. He said he had not been wearing his seat belt. His breathing appeared adequate. There was no visible bleeding present. My general impression was of a male in his 40s who had struck his head on the windshield with enough force to break it.

> *Consider this patient as you read Chapter 18. What would you do to assess and treat his condition?*

INJURIES TO THE HEAD

A head injury may be open or closed. An open head injury is accompanied by a break in the skull, such as that caused by a fracture or an impaled object. It involves direct local damage to tissue accompanied by bleeding. It also can result in brain damage.

A closed head injury does not involve a break in the skull. Even so, the brain can be seriously injured. The skull holds brain tissue, blood, and cerebrospinal fluid. Because the skull is rigid, its capacity is limited. If brain tissue swells or if bleeding occurs, pressure can build up inside the skull causing damage to the brain and possibly death to the patient.

Patient Assessment

The general signs and symptoms of a head injury include:

◆ Altered mental status, from confusion to unresponsiveness.

◆ Irregular breathing.

◆ Open wounds to the scalp.

◆ Penetrating wounds to the head.

◆ Softness or depression of the skull.

◆ Blood or cerebrospinal fluid leaking from the ears or nose (Figure 18-1).

◆ Facial bruises.

◆ Bruising around the eyes ("raccoon eyes") or behind the ears ("Battle's sign").

◆ Abnormal findings in an assessment of pulses, movement, and sensation.

◆ Headache severe enough to be disabling or which appears suddenly.

Figure 18-1 Blood or cerebrospinal fluid may come from the ears and nose of a patient with a head injury.

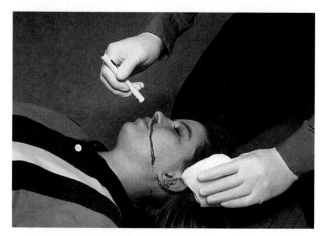

- Nausea, vomiting.
- Unequal pupil size with altered mental status.
- Seizure activity.

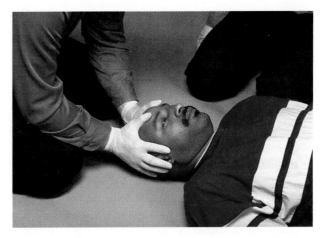

Figure 18-3 Maintain manual stabilization of the head and neck until the patient is completely immobilized.

Suspect spine injury in any patient with a head injury (Figure 18-2). If there is an obvious head injury, if the mechanism of injury suggests a head or spine injury, or if a trauma patient is unresponsive, immediately stabilize the patient's head and neck. Maintain manual stabilization until the patient is completely immobilized (Figure 18-3). If alone with an injured patient, you may be allowed to place a rigid item on each side of the patient's head to prevent movement. Follow local protocols.

During your initial assessment of a patient with a head injury, use a jaw-thrust maneuver to open, assess, and maintain the airway.

During your physical exam, look for open injuries to the head. Closed injuries may present with swelling or depression of the bones of the skull. Check for cerebrospinal fluid, which appears as a clear liquid, possibly tinged pink with blood, leaking from an open head wound or from the ears or nose. Also pay attention to the function of the patient's eyes.

For the head-injured patient, it is especially important to assess pulses, movement, and sensation in the extremities. Note any numbness or tingling, especially of the face, arms, or legs.

When you take the patient's history, be sure to try to find out when the injury occurred, if the patient lost consciousness, and if the patient was moved—or walked around—after the injury occurred. Details about what happened are crucial to his or her medical care. Be sure to document them.

During your ongoing assessment, monitor the patient for any change in level of responsiveness. Keep in mind that change in a patient—not the patient's status at any one time—may be the most important sign of how a patient is doing.

Figure 18-2 Suspect spine injury in any patient with a head injury.

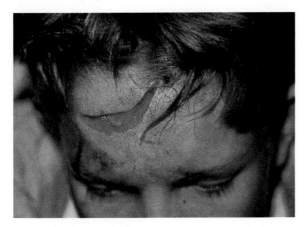

◆ Emergency Care

If you suspect injury to the head, be sure EMS has been activated. After taking BSI precautions and establishing manual stabilization of the patient's head and neck, proceed with emergency care as follows:

1. **Make the airway a top priority.** Oxygen deficiency in the brain is the most frequent cause of death following a head injury. Remember to use a jaw-thrust maneuver to assure an open airway. Monitor the airway and breathing closely, and suction as needed.

Provide high-concentration oxygen, and be prepared to provide ventilations. Follow local protocol.

2. **Control bleeding and dress open wounds.** Scalp wounds often bleed profusely, but are usually easy to control with direct pressure. Never apply direct pressure to a head wound that is accompanied by an obvious or depressed skull fracture. It could drive fragments of bone into brain tissue and cause further injury. Do not try to stop a flow of cerebrospinal fluid. If the fluid is leaking from the ears or a head wound, cover the opening loosely with sterile gauze dressings. If there is a penetrating object, do not try to remove it. Stabilize it with bulky dressings.

3. **Apply a rigid cervical immobilization device,** if trained and allowed to do so. (How to do it is described later in this chapter.) Maintain manual stabilization of the head and neck before, during, and after application and until the patient is immobilized on a long backboard.

4. **Monitor mental status and vital signs closely.** Watch for any sign of deterioration or change. If the patient has convulsions, protect him or her from injury.

5. **Calm and reassure the patient.** Continue to talk with him or her. If you can stimulate the patient, you may be able to prevent loss of consciousness.

Specific Head Injuries

Injuries to the head include skull fracture, injuries to the brain, concussion, and penetrating wounds.

Skull Fracture

The primary function of the skull is to protect the brain from injury. Because of its shape and thickness, the skull usually is broken only by extreme trauma. Suspect skull fracture with any significant trauma to the head, even if the injury is a closed one.

A skull fracture accompanied by brain injury is a serious condition that needs immediate management. Signs and symptoms include (Figure 18-4):

◆ Damage to the skull, visible through lacerations in the scalp.

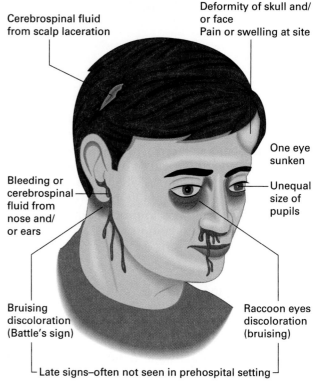

SIGNS AND SYMPTOMS OF SKULL FRACTURE

Cerebrospinal fluid from scalp laceration

Deformity of skull and/or face
Pain or swelling at site

One eye sunken

Unequal size of pupils

Bleeding or cerebrospinal fluid from nose and/or ears

Bruising discoloration (Battle's sign)

Raccoon eyes discoloration (bruising)

Late signs—often not seen in prehospital setting

Figure 18-4 Signs and symptoms of a skull fracture.

◆ Deformity of the skull or face.

◆ Pain or swelling at the injury site.

◆ Clear or pinkish fluid dripping from nose, ears, mouth, or head wound.

◆ Unusual size of pupils.

◆ Purplish bruising under or around the eyes ("raccoon eyes") or behind the ear ("Battle's sign").

Injuries to the Brain

Regardless of whether a head wound is open or closed, brain damage can be extensive. In fact, it is often more severe in closed injuries than in open ones. Severity depends mainly on the mechanism of injury and the force involved. However, consider all suspected head injuries to be serious. Signs and symptoms include:

◆ Changes in mental status, ranging from confusion to unresponsiveness.

◆ Paralysis or flaccidity, usually only on one side of the body.

- Unequal facial movements, squinting, drooping, unequal or unresponsive pupils, disturbances of vision in one or both eyes.
- Headache.
- Ringing in the ears, loss of hearing in one or both ears.
- Rigidity of all limbs (present with severe injury).
- Loss of balance, staggering or stumbling gait.
- Slow, strong heartbeat that gradually becomes rapid and weak (late sign).
- High blood pressure with a slow pulse.
- Rapid, labored breathing or disturbances in the pattern of breathing.
- Vomiting after head injury.
- Incontinence, or loss of bowel or bladder control.

Concussion

A concussion is a temporary loss of the brain's ability to function. There is no detectable damage to the brain. A concussion is classified as mild, moderate, or severe, based on the time interval before return to responsiveness. The key distinguishing factor of concussion is that its effects appear immediately or soon after impact. Then they disappear, usually within 48 hours. If symptoms develop several minutes after impact or do not subside over time, the injury is probably more serious than a concussion.

Signs and symptoms include:

- Momentary confusion, or confusion that lasts several minutes.
- Inability to recall the period just before and after being injured.
- Repeatedly asking what happened.
- Mild to moderate irritability, uncooperativeness, combativeness, verbally abusive.
- Inability to answer questions or obey commands appropriately.
- Headache.
- Persistent vomiting.
- Incontinence.
- Restlessness.
- Seizures.
- Brief loss of consciousness.

Penetrating Wounds

A penetrating wound occurs when an object passes through the skull and lodges in the brain. It often involves bullets, knives, or other sharp objects. Be sure to look for both entry and exit wounds. An extreme emergency, a penetrating wound almost always results in long-term damage.

If the object is impaled in the skull, do not try to remove it. Stabilize it with soft bulky dressings instead. Then dress the area around it with sterile gauze. If an object has penetrated the skull but you cannot see it, cover the wound lightly with sterile dressings. In both cases, permit blood to drain. Apply only enough pressure to control life-threatening bleeding. Never apply overly firm pressure to a head injury that might involve a skull fracture.

INJURIES TO THE NECK

While some injuries to the neck are minor, many can be life threatening (Figure 18-5). They can result from impacts strong enough to cause hidden facial fractures, cervical-spine damage, and skull fractures.

For injuries of the neck, follow the patient assessment and general guidelines for emergency care described at the beginning of this chapter. Be sure the airway stays clear of fragments of teeth, broken dentures, bits of bone, pieces of flesh, and other possible obstructions. Take spinal precautions as appropriate.

Remember to keep the airway a priority. If bleeding into the mouth or throat threatens the airway, maintain control of the cervical spine and roll the patient onto one side to allow for drainage. Be prepared to monitor the airway and breathing constantly. Suction often. Keep in mind that the patient may be very anxious about possible disfigurement. Reassure him, and do your best to help him stay calm.

Common causes of neck injury include hanging (attempted suicide), impact with a steering wheel or windshield, or running into a stretched wire or clothesline. Large wounds may involve injuries to the major vessels in the neck, which can produce massive, even fatal bleeding. If a

Injuries to the Neck

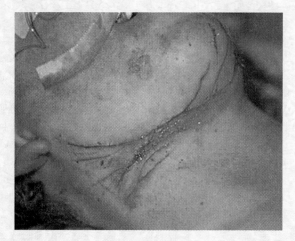

Figure 18-5a Injury to the neck.

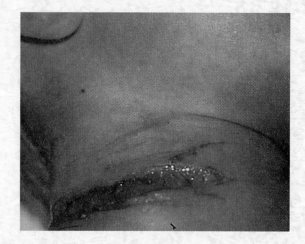

Figure 18-5b Injury to the neck.

SKILL SUMMARY
Severed Neck Veins

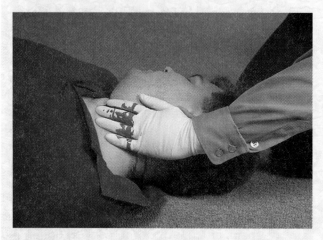

Figure 18-6a Do not delay! Place your gloved palm over the wound.

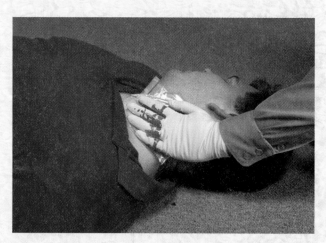

Figure 18-6b Apply moderate pressure with an occlusive dressing.

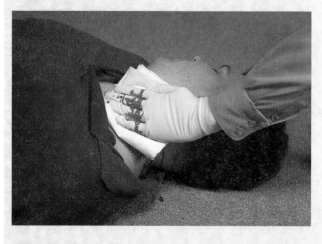

Figure 18-6c Add a bulky dressing.

wound to the neck is left uncovered, air may be sucked into the vessels causing an air embolism.

Signs and symptoms of injuries to the neck include the following:

◆ Airway obstruction.

◆ Obvious lacerations or other wounds.

◆ Deformities or depressions.

◆ Pain or tenderness.

◆ Obvious swelling, which sometimes occurs in the face and chest.

◆ Difficulty speaking, loss of the voice.

◆ Crackling sensations under the skin due to air leaking into the soft tissues (subcutaneous emphysema).

If there is bleeding from a neck wound, apply slight to moderate pressure with an occlusive dressing. Tape down the edges of the dressing to form an airtight seal. Add a bulky dressing over the occlusive one (Figure 18-6). Never apply pressure to both sides of the neck at the same time. Never apply a pressure dressing around the neck. If there is an impaled object in the neck, stabilize it in place with bulky dressings. Do not remove it.

SECTION 3

INJURIES TO THE SPINE

Anatomy of the Spine

The spinal cord lies within the spinal column. It is responsible for sending signals from the brain to the body and for receiving signals from the body and relaying them to the brain. If these signals are interrupted by injury or illness, the patient could lose the ability to move, feel, or even breathe. (Review Chapter 4 for more on the musculoskeletal and nervous systems of the body.)

The spinal column is made up of 33 bones, one stacked on top of another. The vertebrae *articulate,* or fit and move together, so we can bend, turn, and flex.

The spine is divided into five regions: the cervical, thoracic, lumbar, sacral, and coccygeal (Figure 18-7). The cervical spine starts at the base of the skull where the spinal cord begins. Its seven vertebrae not only house delicate nerve tissue, they also support the weight of the head. This makes them especially vulnerable to injury.

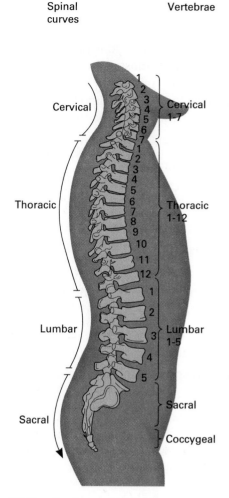

Figure 18-7   Regions of the spine.

The thoracic spine is supported by the rib cage. There are 12 thoracic vertebrae, one for each rib. Because the ribs help protect and support this part of the spine, it is less frequently injured.

The next group of five vertebrae make up the lumbar spine. They carry the weight of most of the body. For this reason, they are heavier and larger. The discs between the lumbar vertebrae are thicker than in other parts of the spine. Sometimes, a disc can shift, slip, or rupture. Injuries to the lumbar spine cost millions of dollars in medical expenses and lost wages every year.

The last two regions of vertebrae are the sacral and coccygeal spines. The sacrum has five fused vertebrae. The coccyx has four. Together they form the posterior portion of the pelvis. Because they are fused, these parts of the spine do not bend easily.

CHAPTER 18 *Injuries to the Head, Neck, and Spine*   **339**

Spine Injuries

During scene size-up, you must identify the mechanism that injured your patient (Figure 18-8). In doing so, you must consider what occurred and what injuries may have resulted. Your index of suspicion for a spine injury should be very high in any of the emergencies described below:

◆ Motor-vehicle collisions.

◆ Motorcycle crashes.

◆ Pedestrian-car crashes.

◆ Falls.

◆ Diving accidents.

◆ Hangings.

◆ Blunt trauma.

◆ Penetrating trauma to the head, neck, or torso.

◆ Gunshot wounds.

◆ Any speed sport accident, such as roller blading, bicycling, skiing, surfing, or sledding.

◆ Any unresponsive trauma patient.

If the mechanism of injury suggests it, proceed as if the patient has a spine injury, even if he says he is not injured at all. The lack of back pain or the ability to walk, move arms and legs, or feel sensation does not rule out spine injury.

Patient Assessment

If you suspect spine injury in your patient, protect the spine from further injury. Immediately upon completing your scene size-up, stabilize the patient's head and neck. Then assess the ABCs. Be sure to use the jaw-thrust maneuver to open and maintain the airway. Remember that a cervical-spine injury can result in severe breathing problems, including respiratory arrest. So, monitor the patient's airway and breathing continuously.

Although there may be no signs or symptoms at all of spine injury, when there are, they typically include one or more of the following:

◆ Respiratory distress.

◆ Tenderness at the site of injury on the spinal column.

◆ Pain along the spinal column with movement. (Do not move the patient or ask the patient to move to test for this pain.)

◆ Constant or intermittent pain without movement along the spinal column or in the lower legs.

◆ Obvious deformity of the spine. (This is rare.)

◆ Soft-tissue injuries to the head, neck, shoulders, back, abdomen, or legs.

◆ Numbness, weakness, or tingling in the arms or legs.

◆ Loss of sensation or paralysis in the upper or lower extremities or below the injury site.

◆ Incontinence, or loss of bowel or bladder control.

◆ Priapism, or a constant erection of the penis (a classic sign of cervical-spine injury).

During the physical exam, do not risk moving the spine by taking off the patient's shirt or coat. Cut off his clothes if necessary. Be sure to ask the patient if and where the spine hurts. If he complains of pain upon palpation of the spine, stop. Continue the assessment in other areas of the body.

Assess pulses, movement, and sensation in all four extremities (Figure 18-9). To assess movement, ask the patient if he can move his hands and feet. Ask him to squeeze both of your hands at the same time. Gauge the patient's strength and decide if it is equal on both sides. Also have the patient push his feet against your hands. Again, gauge strength and equality.

To assess sensation, gently squeeze one extremity and then the other. As you do, ask questions such as: Can you feel me touching your fingers? Can you feel me touching your toes? In the unresponsive patient, apply a painful stimulus to check response (Figure 18-10). Either pinch the webbing between the toes and fingers or apply pressure with a pen across the back of a fingernail. The patient should withdraw from the pain. Note the response to pain in all four extremities.

After assessing the front of the patient, perform a log roll so you can assess the back. However, do so only if you are properly trained and have enough help to do so safely. Details on how to perform a log roll are provided later in this chapter.

Remember that a patient may be uncomfortable, confused, and possibly afraid of paralysis or death. It is important for you to show a caring attitude. As you proceed with the physical exam, for example, be careful how you communicate your findings to your partner. An off-hand remark could terrify the patient. When you speak to his or her family, be honest but do not alarm them unnecessarily.

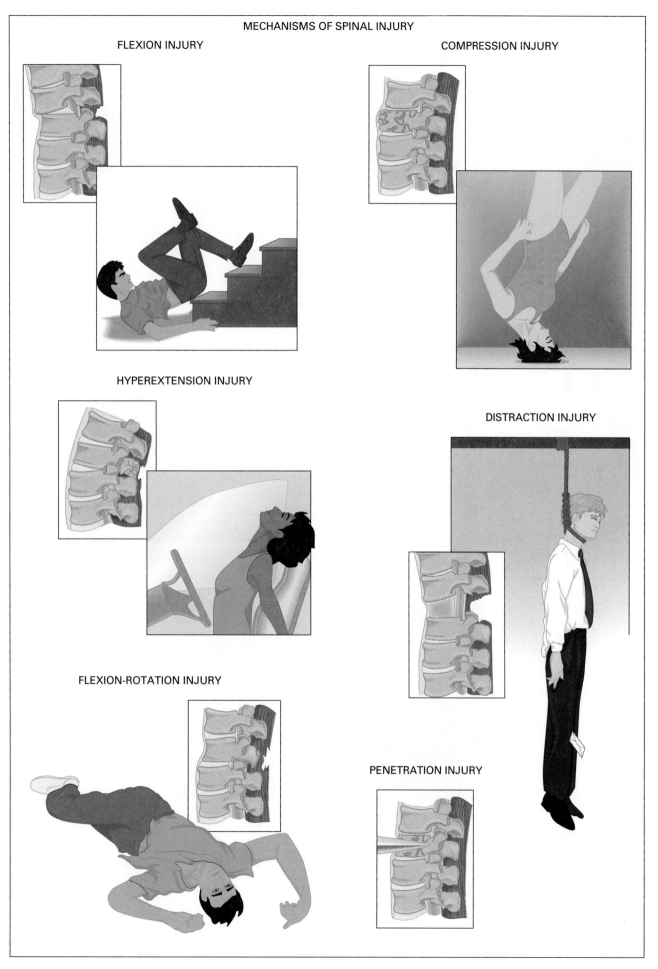

Figure 18-8 Common mechanisms of spine injury.

Assessing Pulses, Movement, and Sensation

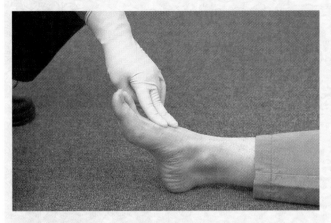

Figure 18-9a *Pulses*—feel for a pulse in all extremities.

Figure 18-9b *Movement*—ask the patient if her hands can move.

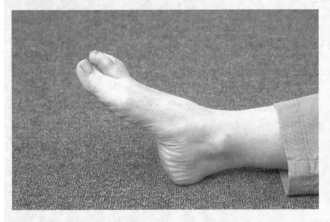

Figure 18-9c *Movement*—ask the patient if her feet can move.

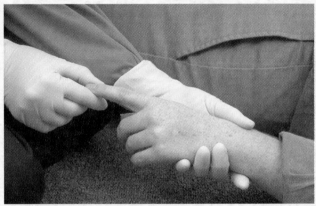

Figure 18-9d *Sensation*—touch the patient's fingers to assess for sensation.

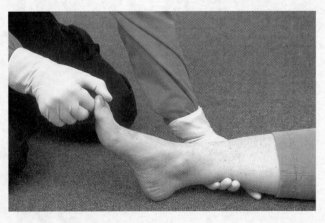

Figure 18-9e *Sensation*—touch the patient's toes to assess for sensation.

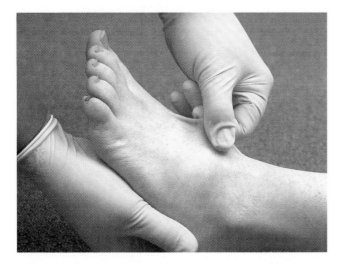

Figure 18-10 In an unresponsive patient, apply painful stimuli to assess response.

◆ Emergency Care

If the mechanism of injury suggests a possible spine injury, immediately stabilize the patient's cervical spine. That is, place your gloved hands just behind the patient's ears. Then hold the patient's head firmly and steadily in a neutral, in-line position. Neutral means the head is not flexed forward or extended back. In-line means the patient's nose is in line with the navel.

If you find that the patient's head is not in-line, you must gently put it there. Stop at once if the responsive patient complains of pain or you feel resistance in the unresponsive patient. Then stabilize the head and neck in the position in which it was found. Manual stabilization may be released only when the patient is immobilized from head to toe on a long backboard. When possible, have another rescuer maintain it, allowing you to care for the patient.

In general, take BSI precautions, identify the mechanism of injury, and then provide emergency care for a suspected spine-injured patient as follows:

1. **Stabilize the patient's head and neck immediately.** Keep the patient from moving.

2. **Perform an initial assessment and provide treatment.** Be sure to open and maintain the airway with a jaw-thrust maneuver. Insert an oropharyngeal or nasopharyngeal airway, if needed. Suction without turning the patient's head.

3. **Provide high-concentration oxygen via nonrebreather mask.** If the patient stops breathing or if breathing is inadequate, assist with artificial ventilation. Maintain neutral, in-line stabilization throughout.

4. **Apply a rigid cervical immobilization device,** if trained and allowed to do so. (How to do it is described later in this chapter.) Maintain manual stabilization of the head and neck before, during, and after application.

5. **Perform a physical exam and provide treatment.** Be sure to monitor the patient's airway and breathing continuously.

6. **Maintain manual stabilization** until the patient is completely immobilized.

Immobilization Techniques

Fire service First Responders may immobilize a suspected spine-injured patient if local protocol allows it. Even if your system does not, you may be called to assist EMTs. Become familiar with the techniques. They include cervical immobilization, long backboard immobilization, rapid extrication, and helmet removal.

Remember: Never attempt to treat or move a spine-injured patient unless you have the proper equipment, training, and personnel.

Cervical Immobilization

After an initial assessment, a rigid cervical immobilization device should be applied to the patient (Figure 18-11). This collar is also called an "extrication collar" or a "c-collar." There are a variety available. However, never use a "soft" collar in

Figure 18-11 If allowed, apply a rigid cervical immobilization device to the patient.

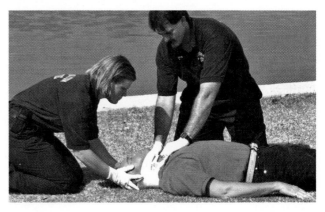

Applying a Rigid Cervical Immobilization Device

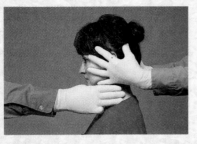

Figure 18-12a

Figure 18-12b

SIZING

It is critical to select a collar that is the correct size. Too tall can overextend the neck, force the jaw closed, and limit access to the airway. Too short can lead to inadequate immobilization. Too tight can impede blood flow. One way to measure collar size is to use your fingers to compare the neck size to the corresponding area of the collar.

NOTE: Do not use soft collars. Only use rigid cervical immobilization devices in the field. Also, do not use the chin piece as an anchoring point for the collar. This may cause hyperextension, which may injure the patient's cervical spine.

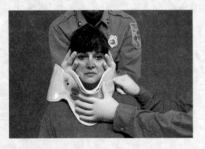

Figure 18-12c

Figure 18-12d

SEATED APPLICATION

The patient's chin must be well supported by the chin piece. To accomplish this, slide the collar up the patient's chest wall. If the collar is pushed directly inward, it may be difficult to position the chin piece and, therefore, to apply the collar tightly enough.

the field. They are nothing more than cotton-covered foam rings, which do not prevent movement of the head and neck.

Use rigid or hard collars in the field. They are designed to prevent the patient from turning, flexing, and extending the head. They can restrict movement by up to 70%. The remaining 30% must be accomplished by manual stabilization.

Follow manufacturer's instructions for applying a collar. Though instructions will vary, all collars are supported at the same points: the maxilla (jaw), shoulders, and clavicles. Note that failure to fit a patient properly can aggravate the injury.

Before application, be sure that jewelry and long hair have been moved away from the area. Also, examine and palpate the patient's neck before the collar is applied.

In general, to apply a rigid cervical collar to a supine patient, follow these steps (Figure 18-12): First, slide the posterior portion of the collar in

Figure 18-12e

TIGHTENING
Grip the trach hole as you tighten the collar. Then check to see that the collar fits according to the manufacturer's instructions.

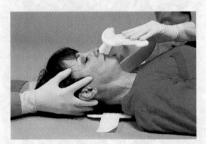

Figure 18-12f

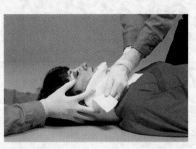

Figure 18-12g

SUPINE APPLICATION
Slip the collar underneath the patient's neck. Then rotate the collar up along the chest until the chin piece is properly positioned.

Figure 18-12h

WARNING
Always check for neutral alignment and proper fit. Improper sizing or application may allow the patient's chin to slip inside the collar. This must be prevented.

the gap under the patient's neck. Then flip the anterior portion over the chin. Finally, secure the collar with the Velcro strap. Be careful not to pull too hard on one end. It could twist the patient's head.

If your patient is sitting, bring the collar up the chest until the chin is trapped. Then slide the posterior portion around the back of the neck and fasten it. Whatever position your patient is in, you must maintain manual stabilization of the head and neck. Release it only when the patient is completely immobilized on a long backboard.

Long Backboard Immobilization

All patients with suspected spine injury must be immobilized onto a long backboard. To immobilize a supine or prone patient, you must first roll the patient onto his side, slip the board under him, and then roll the patient back. This procedure is called a **log roll.**

Three-Rescuer Log Roll

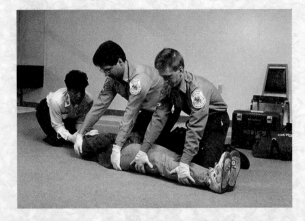

Figure 18-13a Maintain the head and neck in a neutral in-line position.

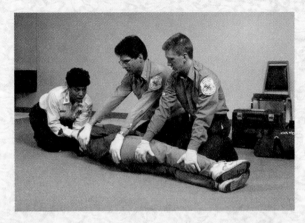

Figure 18-13b Roll the patient onto his or her side.

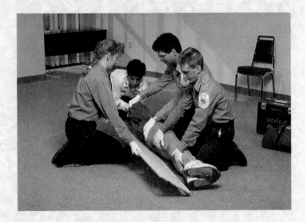

Figure 18-13c A bystander or one of the three rescuers should move the long backboard into place.

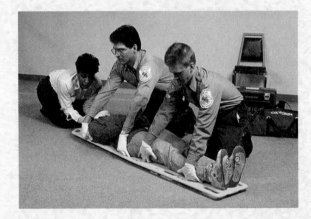

Figure 18-13d Lower the patient onto the long backboard.

To perform a log roll safely, you need at least three rescuers, preferably four, who are trained in the procedure. One should stay at the patient's head to maintain manual stabilization and to coordinate the move. The others should position themselves along one side of the patient's body. Note: If any additional rescuers are present, enlist them to assist with stabilization and rolling. Proceed as follows (Figure 18-13):

1. **Maintain manual stabilization** of the patient's head and neck. Continue to do so until the patient is completely immobilized.

2. **Apply a rigid cervical immobilization device,** if it has not already been applied.

3. **Assess pulses, movement, and sensation** in all four extremities.

4. **Position the patient.** Place the patient's arms straight down by his sides, if possible.

5. **Position the rescuers.** At the signal of the rescuer at the head, the two at the side should reach to the far side of the patient. One rescuer should position his hands on the shoulder and the hip. The second rescuer should position his hands at the thigh and lower leg.

6. **Simultaneously roll the patient onto his side,** on signal from the rescuer at the head. Note that this is a good time to assess the

patient's posterior, if it has not been done already.

7. **Position the board.** A fourth person—another rescuer, a family member, or bystander—should push the board under the patient. If no one else is available, one of the rescuers at the side may lean over the patient, grab the backboard, and pull it under the patient.

8. **Simultaneously roll the patient back down and onto the board,** on signal from the rescuer at the head. If the patient is not in the middle of the board, gently pull the patient down and then up again until he is straight on the board. This is done at the shoulders and hips and by pulling in alignment with the long axis of the spine. Never push a patient over to the middle of the backboard. Note that this down and up, or "V," motion helps to prevent movements to the side, which can cause further injury.

9. **Reassess pulses, movement, and sensation** in all four extremities. Report any change to the incoming EMTs.

Once in place, pad the spaces between the patient and the board (Figure 18-14). In an adult, pad anywhere along the length of the body to maintain neutral alignment and provide comfort.

Your next step is to secure the patient to the long backboard. It should always be done in this order (Figure 18-15):

1. **Immobilize the torso first.**

2. **Immobilize the head next.** The head must always be immobilized after the torso. Take a great deal of care not to lock the jaw in place.

Figure 18-14 Pad the voids between the patient and the board.

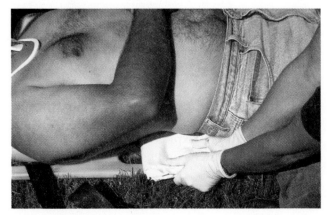

PEDIATRIC NOTE

In the infant and child, also pad the board under the shoulders. This is to keep the relatively larger head of an infant from flexing forward. Take care to avoid any extra movement.

If the patient should have to vomit, he must be able to open his mouth.

3. **Immobilize the legs last.**

4. **Withdraw manual stabilization of the head and neck.**

5. **Reassess pulses, movement, and sensation.** Report any change to the incoming EMTs.

Short Backboard Immobilization

You can use a short backboard to help to immobilize a seated patient. It minimizes the risk of further injury while the patient is being moved to a long backboard.

To apply a short backboard to a seated patient, follow the steps outlined below. Remember to maintain manual stabilization until the patient is completely immobilized. Proceed as follows (Figure 18-16):

1. **Maintain manual stabilization of the head and neck.** If possible, hold the patient's head and neck from behind.

2. **Apply a rigid cervical immobilization device.**

3. **Assess pulses, movement, and sensation** in all four extremities.

4. **Slide the short backboard behind the patient.** Slip it as far down into the seat as possible, but not below the coccyx. The top of the short backboard should be level with the top of the patient's head. The body flaps should fit snugly under the armpits. Try not to jostle the patient or the rescuer who is holding manual stabilization.

5. **Secure the patient to the backboard.** Strap up the torso first. If the device has leg straps, tighten those next. Finally, secure the patient's head. To make sure the head and neck remain in neutral alignment with the rest of the spine, you may need to pad behind them.

To move the patient to a long backboard, position it under or next to the buttocks. Rotate

Securing Patient to a Long Backboard

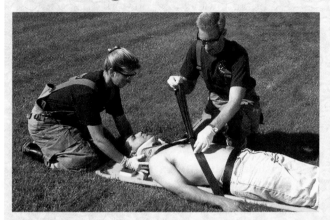

Figure 18-15a Immobilize the patient's torso first.

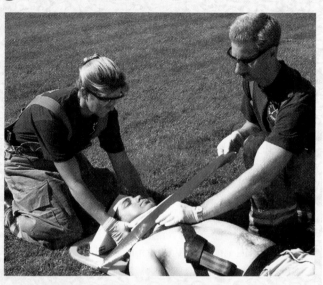

Figure 18-15b Immobilize the head next.

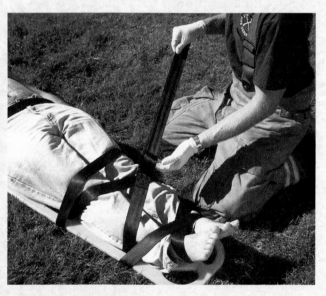

Figure 18-15c Immobilize the patient's legs last.

the patient until his back is in line with it. Then lower him onto the long backboard. Follow the instructions outlined above for securing the patient. Release manual stabilization when the patient is completely immobilized. Release the leg strap of the short board.

Rapid Extrication

In general, rescuers should move a sitting spine-injured patient only after short backboard immo-

bilization. However, in certain emergencies there is not enough time. A rapid extrication may need to be performed instead when:

◆ The scene is not safe. For example, there is a threat of fire or explosion, a hostile crowd, or extreme weather conditions.

◆ Life-saving care cannot be given because of the patient's location or position.

Securing Patient to a Short Backboard

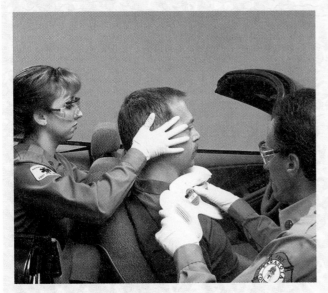

Figure 18-16a Manually stabilize the head and neck. Then apply a rigid cervical collar.

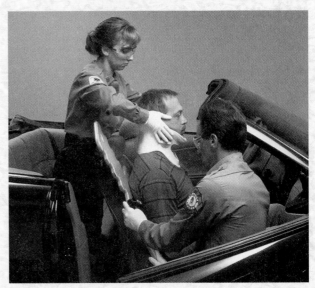

Figure 18-16b Position the short backboard behind the patient.

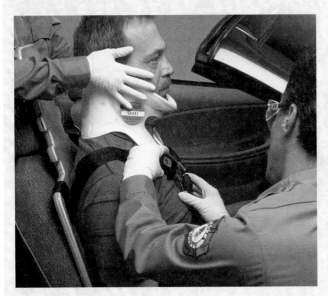

Figure 18-16c Secure the torso to the board.

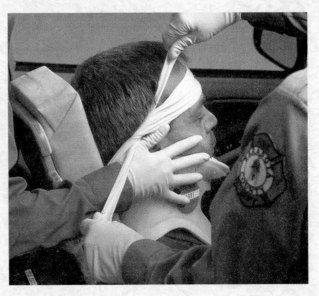

Figure 18-16d Pad behind the head and secure it to the board.

◆ There is an inability to gain access to other patients who need life-saving care.

In general, a rapid extrication must be performed by a team of three or more rescuers. The objective is to move a sitting patient to a long backboard with only manual stabilization of the spine. To do so, proceed as follows (Figure 18-17):

1. **Bring the patient's head into a neutral, in-line position.** This is best done from behind or to the side of the patient.

2. **Apply a rigid cervical immobilization device.**

3. **Rotate the patient into position.** Do so in several, short, coordinated moves until the

Rapid Extrication

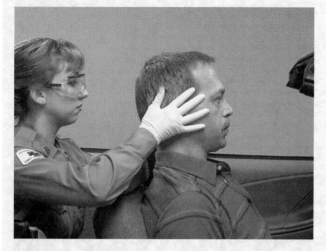

Figure 18-17a Bring the patient's head into a neutral, in-line position.

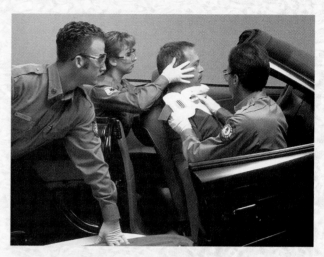

Figure 18-17b Apply a rigid cervical collar.

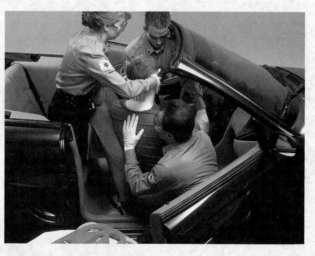

Figure 18-17c Rotate the patient into position.

patient's back is in the open doorway and feet are on the adjoining seat.

4. **Bring the long backboard in line with the patient.** It should rest against the patient's buttocks.

5. **Lower the patient onto the long backboard.** and slide him into position in short, coordinated moves.

6. **Secure the patient to the backboard.** Release manual stabilization only when the

Figure 18-17d Bring the long backboard in line with the patient.

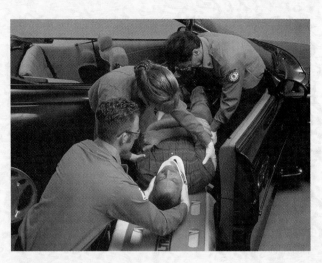

Figure 18-17e Lower the patient onto the long backboard.

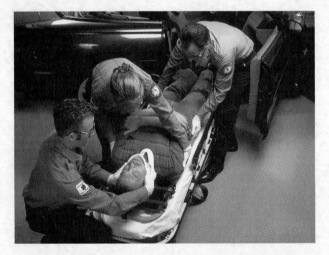

Figure 18-17f Slide the patient into position in small steps, and secure the patient to the backboard.

patient is completely immobilized on a long backboard.

It may be necessary to hand off manual stabilization during the procedure. Be sure it is maintained continuously until the patient is completely immobilized.

Helmet Removal

There are two basic types of helmets: motorcycle helmets and sports helmets such as those worn for football. Typically, a sports helmet has an opening in front that allows easy access to the patient's airway. For many, the face shield can be unclipped

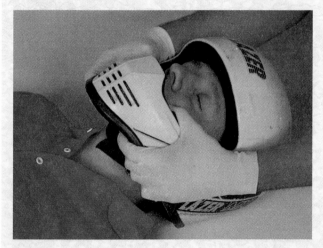

Figure 18-18a Stabilize the helmet, head, and neck to prevent movement.

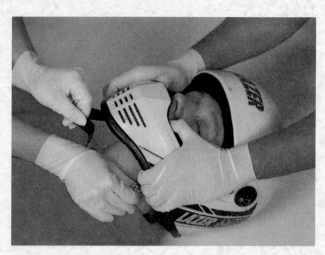

Figure 18-18b Loosen the chin strap, while maintaining manual stabilization.

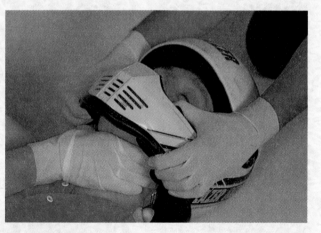

Figure 18-18c Transfer stabilization.

or unsnapped for easy removal. A motorcycle helmet, however, may have a shield that prevents access to the patient's airway.

In general, if your patient can be properly assessed and the airway maintained, a helmet should be left in place. However, remember that a helmet may cause the cervical spine to be in an out-of-line position when the patient is supine.

Do not attempt to remove a helmet alone. Wait for help. If it must be removed, then follow these steps (Figures 18-18):

1. **Stabilize the helmet** to prevent movement. The rescuer at the head holds each side of the helmet. He then places his fingers on the lower jaw.

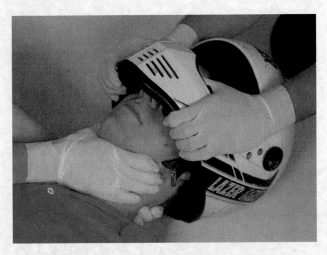

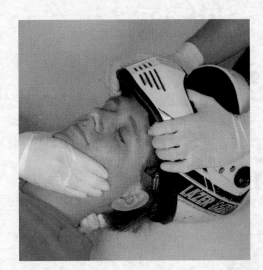

Figure 18-18d Slip off the helmet about half way so your partner can maintain an in-line position of the head.

Figure 18-18e When the helmet is completely removed, transfer manual stabilization to the rescuer at the head.

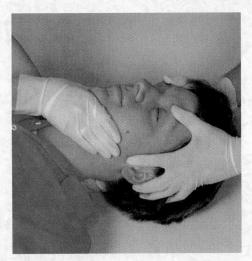

Figure 18-18f Maintain manual stabilization until the patient is completely immobilized.

2. **Loosen the chin strap.** The second rescuer does this, while the first maintains manual stabilization.

3. **Transfer stabilization next.** To do so, the second rescuer places one hand anteriorly on the mandible at the angle of the jaw. He then places the other hand at the back of the head.

4. **Slip off the helmet about half way.** Be sure to spread it so it can clear the ears. The second rescuer then re-adjusts his hands in order to maintain alignment of the head.

5. **Remove the helmet completely.** The rescuer at the head then takes over manual stabilization until the patient is completely immobilized.

At the beginning of this chapter, you read that First Responders were on scene with a male patient who had been in a car crash. To see how the skills in this chapter apply to this emergency, read the following. It describes how the call was completed.

PHYSICAL EXAMINATION

My partner continued to stabilize the patient's head and neck while I conducted a head-to-toe exam. Finding no other injuries, I checked the patient's vital signs and noted they were within normal ranges.

During the exam, the patient asked why we were holding his head and neck. I told him that, because of the way his head impacted the windshield, we were concerned that he may have a neck injury. I added that most of the time in similar situations, this wasn't the case. But, until we could get him to the hospital for an x-ray, we didn't want to take any chances. He seemed relieved by this, and said "OK."

PATIENT HISTORY

I asked our patient if he had any medical conditions, or if he was taking any medications. He told us that he was a diabetic, but he was controlling it with proper diet. He also told us that he was not allergic to any medications that he was aware of. He had eaten about three hours ago.

ONGOING ASSESSMENT

We continued stabilizing the patient's head and neck. Our assessment revealed he was able to wriggle his toes and fingers and continued to respond to questions appropriately. I applied an ice pack to the "goose egg" that had now appeared above his left eye. He told us that the ice made it feel better.

PATIENT HAND-OFF

When the EMTs arrived, I gave them the hand-off report:

"This is Rolando Gomez. He is 38 years old. About 15 minutes ago he was involved in an MVA where he apparently struck an abandoned vehicle in the intersection in front of us. He told us he was traveling about 30 mph when the collision occurred. He denies a loss of consciousness, and states he was wearing no restraints. His chief complaint is neck pain. There also is a contusion above his left eye, which we have been treating with ice. We have noted no other injuries. We have assessed his motor-sensory abilities, and have found him able to wriggle his fingers and toes. The patient states he is a diabetic, but controls it without medication. He denies any allergies. His vital signs are pulse 80, respirations 20, blood pressure 128/88, skin pink, warm, and dry with pupils that are equal and reactive."

The EMTs took over care of the patient and asked if we could help extricate Mr. Gomez from the car using a short backboard. Since we were trained in spinal stabilization procedures, we were glad to assist them. When we finished, the EMTs thanked us for our help and told us that we had done a good job.

We learned when we started our next shift that Mr. Gomez was released from the hospital a couple of hours later with nothing more serious than a bump on the head.

By being the first medically trained rescuer on scene, you have the opportunity to really make a difference in patients suffering from a possible head, neck, or spine injury. Proper assessment and treatment could prevent further injury, permanent disfigurement, and even death.

Chapter Review

Injuries to the head, neck, and spine may require fire service First Responders to take spinal precautions whenever a patient's chief complaint or mechanism of injury suggests the possibility of a spine injury.

Even after a collision with heavy damage to the crash vehicles, there may be patients with a complaint of only minor pain or no pain at all. Note that it is just as important to take spinal precautions with these patients as with patients who complain of pain. Never assume a patient is unin-

jured if the mechanism suggests injury. That is for the hospital physicians to decide.

An EMS instructor wrote this simple but powerful message on the chalkboard during a First Responder class: "Quadriplegia is forever." (Quadriplegia is the inability to use any of the extremities because of a spinal-cord injury.) Use caution with every patient who has a possible spine injury. The consequences of not doing so are extremely serious.

FIRE COMPANY REVIEW

Page references where answers may be found or supported are provided at the end of each question.

Section 1

1. What are the signs and symptoms associated with a head injury? (p. 334)

2. What are the general guidelines for emergency care of a patient with a head injury? (pp. 335–336)

3. Under what conditions should you immediately take spinal precautions with a patient who has a head injury? (p. 335)

Section 2

4. What are the general principles of emergency care of a neck wound? (pp. 337, 339)

5. How would you provide emergency care to a patient with an open and bleeding neck wound? (p. 339)

Section 3

6. Which of the five regions of the spine is most vulnerable to injury? Why? (p. 339)

7. When should you begin manual stabilization of the cervical spine? When may you release it? (pp. 340, 343)

8. What are the signs or symptoms of a patient suffering from spine injury? (p. 340)

9. How would you provide emergency care for a spine-injured patient? (p. 343)

RESOURCES TO LEARN MORE

"Injuries of the Cranium" and "Injuries of the Vertebrae and Spinal Cord" in Moore, E.E., K.L. Mattox, and D.V. Feliciano. *Trauma,* Third Edition. Philadelphia: Appleton-Lange, 1996.

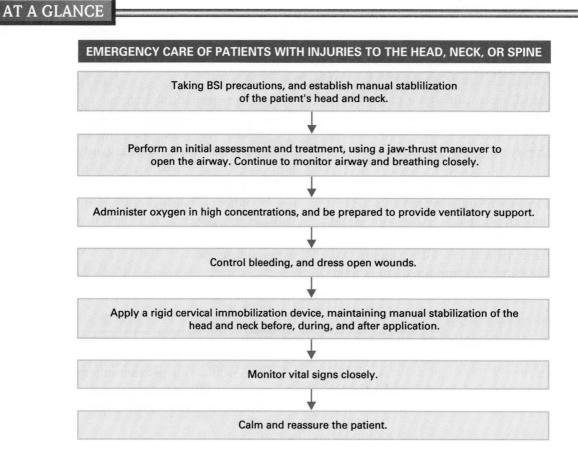

EMERGENCY CARE OF PATIENTS WITH INJURIES TO THE HEAD, NECK, OR SPINE

Taking BSI precautions, and establish manual stablilization of the patient's head and neck.

⬇

Perform an initial assessment and treatment, using a jaw-thrust maneuver to open the airway. Continue to monitor airway and breathing closely.

⬇

Administer oxygen in high concentrations, and be prepared to provide ventilatory support.

⬇

Control bleeding, and dress open wounds.

⬇

Apply a rigid cervical immobilization device, maintaining manual stabilization of the head and neck before, during, and after application.

⬇

Monitor vital signs closely.

⬇

Calm and reassure the patient.

Childbirth

*I*NTRODUCTION *As a fire service First Responder, you may be called to help pregnant patients. Remember, childbirth is a normal, natural process. Only in a few situations will you need to see that the mother receives rapid transport to a hospital.*

Cognitive, affective, and psychomotor objectives are from the U.S. DOT's "First Responder: National Standard Curriculum." Enrichment objectives, if any, identify supplemental material.

Cognitive

6-1.1 Identify the following structures: birth canal, placenta, umbilical cord, amniotic sac. (p. 359)

6-1.2 Define the following terms: crowning, bloody show, labor, abortion. (pp. 359, 361, 368)

6-1.3 State indications of an imminent delivery. (pp. 361–362)

6-1.4 State the steps in the pre-delivery preparation of the mother. (pp. 362–363)

6-1.5 Establish the relationship between body substance isolation and childbirth. (p. 363)

6-1.6 State the steps to assist in the delivery. (pp. 364–366)

6-1.7 Describe care of the baby as the head appears. (pp. 364–365)

6-1.8 Discuss the steps in delivery of the placenta. (pp. 366–367)

6-1.9 List the steps in the emergency medical care of the mother post-delivery. (p. 367)

6-1.10 Discuss the steps in caring for a newborn. (p. 367)

Affective

6-1.11 Explain the rationale for attending to the feelings of a patient in need of emergency medical care during childbirth. (pp. 361–362)

6-1.12 Demonstrate a caring attitude towards patients during childbirth who request emergency medical services. (pp. 361–362)

6-1.13 Place the interests of the patient during childbirth as the foremost consideration when making any and all patient care decisions. (pp. 361–362)

6-1.14 Communicate with empathy to patients during childbirth, as well as with family members and friends of the patient. (pp. 361–362)

Psychomotor

6-1.15 Demonstrate the steps to assist in the normal cephalic delivery. (pp. 364–366)

6-1.16 Demonstrate necessary care procedures of the fetus as the head appears. (pp. 364–365)

6-1.17 Attend to the steps in the delivery of the placenta. (pp. 366–367)

6-1.18 Demonstrate the post-delivery care of the mother. (p. 367)

6-1.19 Demonstrate the care of the newborn. (p. 367)

Enrichment

◆ Discuss specific complications of pregnancy, including toxemia, spontaneous abortion, ectopic pregnancy, placenta previa, and abruptio placenta. (pp. 367–369)

◆ Discuss specific complications of childbirth, including prolapsed umbilical cord, breech birth, limb presentation, multiple births, and premature birth. (pp. 369–371)

ON SCENE

DISPATCH

I was already at the fire station, watching it snow pretty hard, when our training class was cancelled because of the hazardous conditions of the roads.

Just as I was walking out the door, the tone sounded for a 25-year-old woman in labor. As I was getting into our rescue unit, I heard the dispatcher: "Be advised the winter snow advisory has been posted, and all roads are considered hazardous for travel."

My partner and I informed dispatch that we were responding. She reported back that the ambulance could take 20 to 30 minutes, as the paramedics were waiting on a snowplow to meet them at their station.

SCENE SIZE-UP

We arrived at the residence, a farmhouse, and did a quick safety check of the area. My partner advised the dispatcher of the road conditions and best access for the ambulance, while I went up to the door. The husband met me. Once inside the house, he informed me that his wife was due to deliver in a few weeks, but her "water broke" and she was in labor. As I was led to the bedroom where the wife was lying on the bed, I noticed two young children watching TV in the front room. The husband told me that this would be their third child.

INITIAL ASSESSMENT

The patient looked up at me with a grimace and said, "The baby is coming—now!"

> *Consider this patient as you read Chapter 19. How would you proceed?*

SECTION 1
THE PROCESS OF CHILDBIRTH

Anatomy of Pregnancy

The uterus is the organ that contains the developing **fetus,** or unborn infant. (See Figure 19-1.) A special arrangement of smooth muscles and blood vessels in the uterus allows for great expansion during pregnancy and forcible contractions during labor and delivery. It also allows for rapid contraction after delivery, which helps to constrict blood vessels and prevent excessive bleeding.

During pregnancy, the wall of the uterus becomes thin. The **cervix** (the neck of the uterus, which leads into the vagina) contains a mucous plug that is discharged during labor. The expulsion of this plug is known as the **bloody show** and appears as pink-tinged mucus in the vaginal discharge.

The **placenta** is a disk-shaped organ on the inner lining of the uterus. Rich in blood vessels, it provides nourishment and oxygen to the fetus from the mother's blood. It also absorbs waste from the fetus into the mother's bloodstream. The mother's blood and the baby's blood do not mix. The placenta also produces hormones such as estrogen and progesterone, which sustain the pregnancy.

After the baby is delivered, the placenta separates from the uterine wall and delivers as the afterbirth. It usually weighs about a pound or about one-sixth of the infant's weight.

The **umbilical cord** is the unborn infant's lifeline. It is an extension of the placenta through which the fetus receives nourishment. The umbilical cord contains one vein and two arteries. The vein carries oxygenated blood to the fetus. The arteries carry deoxygenated blood back to the placenta. When the baby is born, the cord resembles a sturdy rope about 22 inches (1 050 mm) long and one inch (25 mm) in diameter.

The **amniotic sac,** or *bag of waters,* is filled with a fluid in which the fetus floats. The amount of fluid varies, usually from about one-half to one quart (500 to 1000 milliliters). This sac of fluid insulates and protects the fetus during pregnancy. During labor, part of the sac usually is forced ahead of the baby, serving as a resilient wedge to help dilate (expand) the cervix.

The **birth canal** is made up of the cervix and the **vagina.** The vagina is about 3 to 5 inches (75 to 125 mm) in length. It originates at the cervix and extends to the outside of the body. Its smooth muscle layer stretches gently during childbirth to allow the passage of the infant.

A full-term pregnancy lasts approximately 280 days. Towards the end, the baby usually is in a head-down position, which brings the uterus down and forward. Mothers often can feel the difference and say that the baby has "dropped." This position is most favorable for the baby's passage through the birth canal.

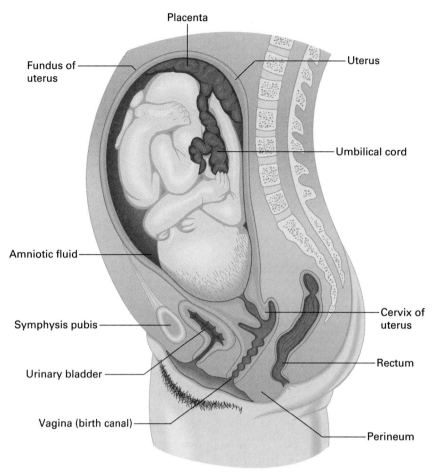

Figure 19-1 Anatomy of pregnancy.

Stages of Labor

Labor is the term used to describe the process of childbirth. It consists of contractions of the uterine wall, which force the baby and later the placenta into the outside world. Normal labor is divided into three stages: *dilation, expulsion,* and *placental* (Figure 19-2). The length of each stage varies greatly in different women and under different circumstances.

First Stage: Dilation

During this first and longest stage, the cervix becomes fully dilated (expanded). This allows the baby's head to progress from the uterus into the birth canal. Through uterine contractions, the cervix gradually stretches and thins until the opening is large enough to allow the baby to pass through.

The contractions may begin as an aching sensation in the small of the back. Within a short time, the contractions become cramp-like pains in the lower abdomen. These recur at regular intervals, each one lasting about 30 to 60 seconds. At first, the contractions usually occur 10 to 20 minutes apart and are not very severe. They may even stop completely for a while and then start again. Appearance of the mucous plug, or bloody show, may occur before or during this stage of labor. Before or during this stage, the amniotic sac may rupture, resulting in a gush of fluid from the vagina. The patient may say something like "my water broke" when this occurs.

Stage one may continue for as long as 18 hours or more for a woman having her first baby. Women who have had a child before may only have two or three hours of labor. When contractions are two minutes apart, birth is very near.

Second Stage: Expulsion

During this stage, the baby moves through the birth canal and is born. As the baby moves down-

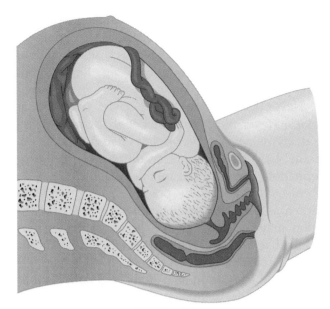

FIRST STAGE:
First uterine contraction to dilation of cervix

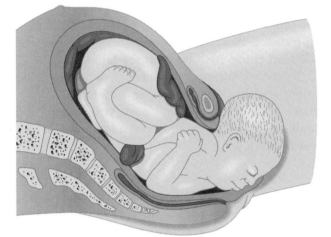

SECOND STAGE:
Birth of baby or expulsion

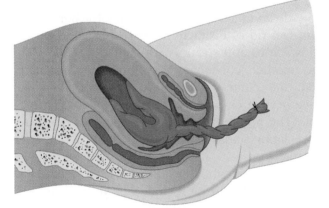

THIRD STAGE:
Delivery of placenta

Figure 19-2 Three stages of labor.

ward, the mother experiences considerable pressure in her rectum, much like the feeling of a bowel movement.

When the mother has this sensation, she should lie down and get ready for the birth of her child. The tightening and bearing-down sensations will become stronger and more frequent. The mother will have an uncontrollable urge to push down, which she may do. There probably will be more bloody discharge from the vagina at this point.

Soon after, the baby's head appears at the opening of the birth canal. This is called **crowning.** The shoulders and the rest of the body follow.

Third Stage: Placental

During this stage, the placenta separates from the uterine wall. Usually, it is then spontaneously expelled from the uterus.

 COMPANY OFFICER'S NOTE

Knowledge of the process of childbirth is critical to having the confidence to deliver a baby. Since delivering babies in the field is a rare event, continuing training and education are important to being able to handle a delivery.

Patient Assessment

Childbirth is a natural, normal process. It is not an illness or disease. It is, however, physically traumatic. Complications can be life threatening to both the mother and the baby.

If you are called to the scene of a childbirth, perform a scene size-up and initial assessment and treatment as you would for any patient. Give the mother calming reassurance. Then assess her condition to see if there will be time for transport to the nearest medical facility or if she will have the baby in her present location. Update incoming EMS units as necessary.

Generally, you should expect to assist in the delivery of the baby on scene if:

◆ You have no suitable transportation.

◆ Delivery of the baby can be expected within five minutes.

Hospital or physician cannot be reached due to a natural disaster, bad weather, or some kind of catastrophe.

Distance to the hospital is so great that the likelihood of delivery would occur before arrival.

To determine if you should have the patient transported, time the contractions. Follow these steps:

1. **Feel for the contractions.** Place your gloved hand on the mother's abdomen, just above her navel. Feel the involuntary tightening and relaxing of the uterine muscles.

2. **Time the contractions in seconds.** Start from the moment the uterus first tightens until it is completely relaxed.

3. **Time the intervals in minutes.** Time them from the start of one contraction to the start of the next.

If the contractions are more than five minutes apart, the mother usually has time to be transported to a hospital safely, as long as traffic or weather conditions are not a problem. If the contractions are two minutes apart, she probably does not have time. Prepare to help deliver the baby where you are.

If the contractions are between two and five minutes apart, you must make a decision based on a number of factors. Ask the mother a few questions and conduct a simple assessment. The mother is usually nervous and apprehensive, so be gentle and kind. Show confidence and support. Ask these questions:

Have you had a baby before? (The birth may take longer in a first pregnancy.) Are you having contractions? How far apart are they? Has the amniotic sac ruptured (or did your water break)? If so, when?

Do you feel the sensation of a bowel movement? (If yes, it is likely that the baby's head is pressing against the rectum and will soon be born. Do not let the mother sit on the toilet.)

Do you feel like the baby is ready to be born?

Examine the mother. She should be on her back with knees bent and legs spread. Inspect the vaginal area, but do not touch it except during delivery and when your partner is present.

Determine if there is crowning. If you can see bulging in the vaginal area, and either the head or other part of the baby is visible, prepare to deliver the baby where you are. Report your findings to EMS.

Never ask the mother to cross her legs or ankles. Never tie or hold her legs together to try to delay delivery. Never delay or restrain delivery in any way. The pressure could result in death or permanent injury to the infant.

Also, be alert to the possibility of a condition known as *supine hypotensive syndrome*. This condition may occur when the pregnant patient lies on her back. The combined weight of the uterus and the fetus presses on the great vein that collects blood from the lower body and delivers it to the heart. This vein is called the *inferior vena cava*. That pressure can limit the blood that must return to the heart, causing the amount of blood that circulates through the body to decrease. You may observe signs of shock in your patient, including reduced blood pressure, increased pulse, and pale skin color. Also be alert for fainting.

To avoid supine hypotensive syndrome, the patient should be lying on her left side, if appropriate, or in a sitting position. If you suspect the condition, position the patient on her left side and treat for shock.

Preparation for Delivery

First, take a deep breath. And remember, childbirth should be a celebration, not an anxiety-ridden process. So, relax. Remember to always act in a professional manner. Reassure the mother. Tell her that you are there to help with the delivery. Provide as much quiet and privacy for her as you can. Get rid of distractions. Hold her hand (if she allows you) and speak encouragingly to her. Help the mother concentrate on breathing regularly with the contractions. Wipe the mother's face. Give her ice chips only if allowed by local protocol. (The mother should not eat or drink anything once labor starts.) The father or another rescuer can help.

At a minimum, the following materials and equipment should be included in your obstetrical (OB) kit. They are:

Sheets and towels, sterile if possible.

One dozen 4 × 4 square gauze pads.

COMPANY OFFICER'S NOTE

If the father is present, and willing, involve him in the process of caring for his wife. Ask him to assist by having him hold his wife's hand. Ask him to wipe his wife's face. If allowed by local protocol, ask him to give ice chips if the mother is thirsty. This can help alleviate the anxiety of the father, mother, and you!

- Two or three sanitary napkins.
- Rubber suction syringe.
- Baby receiving blanket.
- Surgical scissors for cutting the umbilical cord.
- Cord clamps or ties.
- Foil-wrapped germicidal wipes.
- Wide tape or sterile cord.
- Large plastic bags.

All materials used during delivery should be sterile, or at least as clean as possible. This is to protect both the baby and the mother from contamination and infection.

Because delivery results in exposure to a great deal of blood and other body fluids, take BSI precautions. Put on eyewear, a face mask, protective gloves, a disposable gown, and shoe coverings if possible. Handle soaked dressings, pads, and linens carefully. Place them in separate bags that will not leak. Then seal and label the bags. Scrub your arms, hands, and nails thoroughly before and after the delivery, even if you wore gloves.

Other guidelines include:

- Be prepared to provide basic life support to both the mother and the infant, including treatment for shock.
- Help the mother relax with each contraction. Inhaling causes muscles to tighten, so have her exhale with each contraction. Encourage her to keep her breathing slow but comfortable. Tell her not to strain or push during the first stage of labor.
- The amniotic sac may rupture, if it has not already done so. There also may be some blood-tinged mucus. These fluids increase as labor progresses. If you have a clean towel, place it under the mother's buttocks to absorb them. Always wipe the vaginal opening in a down-and-away direction to minimize contamination from the rectal area. Discard soiled towels or sheets used for this purpose. Replace them frequently with clean ones.

- If the patient feels more comfortable sitting, reclining, or in some other position during the first stages of labor, let her do so.

- As contractions become longer and closer together, the patient should lie down on a flat, firm surface that she can push against. It is easiest for you if the mother is on an elevated surface. However, if the floor is the only firm surface available, use it. Pad it with folded sheets, towels, or blankets. Elevate the mother's buttocks about two inches (55 mm) with an additional pad of folded sheets or towels. The pad, which should extend about two feet in front of her, will help to support the slippery baby when it is born.

- When the mother is in position, her feet should be flat on the surface beneath her. Her knees will naturally spread apart because of the size of her abdomen. Do not pull them apart any further. Remove any constricting clothing, or push clothing above the mother's waist.

- Create a sterile field around the opening of the vagina. Place a sterile or clean sheet under the mother's hips. Touching only the corners of the sheet, have the mother lift her hips while you place one fold well under her hips. Unfold it toward her feet. If you have time, place another sheet or towel over the mother's abdomen and legs, leaving the vaginal area uncovered. Direct the best possible light toward the mother's genitals. Do not touch the vagina at any time.

- During the second stage of labor, when the mother bears down, remind her not to arch her back. She should curve it and bring her chin to her chest to avoid excessive straining. Have her hold her breath for 7 to 10 seconds as she bears down. Holding the breath longer can cause too much straining, broken blood vessels, and tearing of the area around the vagina.

Childbirth

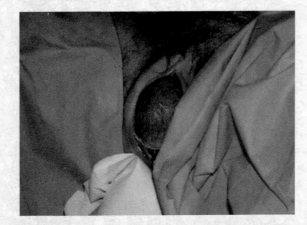

Figure 19-3a Crowning.

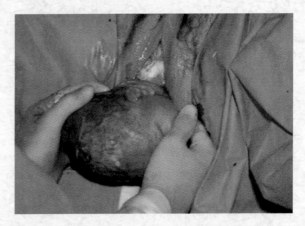

Figure 19-3b Head delivers and turns.

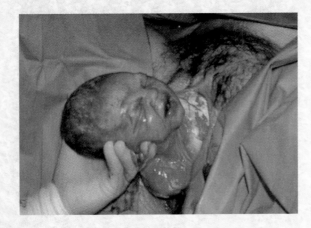

Figure 19-3c Shoulders deliver.

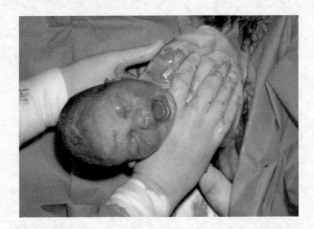

Figure 19-3d Chest delivers.

Delivery of the Baby

(See Figure 19-3 for photographs of the delivery of a baby and the placenta.)

1. **Place the palm of your hand on top of the baby's head.** When it crowns, apply very gentle pressure to prevent an explosive delivery.

2. **Break open the amniotic sac** if it has not already broken. Tear it or pinch it open with your fingers and push it away from the infant's head and mouth. The baby can safely inhale clear amniotic fluid. However, if there is **meconium** staining (greenish or brownish fluid), the baby has had a bowel movement, which could cause pneumonia if inhaled.

In the case of meconium staining, clean the area around the mouth and nose once the head is delivered. Suction the mouth first and then the nose with a rubber suction syringe. Expel all air from the suction bulb prior to placing it in the baby's mouth or nose. Release the bulb to create suction. Suctioning may need to be repeated in order to clear the airway. Note that meconium staining can be a life-threatening event. Consider requesting an advanced life support unit to assist.

3. **Determine the position of the umbilical cord.** When the baby's head delivers, check to see if the umbilical cord is around the baby's neck. If it is, use two gloved fingers

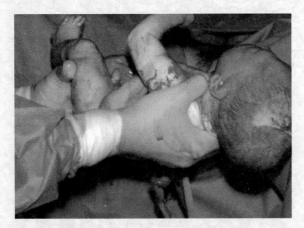

Figure 19-3e Legs and feet deliver.

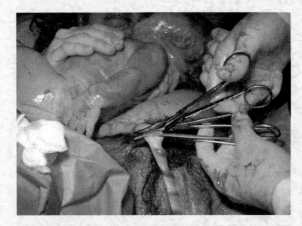

Figure 19-3f Cutting of cord.

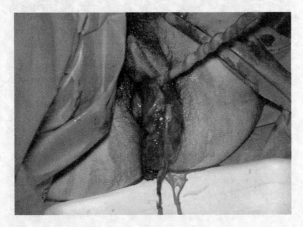

Figure 19-3g Placenta begins delivery.

Figure 19-3h Placenta delivers.

to slip the cord over the shoulder. Only if you cannot dislodge it, attach two clamps a few inches (millimeters) apart. Then, with extreme care, cut the cord between the clamps. Gently unwrap the ends of the cord from around the baby's neck, and proceed with the delivery.

4. **Support the baby's head.** As soon as the baby's head is born, place one hand below it. Spread the fingers of your other hand gently around it. Avoid touching the fontanels (the soft spots at the top of the head). In most normal presentations, the baby's head faces down. Then it turns so that the nose is toward the mother's thigh.

5. **Remove fluids from the infant's airway.** Use a rubber bulb syringe to suction mucus first from the baby's mouth and then from the nose. Make sure you fully compress the syringe before you bring it to the baby's face. Avoid contact with the back of the mouth. Insert the tip no more than an inch (25 mm) into the mouth. Slowly release the bulb to allow fluid to be drawn into the syringe. If a syringe is not available, wipe the baby's mouth and then the nose with gauze.

6. **Support the baby with both hands as the rest of the body is born.** Once the shoulders are delivered, the rest of the body will appear rapidly. Note that you should never

pull the baby from the vagina. Never touch the mother's vagina or anus. Handle the baby's slippery body carefully. Do not put your fingers in the baby's armpits. Pressure on the nerve centers there can cause paralysis of the arms.

7. **Grasp the feet as they deliver.** Do not pull on the umbilical cord.

8. **Position the baby.** He or she should be level with the mother's vagina until the umbilical cord is cut. The neck should be in a neutral position.

9. **Make note of the time of birth.**

10. **Dry, wrap, and reposition the newborn.** Gently dry the infant with towels. Then wrap him or her in a clean, warm blanket. Place the baby on his or her side, head slightly lower than the trunk. Only the face should be exposed.

11. **Clean the newborn's mouth and nose.** Wipe blood and mucus from the baby's mouth and nose with sterile gauze. Suction the mouth first and then the nose. The infant should cry almost immediately.

12. **Assess the baby's breathing.** Provide tactile stimulation, if there is none. Rub the back gently or flick the soles of the feet (Figure 19-4). Administer oxygen as soon as possible. Usually, placing an oxygen mask near the baby's face and allowing the oxygen to blow by is effective. Do not use oxygen tubing without a mask. The force of the oxygen coming out of the tube can be harmful. If you still get no response, start artificial ventilation. (See "Newborn Care" below for details.)

13. **Clamp, tie, and cut the umbilical cord when it stops pulsating,** if your EMS system allows you to do so (Figure 19-5). Place two clamps or ties on it about three inches (75 mm) apart. Position the first clamp about four finger-widths (six inches, or 150 mm) from the infant. Use sterile surgical scissors to cut the cord between the two clamps or ties. Periodically check the end of the cord for bleeding, and control any that occurs.

14. **Record the time of delivery.**

Remember your priorities: make sure the baby's airway is clear and open, and keep the baby warm.

Delivery of the Placenta

After the baby is delivered, observe for the delivery of the placenta:

1. **Feel for contractions.** The contracting uterus should feel like a hard, grapefruit-size ball.

2. **Encourage the mother to bear down as the uterus contracts.** It usually delivers within 10 minutes of the infant, and almost always within 30 minutes. Normally, there will be some bleeding as the placenta separates.

3. **Wrap the placenta after it delivers.** When the placenta appears, slowly and gently guide it from the vagina. Never pull. If you have not cut the cord, wrap the placenta in a sterile towel and place it next to the baby. Wrap the baby and the placenta in a third sheet or blanket. If the cord is cut, place the placenta in a plastic bag to be taken to the hospital where a physician can confirm the delivery is complete.

Figure 19-4 Stimulate breathing by rubbing back or flicking feet.

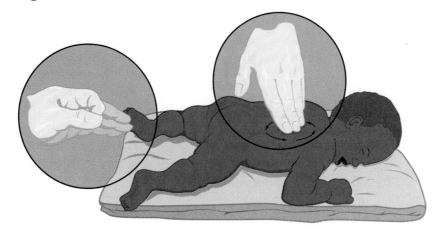

CUTTING THE UMBILICAL CORD

3"

6–9"

Figure 19-5 Cutting the umbilical cord.

After the placenta delivers, check the mother's vaginal bleeding. Up to about one pint (500 cc) of blood loss is normal and usually well tolerated by the mother. Place two sanitary napkins over the opening of the vagina. Touch only the outer surfaces of the pads. Do not touch the mother's vagina.

Ask the mother if she plans to breast feed the infant. If so, now is a good time to encourage the mother to start. Breast feeding helps the uterus to contract, which decreases the size of the uterus and helps to stop bleeding.

Make sure that the mother and the baby are covered and warm. The mother, as well as the infant, can chill easily following birth. Activate the EMS system if you have not already done so. The mother and the baby should be taken together to the hospital for evaluation by a physician.

If after the delivery of the placenta, bleeding appears to be excessive, treat the mother for shock. Arrange for immediate transport. Then massage the uterus to stimulate uterine contractions:

1. **Position your hands.** Place the medial edge of one hand horizontally across the abdomen, just superior to the symphysis pubis. Extend

your fingers. Cup your other hand around the uterus.

2. **Begin the massage.** Use a kneading or circular motion to massage the area. It should feel like a hard ball.

3. **Reassess the bleeding.** If it continues to appear to be excessive, check your massage technique.

Replace any blood-soaked sheets and blankets while waiting for transport. Place all soiled items in a marked infection control bag and seal.

Newborn Care

Perform artificial ventilation on the newborn if any of the following three conditions exist:

◆ Gasping respirations, or no breathing after drying, warming, and tactile stimulation.

◆ Pulse rate is less than 100 beats per minute.

◆ Persistent central cyanosis, or bluish discoloration around the chest, abdomen, and face after 100% oxygen has been administered.

The recommended rate for assisting a newborn's ventilations is between 40–60 breaths per minute. Keep in mind that a baby's lungs are very small and require very small puffs of air. Never use mechanical ventilation on a newborn. A bag-valve-mask device may be used, but it must be the appropriate size for a newborn. Remember to observe for chest rise. Reassess after 30 seconds. For proper positioning of the head, a towel may be placed under the baby's shoulders.

If breathing and pulse are absent or if pulse rate is less than 60 beats per minute, or 60-80 beats per minute and not rising when ventilating with oxygen, then start CPR. The rate of compressions is 120 per minute. The ratio of compressions to breaths for the newborn is 3 to 1.

SECTION
2

COMPLICATIONS OF CHILDBIRTH

Complications of Pregnancy
Trauma During Pregnancy

Out-of-hospital care for a pregnant patient who has been injured is the same as for any trauma patient. By assuring that the pregnant woman has an open airway and adequate breathing and circulation, you will be maximizing the fetus's chances for survival. When you report to EMS, be sure to inform them that your patient is pregnant in order to facilitate rapid transport.

During patient assessment, keep in mind that a pregnant patient's vital signs will differ from those of a woman who is not pregnant. That is, the pulse rate is usually higher and blood pressure is usually lower. Always assume that a woman who has sustained injuries and has an increased heart rate or decreased blood pressure is in shock, and treat accordingly.

Provide normal trauma care, including manual stabilization if needed. However, any pregnant woman with trauma should receive high-flow oxygen and be positioned on her left side. For the pregnant patient who must be immobilized on a long backboard, the entire board can be tilted slightly to the left or a towel can be placed under the patient's right hip. Advanced life support (ALS) should be summoned if the patient is unstable.

Toxemia of Pregnancy

About 5% of women develop toxemia (or "poisoning of the blood") during pregnancy. It occurs most frequently in the last **trimester** (three months). It most often affects women in their twenties who are pregnant for the first time. Women with a history of diabetes, heart disease, kidney problems, or high blood pressure are at greatest risk.

Signs and symptoms of toxemia may include:

◆ High-blood pressure (most common).

◆ Swelling in the extremities (most common).

◆ Sudden weight gain (two pounds or more per week).

◆ Blurred vision or spots before the eyes.

◆ Pronounced swelling to the face.

◆ Decreased urinary output.

◆ Severe and persistent headache.

◆ Persistent vomiting.

◆ Pain in the upper abdomen.

◆ Sudden seizures.

To provide emergency care for toxemia of pregnancy, arrange for immediate transport. Position the patient on her left side to avoid compressing the inferior vena cava. Keep the patient calm. Administer oxygen, if you are allowed to do so.

If the mother suffers a seizure, monitor her breathing closely. When the seizure stops, elevate her head and shoulders and administer high-flow oxygen.

Spontaneous Abortion

Sometimes called a "miscarriage," a spontaneous abortion is the loss of pregnancy before the twentieth week. It occurs naturally, unlike abortions that are deliberately performed. Signs and symptoms include:

◆ Vaginal bleeding, often heavy.

◆ Pain in the lower abdomen that is similar to menstrual cramps or labor contractions.

◆ Passage of tissue from the vagina.

To provide emergency care, arrange for immediate transport. Treat the patient for shock. Save any passed tissue by packaging it in a sealed bag. The bag should then be transported with the patient for evaluation by a physician.

Ectopic Pregnancy

A woman has two **fallopian tubes.** Each one extends up from the uterus to a position near an **ovary.** Each month an egg passes from an ovary into a fallopian tube. The fallopian tube then conveys the egg to the uterus and sperm from the uterus toward the ovary.

In a normal pregnancy, a fertilized **ovum** (egg) is implanted in the uterus. In an ectopic pregnancy, a fertilized ovum is implanted outside the uterus. It could be in the abdominal cavity, on the outside wall of the uterus, on the ovary, or on the outside of the cervix. In 95% of cases, the ovum is implanted in a fallopian tube.

An ectopic pregnancy is a severe medical emergency. The expanding fertilized ovum eventually causes rupture of a blood vessel and severe abdominal bleeding. It is the leading cause of death in pregnant women in their first trimester.

Signs and symptoms include:

◆ Sudden, sharp abdominal pain on one side. If bleeding is extensive, pain will become more diffuse.

◆ Pain under the diaphragm, or pain radiating to one or both shoulders.

◆ Tender, bloated abdomen.

◆ Vaginal spotting or bleeding.

◆ Missed menstrual periods.

- Weakness when sitting.
- Syncope (fainting).
- Signs of shock.

Suspect ectopic pregnancy in any woman of childbearing age when the above signs and symptoms are present. To provide emergency care, arrange for immediate transport. Place the patient on her back with knees elevated. Keep the patient warm. Administer oxygen, if you are allowed.

Placenta Previa

Placenta previa occurs when the placenta is positioned in the uterus in an abnormally low position. So when the cervix dilates (expands), the fetus moves, or labor begins, the placenta separates from the uterus. This puts both the mother and the baby in danger.

Signs and symptoms include severe, usually painless bleeding from the vagina and shock. To provide emergency care, arrange for immediate transport. Elevate the patient's legs. Maintain body temperature. If possible, administer 100% oxygen by mask.

Abruptio Placenta

Another major cause of bleeding during pregnancy is abruptio placenta. It is the leading cause of fetal death after blunt trauma. Life-threatening for both the mother and the baby, it needs to be recognized and treated rapidly.

There are several causes, including toxemia and trauma. Whatever the cause, the normally implanted placenta begins separating from the uterus sometime during the last trimester of pregnancy. Bleeding begins, but it is often behind the placenta and the mother is unaware of it. Shock then develops in the mother, and the baby does not get enough oxygen.

Signs and symptoms include:

- Bleeding from the vagina, not usually in great quantities.
- Severe abdominal pain.
- Rigid abdomen.
- Signs of shock.

To provide emergency care, arrange for immediate transport. Monitor vital signs carefully, and treat for shock. Administer 100% oxygen by mask, if you are allowed.

Complications of Delivery

Prolapsed Umbilical Cord

In some situations, the umbilical cord comes out of the birth canal before the infant. When this happens, the baby is in great danger of suffocating. The cord is compressed against the birth canal by the baby's head, which cuts off the baby's supply of oxygenated blood from the placenta. Emergency care is extremely urgent. Arrange for immediate transport. Follow these steps:

1. **Position the patient.** Have the mother lie down on her left side, if possible. Knees should be drawn to her chest, or her hips and legs should be elevated on a pillow.
2. **Administer high-flow oxygen,** if possible.
3. **Position the infant.** With your gloved hand, gently push the baby up the vagina far enough so that the baby's head is off the umbilical cord. Maintain pressure on the baby's head and keep the cord free until medical help arrives. This is controversial in some areas. Follow local protocol.
4. **Cover the cord.** Use a sterile towel moistened with clean, preferably sterile, water. Do not try to push the cord back into the vagina.

> **FIRE DRILL**
>
> Have you ever had someone pinch shut the low-pressure hose on your SCBA as a prank? How long did it take before you needed to take a breath? This is very similar to what happens to the baby when the umbilical cord is prolapsed.

Breech Birth

In a breech birth, the baby's feet or buttocks are delivered first (Figure 19-6). Whenever possible, the mother should be taken to the hospital for the birth. If that is not possible, then follow these guidelines:

1. **Position and prepare the mother for a normal delivery.**
2. **Let the buttocks and trunk deliver on their own.** Note, never try to pull the baby from the vagina by the legs or trunk.

Figure 19-6 In breech birth, baby's feet or buttocks deliver first.

3. **Support the infant.** Place your arm between the baby's legs. Let the legs dangle astride your arm. Support the baby's back with the palm of your other hand.

4. **Observe the delivery of the head,** which should follow on its own within three minutes. If not, you need to prevent the baby from suffocating. If you do not, the baby's head will compress the umbilical cord, preventing the flow of oxygenated blood from the placenta.

5. **If necessary, form an airway for the baby.** Place the middle and index fingers of your gloved hand alongside the infant's face. Your palm should be turned toward the face. Form the airway by pushing the vagina away from the baby's face until the head is delivered. Hold the baby's mouth open a little with your finger so that the baby can breathe.

Umbilical Cord Around the Neck

If the umbilical cord is wrapped around the baby's neck in the birth canal, try to slip the cord gently over the baby's shoulders or head. If you cannot, and the cord is wrapped tightly around the neck, place clamps or ties three inches (75 mm) apart on the cord. Quickly but carefully, cut between them. Then unwrap the cord from around the neck. Deliver the shoulders and body, supporting the head at all times.

Limb Presentation

If the baby's arm or leg comes out of the birth canal first, it means that the baby has shifted so much in the uterus that a normal delivery is not possible. The baby will have to be delivered by a physician. Delay can be fatal. Never pull on the baby by the arm or leg. The mother must be taken immediately to a hospital. Transport without delay.

Multiple Births

Twins are delivered the same way as single babies, one after the other. In fact, since twins are smaller, delivery is often easier. Identical twins have two umbilical cords coming out of a single placenta. If the twins are fraternal (not identical), there will be two placentas.

The mother may not be aware that she is carrying twins. You should suspect them if one or more of the following conditions exists:

- The abdomen is still very large after one baby is delivered.

- If the baby's size is out of proportion with the size of the mother's abdomen.

- If strong contractions begin again about 10 minutes after one baby is born.

The second baby is usually born within minutes and almost always within 45 minutes. To manage a multiple birth, follow these guidelines:

- After the first baby is born, clamp and cut the cord to prevent bleeding to the second baby. About one-third of second twins are breech.

- If the second baby has not delivered within 10 minutes, the mother should be transported to the hospital for the birth.

- After the babies are born, the placenta or placentas will be delivered normally. You can expect bleeding after the second birth.

- Keep the babies warm. Twins are often born early and may be small enough to be considered premature. Guard against heat loss until they can be taken to a hospital.

Premature Birth

If a woman gives birth before the thirty-sixth week of pregnancy, or if the baby weighs less than five and one-half pounds, the baby is considered to be premature. Premature babies are smaller and redder. They have heads that are proportionately larger than full-term babies. Because they are very vulnerable to infection, special care must be taken.

- Keep the baby warm with a blanket or swaddle. If you lack other supplies, use aluminum foil as an outer wrap.

- Keep the baby's mouth and nose clear of fluid by gentle suction with a bulb syringe.

- Prevent bleeding from the umbilical cord. A premature infant cannot tolerate losing even a little blood without being at risk for shock.

- Administer oxygen, if allowed, by blowing it gently across the baby's face. Never blast oxygen directly into the face.

- Since premature babies are so vulnerable to infection, do not let anyone breathe into the baby's face. Do everything you can to prevent contamination.

ON SCENE FOLLOW-UP

> At the beginning of this chapter, you read that fire service First Responders were on the scene of a childbirth. To see how chapter skills apply to this emergency, read the following. It describes how the call was completed.

INITIAL ASSESSMENT (CONTINUED)

I introduced myself to the patient and assured her that I had been trained to handle this type of situation. As we talked, I determined that she was alert and oriented with a good airway and adequate respiration. Her pulse was strong and regular, and her skin warm and slightly sweaty. There was no evidence of bleeding.

I moved to a position beside the patient and explained that the ambulance was on the way. At this point, I told both parents what needed to be done and how they could help until the paramedics arrived. I also congratulated each of them on the imminent arrival of the new member of their family. Meanwhile, my partner updated dispatch.

PATIENT HISTORY

The patient, Mrs. Jabari Billings, told me that this was her third pregnancy and that there had been no problems with her first two babies. Her doctor assured her that this pregnancy was progressing normally. She reported that her water broke 45 minutes before and contractions were two minutes apart with about a 45-second duration. She said she felt pressure on her rectum, as if she had to move her bowels.

PHYSICAL EXAMINATION

I asked Mrs. Billings if I may examine her for crowning and to prepare her for delivery. She said yes. Acting quickly, my partner and I donned all appropriate personal protective equipment. Her husband stayed next to her, holding her hand while coaching her breathing.

The baby's head had crowned. I placed my hands gently on the head. The shoulders and the rest of the baby followed rapidly. Grabbing the bulb syringe from my partner, I suctioned the mouth first and then the nose. The baby cried loudly. It was a boy. Wrapping him in a warm towel and placing him on his mother's belly, I couldn't help but smile at the joy evident in the faces of both parents as they gazed upon their new son. As my partner clamped the umbilical cord, I looked up at the time.

ONGOING ASSESSMENT

Not long after the birth, Mrs. Billings delivered the placenta. I placed it in a container for the paramedics to transport with the patients. Although the mother was tired, she was in good spirits and had no unusual complaints or distress. Her mental status was normal and vitals were stable.

After being cleaned up a bit, the baby's color also appeared to be normal. His respirations were 44. Pulse was 142 and regular. He was actively moving around and had a good strong cry.

PATIENT HAND-OFF

When the paramedics arrived, I gave them the hand-off report. They quickly packaged the two patients and moved them to the ambulance. They had gotten to the residence quickly, considering the weather. The snowplow driver was my new hero. Even though there were no complications—you could say it had been a "textbook" delivery—my heart was still racing as I watched them drive away.

Childbirth is a rare and exciting event for a fire service First Responder. Remember that birth is a natural process. You are there to assist the mother and then care for her and the baby after delivery. Your top priorities are the same as for any call: scene safety and the ABCs of your patients, both mother and child.

Chapter Review

Calls for emergency childbirth are rare when compared to other types of calls. Unlike other calls, childbirth is not an injury or illness. If you are called to the scene where the delivery of a baby is imminent, remain calm. Remember what you have learned, and keep in mind that you are assisting a natural process. Also, remember that this is one of those few occasions in EMS where we are requested to assist in a joyous event—the beginning of life!

Become familiar with procedures and protocols in your area. In the event of complications, care for the infant and the mother until they are turned over to EMTs for transport to a hospital.

FIRE COMPANY REVIEW

Page references where answers may be found or supported are provided at the end of each question.

Section 1

1. What is the function of each of the following: placenta, umbilical cord, amniotic sac? (p. 359)

2. What happens during the first stage of labor? The second stage? The third stage? (pp. 360–361)

3. What information must you have in order to decide if a birth is imminent? (pp. 361–362)

4. How can you determine how fast and how close together contractions may be? (p. 362)

5. What part of the baby usually presents first in a normal delivery? (p. 361)

6. What can you do to assist a mother in the delivery of the baby? Describe the process briefly. (pp. 364–366)

7. When does the placenta usually deliver? What should you do with it when it does? (pp. 366–367)

Section 2

8. What would you observe in a breech birth? A prolapsed cord? A limb presentation? (pp. 369–370)

9. How does a multiple birth differ from a single birth? (pp. 370–371)

RESOURCES TO LEARN MORE

"Emergency Aspects of Obstetrics" in Harwood-Nuss, A.L., et al. *The Clinical Practice of Emergency Medicine,* Second Edition. Philadelphia: Lippincott-Raven, 1995.

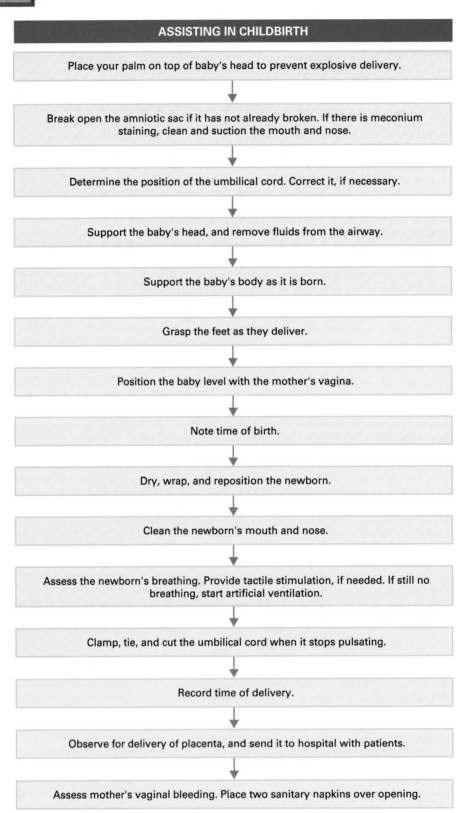

ASSISTING IN CHILDBIRTH

Place your palm on top of baby's head to prevent explosive delivery.

↓

Break open the amniotic sac if it has not already broken. If there is meconium staining, clean and suction the mouth and nose.

↓

Determine the position of the umbilical cord. Correct it, if necessary.

↓

Support the baby's head, and remove fluids from the airway.

↓

Support the baby's body as it is born.

↓

Grasp the feet as they deliver.

↓

Position the baby level with the mother's vagina.

↓

Note time of birth.

↓

Dry, wrap, and reposition the newborn.

↓

Clean the newborn's mouth and nose.

↓

Assess the newborn's breathing. Provide tactile stimulation, if needed. If still no breathing, start artificial ventilation.

↓

Clamp, tie, and cut the umbilical cord when it stops pulsating.

↓

Record time of delivery.

↓

Observe for delivery of placenta, and send it to hospital with patients.

↓

Assess mother's vaginal bleeding. Place two sanitary napkins over opening.

Infants and Children

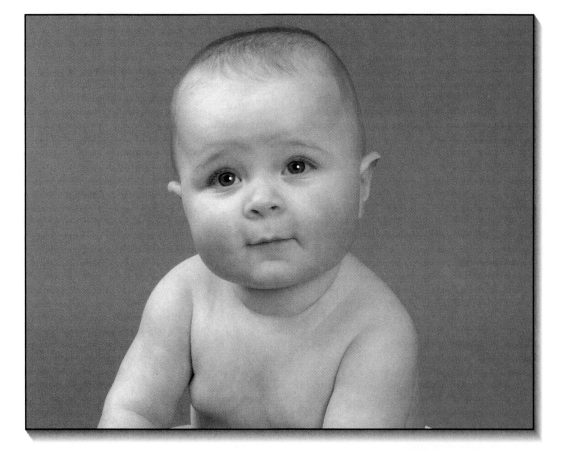

*I*NTRODUCTION   *As a fire service First Responder, you will encounter injured or sick infants and children—the pediatric patient—all too often. Motor-vehicle crashes, followed by burns and drowning, are the most frequent causes of injuries in children. While your assessment approach to the ill or injured child is somewhat different from your approach to an adult, your patient care plan is the same. This chapter focuses on the particular needs of the pediatric patient and how the fire service First Responder should address those needs in emergencies.*

Cognitive, affective, and psychomotor objectives are from the U.S. DOT's "First Responder: National Standard Curriculum." Enrichment objectives, if any, identify supplemental material.

Cognitive

6-2.1 Describe differences in anatomy and physiology of the infant, child, and adult patient. (pp. 380–385, 388)

6-2.2 Describe assessment of the infant or child. (pp. 380–385)

6-2.3 Indicate various causes of respiratory emergencies in infants and children. (pp. 388–390)

6-2.4 Summarize emergency medical care strategies for respiratory distress and respiratory failure/arrest in infants and children. (pp. 388–389)

6-2.5 List common causes of seizures in the infant and child patient. (p. 390)

6-2.6 Describe management of seizures in the infant and child patient. (pp. 390–391)

6-2.7 Discuss emergency medical care of the infant and child trauma patient. (pp. 386–388)

6-2.8 Summarize the signs and symptoms of possible child abuse and neglect. (pp. 392–394)

6-2.9 Describe the medical-legal responsibilities in suspected child abuse. (pp. 392–394)

6-2.10 Recognize need for First Responder debriefing following a difficult infant or child transport. (pp. 392, 394)

Affective

6-2.11 Attend to the feelings of the family when dealing with an ill or injured infant or child. (pp. 377–379)

6-2.12 Understand the provider's own emotional response to caring for infants or children. (pp. 392, 394)

6-2.13 Demonstrate a caring attitude towards infants and children with illness or injury who require emergency medical services. (pp. 377–378, 385–386)

6-2.14 Place the interests of the infant or child with an illness or injury as the foremost consideration when making any and all patient care decisions. (pp. 378–379)

6-2.15 Communicate with empathy to infants and children with an illness or injury, as well as with family members and friends of the patient. (pp. 377–379)

Psychomotor

6-2.16 Demonstrate assessment of the infant and child. (pp. 379–386)

Enrichment

◆ Describe characteristics associated with the stages of infant and child development. (pp. 377, 378)

◆ Discuss common respiratory emergencies in infants and children, including croup, epiglottitis, and asthma. (pp. 388–390)

◆ Identify the signs and symptoms of shock in cases of trauma or dehydration in infants and children. (pp. 387–388)

◆ Discuss the management of a patient with suspected sudden infant death syndrome (SIDS). (pp. 391–392)

ON SCENE

DISPATCH

My partner and I had just finished lunch as tones came over the speaker in the squad room. The dispatcher's voice followed. "Squad Two, respond to a child struck by a car at the intersection of Southport and Roscoe. Be advised that an ambulance is being dispatched. Time of call is 1423 hours."

En route to the call, we were informed that the child's condition was unknown and the ambulance would be delayed.

SCENE SIZE-UP

Upon arrival, I saw that the police were already present on scene. My partner and I took BSI precautions as we were led through the crowd by an officer. He led us to a grassy area, where a child was lying on her left side. A very upset father was beside the child. While my partner tried to introduce herself to the father and calm him down, I questioned a witness, who told me that the car had been traveling at about 15 mph when it hit the child. She said the child appeared to glance off the left front of the vehicle, fall to the ground, and roll onto the grass.

By that time, my partner had calmed the father down and gotten consent to provide emergency care.

INITIAL ASSESSMENT

The patient was on her left side and not moving. The father said she was six years old. Checking her responsiveness, I noted she responded only to pain. She also had snoring respirations.

> *Consider this patient as you read Chapter 20. Will assessment and treatment of the six-year-old be different or the same as for an adult?*

THE PEDIATRIC PATIENT

Determining the age of a pediatric patient can be difficult. Although age is a common benchmark for certain types of treatments, not all young patients physically mature at the same pace. Use your best judgment when the exact age of a patient is unknown.

Developmental Characteristics

Knowing the characteristics of children at each age can help you in an emergency. You will have a good idea of what to expect from them and how best to communicate. (See Table 20-1 for a summary.)

As a fire service First Responder, your encounters with pediatric patients will be when they are ill or injured. Frightened before your arrival, their perception of you as an unfamiliar person will add to their fear.

Children pick up on anxiety easily. Similarly, they often take their cues from what they observe. So, it is very important to stay calm. Parental involvement is often helpful. If all the adults on scene stay calm, a pediatric patient is likely to stay calm, too.

When dealing with younger children, it will help to get down to their eye level. Do not stare. Instead, include them in your conversation. Move slowly and deliberately when performing an assessment or providing emergency care. If a child is old enough to understand, ask permission to remove a piece of clothing or to touch his or her body. If a child holds out a hand or allows you to examine some part, seize the moment. The rule in pediatric care is to examine and palpate what you can when the opportunity presents itself.

With adolescents, it is important to be sensitive to feelings of modesty. In many ways, they are young adults. Permanent disfigurement may be a major concern. Also, they are very sensitive to peer pressure, requiring reassurance that what they tell you will be held in confidence. Of course, you would still be free to include relevant information in your prehospital care report or in any report to emergency medical providers who have a need to know.

Dealing with Caregivers

With pediatric patients, it is important to understand that caregivers, especially parents, may be very upset and concerned. In fact, when a child is ill or injured, the fire service First Responder should take a holistic approach, one involving not only the patient, but the family as well.

Anticipate a variety of responses from caregivers. A few of the more common include crying, emotional outbursts, anger, guilt, and confusion, which may be directed at fire department and other EMS personnel. Do not take it personally. Caregivers need support and understanding just

Table 20-1

Childhood Development by Age

Common Term	Age	Characteristics and Behaviors
Infant	Birth to 1 year	Knows the voice and face of parents. Crying may indicate hunger, discomfort, or pain. Will want to be held by a parent or caregiver. Difficult to identify the precise location of an injury or source of pain.
Toddler	1–3 years old	Very curious at this age. Into everything. Be alert to the possibility of poison ingestion. May be distrustful and uncooperative. Usually does not understand what is happening, which raises level of fear. May be very concerned about being separated from parents or caregivers. A stuffed toy may be helpful in gaining trust.
Preschooler	3–5 years old	Able to talk, but appears not to listen well because may not understand what is being said. Use simple words. May be scared and believe what is happening is own fault. Sight of blood may intensify response. Sometimes a cartoon-character bandage helps.
School Age	6–12 years old	Should cooperate and be willing to follow the lead of parents and EMS provider. Has active imagination and thoughts about death. Continual reassurance is important.
Adolescent	13–18 years old	Acts like adult. Able to provide accurate information. Modesty is important. Fears permanent scarring or deformity. May become involved in "mass hysteria." Be tolerant but do not be fooled by it.

COMPANY OFFICER'S NOTE

For the fire service First Responder, empathy skills—including listening with your ears and your heart—are just as important as clinical skills with every patient. Multiply this by 100 when dealing with the parents or caregivers of an injured or sick infant or child. So, it may be appropriate for you, as the company officer, to deal with others on scene while the First Responder is providing emergency care.

For example, a parent can hold an oxygen mask near the infant's face (Figure 20-1).

Occasionally you may encounter a parent who will not let you help an ill or injured child. That parent may insist on remaining in control. Avoid becoming defensive. The parent's behavior has nothing at all to do with you. Remember, even though they may be coping in the only way they know how, ultimately the child is the patient. If you were called, you must make sure the patient is not in need of emergency care.

The following techniques may help in situations where parents are especially anxious:

◆ Your first priority is to protect the health and safety of the patient. If parents are making unsafe demands, try an approach of reserved confrontation. Explain that your opinion and procedures are based on sound medical knowledge and their demands are obstructing

as much, if not more than the injured or sick infant or child does.

As the assessment of the patient progresses, explain what you are doing. If time permits, tell them why. If appropriate, ask for their assistance.

Figure 20-1 "Blow-by" oxygen using tubing and a paper cup. Never use Styrofoam.

what is considered appropriate and in the child's best interest.

- Realize that the parents may be correct. They usually know their children extremely well. A parent of a chronically ill child probably has a good grasp of what is going on.

- Regardless of how parents behave, treat them with courtesy, respect, and understanding. Avoid raising your voice. Tell them that you know they want help.

- Let parents stay as close to the patient as possible, as long as they are not interfering with care. If medically appropriate, a child can be held by a parent. Also, consider letting the parents do something for the child in order to help focus their attention away from you.

- Whatever you do, do not react with anger.

> **COMPANY OFFICER'S NOTE**
>
> If you believe an infant or child needs immediate medical attention but the parents or caregivers refuse it, call for help. That may be the police department or an agency within your jurisdiction charged with protecting the welfare of infants and children (child protective services).

Using Appropriate Equipment

Having the correct equipment in the correct size is extremely important to providing quality care to the pediatric patient. The following list is an example of some of the EMS equipment appropriate for the different needs of infants and children:

- Adjunct airways in pediatric sizes.
- Face masks, oxygen masks, and nasal cannula in pediatric sizes.
- Pediatric bag-valve-mask (BVM) resuscitator with oxygen enrichment attachment. If the BVM has a pop-off valve, it must be able to be closed.
- Bulb syringe for suctioning.
- Blood pressure cuffs in pediatric sizes.
- Pediatric stethoscope.
- Cervical immobilization devices in various pediatric sizes.
- Backboards made especially for infants and children.
- New, clean stuffed animal to be used to comfort or distract the patient.
- Cartoon-character bandages.

Patient Assessment

Scene Size-up

When entering an emergency scene involving an infant or child, many EMS rescuers say that 90% of the assessment is done from the doorway. Although not totally true, it is possible to get a fairly comprehensive "snapshot" as you enter the scene. In addition to your usual scene size-up, observations made from the doorway include the interaction between the patient and parents, general appearance of the environment, and signs of possible **non-accidental trauma** (child abuse).

Notice how caregivers are reacting and, if appropriate, consider how best to involve them in assessment and treatment of the patient. Note that you may be called to treat infant or child patients at a location other than where they were injured or became ill. It is not uncommon for adults to pick up a child in order to provide comfort or assistance. Understand they mean no harm, even in a situation in which a First Responder would not normally want to move a child. Ask caregivers specifically:

- Why was EMS called?
- What is the chief complaint?
- Has the child been moved? If so, where did the incident occur?

Be alert to the possibility of poison ingestion or a possible fall.

Initial Assessment

When assessing an infant's or child's level of responsiveness, determine from the caregiver what is "normal." If a patient appears to be responsive but is unable to recognize his or her own parents, consider this a sign of a serious medical emergency.

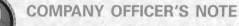

COMPANY OFFICER'S NOTE

One way to help assess an infant or child's level of responsiveness is to ask a question pertinent to the age of the patient. For example, you might ask a four-year-old: "Who is your favorite Sesame Street character?" You can ask the parents for the title of the last movie the child watched. If it was, for example, "Beauty and the Beast," you might ask the child: "What was the name of the beautiful woman who fell in love with the Beast?"

Although the anatomy of an infant or child is not the same as an adult's (Figure 20-2), assessment and treatment of the ABCs are just as critical. Remember, your single most important task is to ensure an open airway. When trauma is suspected, always use a jaw-thrust maneuver. Never hyperextend the head and neck. One way to keep the head in a neutral position is to place a folded towel under the patient's shoulders.

Common signs in infants and children that indicate respiratory distress are as follows (Figure 20-3):

◆ Noisy breathing, such as stridor, crowing, grunting.

◆ Cyanosis.

◆ Flaring nostrils.

◆ Retractions (drawing back) between or below the ribs.

◆ Use of accessory muscles to breathe.

◆ Breathing with obvious effort.

◆ Altered mental status.

◆ Decreased heart rate.

Immediately provide oxygen if any of these signs are evident. Continually monitor for signs of respiratory distress. If an infant's respirations are less than 20 per minute or a child's are less than 10, assist ventilations. One way to do this is to use a pediatric BVM with supplemental oxygen.

Assess circulation by palpating the infant's brachial pulse. Palpate the unresponsive child's carotid or femoral pulse. Palpate the responsive child's radial or brachial pulse.

FIRE DRILL

New firefighters must learn where all the equipment is kept on the apparatus they are assigned to. For example, they must be able to locate the axe, pike pole, and Haligan tool. Learning where a tool is kept makes it much easier to find in the stressful environment of the fireground.

Similarly, it is essential for the fire service First Responder to practice locating an infant's or child's pulse. Practice on your own baby, for example, or on a family member's infant or child. By doing so, you will find you will be able to take a pulse with skill and confidence whenever you are in the stressful environment of a medical emergency.

When you assess the pediatric patient's circulation, remember that inadequate oxygen can slow the heart. Provide oxygen as soon as you detect a slow pulse, and use a mask that is the correct size for the patient. When the patient is not breathing and has no gag reflex, an oropharyngeal airway should be inserted to assist in maintaining an open airway. (See Chapter 5 for review of how to insert adjunct airways.)

Control external bleeding immediately. Remember, a child has the ability to compensate for blood loss longer than an adult can. As a result, decompensation (failure of the heart to maintain sufficient circulation of blood) occurs very rapidly. Therefore, it is imperative to monitor pediatric patients constantly.

After initial assessment and treatment, update other EMS providers who may be responding to the scene. For those patients with significant airway problems, respiratory or cardiac arrest, or the possibility of shock, rapid transport is indicated.

NOTE: The most important care you can give a pediatric patient is done in the initial assessment. Never stop required airway care to perform a physical exam or gather a patient history.

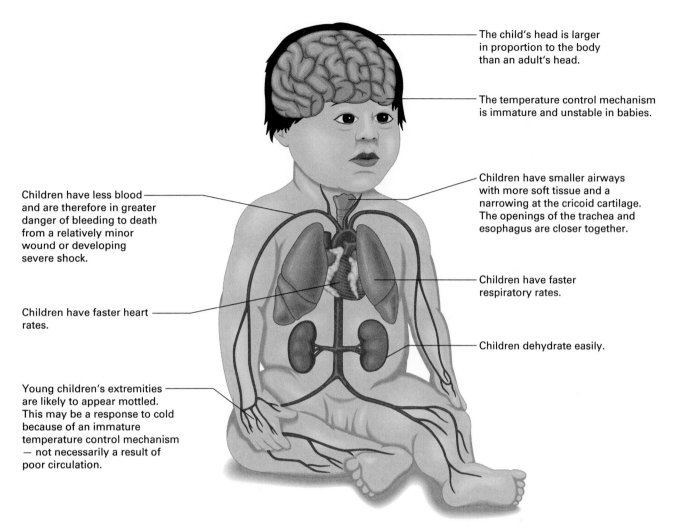

The child's head is larger in proportion to the body than an adult's head.

The temperature control mechanism is immature and unstable in babies.

Children have smaller airways with more soft tissue and a narrowing at the cricoid cartilage. The openings of the trachea and esophagus are closer together.

Children have faster respiratory rates.

Children dehydrate easily.

Children have less blood and are therefore in greater danger of bleeding to death from a relatively minor wound or developing severe shock.

Children have faster heart rates.

Young children's extremities are likely to appear mottled. This may be a response to cold because of an immature temperature control mechanism — not necessarily a result of poor circulation.

Figure 20-2 The anatomy of an infant or child is not the same as an adult's.

Patient History

If time permits, gather a pertinent medical history. Remember that anxious, upset, screaming parents and children can be unnerving. If possible, talk to the child and involve the caregivers. Interview witnesses. Avoid asking questions that require only yes-or-no answers. If responses to questions seem inconsistent or if they do not correspond to your initial assessment, try rephrasing the questions.

When you take a SAMPLE history for a medical patient, include questions such as: When did the signs and symptoms develop? How have they progressed? Is the problem a recurring one? If so, has the child been seen by a physician? Is the specific diagnosis known? What treatment was received?

When you take a SAMPLE history for a trauma patient, include the details of the emergency, such as the time it occurred, mechanism of injury, and emergency care already given.

Physical Exam

It is difficult to assess pain in children. They may lack the body awareness and vocabulary necessary to describe it. Children also may not be able to separate the fear they feel from their physical condition. Ask the parent, if possible, how the child usually responds to pain. This may give you some idea of how the present condition compares.

The bodies of infants and children can hide injury for some time. Only after their compensatory abilities fail will you see changes in vital signs. Changes can occur very quickly, and the patient's condition may deteriorate very fast. Take the vital signs of infants and children more frequently than

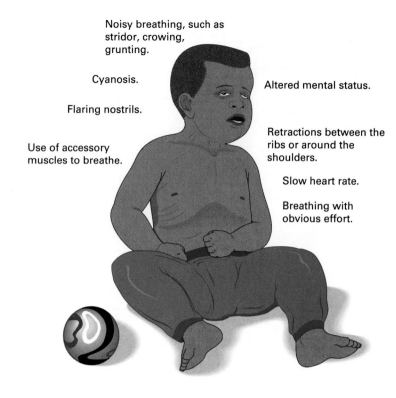

Noisy breathing, such as
stridor, crowing,
grunting.

Cyanosis.

Flaring nostrils.

Use of accessory
muscles to breathe.

Altered mental status.

Retractions between the
ribs or around the
shoulders.

Slow heart rate.

Breathing with
obvious effort.

Figure 20-3   Signs of respiratory problems.

you would for an adult. (See Table 20-2.) Also, pay close attention to your overall impression of how the patient looks and acts. Your observations and intuitive feelings may tell you more about the status of the child than any one vital sign.

When you do assess a pediatric patient's vital signs, keep the following in mind:

◆ *Pulse.* Use the brachial pulse in an infant and the radial pulse in a child. If the radial pulse is not clear, check the brachial. If the pulse is too rapid or too slow, immediately examine the patient for problems such as signs of respiratory distress, shock, or head injury.

Rapid pulse may indicate hypoxia (oxygen deficiency), shock, or fever. It also may be normal in scared or overly excited children. A slow pulse in a child must first be presumed to be a sign of hypoxia. Other causes of slow pulse include pressure within the skull, depressant drugs, or a rare medical condition.

◆ *Respiration.* Children sometimes breathe irregularly. Therefore, monitor respirations for a full minute when determining rate. Do so frequently. Note that you may need to place your hand on a "belly breather's" abdomen to get an accurate breathing rate. Rates in children alter easily due to emotional and physi-

cal conditions. An increase over a previous rate can be significant.

The quality of breathing also is important. Determine if it is adequate. Observe to see if the child is working to breathe and using accessory muscles. Look for retractions. Notice if breathing is noisy. Shortness of breath may indicate the need for you to assist ventilations.

◆ *Blood pressure.* A falling or low blood pressure in a pediatric patient can be a late indicator of shock. Be sure to use a blood pressure cuff that is the correct size for your patient. Use a pediatric stethoscope, if available. Do not take a blood pressure in children under three years of age.

◆ *Temperature.* Feel the arms and legs of infants and children to see if they are cold. The torso may be warm in comparison. Cold hands and feet may indicate shock, especially when your patient is not in a cold environment.

◆ *Skin condition and capillary refill.* Always look at the pediatric patient's skin for signs of injury. Notice skin color. Newborns may have a mottled color or bluish discoloration on their hands and feet. Children should not. Be alert. Assess capillary refill (Figure 20-4). If it takes more than two seconds, the patient may be in shock. A delayed capillary refill

Table 20-2

Normal Vital Signs for Infants and Children

Normal Pulse Rates (Beats per Minute, at Rest)

Newborn	120 to 160
Infant 0–5 months	90 to 140
Infant 6–12 months	80 to 140
Toddler 1–3 years	80 to 130
Preschooler 3–5 years	80 to 120
School-age 6–10 years	70 to 110
Adolescent 11–14 years	60 to 105

Normal Respiratory Rates (Breaths per Minute, at Rest)

Newborn	30 to 50
Infant 0–5 months	25 to 40
Infant 6–12 months	20 to 30
Toddler 1–3 years	20 to 30
Preschooler 3–5 years	20 to 30
School-age 6–10 years	15 to 30
Adolescent 11–14 years	12 to 20

Normal Blood Pressure Ranges

	Systolic	Diastolic
Preschooler 3–5 years	Average 99 (78 to 116)	Average 65
School-age 6–10 years	Average 105 (80 to 122)	Average 69
Adolescent 11–14 years	Average 114 (88 to 140)	Average 76

Note: A high pulse in an infant or child is not as great a concern as a low pulse. A low pulse may indicate imminent cardiac arrest. Blood pressure is usually not taken in a child under 3 years of age. In cases of blood loss or shock, a child's blood pressure will remain within normal limits until near the end, and then it will fall swiftly.

should be considered an emergency in the pediatric patient.

Many children with head or spine injuries suffer nervous system damage as well. In fact, it is the cause of traumatic death at least two-thirds of the time. If the patient's history suggests trauma or if there is a significant mechanism of injury, manually stabilize the head and neck immediately. Remember that children do not have the same verbal skills as adults, and their response to stimuli may be different. Parents may be able to describe a normal response.

Assess for damage to the nervous system as follows:

◆ Determine the level of responsiveness. If the patient appears to have an altered mental status, ask parents to describe the normal level.

◆ Check the pupils. Find out if they are of equal size and, if practical, how they respond to light.

Figure 20-4 Assess capillary refill in patients who are less than six years old.

- Examine the head, neck, and spine for signs of injury.
- Check to see if the patient responds to verbal and painful stimuli. Pinch the skin between the patient's thumb and forefinger.
- Check the ability to recognize familiar objects and people.
- Check the ability to move arms and legs purposefully.
- Check to see if there is clear or bloody fluid draining from the ears or nose.

If you suspect damage to the nervous system, keep your patient as still and calm as possible. Provide in-line stabilization of the patient's head and neck. Apply a cervical immobilization device if you are allowed to do so. *Do not allow untrained people to move the patient.* Decide on the appropriate resources necessary for transport to the hospital. Continue to assess vital signs.

When caring for an infant or child, the following special conditions and situations should be considered by the fire service First Responder (Table 20-3):

- Monitor an infant's breathing continually. Infants have proportionately larger tongues, which if relaxed can easily block the airway.
- Head injuries are more likely in infants and children. When you suspect an injury above the clavicles or when the mechanism of injury is unknown, provide in-line stabilization. Apply a cervical stabilization device, if possible.
- Always support the head when you lift an infant. Before the age of nine months, an infant cannot fully support his or her own head.
- During the physical exam, check the anterior fontanel (soft spot) on top of an infant's skull. If you see a bulge, there may be pressure inside the skull. If you see a depression, the infant may be dehydrated and in shock. (The fontanel stays open for up to two years after birth.)
- Injuries to the extremities can damage the growth plates with long-term effects. Carefully assess extremities, including pulses, capillary refill, and skin condition. Be careful not to cause additional pain. Follow local protocols for splinting.
- Stop bleeding as quickly as possible. A comparatively small blood loss in an adult would be major for a child. Be alert to open fractures, which can bleed profusely.
- A child's skin surface is large compared to body mass. This makes children more susceptible to hypothermia. Response to burns also can be more severe. Watch for signs of shock.
- Make sure cervical stabilization devices fit correctly. Note that some children have very short necks. Such a device may not work on them. Use a rolled towel or blanket instead.
- Children often get arms, legs, hands, feet, or heads trapped under or in rigid structures. The child is sometimes in pain and is usually panicky. Calm the child first. Then see if he or she can move independently. Consider calling

Table 20-3

Anatomical Differences in Infants and Children

Anatomical Differences	Impact on Assessment and Treatment
Larger tongue.	Can block airway.
Reduced size of airway.	Can become easily blocked.
Abundant secretions.	Can block airway.
"Baby" teeth.	Can easily dislodge and block airway.
Flat nose and face.	Difficult to obtain good airway seal with face mask.
Proportionally large head.	Must maintain neutral position to keep airway open. Higher potential for head injuries in cases of trauma.
"Soft spots" (fontanels) on head.	Bulging "soft spots" may indicate intracranial pressure; sunken ones may indicate dehydration.
Thinner and softer brain tissue.	Consider head injury more serious than in adults.
Short neck.	Difficult to stabilize and immobilize.
Shorter and narrower trachea, with more flexible cartilage.	Can close off trachea with overextension of the neck.
Faster respiratory rate.	Muscles fatigue easily, which can lead to respiratory distress.
Primarily nose breathers (newborns).	Airway more easily blocked.
Abdominal muscles used to breathe.	Difficult to evaluate breathing.
More flexible ribs.	Lungs are more easily damaged. May be significant injuries without external signs.
Heart can sustain faster rate for longer period of time.	Can compensate longer before showing signs of shock and usually "crashes" more quickly than an adult.
More exposed spleen and liver.	Significant abdominal injury more likely. Abdomen more often a source of hidden injury.
Larger body surface.	Prone to hypothermia.
Softer bones.	Can easily bend and fracture.
Thinner skin.	Consider burns to be more serious than in an adult.

for additional resources. Lubricate skin surface with baby oil or a water-soluble lubricant. Make sure it is not applied near the patient's mouth, nose, or eyes. If sawing or cutting is needed, make sure the child is protected with a heavy, fire resistant cover. Someone may talk to the child to provide encouragement. After the child is freed, perform a complete patient assessment.

Following are helpful hints for conducting a physical exam:

◆ If possible, assess the child while he or she is on the parent's lap (Figure 20-5).

◆ Prepare yourself so you can radiate confidence, competence, and friendliness. Remember that children between one and six seldom like strangers.

Figure 20-5 Having the child sit on a parent's lap can have a calming influence.

◆ Get as close as you can to a child's eye level. Sit next to the child if possible. When it comes time for a hands-on assessment, do it in the least threatening way. Consider starting at the toes and working your way up to the head.

◆ Explain what you are doing in terms a child can understand. Follow up on the child's questions. Maintain eye contact, but do not stare. Speak in a calm, quiet voice. Even infants will respond to a calm voice, and an apparently unresponsive child may absorb much of what you say.

◆ Younger children tend to take statements literally. For example, if you say you want to take a pulse, they may think you mean to take something away from them. Watch your phrasing. Also, older children do not like being talked about. Talk with them directly.

◆ Be gentle. Do everything you can to reduce the amount of pain that a child must endure. However, when there will be pain, be honest. "It will hurt when I touch you here, but it will last only a second. If you feel like crying, it's okay." Children can tolerate pain, if they are prepared for it and are given adequate support. With children under school age, keep the most painful parts of the assessment for the end. It will help if the child is kept on a parent's lap.

◆ Do not lie to a child. Always be honest. This does not mean you need to explain everything that is going on, but when you do answer questions or perform a procedure that could be uncomfortable, be candid.

◆ If a child is not calm enough to be treated, you must restrain him. But be sure it is absolutely necessary. Only when care is compromised should the child be separated from a parent.

◆ A stuffed animal may help to win the confidence of a young child. It also may help to distract the child during assessment and treatment. Also, very shy children sometimes "talk through" a stuffed animal. For example, they may not tell you where they hurt but might tell where the animal hurts, thereby giving you information about themselves.

Ongoing Assessment

The ongoing assessment is the same for infants and children as it is for adults. It never ends as long as you are caring for the patient.

Patient Hand-off

Just as you would for adult patients, include in your infant or child hand-off report:

◆ Age and sex of the patient.

◆ Chief complaint.

◆ Level of responsiveness (AVPU).

◆ Airway and breathing status.

◆ Circulation status.

◆ Physical exam findings.

◆ SAMPLE history.

◆ Treatment, interventions, and the patient's response to them.

SECTION 2

COMMON PEDIATRIC EMERGENCIES

Trauma

Injuries, especially blunt ones, are the leading cause of death in infants and children. Emergency care for these patients is similar to care provided to adults. Treat as you go. That is, as you assess the patient's ABCs, take care of any problem you find. Remember that children may take longer to go into shock, but when they do, it can develop very rapidly.

Always suspect trauma when infants and children are passengers in motor-vehicle collisions. If they are unrestrained, suspect head and neck injuries. Suspect blunt trauma to the abdomen if the child was wearing a lap belt without a shoulder belt. Fully restrained passengers often have abdominal and lower spine injuries. If the infant's car seat is fastened improperly, suspect head and neck injuries.

In some EMS systems, you may be allowed to immobilize a child right in his or her car seat. This is done by applying a cervical stabilization device to the patient and padding between the child and seat to prevent movement. This procedure should not be used if the child has or may have serious injuries, since the car seat can prohibit airway care and other important emergency care.

If a child is immobilized in a car seat, never tip the seat back. The position may impair breathing by putting pressure on the child's diaphragm. Always make sure that the seat remains upright.

If a child riding a bicycle is struck by a car, suspect head, spine, and abdominal injuries. If the child was walking when struck, suspect head injury, abdominal injury, and pelvis or femur injury.

Shock

A major cause of shock in children is dehydration due to vomiting and diarrhea related to infection. Another cause is blood loss due to trauma.

Children tend to compensate more efficiently for shock. When they decompensate, it is a rapid process that can be devastating. Blood pressure may drop so far and so fast that the patient may go into cardiac arrest. It is critical to constantly monitor the injured patient for signs and symptoms of shock. Take his or her vital signs constantly, including capillary refill.

Signs and symptoms of shock in the pediatric patient include (Figure 20-6):

- Altered mental status, from anxiety to unresponsiveness.
- Apathy or lack of vitality. This may present as the inability of the child to identify a parent, an ominous sign.
- Delayed capillary refill.
- Rapid or weak and thready pulse.
- Pale, cool, clammy skin.
- Rapid breathing.
- Falling or low blood pressure (a late sign).
- Absence of tears when crying.

Hypothermia can intensify shock in infants. Since they cannot shiver to warm themselves, you must keep them warm. Remember especially to cover an infant's head. If you are waiting for advance EMS providers to arrive on scene, the following steps for emergency care of an infant in shock should be taken:

1. **Position the patient.** Have the infant or child lie flat. Make sure that the position does not interfere with breathing.
2. **Monitor the airway and breathing.** Be prepared to assist ventilations. Provide oxygen, if you are allowed.
3. **Keep the patient warm and as calm as possible.**
4. **Monitor vital signs often.**

Remember that a relatively small blood loss in a pediatric patient can be very dangerous. For example, a newborn usually has less blood than the contents of a soda can. When there is

Figure 20-6 Signs of shock in an infant or child.

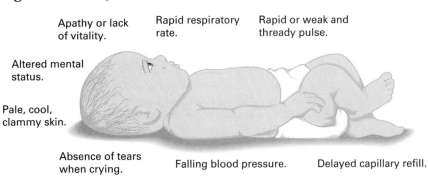

Apathy or lack of vitality.

Rapid respiratory rate.

Rapid or weak and thready pulse.

Altered mental status.

Pale, cool, clammy skin.

Absence of tears when crying.

Falling blood pressure.

Delayed capillary refill.

significant visible blood loss or suspected internal bleeding, rapid transport and constant monitoring are essential.

Respiratory Emergencies

Nothing is more critical than controlling the airway and ensuring adequate breathing in a pediatric patient. The National Pediatric Trauma Registry has reported that 30% of all pediatric trauma deaths are related to inappropriate management of the airway. The American Heart Association reports that more than 90% of pediatric deaths from foreign body airway obstruction occur in children under five years of age. Of those, 65% are infants. It is estimated that most could be saved by early detection.

Note that there are some differences in the pediatric patient's airway (Figure 20-7) and how you manage it:

◆ Infants and children are more susceptible than adults to respiratory problems. Having smaller air passages and less reserve air capacity, airway compromise from trauma or infection is more likely than in the adult patient. Be especially attentive to ensure a clear and open airway.

◆ Airway structures of infants and children are not as long or as large as an adult's. An infant's or a child's can close off if the neck is flexed or extended too far. The best position is a neutral or slightly extended position. If the patient is flat on the back, place a thin pad or towel under the shoulders to keep the head and neck properly aligned. Monitor signs of breathing.

◆ Because of immature accessory muscles, infants and children use their diaphragms to breathe. If there are no reasons to prevent you from doing so, place the pediatric patient in a position of comfort, such as sitting.

◆ Infants and children have a large tongue that can block the airway. Make sure the tongue is forward. If a jaw-thrust maneuver is used, be sure your hand stays on the bony part of the chin (Figure 20-8). If it falls below, the tongue could be pushed back to block the airway.

◆ Infants and children tend to breathe through their noses. They also have abundant secretions. So be prepared to suction often to make sure the nose stays clear (Figure 20-9). (You may wish to refer to Chapter 19 for directions on the proper use of a bulb syringe.)

◆ Apply oxygen, if you are allowed, by way of a mask. Humidified oxygen is preferred. But never withhold or delay oxygen in order to have it humidified. If an infant or child will not tolerate a mask, then hold it slightly away from the patient's face.

Figure 20-7   Comparison of airways of an adult and an infant or child.

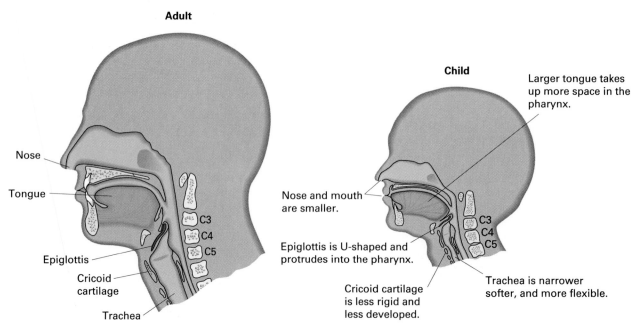

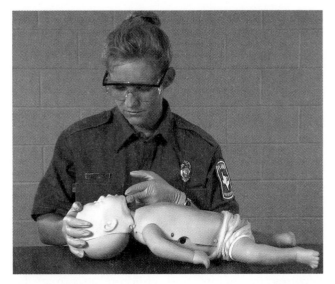

Figure 20-8 Airway obstruction may be relieved in an infant or child by moving the head to a neutral or slightly extended position.

◆ If an infant or child is having a respiratory emergency, notify the incoming EMS unit and be sure an advanced life support (ALS) team is en route, if available.

◆ Even if the signs and symptoms of a respiratory emergency subside, it is still important for the infant or child to be transported to a hospital.

For emergency care of infants and children with respiratory emergencies, see Chapter 5. For

Figure 20-9 Use a bulb syringe for suctioning an infant.

more information about respiratory problems in all patients, see Chapter 10.

Croup

Croup is a common viral infection of the upper airway characterized by swelling that progressively narrows the airway. It is most common in children between the ages of one and five. As the child breathes, he or she may produce strange whooping sounds or high-pitched squeaking. There may be hoarseness, with the child's cough typically described as a "seal bark." Episodes of croup occur more commonly at night. As the child gets worse, he or she may experience the following signs of respiratory distress:

◆ Breathing with effort, including nasal flaring and retractions.

◆ Rapid breathing.

◆ Rising pulse initially, then a slow pulse as hypoxia develops.

◆ Paleness or cyanosis.

◆ Restlessness or altered mental status.

Severe attacks of croup can be dangerous. About 10% of all children with croup need to be hospitalized. Treat a child with croup the same way you would treat any respiratory emergency. Arrange for transport to the nearest hospital as quickly as possible.

Epiglottitis

While croup is usually viral in nature, epiglottitis is caused by a bacterial infection that inflames the epiglottis. It may resemble croup, but it is more serious. Left untreated, epiglottitis can be life threatening. Signs and symptoms may include:

◆ Occasional noise while inhaling.

◆ Anxious concentration on breathing. The child may try to stay very still.

◆ Sitting up and leaning forward, usually with chin thrust outward.

◆ Pain on swallowing and speaking.

◆ Drooling.

◆ Changes in voice quality.

◆ High fever (usually above 102°F or 38.9°C).

If you suspect epiglottitis, do not ask the child to open his or her mouth. Do not try to examine the child's throat or place anything in the mouth.

Touching the larynx can cause the airway to close completely. In all other ways, treat the child as you would for any respiratory emergency. Arrange for transport to the nearest hospital as quickly as possible.

Asthma

Asthma is common among children, especially those with allergies. Always consider it to be a serious medical emergency.

An acute asthma attack occurs when the bronchioles (small airway passages) spasm and constrict. This causes the bronchial membranes (tissue surrounding the small airway passages) to swell and become congested with mucus, further constricting the airways. As a result, the ability to exhale is affected. Air gets trapped in the lungs, the chest gets inflated, breathing becomes impaired, and oxygen deficiency occurs.

Especially critical signs and symptoms include:

◆ Rapid irregular breathing, especially in younger children.

◆ Exhaustion.

◆ Sleepiness and changes in level of responsiveness.

◆ Cyanosis.

◆ Rapid pulse, followed by a slow pulse as hypoxia develops.

◆ Wheezing (high-pitched breathing sounds).

◆ Quiet or silent chest. (In the late stages of respiratory distress, respirations may become so shallow that they no longer cause noise. Do not be fooled into believing that the child has gotten better. The condition has worsened.)

Emergency care is the same as for any respiratory emergency. Be sure to monitor the airway and breathing constantly. Arrange for transport to the nearest hospital as quickly as possible.

Cardiac Arrest

In infants and children, most cardiac arrests result from airway obstruction and respiratory arrest. So it is very important to assure an open airway and adequate breathing in these patients.

In cases of cardiac arrest, provide CPR and make sure advanced EMS providers are on the way. Your goal is to keep your patient's brain alive. In cases of hypothermia and drowning, especially in cold water, do not stop or interrupt CPR. Children have been known to survive after prolonged submersion, so keep at it. Arrange for transport to the nearest hospital as quickly as possible.

CPR techniques vary for adults, children, and infants. See Chapter 6 for the details of emergency care.

Seizures

Febrile seizures are caused by high fever, or more specifically by a rapid rise in an infant's or child's body temperature. Febrile seizures are the most common type of seizure in pediatric patients. Seizures in infants and children also may be caused by infection, poisoning, trauma, decreased levels of oxygen, epilepsy, hypoglycemia, inflammation to the brain, or meningitis. They also may have unknown causes. However, all seizures, including febrile seizures, should be considered potentially life threatening.

During most seizures, an infant's or child's:

◆ Arms and legs become rigid.

◆ Back arches.

◆ Muscles may twitch or jerk in spasm.

◆ Eyes roll up and become fixed, with dilated pupils.

◆ Breathing is often irregular or ineffective.

◆ Bladder and bowels lose control.

The infant or child may be completely unresponsive. If the seizure is prolonged, the patient will show signs of cyanosis. The spasms will prevent swallowing. He or she will push the saliva out of the mouth, which will appear to be frothing. If saliva is trapped in the throat, the infant or child will make bubbling or gurgling sounds. This may mean the airway needs suctioning when the seizure has ended. After the seizures, the patient often appears to be extremely sleepy.

To obtain a patient history, ask the parents the following questions:

◆ Has the patient been sick, with a rash, or running a fever? Has the patient had a head injury, a stiff neck, or a recent headache? Does he or she have diabetes?

◆ Has the patient had seizures before? How often? Is this his or her normal seizure pattern?

Have the seizures always been associated with fever or do they occur when the patient is well? Did others in the family have seizures when they were children?

◆ What did the seizure look like? Did it start in one part of the body and progress? Did the eyes go in different directions?

◆ How many seizures has the patient had in the last 24 hours? What was done to help?

◆ Is the patient taking a seizure medication? Could he or she have ingested any other medications?

Emergency care of childhood seizures includes the following:

1. **Maintain an open airway.** During the seizure, the tongue may relax and shift backward, decreasing the size of the air passage. To prevent this, as well as to encourage draining of mucus and frothing, place the patient in the recovery position. But do so only if there is no possibility of spine injury. Do not put anything in his mouth.

2. **Protect the patient from injury.** Do not restrain him during the seizure, but place him where he cannot fall or strike something. An open space on the floor with furniture and other objects moved away is fine. If he is on a bed that does not have sides, move him to prevent a fall, if necessary.

3. **Loosen any clothing that is tight and restricting,** especially around the neck or face.

4. **Assure an open airway.** Be prepared to suction.

5. **Administer high-concentration oxygen,** if local protocol allows. Hold the mask slightly away from the patient's face until the seizure is completely over. If breathing is diminished or absent and the airway is clear, assist ventilations with a bag-valve-mask (BVM) or pocket face mask with oxygen enrichment attachment. Follow local protocol.

6. **Assess for injuries,** which may have occurred during the seizure. Also, avoid overbundling the patient, especially if the seizure may have been due to a high fever.

If the seizure lasts longer than a few minutes or recurs without a recovery period, then the

COMPANY OFFICER'S NOTE

Seizures occur due to over-stimulation of brain cells. These cells become overly excited for a number of reasons (fever, infection, trauma, and so on). As a result, their requirements for oxygen can increase as much as 300%. In a generalized seizure, during which the infant or child may not be breathing, hypoxia is a real concern. Therefore, it is essential that pediatric patients who are actively seizing receive high-concentration oxygen.

seizure may be *status epilepticus.* This condition is a true medical emergency. Notify the incoming EMS unit immediately. (See Chapter 11 for more information about seizures.)

Sudden Infant Death Syndrome (SIDS)

Sudden infant death syndrome (SIDS) is defined as the sudden death of infants in the first year of life. It used to be more commonly known as "crib death" or "cot death."

SIDS is still not completely understood. It almost always occurs while the infant is sleeping. The infant is typically healthy, born prematurely, and between the ages of four weeks and seven months when he or she suddenly dies. No illness has been present, though there may have been recent cold symptoms. There is usually no indication of struggle.

Managing the SIDS Call

Unless the infant has *rigor mortis* (stiffness), immediately begin basic life support. Have your partner inform the EMTs of what is occurring.

The extreme emotional state of the parents makes them victims as much as the baby. They will be in agony from emotional distress, remorse, and guilt. Avoid any comments that might suggest blame. Help them feel that everything possible is being done, but do not offer false hope. If possible, have a member of your team stay with the parents and assist them in any way possible. Always follow local protocol.

If possible, obtain a brief medical history of the infant. However, this should not delay life-support efforts. If necessary, have other medical personnel find out the following:

◆ When was the infant put in the crib?

◆ What was the last time the parents looked in on the baby?

◆ What was the position of the baby in the crib?

◆ What else was in the crib?

◆ Is there medication present (even if it is for the adults)?

◆ What is the general health of the infant, recent illnesses, medications, or allergies?

Note that it is common for fire service First Responders to feel anxiety, guilt, or anger after a SIDS call. Ignoring these feelings will not cause them to go away. They may even have serious, negative impact on your mental health. After a SIDS case, a critical incident stress debriefing may be helpful. Refer to Chapter 2 for a review of the overall CISM process.

Child Abuse and Neglect

The definition of **abuse** is improper or excessive action so as to injure or cause harm. The term **neglect** refers to giving insufficient attention or respect to someone who has a claim to that attention and respect.

Child abuse, or non-accidental trauma, occurs in all parts of our society. The number of children who are abused or neglected in the U.S. is staggering. Estimates range from 500,000 to 4 million cases annually with thousands dying. In fact, child abuse has been the only major cause of infant and child death to increase in the last 30 years. These are numbers that are cause for alarm.

During an emergency call, the adult (usually a parent) who abused the child often behaves in an evasive manner. He or she may volunteer little or contradictory information about what happened. Even so, a call for child abuse may be a call for help. Do not be judgmental. Focus on the child.

Some forms of abuse are difficult to recognize on scene. For example, broken bones at various stages of healing can be identified only at the hospital. However, you may be able to recognize some forms of physical abuse and neglect (Figure 20-10). Signs and symptoms may include:

◆ Multiple bruises in various stages of healing.

◆ Injury that is not consistent with the mechanism of injury described by the caregivers.

◆ Patterns of injury that suggest abuse, such as cigarette burns, whip marks, or hand prints.

◆ Fresh burns such as scalding in a glove or dip pattern.

◆ Burns not consistent with the history presented by the caregivers.

◆ Untreated burns.

◆ Unusual circumstances surrounding a cardiac arrest.

Also suspect abuse when repeated calls to the same address occur, when the caregivers seem unconcerned or give conflicting stories, and when the child seems afraid to discuss how the injury occurred. If an infant or child presents with unresponsiveness, seizures, or signs of severe internal injuries but no external signs, suspect central nervous system injuries, or *shaken baby syndrome.*

COMPANY OFFICER'S NOTE

Shaken baby syndrome was first defined by Dr. John Caffey in 1972. He reported that retinal hemorrhages and intracranial bleeding were common injuries resulting from an indirect force. The causes of these indirect forces—the results of shaking—are from both innocent behavior, such as repeated playful episodes of a child being tossed in the air, as well as shaking by a caregiver attempting to unsuccessfully deal with an inconsolable child. Initial signs of a severe injury can mimic a virus—vomiting, drowsiness, or poor feeding. More severe signs include apnea (temporary cessation of breathing), bradycardia (slow heartbeat), seizures, a bulging fontanel, and lethargy.

Signs and symptoms of neglect include the following:

◆ Lack of adult supervision.

◆ Appearance of malnourishment.

◆ Unsafe living conditions.

◆ Untreated chronic illness, such as no medication for asthma.

◆ Delay in reporting injuries.

Child Abuse and Neglect

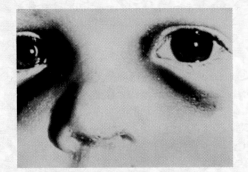

Figure 20-10a Child physical abuse.

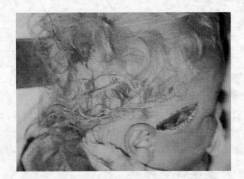

Figure 20-10b Child physical abuse.

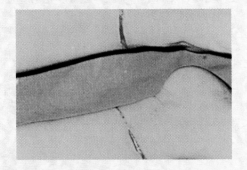

Figure 20-10c Child neglect—injury due to lack of appropriate medical care.

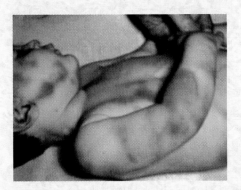

Figure 20-10d Child abuse death from multiple injuries.

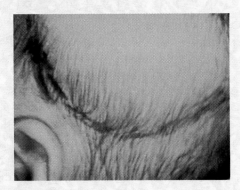

Figure 20-10e Physical abuse—restraining by tying.

Figure 20-10f Physical abuse—burns from hand held on an electric stove.

If you suspect abuse or neglect, first and foremost make sure that the environment is safe for the patient. Provide all necessary emergency care. As time permits, observe the child and the caregivers. Look for objects that might have been used to hurt the child. Look for signs of neglect in the general appearance of the child. Sometimes abuse and neglect are well hidden.

If you find yourself in a position of giving emergency care to a possible victim of child abuse or neglect, follow these guidelines:

◆ Gain entry to the home and access to the child, if it can be done safely. If the parents placed the emergency call, you will probably get in without any difficulty. If someone else called, the parents may resist, and you may need to call the police. Remember, your safety and the safety of your team come before the safety of anyone else.

◆ If you are asked to help the child, calm the parents. Tell them that if the child needs care, you will provide it. Tell them that is the only reason you are there. Speak in a low, firm voice.

◆ Focus attention on the child while you provide emergency care. Speak softly. Use the child's first name. Do not ask the child to re-create the situation while parents are present.

◆ Do a full patient assessment. Treat as you go. Note any suspicious wounds and evidence of internal injury. Look also for signs of head injury. Remember that you are there to provide emergency care, not to determine child abuse.

◆ Update the responding EMTs in the same way you would for any other child in need of care and transport.

◆ You are not expected to deal with child abuse issues on the spot unless the child is in danger. In all suspected cases, the child should be transported.

◆ Never confront parents with a charge of child abuse. Being supportive and nonjudgmental will help them be more receptive to others providing emergency care.

◆ Accusations can delay transport. Instead, report objective information to the transporting unit's crew. Report only what you see and what you hear. Do not comment on what you think.

◆ Maintain total confidentiality. Do not discuss the incident with your family or friends.

It is important for you to report any suspicions of child abuse to the proper authorities. Therefore, it is critical to understand the reporting laws in your area and the reporting protocols for your EMS system. Specifically, determine:

◆ Who must report the abuse.

◆ What types of abuse and neglect must be reported.

◆ To whom the reports must be made.

◆ What information a First Responder must give.

◆ What immunity a First Responder is granted.

◆ Criminal penalties for failing to report.

Taking Care of Yourself

Almost half of the children in the U.S. who die from accidents are pronounced dead either at the scene or at the hospital. The sudden and violent death of a child is emotionally wrenching, whether it occurs before you arrive or while you are giving care.

Recognize your reactions. Feelings of fear, rage, helplessness, anxiety, sorrow, and grief are common. It also is common for rescuers to feel shame and guilt, even if they did everything possible to help the child. These feelings are particularly intense if the child dies. Remember, some children will die despite your best efforts.

As a fire service First Responder, you need to control your emotions while you are treating the child. In this way, you can render the best assistance possible. After the case is over, however, you need to deal with your feelings. Talk them out. Use any critical incident stress debriefing (CISD) team that you have access to. CISD is an excellent way to help prevent or minimize long-term problems.

Not finding a way to talk through your feelings and resolve them can have a serious negative impact on your mental health. In addition to formal debriefing, it may be helpful to find a trusted friend who will listen. It is important to deal with the feelings and issues that can result from pediatric calls. Do not delay!

COMPANY OFFICER'S NOTE

Many fire service EMS providers find comfort by spending time with their loved ones immediately following a stressful call involving infants and children. Hugging and kissing your own children as soon as possible, for example, is a powerful drug for healing emotional injuries caused by the death of an infant or child.

ON SCENE FOLLOW-UP

At the beginning of this chapter, you read that First Responders were on scene with a six-year-old patient who had been hit by a car. To see how chapter skills apply to this emergency, read the following. It describes how the call was completed.

INITIAL ASSESSMENT (CONTINUED)

We found the patient responsive only to pain with snoring respirations. My partner and I knew what we had to do in one word: airway. Getting in position, we applied manual stabilization, turning her for better airway assessment and control. While my partner held her head and neck in line, I applied a cervical stabilization device.

A jaw-thrust stopped the snoring. Further examination showed that there was blood in the mouth. As we suctioned, her breathing appeared to become adequate. I applied oxygen by way of a pediatric nonrebreather at 10 liters per minute.

Dispatch informed us that the transport unit would be on scene in about four to five minutes. We reported our general impression of the patient: "Six-year-old female was struck by vehicle. The airway is open, and she is breathing on her own. The patient is responsive to pain only." We advised the ambulance should continue to respond with red lights and sirens, priority one.

PHYSICAL EXAMINATION

Covering the patient with a blanket to maintain body heat, we proceeded with a head-to-toe exam. Our findings included swelling with discoloration on the left forehead above the eye. Pupils were reactive but sluggish. The lower left arm was observed to be deformed and swollen. We found her respirations were 12 and shallow, which was slower than our last assessment, so we assisted ventilations with our BVM and supplemental oxygen flowing at 12 liters per minute.

PATIENT HISTORY

The father reported that the child was in good health and had no allergies.

ONGOING ASSESSMENT

We continued to assist ventilations until the ambulance crew arrived.

PATIENT HAND-OFF

When the paramedics arrived on scene, I gave them the hand-off report:

"This is Neena Chong, six years old. She was hit by a car traveling approximately 15 mph about 20 minutes ago. While stabilizing her cervical spine, we logrolled her onto her back, opened her airway with a jaw-thrust maneuver, and noted that the snoring respirations stopped. We applied a cervical collar and administered oxygen. She has an injury to the left forehead and a swollen, deformed left arm. Her vitals are pulse 60, strong, regular; respirations 12 and shallow. We assisted ventilations with our BVM and supplemental oxygen flowing at 12 liters per minute. There is no medical history of allergies."

A few days later we went to the trauma center and asked about this patient. We were told that her head injury required surgery, but she was doing well and was expected to make a full recovery. One of the doctors asked what had happened, and we gave him all the details. He listened intently. When we were done, he said that opening the airway and alerting the paramedics to the slowing respirations may have made the difference in the patient's outcome.

Learn the unique aspects of providing emergency care to pediatric patients. One of the most important is that you are not treating the patient only. A calm, professional, reassuring approach can help to minimize the impact of the emergency on both the patient and the family.

Chapter Review

Some points to remember about the assessment and emergency care of pediatric patients follow:

◆ Maintaining a good airway and ensuring quality breathing are the two most important concerns for a First Responder when dealing with a pediatric emergency.

◆ When assisting ventilations, be sure to watch for chest rise. This is a good indicator of whether or not your breaths are effective.

◆ If you do not have an airway, you will not have a patient.

◆ With pediatric patients, all "roads" lead to the ABCs.

◆ Be alert. Infants and children have an amazing ability to maintain vital signs until they are deep in shock and near death.

◆ Infants and children are prone to hypothermia. Keep them warm.

◆ When performing a patient assessment, take what you can when you can get it. For example, if an infant or child holds out a hand, it is a good time to check the pulse or capillary refill.

◆ When the age of an infant or child is unknown, use your best judgment based on the patient's size.

◆ In cases of abuse or neglect, EMS personnel may be the only advocates an infant or child has.

◆ When you treat a traumatized infant or child, you are treating a family.

◆ The interests of the infant or child must always be placed as the foremost consideration when making any and all patient care decisions.

FIRE COMPANY REVIEW

Page references where answers may be found or supported are provided at the end of each question.

Section 1

1. How is scene size-up for an emergency involving an infant or child different from that involving an adult? (pp. 379–380)

2. How is the initial assessment different? (p. 380)

3. How is taking vital signs in an infant or child different from taking signs in an adult? (pp. 381–384)

4. Why must you be more concerned about a small amount of blood loss in an infant or child than you would be about a small blood loss in an adult? (p. 384)

5. What are some techniques you can use to help make the physical exam less threatening to the pediatric patient? (pp. 385–386)

Section 2

6. What are the signs and symptoms of shock in the pediatric patient? What is the emergency care? (pp. 387–388)

7. What are some of the differences in the way you manage the airway of a pediatric patient? (pp. 388–389)

8. What are some of the common causes of seizures in the infant and child? Which ones are considered to be life-threatening? (pp. 390–391)

9. How would you manage a sudden infant death syndrome (SIDS) call? (pp. 391–392)

10. What are the signs and symptoms of possible child abuse and neglect? (pp. 392–394)

RESOURCES TO LEARN MORE

American Heart Association and American Academy of Pediatrics. *Textbook of Pediatric Advanced Life Support.* Dallas: American Heart Association, 1998.

Eichelberger, M.R., et al. *Pediatric Emergencies: A Manual for Prehospital Care Providers,* Second Edition. Upper Saddle River, NJ: Brady/Prentice-Hall, 1998.

Freisher, G.R., and S. Ludwig. *Textbook of Pediatric Emergency Medicine,* Third Edition. Baltimore: Williams & Wilkins, 1993.

AT A GLANCE

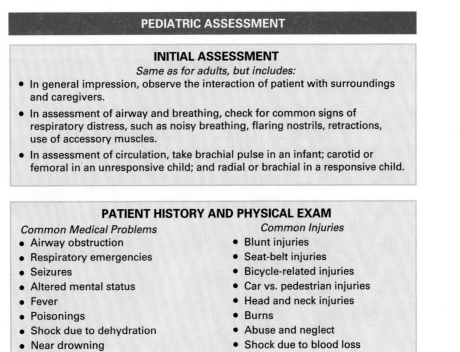

PEDIATRIC ASSESSMENT

INITIAL ASSESSMENT

Same as for adults, but includes:

- In general impression, observe the interaction of patient with surroundings and caregivers.
- In assessment of airway and breathing, check for common signs of respiratory distress, such as noisy breathing, flaring nostrils, retractions, use of accessory muscles.
- In assessment of circulation, take brachial pulse in an infant; carotid or femoral in an unresponsive child; and radial or brachial in a responsive child.

PATIENT HISTORY AND PHYSICAL EXAM

Common Medical Problems
- Airway obstruction
- Respiratory emergencies
- Seizures
- Altered mental status
- Fever
- Poisonings
- Shock due to dehydration
- Near drowning
- Sudden infant death syndrome (SIDS)

Common Injuries
- Blunt injuries
- Seat-belt injuries
- Bicycle-related injuries
- Car vs. pedestrian injuries
- Head and neck injuries
- Burns
- Abuse and neglect
- Shock due to blood loss

Lifting and Moving Patients

*I*NTRODUCTION *After receiving emergency care, a patient may need to be handled or transported. If done improperly, the patient may suffer unnecessary pain and further injury. It is part of your responsibility to see that this does not occur. Each EMS system defines if and when fire service First Responders may move patients. Generally, you may only move patients who are in immediate danger. You may position patients to prevent further injury. You also may assist other EMS workers in moving patients. Learn and follow your local protocols.*

Cognitive, affective, and psychomotor objectives are from the U.S. DOT's "First Responder: National Standard Curriculum." Enrichment objectives, if any, identify supplemental material.

Cognitive

1-5.1 Define body mechanics. (p. 400)

1-5.2 Discuss the guidelines and safety precautions that need to be followed when lifting a patient. (pp. 400–401)

1-5.3 Describe the indications for an emergency move. (pp. 403–404)

1-5.4 Describe the indications for assisting in non-emergency moves. (p. 405)

1-5.5 Discuss the various devices associated with moving a patient in the out-of-hospital arena. (pp. 410–412)

Affective

1-5.6 Explain the rationale for properly lifting and moving patients. (pp. 400–401)

1-5.7 Explain the rationale for an emergency move. (pp. 403–404)

Psychomotor

1-5.8 Demonstrate an emergency move. (pp. 403–404)

1-5.9 Demonstrate a non-emergency move. (p. 405)

1-5.10 Demonstrate the use of equipment utilized to move patients in the out-of-hospital arena. (pp. 410–412)

Enrichment

◆ Describe the power lift and the power grip. (pp. 401–402)

◆ Explain how good posture and physical fitness can contribute to your well being as an EMS provider. (pp. 402–403)

ON SCENE

DISPATCH

My partner and I were returning from a call in our first response vehicle when the sport utility vehicle in front of us suddenly swerved off the road. Hitting the soft shoulder, the vehicle veered first to the left, then back to the right.

"It's gonna flip!" Sonya exclaimed. My partner was right. The vehicle rolled over twice, landing upright on its wheels.

As I grabbed the radio to call for a medic unit, Sonya parked our vehicle a safe distance from the scene.

SCENE SIZE-UP

As soon as I made my call, I saw smoke billow out from under the hood of the crashed vehicle. The fire must have been fueled by grease. Sonya grabbed a portable fire extinguisher off our rig, while I called the dispatcher to notify the fire department that we were going to need a pumper at the scene.

We had turnout gear on, so we carefully approached the car. A woman was in the driver's seat. She was wearing a shoulder harness and lap belt. I noticed the airbag had deployed. She seemed dazed. Smoke was filling the car, the windshield was turning black, and then we saw flames. I was sure that before long the passenger compartment would be on fire.

> *Lifting and moving patients is an important responsibility. Many lifts and moves are routine while others require quick thinking and skill. Consider this patient as you read Chapter 21. Is it within the First Responder's scope of care to move her? If so, how can it be done without causing further injury?*

BODY MECHANICS

Basic Principles

As a firefighter, you are often asked to lift and carry heavy objects. Similarly, as a First Responder, you may be asked to lift and carry patients. Done improperly, lifting can cause you to suffer an injury and life-long pain. With planning, good health, and skill, you can do your job with minimum risk to yourself.

Practice the principles of proper lifting and moving every day. By doing so, you can make them automatic, a habit that increases your safety and performance, even in the most stressful emergency situations.

The term **body mechanics** refers to the safest and most efficient methods of using your body to gain a mechanical advantage. For example:

◆ *Use your legs to lift, not your back.* To move a heavy object, use the muscles of your legs, hips, and buttocks, plus the contracted muscles of your abdomen. These muscles let you safely generate a lot of power. Never use the muscles in your back to help you move or lift a heavy object.

◆ *Keep the weight of the object as close to your body as possible* (Figure 21-1). Reach a short distance to lift a heavy object. Back injury is much more likely to occur when you reach a long distance to lift an object.

◆ *Align shoulders, hips, and feet.* That is, visualize your shoulders stacked on top of your hips, and your hips on top of your feet. Then move as a unit. If your shoulders, hips, and feet are not aligned, you could create twisting forces that can harm your lower back.

◆ *Reduce the height or distance you need to move an object.* Get closer to the object, or reposition it before you try to lift. Lift in stages if you need to.

Apply the principles of body mechanics to lifting, carrying, moving, reaching, pushing, and pulling. Key to preventing injury during all those tasks is correct alignment of the spine. Keep a normal inward curve in the lower back. Keep wrists and knees in normal alignment, too. Whenever possible, use equipment to do the lifting for you.

In an emergency, teamwork is essential. Just as a football coach positions players according to ability, rescuers should too. It can help capitalize on their abilities to ensure the best outcome in any emergency.

All members of a team should be trained in the proper techniques. Problems can occur when teams members are greatly mismatched. The stronger partner can be injured if the weaker one fails to lift. The weaker one can be injured if he tries to do too much. Ideally, partners in lifting and moving should have adequate and equal strength and height. Know your physical ability and limitations. Respect them. Consider the weight of the patient and recognize the need for help.

Team members also need to communicate during a task, clearly and frequently. Use commands

Figure 21-1   Keep weight close to the body as it is lifted.

Power Lift

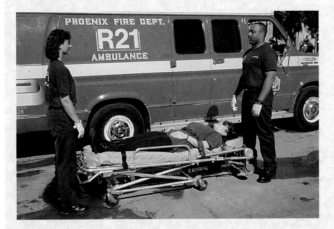

Figure 21-2a Get into position.

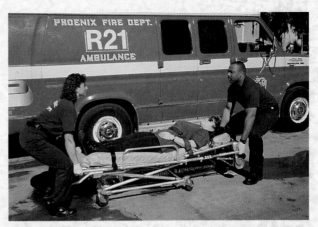

Figure 21-2b Lift in unison, keeping your back locked and feet flat.

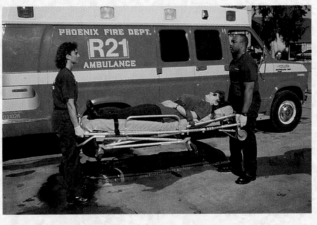

Figure 21-2c Stand, making sure your back remains locked.

that are easy for team members to understand. Verbally coordinate each lift from beginning to end.

The Power Lift

The power lift is a technique that offers you the best defense against injury. It also protects the patient on a stretcher with a safe and stable move. It is especially useful for rescuers who have weak knees or thighs. Remember that in performing the power lift, keep your back locked and avoid bending at the waist (Figure 21-2). Follow these steps:

1. **Place your feet a comfortable distance apart.** For the average-size person, this is

COMPANY OFFICER'S NOTE

The spine works on a principle of leverage. The discs in the back act as the fulcrum. The amount of pressure on the discs is related to the distance the weight is away from the spine (fulcrum). The solution for proper lifting can be said in three words: bend your knees. By doing so, the load is brought closer to the spine (fulcrum), reducing the pressure on the discs. Loads that cannot be lifted by bending the knees cannot be lifted safely!

usually about shoulder width. Taller rescuers might prefer a little wider stance.

2. **Turn your feet slightly outward.** Most people find that this helps them feel more comfortable and more stable.

3. **Bend your knees,** which will bring your center of gravity closer to the object. You should feel as though you are sitting down, not falling forward.

4. **Tighten the muscles of your back and abdomen** to splint the vulnerable lower back. Your back should remain as straight as you can comfortably manage, with your head facing forward in a neutral position.

5. **Keep your feet flat,** with your weight evenly distributed and just forward of the heels.

6. **Position your hands.** Place them a comfortable distance from each other to provide balance to the object as it is lifted. This is usually at least 10 inches (250 mm) apart.

7. **Always use a power grip to get maximum force from your hands.** That is, your palms and fingers should come in complete contact with the object and all fingers should be bent at the same angle. (See Figure 21-3.)

8. **As lifting begins, raise your upper body before your hips.** Remember to keep your back locked as the force is driven through the heels and arches of your feet.

9. **Reverse these steps to lower the object.**

Posture and Fitness

Posture is a much overlooked part of body mechanics. When people spend a great deal of time sitting or standing, poor posture can easily tire back and stomach muscles. This can only make back injury more likely.

One extreme of poor posture is the *slouch* (Figure 21-4). In it the shoulders are rolled forward, putting increased pressure on every region of the spine. Another is the *swayback*. In it the stomach is too far forward and the buttocks too far back, causing extreme stress on the lower back.

Be aware of your posture. While standing, your ears, shoulders, and hips should be in vertical alignment. Your knees should be slightly bent, and your pelvis slightly tucked forward (Figure 21-5).

When sitting, your weight should be evenly distributed on both ischia (the lower portion of your pelvic bones). (See Figure 21-6.) Your ears, shoulders, and hips should be in vertical alignment. Your feet should be flat on the floor or crossed at the ankles. If possible, your lower back should be in contact with the support of the chair.

Finally, proper body mechanics will not protect you if you are not physically fit. A proactive,

Figure 21-4 Slouch and swayback are extremes of poor posture.

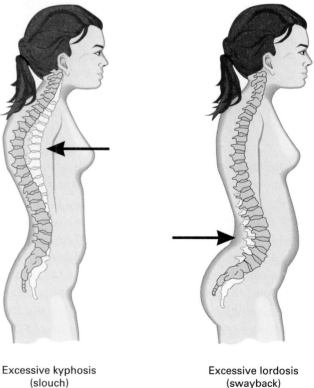

Excessive kyphosis (slouch)

Excessive lordosis (swayback)

Figure 21-3 The power grip.

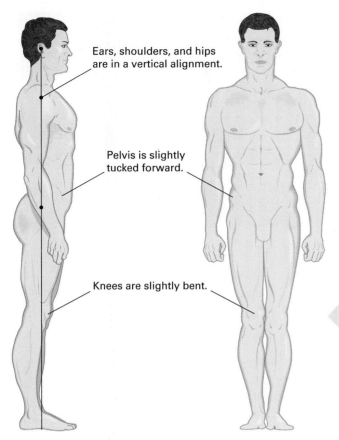

Figure 21-5 Proper standing position.

Ears, shoulders, and hips are in a vertical alignment.

Pelvis is slightly tucked forward.

Knees are slightly bent.

Figure 21-6 Proper sitting position.

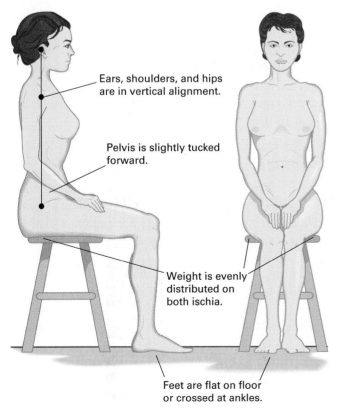

Ears, shoulders, and hips are in vertical alignment.

Pelvis is slightly tucked forward.

Weight is evenly distributed on both ischia.

Feet are flat on floor or crossed at ankles.

well balanced physical fitness program should include flexibility training, cardiovascular conditioning, strength training, and nutrition.

> **COMPANY OFFICER'S NOTE**
>
> A recent study conducted by the National Institute of Occupational Safety and Health found that 10% of today's employees are not physically capable of performing their jobs and that 75% of back injuries occur within that same group.

SECTION 2

PRINCIPLES OF MOVING PATIENTS

Emergency Moves

The top priority in emergency care is to maintain a patient's airway, breathing, and circulation. However, if the scene is unstable or poses an immediate threat, you may have to move the patient first. Follow local protocols.

In general, when there is no threat to life, provide emergency medical care in place and wait for the EMTs to move the patient. Make an **emergency move** only when there is an immediate danger to the patient. Examples of situations in which you may make an emergency move are:

◆ *Fire or threat of fire.* Fire should always be considered a grave threat, not only to patients but also to rescuers.

◆ *Explosion or the threat of explosion.*

◆ *Inability to protect the patient from other hazards at the scene.* Examples of hazards include an unstable building, a rolled over car, spilled gasoline and other hazardous materials, an unruly or hostile crowd, and extreme weather conditions.

◆ *Inability to gain access to other patients who need life-saving care.* This may occur at the scene of a car crash involving two or more patients, for example.

◆ *When life-saving care cannot be given because of the patient's location or position.* For instance, a patient in cardiac arrest must be supine on a flat, hard surface in order for you to perform CPR properly. If that patient is sitting on a chair, an emergency move must be made in order for you to provide life-saving care.

The greatest danger to the patient in an emergency move is the possibility of making a spine injury worse. Therefore, provide as much protection to the spine as possible by pulling the patient in the direction of the long axis of the body.

It is impossible to move a patient from a vehicle quickly and, at the same time, completely protect the spine. So move a patient from a vehicle immediately only if one of the five conditions described above exist.

If the patient is on the floor or the ground, use one of the following emergency moves. But be sure never to pull the patient's head away from the neck and shoulders. Always pull along the long axis of the body. If time permits, you may wish to bind the patient's wrists together with a cravat or gauze. This will make the patient easier to move, and it will help protect the hands and arms from injury.

Shirt Drag

To perform a shirt drag, do the following (Figure 21-7). First, fasten the patient's hands or wrists loosely with a cravat or gauze to protect them during the move. Then grasp the shoulders of the patient's shirt (not a tee shirt). Pull the shirt under the patient's head to form a support. Then, using the shirt as a "handle," pull the patient toward

Figure 21-7 Shirt drag.

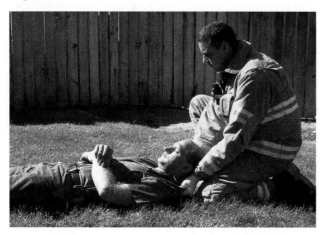

Figure 21-8 Blanket drag.

you. Be careful not to strangle the patient. The pulling should engage the patient's armpits, not the neck.

Blanket Drag

To perform a blanket drag, do the following (Figure 21-8). First, spread a blanket alongside the patient. Gather half of it into lengthwise pleats. Roll the patient away from you onto his side, and tuck the pleated part of the blanket as far under him as you can. Then roll the patient back onto the center of the blanket, preferably on his back. Wrap the blanket securely around the patient. Grabbing the part of the blanket that is under the patient's head, drag the patient toward you.

If you do not have a blanket, you can use a coat in the same way.

Shoulder or Forearm Drag

To perform a shoulder drag, do the following (Figure 21-9). First, stand at the patient's head. Then slip your hands under the patient's armpits from the back. If you must drag the patient a long distance and need a better grip, perform a forearm drag. That is, position yourself as you would in a shoulder drag. After you slip your hands under the patient's armpits, grasp the patient's forearms and drag the patient toward you. Use your own forearms as a support to keep the patient's head, neck, and spine in alignment.

Figure 21-9 Shoulder drag.

Other Emergency Moves

Other emergency moves include the piggyback carry, one-rescuer crutch, one-rescuer cradle carry, firefighter's drag, and others. (See Figures 21-10 and 21-11.)

Non-Emergency Moves

Non-emergency moves, or non-urgent moves, are generally performed with other rescuers. They require no equipment and may take less time than moves such as a blanket drag. However, do not use them with possible spine-injured patients, since they offer no spinal protection.

Non-emergency, or non-urgent, moves include the direct ground lift and extremity lift.

Direct Ground Lift

The direct ground lift requires two or three rescuers. It is valuable when the patient cannot sit in a chair and when a stretcher cannot be brought close to the patient. It is difficult if the patient weighs more than 180 pounds (80 kg), is on the ground or some other low surface, or is uncooperative.

Position the stretcher as close to the patient as possible. Undo the stretcher straps, lower the railings, and clear any equipment off the mattress. Tell the patient what you are going to do. Then

warn him that he must remain still in order to protect your balance. Place the patient's arms on his chest if possible.

To perform a direct ground lift, follow these steps (Figure 21-12):

1. **Get in position.** Line up on one side of the patient. If at all possible, line up on the least injured side. Kneel on one knee, preferably the same knee for all rescuers.

2. **Position your hands.** Have the first rescuer cradle the patient's head by placing one arm under the neck and shoulder. He must place his other arm under the patient's lower back.

   Have the second rescuer place one arm under the patient's knees, and the other arm above the buttocks.

   If a third rescuer is available, have him place both arms under the patient's waist. The other two rescuers should slide their arms up to the middle of the back and down to the buttocks as appropriate.

3. **On signal, all rescuers should lift the patient to their knees as a unit.**

4. **Slowly turn the patient toward you.** Using a gently rocking motion, roll the patient as a unit toward your chests, until he is cradled in the bends of your elbows. Then tuck the patient's head in toward your chests.

5. **Rise to a standing position in unison,** and carry the patient to the stretcher.

6. **Reverse the steps to lower the patient onto the stretcher.**

Extremity Lift

Do not use the extremity lift if the patient has injuries to his arms or legs. Use this lift to move an unresponsive patient from a chair to the floor. Two rescuers are needed to perform the lift (Figure 21-13):

1. **Get in position.** Take a position at the patient's head. The other rescuer should kneel at the patient's side by the knees.

2. **Position your hands.** Place one hand under each of the patient's shoulders, reaching through to grab the patient's wrists. The second rescuer should slip his hands under the patient's knees.

Figure 21-10a Sheet drag.

Figure 21-10b Piggyback carry.

Figure 21-10d Cradle carry.

Figure 21-10c One-rescuer crutch.

Figure 21-10e Firefighter's drag.

Firefighter's Carry

Figure 21-11a Grasp the patient's wrists.

Figure 21-11b Stand on the patient's toes and pull.

Figure 21-11c Pull the patient over a shoulder.

Figure 21-11d Pass an arm between the legs and grasp the arm nearest you.

3. Move the patient. On signal, both of you then can move the patient to the desired location.

Positioning the Patient

How you position a patient depends on the patient's condition. General guidelines include:

◆ An unresponsive patient who is not injured should be placed in the recovery position.

This is done by rolling the patient onto his or her side, preferably the left side.

◆ Unless there is a life-threatening emergency, a patient who has been injured should not be moved. The EMTs will evaluate, stabilize, and move the patient as necessary.

◆ A patient who shows signs of shock may be placed in the shock position. If it will not

Figure 21-12a   Kneel on one knee on the least injured side.

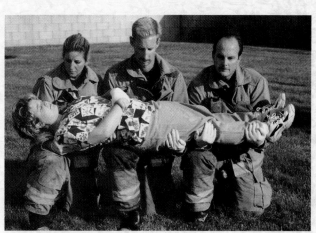

Figure 21-12b   In unison, lift the patient to knee level.

Figure 21-12c   In unison, slowly turn the patient toward you. Then rise to a standing position.

aggravate injuries to the legs or spine, this is done by elevating the supine patient's legs 8–12 inches (200 to 300 mm).

◆ A patient who has pain or breathing problems may get in any position that makes him most comfortable, unless his injuries prevent it. Generally, a patient who has breathing difficulties will want to sit up. A patient with abdominal pain will want to lie on his side with knees drawn up.

◆ A responsive patient who is nauseated or vomiting should be allowed to remain in a position of comfort. However, you should always be positioned so you can manage the patient's airway if needed.

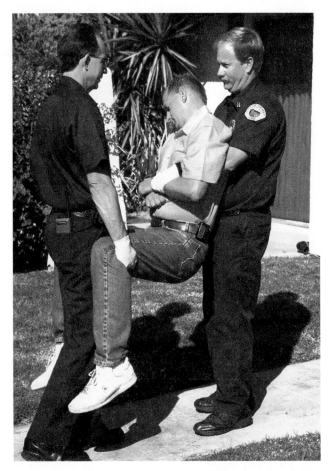

Figure 21-13 Extremity lift.

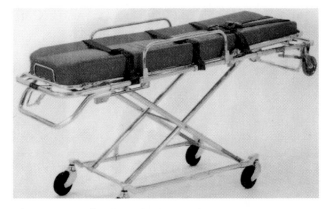

Figure 21-14 Standard stretcher.

3 EQUIPMENT

Become completely familiar with the equipment used to move patients in your jurisdiction. To decide which to use, base your decision on the patient's condition, the environment in which he is found, and the resources available. Generally, the best way to move a patient is the easiest way that will not cause injury or pain.

Let your equipment do the work whenever possible. Drag or slide the patient (not lift), whenever you can. If you must lift a patient, do it with a device designed for that purpose. As a rule, carry a patient only as far as absolutely necessary. Make sure you have adequate help. If you do not, get it. Never risk injuring yourself.

Typical equipment used in EMS includes: various types of stretchers, the stair chair, and backboards.

Stretchers

A standard stretcher, or cot, has wheeled legs. It also has a collapsible undercarriage that makes it possible to load it into an ambulance. (See Figure 21-14.)

A portable stretcher is lightweight, folds compactly, and is easy to clean. It does not have an undercarriage and wheels. It is comfortable to rest on, especially if the head is padded. It is valuable when there is not enough space for a standard stretcher or when there are multiple patients. There are a variety of styles. The most common has an aluminum frame with canvas fabric. (See Figures 21-15 and 21-16.)

A scoop, or orthopedic, stretcher splits in two or four sections (Figure 21-17). Each section can be fitted around a patient who is lying on a relatively flat surface. It is used in confined areas where larger stretchers will not fit. Once secure in a scoop stretcher, a patient can be lifted and moved to a standard one. To operate a scoop

Figure 21-15 Pole stretcher.

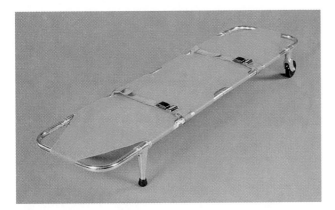

Figure 21-16 Portable ambulance stretcher.

stretcher, split it apart lengthwise. Carefully slide it under the patient from both sides. Then lock the brackets at each end, and lift the patient.

A stretcher also can be improvised with a blanket, canvas, brattice cloth, or a strong sheet and two 7- to 8-foot pike poles. To improvise a stretcher with two pike poles and a blanket, follow these steps (Figure 21-18):

1. **Position the first pole.** Place it one foot from the center of the unfolded blanket.

2. **Fold the short side of the blanket over the pole.**

3. **Position the second pole.** Place it on top of the two folds of blanket. It should be about two feet (60 cm) from the first pole and parallel to it.

4. **Fold the remaining side of the blanket over the second pole.** When the patient is placed on the blanket, the weight of the body secures the poles.

Cloth bags or sacks may be used as stretchers. Make holes in the bottoms of bags or sacks so that the poles pass through them. Enough bags should be used to provide the required length.

A stretcher also can be made from three or four coats or jackets. First, turn the sleeves inside out. Then fasten the jacket with the sleeves inside the coat. Place a pole through each sleeve.

Stair Chair

Moving patients up or down stairs dramatically increases the potential for rescuers to be injured. The safest way to do it is with a stair chair. A stair chair is a lightweight folding device. It has straps to confine the patient, wheeled legs, a grab bar below the patient's feet, and handles that extend behind the patient's shoulders.

When you use a stair chair, make sure as many people as necessary are helping. A "spotter" is needed to help maneuver the stair chair down stairs. He should continually tell how many stairs are left and what conditions are ahead. A spotter also can place his hand on the back of the rescuer who is moving backward to help steady him (Figure 21-19).

Rescuers carrying a patient in a stair chair should keep their backs in a locked position. They should flex at the hips instead of at the waist, bend at the knees, and keep arms (and the weight of the chair) as close to their bodies as possible. Once off the stairs, the patient can be transferred to a more conventional stretcher.

Stair chairs work well for patients in respiratory distress who must be moved up or down stairs. The sitting position does not worsen the patient's breathing problems.

Figure 21-17 Scoop or orthopedic stretcher.

Improvising a Stretcher

Figure 21-18a Position the first pole. Then fold the blanket over it.

Figure 21-18b Position the second pole.

Figure 21-18c Fold the remaining side of the blanket over the second pole.

Figure 21-18d When the patient is placed on the blanket, the weight of the patient should secure the poles.

Backboards

There are both long and short backboards (Figure 21-20). A long backboard may be about 6 to 7 feet (2 m) long, which means it can stabilize the patient's entire body. It is used for patients with suspected spine injury who are lying down.

A short backboard is about 3 to 4 feet (1 m) long and can stabilize the patient down to the hips. It is used to stabilize a patient with suspected spine injuries who is in a sitting position.

Both the long and short backboards feature handholds and straps. Most are made of synthetic material that will not absorb blood and is easy to clean. Regardless of whether you use a long or short backboard, always maintain manual support of the patient's head and neck in the normal anatomical position. Maintain that support until the patient is fully secured to the backboard.

Using a Stair Chair

Figure 21-19a Stair chair.

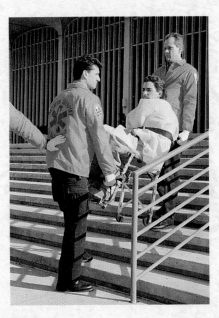

Figure 21-19b Moving a patient down steps with a third rescuer as spotter.

Backboards

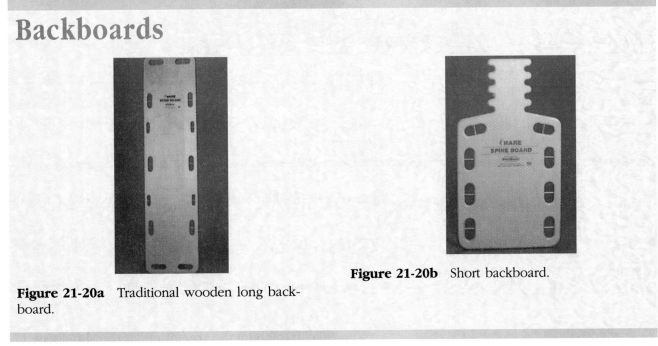

Figure 21-20a Traditional wooden long backboard.

Figure 21-20b Short backboard.

(continued)

Backboards (continued)

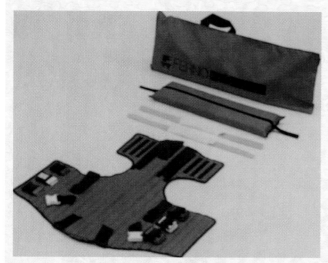

Figure 21-20c   Vest-type immobilization device.

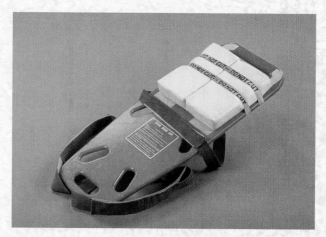

Figure 21-20d   Short backboard.

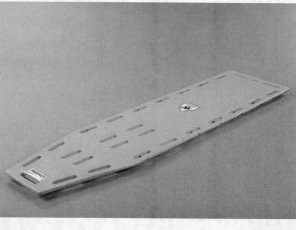

Figure 21-20e   Long backboard.

◆ *ON SCENE FOLLOW-UP*

At the beginning of this chapter, you read that fire service First Responders were at the scene of a collision, with a patient having unknown injuries and in serious danger from smoke and fire. Read the following to see how chapter skills apply to this emergency.

SCENE SIZE-UP (CONTINUED)

My partner immediately went to the main body of the fire and applied the fire extinguisher, but

it had minimal effects on the flames. Within seconds, the fire was as intense as before.

Sonya and I looked at each other. We both knew the patient was in mortal danger and had to be moved immediately. While we were concerned about her possible injuries, we knew we could not leave her in the smoke-filled, flaming car.

Grabbing the patient under the shoulders, I began the move. I cradled her head in my arms to help minimize movement of her spine. Sonya grabbed her legs. We moved her a safe distance from the car, and set her down carefully.

INITIAL ASSESSMENT

Sonya stabilized the patient's head and monitored her airway, while I assessed responsiveness. The patient moaned when I spoke loudly. She was breathing adequately and had no foreign material in her mouth, but I knew that could change quickly. She had no signs of external bleeding.

Our general impression was that the patient was about 55 years old and in potentially serious condition. She may have had injuries from the crash and from the smoke. We applied a nonrebreather mask at 15 liters per minute. We updated the incoming units of our suspicions and our treatment so they could be prepared when they arrived on scene.

PHYSICAL EXAMINATION

Based on the mechanisms of injury, any type of injury was possible. The patient was not alert and could not tell us what hurt. Palpation of her head and neck were negative for signs of injury. I palpated her chest and she groaned, indicating that she felt pain from an injury there. Listening to her chest with my stethoscope, I detected unequal breathing sounds—diminished on the left—which is where she had the pain. She had some abrasions on her lower legs, but no other injuries that I could find. As I finished the exam, Sonya informed me that respirations were becoming inadequate. Removing the nonrebreather, she began assisting ventilations with the BVM flowing oxygen at 15 liters per minute.

I checked the patient's pulse, which was 112, regular and weak. Her respirations were 28 and shallow.

PATIENT HISTORY

I looked, but didn't find any medical information devices on the patient. She couldn't tell us anything about her condition and no family members were present.

ONGOING ASSESSMENT

We continued to assist her ventilations. They remained shallow and rapid. Her pulse increased slightly to 120 and remained weak. We concentrated on ventilating the patient until the EMTs arrived.

PATIENT HAND-OFF

We told the EMTs about the crash we witnessed. They agreed that the patient had to be moved fast. We told them:

"This is a female, about 55 years of age, involved in a motor-vehicle crash. This was a one-vehicle rollover, speed approximately 45-50 mph. It rolled over twice. We had to move the patient rapidly away from the vehicle, due to the fire. We could not determine if the patient had been properly restrained or not. She responds only by moaning. We are currently assisting ventilations because of inadequate breathing. We are stabilizing the spine because of the mechanism of injury. We noted pain in her chest while we palpated. Breath sounds on the left are diminished compared to the right. Her pulse increased from 112 to 120 and is weak. Her respirations are shallow and 28. We have no information on history. We'll give you a hand with the backboarding so you can get off the scene quickly."

It turns out the patient experienced a collapsed lung in the crash. The smoke from the fire really didn't help. The EMTs told us that the emergency department staff corrected the lung problem and the patient improved dramatically. She was expected to recover fully. Our actions at the scene were an important part of her doing so well.

> *When to move a patient is determined by the patient's condition and the environment in which he or she is found. How to move a patient is determined by considering his or her condition, location, and resources. Remember, in general, a fire service First Responder does not move a patient unless an emergency move must be made or the EMTs ask for assistance. When you do move a patient, be sure to follow the rules of good body mechanics.*

Chapter Review

You will find most patients are in safe locations. Others have problems that would be worsened by movement. So in most cases, you will not move a patient until the EMTs arrive on scene.

However, when moving a patient means the difference between life and death, your ability to do so safely and properly is key. For example, a patient in a burning vehicle must be moved. If you have a sound knowledge of lifting and moving skills, you could move that patient without causing further injury to the spine or other parts of his or her body.

Study and practice the skills presented in this chapter. You will use them more than you think.

FIRE COMPANY REVIEW

Page references where answers may be found or supported are provided at the end of each question.

Section 1

1. Why should you follow the principles of body mechanics? (p. 400)

Section 2

2. What are five situations in which you should perform an emergency move of a patient? (pp. 403–404)

3. Who should move a patient when there is no immediate threat to the patient's life? (p. 405)

4. How would you perform a shirt drag? Blanket drag? Shoulder drag? (p. 404)

5. How would you perform a direct ground lift? Extremity lift? (p. 405)

6. What is the preferred position for an unresponsive patient? A patient with signs of shock? A patient who has pain or breathing problems? A patient who is nauseated or vomiting? (pp. 408–409)

Section 3

7. What is a portable stretcher? Describe it. (p. 410)

8. What is a scoop, or orthopedic, stretcher? Describe it. (pp. 410–411)

9. How should you use a stair chair? Describe the procedure. (p. 411)

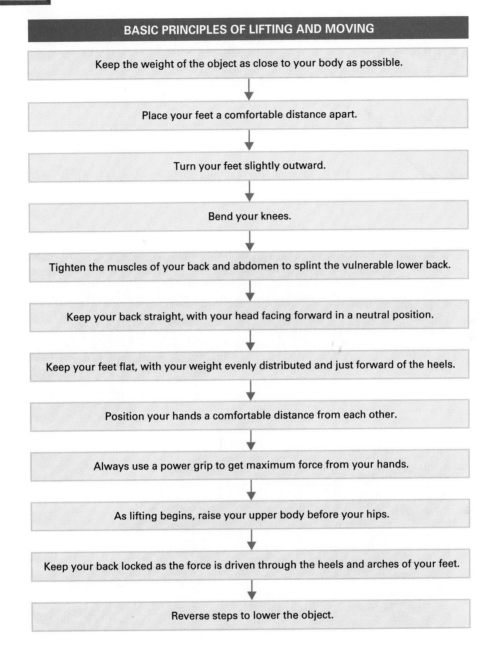

BASIC PRINCIPLES OF LIFTING AND MOVING

Keep the weight of the object as close to your body as possible.

Place your feet a comfortable distance apart.

Turn your feet slightly outward.

Bend your knees.

Tighten the muscles of your back and abdomen to splint the vulnerable lower back.

Keep your back straight, with your head facing forward in a neutral position.

Keep your feet flat, with your weight evenly distributed and just forward of the heels.

Position your hands a comfortable distance from each other.

Always use a power grip to get maximum force from your hands.

As lifting begins, raise your upper body before your hips.

Keep your back locked as the force is driven through the heels and arches of your feet.

Reverse steps to lower the object.

CHAPTER 22

Multiple-Casualty Incidents and Incident Management

*I*NTRODUCTION   *Multiple-casualty incidents range from a car crash to hurricanes, floods, earthquakes, bombings, and terrorism incidents. This chapter will introduce you to ways in which EMS personnel respond to such emergencies. It also will give you an overview of your role as a fire service First Responder, including how you can provide the best emergency care to the greatest number of patients.*

Cognitive, affective, and psychomotor objectives are from the U.S. DOT's "First Responder: National Standard Curriculum." Enrichment objectives, if any, identify supplemental material.

Cognitive

7-1.8 Describe the criteria for a multiple-casualty situation. (p. 420)

7-1.9 Discuss the role of the First Responder in the multiple-casualty situation. (pp. 423–430)

7-1.10 Summarize the components of triage. (pp. 426–430)

Affective

No objectives are identified.

Psychomotor

7-1.2 Given a scenario of a mass casualty incident, perform triage. (pp. 426–430)

Enrichment

◆ Describe the role of Incident Command in a multiple-casualty incident. (pp. 421–422)

◆ Identify communications as a key component of an MCI. (p. 423)

◆ Describe commonly used EMS sector functions. (p. 425)

◆ Identify a triage tag. (pp. 428–429)

◆ Discuss ways to reduce the psychological impact of disasters on patients and rescuers. (p. 430)

ON SCENE

DISPATCH

Last week while on duty for the fire department, I was assigned to the medical first response unit. Shortly after checking out our equipment and supplies, my partner and I were dispatched to an explosion with fire at a small factory.

SCENE SIZE-UP

Our response time was five minutes. As we approached the scene, we requested an assignment from the incident commander. We were told where to park our vehicle. I was then instructed to establish an EMS division/group and determine the number and extent of injuries. My partner and I immediately donned our identification vests. I wore the EMS division/group vest. She put on the triage sector vest. She commented on how lucky it was that we had participated in a multiple-casualty incident drill just two weeks ago.

A plant security guard approached our sector and informed me that nine people were working in the area when the explosion occurred. Eight people escaped with their lives. One was still unaccounted for. I radioed this information to the incident commander.

Emotions were running high among the patients. I knew that many patients follow the rescuer's lead, so calm was the rule. I remembered something I had been taught in my incident division/group training class: There's no reason to get excited. After all, you didn't cause this disaster. You're there to make sure it doesn't get worse!

Consider this situation as you read Chapter 22. What can be done to assure that all the patients at this incident get timely and appropriate emergency medical care?

MULTIPLE-CASUALTY INCIDENTS

One of the most challenging emergencies for a fire service First Responder, or any emergency provider for that matter, is a **multiple-casualty incident (MCI).** An MCI also is called a multiple-casualty situation (MCS) or a mass-casualty incident (MCI). An MCI is an event that places great demands on the personnel and resources of an EMS system. Different jurisdictions define MCIs in different ways. Some systems define it as any incident involving three or more patients. Other jurisdictions set the level for an MCI at five, seven, or more patients. The most common MCI is an automobile crash with three or more patients. You will likely respond to many incidents with three to 15 potential patients. Incidents with large-scale casualties are rare and apt to be "once in a career" events.

The demands of an MCI usually go beyond what a single emergency unit can handle. In fact, the resources of multiple agencies—fire, EMS, and police—are often required to handle an MCI. With many people and different agencies involved in an incident, a clear plan for handling the situation is needed. Otherwise, confusion is likely to result. Personnel may fail to perform important tasks, or crews may not be where they are needed most at the proper times.

Coordination of personnel is the key to safe and efficient functioning at MCIs. To ensure this coordination, every jurisdiction should have a disaster plan. This is a predefined set of instructions that tells a community's various emergency responders what to do in different types of emergencies. No plan can predefine responses for all conditions that might arise. Nevertheless, a very good disaster plan should be:

◆ *Flexible.* The plan must be adaptable to a variety of incidents. It should be expandable to cover MCIs involving three patients to those involving 15 or more.

◆ *Written to address the events conceivable for a particular situation.* This means that a plan for a town in Kansas should address tornadoes, not hurricanes.

◆ *Well-publicized.* Personnel in all responding agencies should know about the plan and how it operates.

◆ *Realistic.* The plan should be based on the actual resources available.

◆ *Rehearsed.* Agencies covered by the plan should practice using it.

Most well-trained emergency responders can handle small-scale MCIs. In theory, large-scale MCIs are managed using the same principles. However, the scope of the larger incidents can overwhelm responders. To handle these incidents, the jurisdiction must have a good plan and emergency personnel must know it and practice it.

INCIDENT MANAGEMENT SYSTEM

There are many systems for organizing responses to large-scale emergency incidents. Historically, there have been two predominant incident management systems used in the fire and EMS fields. They were the FIRESCOPE Incident Command System (ICS) and the Fire Ground Command (FGC) systems. ICS was developed in Southern California in the early 1970s and was adopted nationwide in the U.S. through the teaching efforts of the National Fire Academy in Emmitsburg, Maryland. FGC was developed by the Phoenix, Arizona, Fire Department and is used in many fire departments in North America.

In recent years, these two systems have merged into a common system called the **National Fire Service Incident Management System (IMS).** The International Fire Service Training Association (IFSTA) officially adopted the IMS system for use in all of its materials. This system is adaptable to all types of emergency incidents. In addition, it is mandated by law for the management of some types of incidents, such as those involving hazardous materials. IMS provides procedures for controlling personnel, facilities, equipment, and communications. It is designed for flexibility in both small-scale and large-scale incidents.

IMS has five major functional areas:

◆ *Command*—responsible for direction and oversight of all incident activities.

◆ *Operations*—responsible for carrying out the goals of the mission as set by Command.

- *Planning*—responsible for gathering and evaluating information about the incident, including the status of resources.
- *Logistics*—responsible for providing facilities, services, and materials to support response at an incident.
- *Finance/administration*—responsible for costs and financial aspects of an incident.

Not all of these components will be required at every incident. For example, finances will only be a concern at really large-scale, long-term incidents. The most commonly used components at MCIs are Command and Operations.

Command

Command must be established at all MCIs. **Command** is the person who assumes responsibility for management. This individual stays in command unless that role is transferred to another person or until the incident is brought to a conclusion.

IMS operates on the assumption that a manageable span of control is three to seven people or units, with five being the optimum number. As an incident escalates and becomes more complex, the number of people involved grows and becomes too great for one person to control. At this point, Command names **Sector Supervisors** (also called Division or Group Supervisors) to help. These Sector Supervisors oversee specific functions at an incident, such as transportation or rehab. Command retains overall responsibility for directing the incident. If the incident expands still further, it may be necessary to implement a medical branch to supervise the various sectors.

There are two basic methods of Command—singular and unified. In singular command, one agency controls all resources and operations at an incident. As you know, EMS is often managed by a fire department. Singular command is often used at fire and rescue incidents.

There are incidents, however, that will involve fire, police, additional EMS providers, and other agencies. With such incidents, unified command is more appropriate. In unified command, several agencies work independently but cooperatively, rather than one agency assuming overall control. Unified command is a highly effective means of incident management (Figure 22-1). It recognizes that large-scale incidents are complex and that the right agency must take the lead at the right time with the cooperation and support of the other agencies.

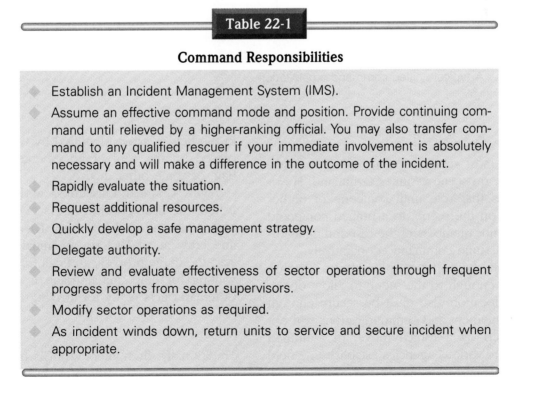

Table 22-1

Command Responsibilities

- Establish an Incident Management System (IMS).
- Assume an effective command mode and position. Provide continuing command until relieved by a higher-ranking official. You may also transfer command to any qualified rescuer if your immediate involvement is absolutely necessary and will make a difference in the outcome of the incident.
- Rapidly evaluate the situation.
- Request additional resources.
- Quickly develop a safe management strategy.
- Delegate authority.
- Review and evaluate effectiveness of sector operations through frequent progress reports from sector supervisors.
- Modify sector operations as required.
- As incident winds down, return units to service and secure incident when appropriate.

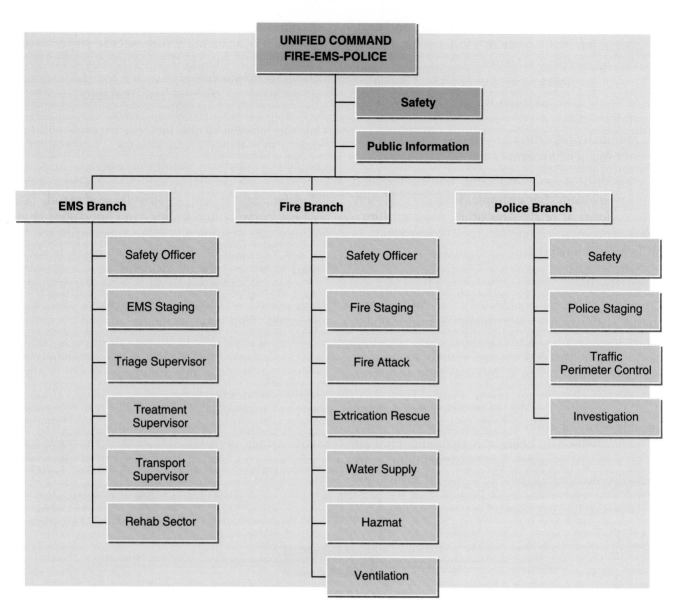

Figure 22-1 A typical unified command organization.

Command Functions

Initially, the most senior member of the first service to arrive on scene assumes Command. He or she performs that role until someone of higher rank arrives on the scene. In a unified command structure, senior members of the responding agencies would assume command cooperatively.

Command should be positioned at a location that is close enough to the scene to allow observation. The location also must be secure enough so that management and communications will not be interrupted by a forced move. In a unified command, the various agencies establish one field command post and stay there. The Command post

in the field is often identified by two traffic cones placed on the roof of an emergency vehicle, a flag, or some other device.

With Command assumed and a post established, the next priorities are scene size-up/triage and organization/delegation. That is, Command and crew at an MCI do an initial scene size-up, start the triage process, and call for backup. While waiting for help, initial triage is completed and Command gets ready for arriving resources.

Scene Size-Up

Traditionally, firefighters at MCIs have carried out fire suppression, rescue, and extrication duties. As a fire service First Responder, you will probably be assigned to a role involving patient assessment and treatment.

If your unit is the first on scene, perform a scene size-up. Remember that at this stage your job is to get an overall picture of the incident. Even though many patients may be in obvious distress, perform the scene size-up and triage before turning to a detailed assessment and treatment of individual patients. Only by doing this will you assure the rapid arrival of adequate resources to help the largest number of patients.

After arriving on scene and establishing command, perform a scene size-up as follows:

◆ Do a quick walk through or, if the scene involves hazardous materials, observe from a safe distance. You may assign the remainder of your crew to perform this task and report back to you. Assess the scene for the following information:

   ◆ Number of patients, including the "walking wounded."

   ◆ Scene hazards.

   ◆ Apparent patient priorities.

   ◆ Need for extrication.

   ◆ Number of ambulances required.

   ◆ Other factors affecting the scene and resources needed to address them.

   ◆ Areas to stage resources.

◆ Radio in an initial scene report with your request for additional resources.

Communications

Once the scene size-up is completed, an initial scene report should be made to the communications center. This report should be short and to the point. It must also make clear the severity of the situation and the need for additional resources. When making the report, Command should have a unique name to distinguish it from other units that may be using the same communications system. Command's location also should be a part of the initial report. One example of an initial report is as follows:

Firecom, this is Engine 62. We are on the scene of a two-car MVC with severe entrapment of four Priority-1 patients. Dispatch a rescue company and three paramedic ambulances. I will be establishing Central Avenue Command. Police are needed at the scene to assist with traffic and crowd control as soon as possible.

If the jurisdiction's disaster plan is to be put into effect, make sure that all responding units are informed of this. They must follow the same organizational chart for the incident that Command is using to assure that rescue efforts are coordinated. Command must tell responding units as early as possible what equipment to bring, how to access the scene, where to stage, and what they should do upon arrival.

As help begins to arrive, control of on-scene communication is important. Use as much face-to-face communications as possible, especially between Command and Sector Supervisors and between Sector Supervisors and subordinates. This will help to reduce radio channel crowding. Command may have to designate a radio aide if radio communications take up too much time.

Basically, the flow of communications on the scene should correspond to the organizational chart being used for the incident. Accordingly, the only unit talking to the communications center and requesting resources is Command. The only ones who report to Command are the Sector Supervisors. Other personnel report only to their assigned Sector Supervisors.

Organization

Getting organized early and aggressively is very important. If you are serving as initial Command at an incident, you must have a plan to deploy resources when they arrive. You must have decided what Sector Supervisors will be needed and where resources will be placed.

A command mistake is to underestimate the resources needed. Somehow, new patients not found during size-up have a way of appearing. Think big. Order big. Put resources in the staging area if they are not needed right away. In urban/suburban incidents, backup can be fast and overwhelming. Think about supply and staffing areas early or you will risk being overrun.

Another of the responsibilities of Command is to prevent "freelancing." Freelancing is uncoordinated or undirected activity at the emergency scene. Given the opportunity, most responders will arrive on scene and begin setting their own priorities. Command can prevent this problem. When Command is established early in the incident,

COLONIE EMS — Incident Tactical Worksheet

Location _____
Med. Command _____

____ Establish unified command with fire & police
____ Place 2 cones on command vehicle
____ Put bib on
____ Designate triage office

____ Advise inbound units where to stage
____ Advise crews to stay with units until given instructions
____ Advise units to switch to EMS Admin., 265 or 715

LEVEL 1 (3-10 Patients)

____ Declare MCI
____ EMS All Call
____ Request # of Units Needed
____ Cover Town/Sr. Medic Act 615
____ Roll Call Hospitals
____ Transport Officer?

(2-5 Amb. Needed)

LEVEL 2 (11-25 Patients)

____ Declare MCI
____ EMS All Call
____ Request # of Units Needed
____ Cover Town/Sr. Medic Act 615
____ Roll Call Hospitals
____ Get Mutual Aid Units
____ Designate Treatment Officer
____ Designate Transport Officer
____ Designate Staging Officer
____ REMO MD to Scene
____ Consider Rehab & CISD

(6-13 Amb. Needed)

LEVEL 3 (over 25 Patients)

____ Declare MCI
____ EMS All Call
____ Request # of Units Needed
____ Cover Town/Sr. Medic Act 615
____ Roll Call Hospitals
____ Get Mutual Aid Units
____ Designate Treatment Officer
____ Designate Transport Officer
____ Designate Staging Officer
____ REMO MD to Scene
____ Request Bus to Scene

(over 13 Amb. Needed)

FIRE

____ Assess # of Units Needed
____ EMS All Call Req. 619
____ Designate Triage
____ Set up Rehab at Air Bank
____ Use 619 as ALS Unit

RESCUE

____ Establish Perimeter
____ Request Speciality Units
____ Triage Officer Handles Inner
____ Circle

HAZ-MAT

____ Req. # of Units Needed
____ EMS All Call
____ Est. Command in Cold Zone
____ Designate Triage
____ Identify Agent
____ Research Decontamination
____ Research Med.

____ Medical Baseline Assessment of Team
____ Don Protective Barriers
____ Assist With Decontamination
____ Rehabilitate

HOSPITAL ROLL CALL	AMCH	St. PETERS	MEMORIAL	VA	ELLIS	St. CLARE'S	LEONARD	St. MARY'S	SAMARITAN
CAN TAKE									
# PATIENTS SENT									

UNITS RESPONDING

620 621 622
630 631 632
640 641 642
650 651 652
610 611 605
TSU-1 TSU-2
619 ____
Guild. ____
CPHM ____
Albany ____
Mohawk ____
Empire ____

UNITS IN STAGING

620 621 622
630 631 632
640 641 642
650 651 652
610 611 605
TSU-1 TSU-2
619 ____
Guild. ____
CPHM ____
Albany ____
Mohawk ____
Empire ____

OF PATIENTS BY PRIORITY

1 (Red)	2 (Yellow)	3 (Green)	0 (Black)	TOTALS

Figure 22-2 An example of an incident tactical worksheet.

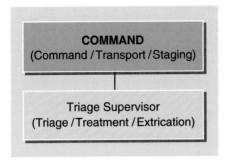

Figure 22-3 An organization for a smaller incident.

people and crews can be assigned to tasks as they arrive.

Some tools are available to help in the organization of Command. For example, many agencies list the main points of their disaster plans on a "tactical worksheet" that can be used in the field (Figure 22-2). Such worksheets are useful aids for remembering all the steps that must be carried out during an incident.

EMS Sector Functions

In smaller MCIs, Command may be able to manage all aspects of the incident. As incidents grow in scale, they come to involve more information, resources, and personnel than a single person can effectively manage. In larger incidents, Command should designate Sector Supervisors to manage different aspects of the emergency operation. The number of sectors will vary depending on the size and complexity of the incident (Figures 22-3 through 22-5). In MCIs, some of the common EMS sectors, in addition to Command, are:

◆ *Staging.* This sector, set up a safe distance from the emergency, holds, monitors, and inventories ambulances and emergency vehicles and their crews and releases them to other sectors as needed.

◆ *Supply.* This sector monitors, inventories, and distributes patient-care equipment.

◆ *Extrication.* This sector oversees and directs rescue teams who are responsible for freeing patients from wreckage.

◆ *Triage.* This sector is responsible for sorting patients to determine the order in which they will receive medical care or transportation to definitive care.

◆ *Treatment.* This sector provides field care for patients. Triage also continues in this sector.

◆ *Transportation.* This sector is responsible for moving patients from treatment sectors to hospitals. The Transportation Sector Supervisor must communicate often with Command and with Sector Supervisors in staging, triage, and treatment to assure the smooth movement of patients out of the emergency scene. The Transportation Supervisor must know the priorities, identities, and destinations of all patients leaving the scene. The Transportation Supervisor may assign a Medical Communications Aide to communicate with the receiving hospitals. He or she will continually monitor resource availability at the receiving hospitals to assure optimum patient care.

◆ *Rehab.* This sector is responsible for monitoring the health and well-being of rescue workers and providing them with rest, food, and rehydration as needed.

Individuals and agencies on scene will be assigned roles in one or more sectors. Most systems use brightly colored reflective vests that can be worn over protective clothing to make each Sector Supervisor easy to identify (Figure 22-6). Emergency personnel arriving on scene after Command has been established would be expected to report to a Sector Supervisor for

Figure 22-4 An organization for a medium-sized incident.

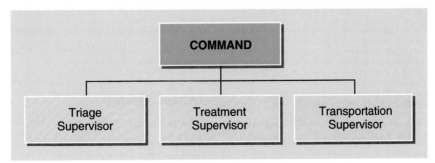

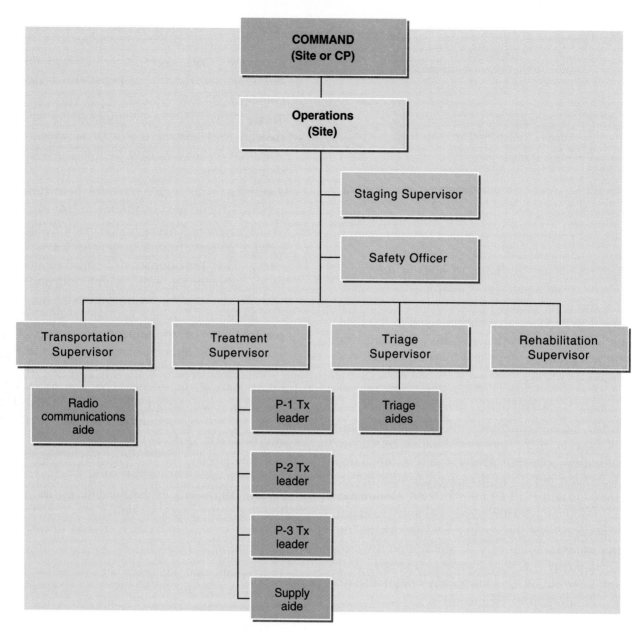

Figure 22-5 An organization for a major incident.

assignment to specific duties. When reporting in, identify yourself, give your level of training, and ask for instructions. Be prepared to care for patients or to support rescue personnel as directed. Once assigned a task, an emergency worker should complete it and then report back to the Sector Supervisor.

Triage Sector

Once EMS Command has been established, the next task is to quickly assess all the patients and assign each a priority for receiving emergency care or transportation to definitive care. This

process is called **triage,** which comes from a French word meaning "to sort." The most knowledgeable EMS provider becomes the Triage Supervisor. The Triage Supervisor calls for additional help if needed, assigns available personnel and equipment to patients, and remains at the scene to assign and coordinate personnel, supplies, and equipment.

Initial Triage

When faced with more than one patient, your goal must be to afford the greatest number of people the greatest chance of survival. To accom-

Figure 22-6 Using brightly colored reflective bibs make personnel readily identifiable.

plish this, you must provide care to patients according to the seriousness of their illnesses or injuries. While you provide care, you must keep in mind that spending a lot of time trying to save one life may prevent a number of other patients from receiving needed treatment.

In performing triage on a group of patients, it is necessary to quickly assess them and assign them to one of four categories:

◆ *Priority 1: treatable life-threatening illnesses or injuries.* This is the highest priority, which includes airway and breathing difficulties, uncontrolled or severe bleeding, decreased mental status, severe medical problems, shock (hypoperfusion), and severe burns.

◆ *Priority 2: serious but not life-threatening illnesses or injuries.* This is the second priority, which includes burns without airway problems, major or multiple bone or joint injuries, back injuries with or without spinal-cord damage.

◆ *Priority 3: "walking wounded."* This includes minor musculoskeletal injuries and minor soft-tissue injuries. In some areas, the "walking wounded" are in the same category as the dead and are considered the lowest priority.

◆ *Priority 4 (or Priority 0): dead or fatally injured.* This includes exposed brain matter, cardiac arrest (no pulse for over 20 minutes except in cases of cold-water drowning, severe hypothermia, or lightning injuries), decapitation, severed trunk, incineration. Patients in

arrest must be considered Priority 4 when resources are limited. The time that must be devoted to rescue breathing or CPR for one person is not justified when there are many patients who need attention. Once ample resources are available, patients in arrest become Priority 1.

How triage is performed depends on the number of injuries, the immediate hazards to personnel and patients, and the location of backup resources. Your local operating procedures will supply guidance on the exact methods of triage to follow in a given situation. Some basic, standard triage procedures are described below.

If you are assigned to triage, begin the initial assessment. You can first request—using a bullhorn or loud voice—all patients who can work to move to a particular area. This will identify the "walking wounded" patients who have adequate airway and breathing. Then, quickly assess the ABCs of remaining patients as follows:

1. **If a patient's airway is not open,** open it with a manual maneuver. If the patient responds appropriately, move on to the next patient.

2. **If the patient is unresponsive,** check for breathing and pulse. If there is no breathing or pulse, do not provide care for that person. Move on to the next patient.

3. **If you feel a pulse,** check for severe bleeding. If there is severe bleeding, quickly apply a pressure dressing to the wound and move on to the next patient.

As long as there are other patients waiting to be triaged, the only initial treatment you should provide is airway management and control of severe bleeding. If there is an immediate danger to patients, begin moving them regardless of injuries. Your "walking wounded" may be able to assist in airway management, bleeding control, and moving patients.

Patient Identification

For triage procedures to work effectively, a system of identifying patients with the priorities assigned to them must be in place. Arriving EMS personnel must be able to look at patients and determine at a glance their treatment and transportation priorities or whether they have been assessed or not.

Different jurisdictions use different systems for patient identification. One of the most common is the color-coding method:

◆ Red = Priority 1
◆ Yellow = Priority 2
◆ Green = Priority 3
◆ Black (or gray) = Priority 4

Many systems use triage tags coded with these colors to identify patients (Figure 22-7). EMTs carrying out the initial assessment attach the tags to patients as they prioritize them. Often tags have space on them where basic medical information about a patient can be recorded.

Secondary Triage and Treatment

As more personnel arrive at the incident scene, they should be directed to assist with the completion of initial triage. If triage has been completed, they can instead begin treatment.

Figure 22-7 Front and back of a triage tag. *(Source:METTAG)*

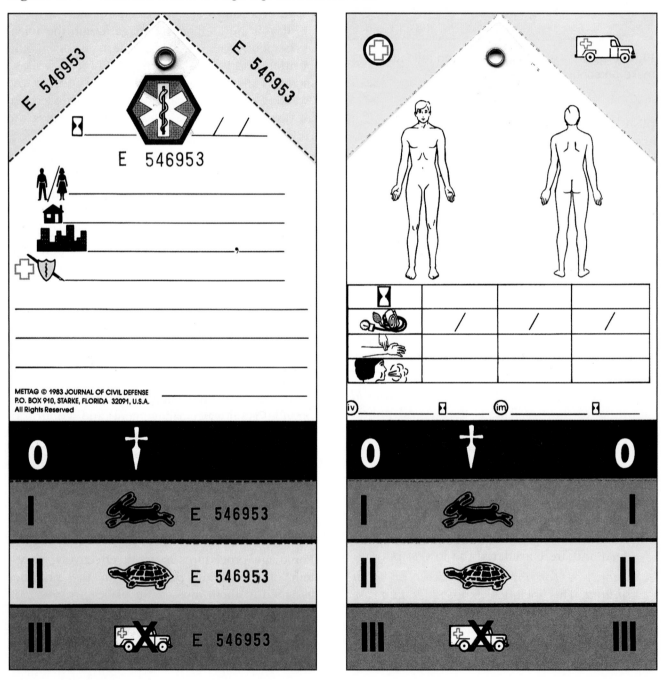

Secondary triage generally begins at this point. In ideal triage systems, patients are gathered into a triage sector and, under the direction of the Triage Supervisor, are physically separated into treatment groups based on their priority levels as indicated on their triage tags.

An area to which patients are removed is referred to as a Treatment Sector. Each Treatment Sector should have its own Treatment Supervisor and an EMT responsible for overseeing the triage and treatment within that sector.

The Treatment Supervisor should again triage the patients within that sector to determine the order in which they will receive treatment. Patients' conditions may improve or deteriorate. If they do, it may be necessary to reassign a patient a higher or lower priority. The patient would then be moved to the appropriate Treatment Sector. Some systems use a different disaster tag during secondary triage on which more detailed information about the patient may be recorded (Figure 22-8). Some systems use colored tarps to identify

Figure 22-8 Front and back of a disaster tag. *(Source: EMS Associates)*

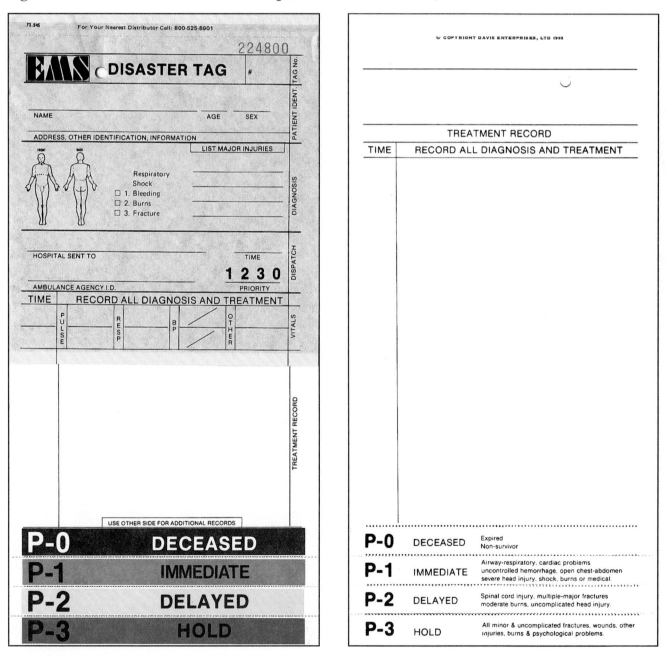

Treatment Sectors in multiple-casualty and large-scale incidents.

Treatment Sector personnel will need supplies and equipment from ambulances such as bandages, blood-pressure cuffs, and oxygen.

Transportation and Staging

Once patients have been properly prioritized and treatment has begun, consideration must be given to the order in which to transport them to hospitals. Again, this is done according to triage priority.

Ambulances for transport should be assembled in the Staging Sector. The Staging Supervisor is responsible for having vehicles and their operators ready to roll. No ambulance should proceed to a Treatment Sector without having been requested by the Transportation Supervisor and directed by the Staging Supervisor.

The Staging Supervisor should determine from each Treatment Sector the number and priority of patients in that sector. This information can then be used by the Transportation Supervisor to arrange for patients to be taken from the scene to the hospital as efficiently as possible.

An ambulance should not transport any patient without the approval of the Transportation Supervisor. This is because the Transportation Supervisor is responsible for maintaining a list of patients and the hospitals to which they are transported. This information is relayed from the Transportation Supervisor (or the designated Medical Communications Aide) to each receiving hospital. In this way, the hospitals know what to expect and receive only those patients they are capable of handling. The EMTs on board the transporting ambulances must follow the Transportation Supervisor's directions. Failure to do so may result in patients being transported to the wrong facilities and delays in their treatment.

Once an ambulance has completed a run to a hospital, it will probably be directed to return to the staging area, perhaps bringing needed supplies. It should wait in the staging area for its next instructions from the Staging Supervisor.

Communicating with Hospitals

Receiving medical facilities must be alerted about an MCI or disaster as soon as the scope of the incident is known. This allows them to call in additional personnel and otherwise prepare to receive the anticipated numbers of patients.

In an MCI, communications channels will be in heavy use. It is crucial that only the Transportation Supervisor communicate with hospitals. This will help eliminate unnecessary radio traffic. It also will ensure that proper patient information is recorded at both ends of the ambulance ride.

In large-scale MCIs, individual patient reports are not necessary. Treating and transporting personnel are usually different. Also, there are usually too many patients to permit a radio report on each one. In these circumstances, hospitals may only be provided with the most basic information, such as they are receiving a Priority-1 patient with respiratory problems.

Psychological Considerations

By their nature, MCIs are extremely stressful situations. This is true for both patients and rescuers involved in them. While they may outwardly show few signs of injury or emotional stress, people at MCIs face devastating circumstances with which they are normally unprepared to cope.

Proper early management of a psychologically stressed patient can support later treatment and help ensure a faster recovery. This management may require you to administer "psychological first aid." This may mean talking with a terrified patient, child, or witness. This does not mean that you should try to perform psychoanalysis. Nor should you lie to a patient in an attempt to calm him or her. Instead, present a caring, honest demeanor. Listen to the patient and show you are listening by acknowledging his or her fears and problems. This may be the only "psychological first aid" necessary.

Rescuers, too, face stress during MCIs. Those who become emotionally incapacitated should be treated as patients and removed to the Rehab Sector. They should be monitored there by a clinically competent provider of care. These rescuers should not return to duty without first being evaluated by someone professionally trained to do so.

A critical incident stress debriefing (CISD) can be invaluable after an MCI or disaster. You can review critical incident stress management (CISM) in Chapter 2.

At the beginning of this chapter, you read that fire service First Responders were on the scene of an MCI. To see how chapter skills apply to the emergency, read the following. It describes how the call was completed.

INITIAL ASSESSMENT AND TRIAGE

Janet, my partner, started triage. When another fire service First Responder arrived, I assigned him to establish a transport sector and gave him a vest. I reminded him to make sure that he identified a safe area where ambulances could enter and exit quickly and safely.

Janet soon reported that two patients were Priority 1, three were Priority 2, and three were Priority 3. I updated the incident commander and requested ambulances, three of which were to be equipped for advanced life support (ALS). I also requested that he alert the trauma center.

A minute or so later it was confirmed that one patient was still inside the plant. We believed that patient to be a Priority 0. At that moment, the paramedic EMS supervisor arrived. I provided a full report and turned over EMS division/group to him. I was then assigned to assist in the treatment area.

Over the next half hour, all eight patients were transported to a hospital. A police officer was assigned to guard the one dead body until the coroner arrived. Triage and transport for the injured patients had taken 45 minutes.

When the EMS division/group was terminated, the incident commander told us that a critical incident stress debriefing would be organized for all rescue personnel involved.

For you to be an effective member of EMS response to an MCI, you must learn your local plans and protocols. Review them often. Practice them whenever you are given the opportunity.

Chapter Review

When most people think of multiple-casualty incidents, they think of plane crashes and commuter train wrecks. While these certainly qualify, the most common MCI is a two-car collision, each car with three injured passengers. Most EMS systems would have to stretch resources to take care of these six patients. Learn what your EMS system expects of you.

In the first few minutes of an MCI, the most important task that the first trained person on the scene can perform is to plan, not rush in.

FIRE COMPANY REVIEW

Page references where answers may be found or supported are provided at the end of each question.

Section 1

1. What is a multiple-casualty incident? (p. 420)

Section 2

2. What is the incident management system (ICS)? (pp. 420–421)

3. What are some common EMS sector functions? (p. 425)

4. What should you include in a size-up of an MCI scene? (pp. 422–423)

5. What are four categories of a triage system? Describe each one. (p. 427)

RESOURCES TO LEARN MORE

Christen, H., and P.M. Maniscalco. *The EMS Incident Management System: EMS Management for High-Impact and Mass-Casualty Incidents.* Upper Saddle River, NJ: Brady/Prentice-Hall, 1998.

Dickinson, E., and M.A. Wieder. *Emergency Incident Rehabilitation.* Upper Saddle River, NJ: Brady/Prentice-Hall, 2000.

National Fire Service Incident Management System Consortium. *IMS Model Procedures Guide for Emergency Medical Incidents.* Stillwater, OK: Fire Protection Publications, 1996.

INITIAL RESPONSIBILITIES AT AN MCI

First arriving EMS personnel establishes EMS Command, and dons proper identification.

↓

Walk through the scene (or observe from a safe distance). Assess for:
- Number of patients
- Scene hazards
- Apparent patient priorities
- Need for extrication
- Number of ambulances needed
- Other factors affecting the scene
- Resources needed to address them
- Areas in which to stage resources

↓

Radio in an initial scene report with request for additional resources.

↓

Organize, deciding where to place resources and what sectors will be needed. Sectors may include:
- Staging
- Supply
- Extrication
- Triage
- Treatment
- Transportation
- Rehab

↓

Begin initial triage of patients, if appropriate.

↓

Assign incoming units and personnel to appropriate sectors.

23

EMS Operations

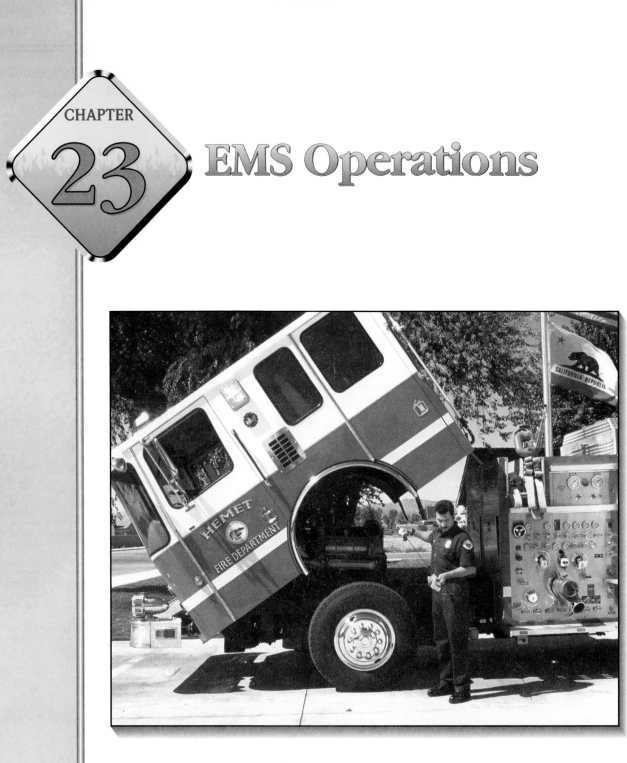

*I*NTRODUCTION   *The following chapter will provide you with a brief overview of the basic operational aspects of out-of-hospital emergency care. Learn the six basic phases of an emergency response. Become familiar with the medical and non-medical equipment used on scene.*

   Even if you are not required to drive an emergency vehicle, become familiar with emergency vehicle safety. Since there may be a situation in which you are asked to travel in an ambulance, this chapter also provides related basic safety precautions.

Cognitive, affective, and psychomotor objectives are from the U.S. DOT's "First Responder: National Standard Curriculum." Enrichment objectives, if any, identify supplemental material.

Cognitive

7-1.1 Discuss the medical and non-medical equipment needed to respond to a call. (p. 436)

7-1.2 List the phases of an out-of-hospital call. (p. 437)

Affective

7-1.11 Explain the rationale for having the unit prepared to respond. (p. 437)

Psychomotor

No objectives are identified.

Enrichment

◆ Discuss ways of driving an emergency vehicle safely, including how to use seat belts, lights, and sirens properly. (p. 440)

◆ Describe how to stay safe in traffic on foot, including how to channel traffic away from a scene. (pp. 441–443)

◆ Discuss safety tips for traveling in the passenger compartment of an ambulance. (pp. 443–444)

 ON SCENE

DISPATCH

Our fire department rescue truck was dispatched to the intersection of Highway 82 and Route 90 for a multiple-vehicle collision. Some of the worst incidents I've scene have occurred at that intersection.

SCENE SIZE-UP

With a 15-minute response time, my partner and I discussed what we might expect when we arrived. Both of us knew we must put on BSI gear and wear our reflective vests. Also, we talked about where we would position our rescue vehicle to afford us the most protection. We knew from experience that this was a 45-mph intersection. Yet most people drive through it at 65 or faster. There's only a flashing yellow light warning motorists on the north-south highway and a red flashing light on the east-west road. Drivers who stop typically misjudge the speed of the vehicles at that flashing yellow light. As a result, they tend to pull out, causing the other drivers to slam on their brakes. We were aware that a collision at this location could mean severe injuries, maybe even fatalities.

Consider the scene as you read Chapter 23. What can fire service First Responders do to keep it safe for themselves, for other rescuers, and for the patient?

RECOMMENDED EMS EQUIPMENT

When you are on duty, you should have EMS equipment at your disposal. This equipment will include the following items (Figure 23-1):

- Equipment for airway and breathing:
  - Adjunct airways.
  - Suction devices.
  - Pocket masks or other artificial ventilation devices.
- Equipment for bleeding control and bandaging:
  - Dressings of various types and sizes.
  - Bandages of various types and sizes.
  - Materials to stabilize impaled objects.
  - Sterile saline.
  - Scissors.
  - Adhesive tape.
- Equipment for patient assessment:
  - Stethoscope.
  - Wristwatch.
  - Pen light.
  - Sphygmomanometer (optional).
  - Prehospital care report forms.
  - Pen and notebook.
- Miscellaneous equipment:
  - OB (obstetric) kit.
  - Blankets.

- Triage tags.
- Chemical cold and heat packs.
- BSI equipment, such as disposable gloves, masks (including HEPA or N-95 respirator), eye wear, and gowns.
- Antiseptic liquid or wipes, waterless hand-washing solution, and bags or containers for contaminated materials.
- Turnout gear, heavy-duty and puncture-proof gloves, shatter-resistant eye protection, and other clothing—such as water-proof and reflective clothing—that will protect you from environmental hazards.
- Flares, cones, or reflective triangles for protection against traffic.
- Fire extinguisher.
- Flashlight and spare batteries.
- Local street maps.
- Binoculars.
- Latest edition of DOT's *North American Emergency Response Guidebook* for hazardous materials situations.
- Personal flotation device.
- Optional equipment—depending on the skills taught in your area, you may be trained to use some or all of the following:
  - Oxygen administration equipment.
  - Splints.
  - Backboards.
  - Automated external defibrillator (AED).
  - Body armor.

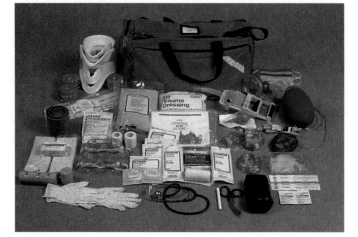

Figure 23-1 A basic First Responder jump kit.

Even when a fire service First Responder is off duty, he or she may come upon the scene of an injury or illness. That is why so many First Responders carry their own personal protective equipment in their vehicles and in their homes. Be sure you carry only equipment you are authorized to use.

Always use extra caution when off duty. You may not have radio contact or all the protective clothing you are used to having. You also may not be wearing clothing that identifies you as an EMS responder. Take extra time to explain to the patient who you are and that you are trained to help.

Figure 23-2a

Figure 23-2b If you drive a EMS vehicle while on duty, inspect it daily.

SECTION 2

PHASES OF A RESPONSE

There are six generally accepted phases of an EMS response. They are preparation, dispatch, en route to the scene, arrival on scene, transfer of care, and post-run activities.

Preparation

It is in the preparation phase of an EMS response that you report for duty and remain available for calls. This part of your job may be the most important. That is because it also is the time when you must check and ready your equipment for service. Supplies should be checked daily. They also should be restocked, cleaned, or maintained after each run.

If you drive an EMS vehicle while on duty, inspect it daily (Figure 23-2). Your employer or volunteer organization should have a clear protocol for performing regular service and maintenance, reporting vehicle problems, and taking vehicles out of service when they are unsafe. Legally, you may be liable for damage caused by a malfunctioning vehicle if you were aware of the problem. Therefore, you also may be within your rights to refuse to use a vehicle you have reason to believe is unsafe.

Dispatch

Dispatch is the formal beginning of an EMS response. Dispatchers get important information from callers who report an emergency. That information includes:

- Nature of the call.
- Name, exact location, and call-back number of the caller.
- Location of the patient.
- Number of patients and the severity of the patient's problem.
- Any other special problems or considerations that may be pertinent.

As a fire service First Responder, you may sometimes be the one who activates EMS by calling dispatch (Figure 23-3). Other times, a witness to the emergency will dial 9-1-1. He or she will provide the information to the dispatcher, who will in turn give it to you. Write the information down so you can refer to it en route. Use it to prepare yourself physically and mentally for the call. Do not hesitate to ask the dispatcher to repeat or restate information if anything is unclear.

Note that while you are en route to the scene, an emergency medical dispatcher may give specific, life-saving instructions to the caller to perform until you arrive.

En Route to the Scene

Traveling to the scene involves much more than speed. Responses also must be safe. Excess speed or carelessness will result in a crash, which at least will prevent you from helping the people who need it. In addition, crashes involving emergency vehicles are a source of legal liability and disciplinary action.

When responding, be sure to know the exact location of the emergency. Have a route planned for your response. Wear your seat belts at all times in moving vehicles. Notify the dispatcher when you have begun your response, so that other emergency teams will be aware you are responding and the time may be logged.

Arrival on Scene

Notify dispatch when you arrive on scene. Then approach cautiously. If you have driven to the

scene, park your vehicle in a safe place. Complete your scene size-up in a rapid, organized, and efficient manner. Once you have determined that the scene is safe to enter, proceed with your patient assessment plan.

Transfer of Care

By the time the responding EMTs arrive on scene, you may already be performing an ongoing assessment and beginning to package your patient. (The term **package** refers to getting the patient ready to be moved and includes procedures such as stabilizing impaled objects and immobilizing injured limbs.)

Be prepared to give a concise and accurate patient hand-off report. Also be ready to assist the EMTs with packaging or lifting and moving, if they request your help (Figure 23-4).

Post-Run Activities

Report to dispatch when you return to your station. Then, after you turn in your prehospital care report, prepare for your next call. This means cleaning and disinfecting any equipment that may have become soiled as per your local protocols.

Figure 23-4 After hand-off, be prepared to assist the EMTs.

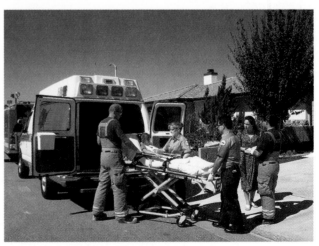

Figure 23-3 Sometimes you may be the one to activate the EMS system by calling 9-1-1.

Figure 23-5 Post-run activities include cleaning and disinfecting equipment, replacing disposable supplies, and performing vehicle maintenance according to local protocols.

Replace any disposable supplies. Change any soiled clothing. Fuel your vehicle, if necessary. Clean the exterior of the unit (Figure 23-5). Then notify dispatch that you are in service and ready for another call.

EMERGENCY VEHICLE SAFETY

Many rescuers spend a lot of time in traffic, both driving and moving around on foot at emergency scenes. If you have not had a basic safety course, check into the possibility of enrolling in one. It is a good idea to take advantage of refresher courses, too.

Driving Safely

Statistics show that haste is unnecessary in about 95% of all emergency runs. The experts maintain that only about 3% to 5% of all runs are true life-or-death situations. Yet, according to national data, approximately one in every 10 ambulances is involved in a collision every year. Excess speed makes an emergency vehicle less stable and poses a greater risk to you and others.

Much of driver safety depends on common sense and good judgment. The following basic tips can improve driver safety:

◆ Learn all local and state guidelines related to driving emergency vehicles before you drive one. By law you must always exercise due regard for the safety of others, and that includes yourself.

◆ When you can, travel in pairs. For example, the rescuer who is not driving can help the driver find the route, clear right-hand intersections, watch the road, and in case of litigation, act as a witness who can substantiate the record.

◆ If you need to back up your vehicle, do so slowly and carefully. Use all available mirrors. Have your partner take a position near the rear of the vehicle and act as a spotter (Figure 23-6).

◆ Know your territory. Take alternate routes whenever possible to avoid potential problems such as tunnels, bridges, traffic jams, and railroad crossings. Besides the obvious advantage of arriving on scene more quickly, when you know your territory you also can avoid collisions caused by trying to read a map while you drive.

◆ Exercise extra caution when traveling in congested traffic, such as rush-hour traffic in urban areas or areas just around industrial centers at shift change.

◆ Plan for stopping distance. Sudden braking is especially dangerous at high speeds. Remember that stopping time increases dramatically as speed increases. Plan for it.

◆ Practice special caution on curves and hills. Brake to a safe and comfortable speed before

Figure 23-6 The rescuer who is not driving can help make sure patient and vehicle arrive safely.

you enter a curve. Then stay on the outside of it. Speed up carefully, gradually, and steadily as you leave the curve. When traveling down hills, use a lower gear instead of the brakes to maintain control of the vehicle.

Seat Belts

When in an emergency vehicle, always wear proper safety restraints, including lap and shoulder belts. Fasten them before you start the ignition. Keep them on until you have turned off the ignition. Do not remove your seatbelt until you have come to a complete stop. Even though seatbelts are the law in many states, a three-year study of ambulance collisions in one state showed a shocking rate of noncompliance. As many as 50% of those driving an ambulance failed to wear their seatbelts.

Never take off safety restraints as you approach the emergency scene. Research shows that the last two intersections before arriving on scene are especially dangerous to rescue drivers, who try to save time by disengaging restraints.

Lights and Sirens

Whenever you respond to an emergency in a vehicle, use headlights and emergency lights even during the daytime. They can help to alert other drivers in case emergency lights blend in with traffic lights, the tail lights of cars traveling in opposite directions, the color of buildings, or holiday decorations. Use minimal lighting in heavy fog. Turn off headlights when you park. If you need to alert oncoming traffic while parked, leave the emergency lights on.

Always use emergency lights and sirens as required by your local protocols. Note, however, that a siren may have a bizarre effect on the driver of an emergency vehicle. Some drivers are easily hypnotized by a siren and lose the ability to negotiate curves and turns. A siren also can disorient or panic drivers. (Never pull up behind another driver and blast the siren.) Finally, a siren may not be effective, because often it is not heard. Insulation in newer vehicles can mask the sound of an approaching siren, as can conversation, pelting rain, thunder, dense shrubbery, trees, and buildings. To alert others of your presence, use your vehicle's horn in conjunction with emergency lights and siren.

> **COMPANY OFFICER'S NOTE**
> Always assume that people driving other vehicles do not see you, especially when you have your lights and siren on.

A significant safety advantage is to turn off your lights and siren as you approach the scene of an injury or illness. Lights and sirens attract crowds and can add to the chaos that already may exist. If there are hostile people on scene, you can take the first step toward calming and controlling the scene by arriving discreetly.

Remember that once you turn off your lights and siren, you are no longer driving an "authorized" emergency vehicle. You are subject to all the laws meant to govern regular traffic.

Hearing Protection

Chronic exposure to loud noises poses a danger to hearing. The trauma associated with repeated loud noise can cause permanent hearing loss. If your EMS system allows fire service First Responders to drive or ride in emergency vehicles, take the following precautions to protect your hearing:

- Keep the windows closed while the siren is in use.
- Wear ear plugs or other device to protect your ears. Do not use a device that will prevent you from hearing other emergency vehicles as they approach. Use caution and always follow local protocol.
- If you can, move the siren speakers from the top of the cab to the front grille. This move can reduce the decibel level by about 5%.

Avoid prolonged or repeated loud noise off the job, too. Protect yourself against loud music, high-volume radios and televisions, loud chain saws, lawn mowers, hydraulic tools, generators, and other noise.

Driving an Ambulance

As a fire service First Responder, there may be occasions when you are asked to drive the ambulance. If so, follow these guidelines to help assure a safe trip:

- Except in the most critical situations, do not exceed the posted speed limit. Excess speed is unsafe for everyone in the ambulance, especially for the patient. Speed poses special hazards at intersections and on curves.

- Start and stop smoothly. Make the transition from one speed to another gradually to avoid aggravating the patient's illness or injury.

- Drive at a steady but safe speed. If you keep your speed even, you often can time your approach to intersections and travel through with the green light. Also avoid weaving through traffic. It can compromise the patient.

- Whenever you can, avoid rough dirt roads, roads with speed bumps, potholes, and other hazards that can jostle the patient. The inner two lanes of a four-lane highway are the smoothest. The lane closest to the gutter on city streets is the roughest. If you see that you are going to drive over a bump, railroad tracks, or a stretch of rough road, warn those in the back so they can protect themselves.

As part of a driver training program, many drivers are required to lie on the stretcher while the instructor drives the ambulance. Even at low to moderate speeds, the experience from the point of view of the patient is often enlightening. (Do not use the lights or sirens or drive at increased speeds for training purposes unless you are at an approved training facility.)

Staying Safe in Traffic on Foot

Upon exiting your vehicle at the scene, you are at risk from oncoming traffic. Even while your focus is on the patient, or the scene itself, you must never compromise your own safety.

Park for Maximum Safety

An important part of ensuring your safety is to scan the scene visually as you approach. Notice areas of vulnerability or potential danger. Pinpoint places where you could seek concealment or protection. If you are with a partner, plan how you will approach before doing so. You may decide that your partner will go to the passenger who is still in the green car, for example, and you will go to the driver who is lying on the gravel.

Protect yourself from environmental hazards on scene by parking at a safe distance. Park a safe distance from a burning vehicle and from an incident involving hazardous materials. Park uphill and upwind of any hazardous material or fire. Avoid parking on or driving over spilled liquids and broken glass.

Turn off your headlights as soon as you park to prevent blinding oncoming drivers. If necessary, leave parking lights on or leave the ambulance warning lights flashing to warn oncoming traffic. Follow local protocols.

Note that proper parking allows the ambulance the best and safest position for patient loading. When possible, try to park your vehicle off the street, such as in a driveway, parking lot, or yard. This will eliminate hazards associated with oncoming traffic. However, make sure the surface you park on is stable enough to support the weight of the vehicle. If it is not possible to park off the street, work together with police to use your vehicle as a shield between the work area and oncoming traffic. Try to park larger vehicles (such as a pumper) between smaller vehicles and oncoming traffic. Guard the patient loading area of the ambulance by shielding it with another vehicle. If possible, position traffic cones so that oncoming traffic will be directed away from the work area.

Exiting a Vehicle Safely

A specific transition takes place as soon as you leave your vehicle. You move from your "sanctuary" to someone else's turf, and that makes you vulnerable. Follow these tips for the greatest protection:

- Before you open the door, check the rearview mirror to determine how much traffic is approaching from behind. If you can, wait a minute or two until traffic passes.

- Open your door slowly to alert passing motorists that you are getting out (Figure 23-7). Move carefully but quickly away from passing or oncoming traffic.

- If you have passengers in the rear compartment, have them get out through the rear doors instead of a side door, which might open into passing traffic.

- Be especially cautious about hazards at the scene, such as broken glass, twisted metal, or spilled gasoline. Immediately assess any crash involving trucks for hazardous materials.

Figure 23-7 Open the door slowly and cautiously when exiting your vehicle.

◆ If the scene is safe to enter, walk purposefully to the patient. Running is a signal to others that you are out of control, and it causes your heart to race. The boost to your pulse and adrenaline levels can hinder your ability to treat the patient effectively.

Wear Protective Equipment

Plenty of rescues take place outdoors in the dark and in bad weather. An essential for every rescuer is reflective clothing. That includes reflective tape, at least, and a reflective vest or other gear if you have access to it. If you are channeling traffic away from the crash scene while waiting for police, it is essential that you wear as much reflective gear as possible. It will help you to be visible to drivers who may have pitted windshields, frayed windshield wipers, or drug or alcohol impairment.

Depending on the situation, consider the following protective gear:

◆ In crashes involving hazardous materials, protect yourself with masks, gowns, and gloves as dictated by local protocol. In incidents involving grain, cement, or similar materials, wear a dust respirator.

◆ If there is any risk of falling debris, wear an impact-resistant protective helmet with reflective tape and a strap under the chin.

◆ In situations where splashing may occur (including splashing of blood and body fluids), wear protective eyewear or a face shield specified for work with power equipment. If

you usually wear eyeglasses, get clip-on side shields. All protective eyewear must meet American National Standards Institute (ANSI) standards.

◆ To protect yourself against the cold, wear gloves, a warm hat, long underwear, and several layers of medium-weight clothing.

◆ Depending on the rescue situation, you may need rubber or waterproof boots and slip-resistant waterproof gloves.

Channeling Traffic Away from the Scene

While waiting for the arrival of law enforcement, you need to channel traffic away from the scene. This is not only for the safety of patients and bystanders, but it is also for your safety and the safety of other rescuers. Your goals should be:

◆ To channel the regular flow of traffic around the scene, preventing additional collisions and injuries.

◆ To monitor traffic so that there is a minimum of disruption.

◆ To clear the scene so that other emergency vehicles can reach the patients quickly.

Unless there is a distinct hazard that dictates otherwise, keep traffic moving. Even if the roadway is blocked, try to reroute traffic to an alternate road rather than bring traffic to a standstill. To effectively channel traffic, you need additional rescue personnel and attention-getting devices such as flares, chemical lights, or reflective cones. Follow these general guidelines:

◆ Make sure all those who are channeling traffic are wearing adequate reflective clothing or tape so that they can be clearly seen by approaching drivers.

◆ Visual signals given by rescuers must be clear. Approaching drivers need to understand quickly and exactly what you want them to do.

◆ Place flares, chemical lights, or reflective cones 10 to 15 feet (3 m to 5 m) apart and approximately 100 feet (30 m) toward the oncoming traffic.

If a collision is on a two-lane highway, the flares or cones should be placed in both directions. On a curve or hill, place them at the beginning of the curve or at the crest of the hill. They should direct motorists around the crash scene, at least 50 feet (15 m) from wrecked cars. Use this general rule: the flares or cones should begin far enough from the scene so that a car can safely stop before it hits the scene, even if the driver did not notice the flares or cones from a distance.

Many EMS systems recommend the use of reflective cones instead of flares because of possible burns while lighting or using flares, the need to keep lighting new flares as old ones extinguish, the difficulty of keeping flares working in bad weather, and the possibility of toxins from the smoke. Follow local protocol. Remember that flares are an ignition source for spilled fuel or brush. Use caution.

The Patient Compartment

If you assist other EMS providers, you face tremendous hazards every time you climb into the patient compartment. It is normal and necessary to move around in the compartment as you treat the patient. However, too few rescuers remember to protect their own safety. Using appropriate restraints and learning to position yourself can help to prevent injury en route to the hospital.

Whenever you do not need to move around in the patient compartment, wear proper restraints.

Hanging on and Bracing

The principle of hanging on is one borrowed from rock climbers. You have four possible points of contact with the ambulance: two hands and two feet. At any one time, you must maintain at least three-point contact for optimum safety and stability. In other words, you should never have more than one hand or one foot at a time off a stable surface in the patient compartment. Do not forget your fifth point of contact—the seat of your pants. Follow these tips:

◆ To maintain the greatest stability, keep a wide base of support. That is, keep your feet at about shoulder width. Do not place your hands too close together.

◆ If you need to reach for something, keep both feet planted firmly on the floor. Grasp a stable object, such as the overhead bar, with your free hand.

◆ If you need to walk, even a single step, hold on to a stable object with both hands. Slide your hands along as you walk instead of moving arm-over-arm. If you need equipment, try putting it down and sliding it.

◆ Even when you are sitting, hook your feet under the stretcher bar to give yourself increased stability.

Bracing is exerting an opposing force against two parts of the ambulance with your body (Figure 23-8). It provides additional stability and protection. You can use your hands, feet, knees, or any combination of hip, shoulder, knees, and hands. You can exert yourself against any solid surface in the patient compartment, such as the squad bench or an interior wall. For greatest stability and safety, keep your center of gravity low.

Securing the Patient

There are two reasons to secure the patient firmly in the compartment. First, you want to protect any

Figure 23-8 Hanging on and bracing help you to achieve optimum safety and stability inside the patient compartment.

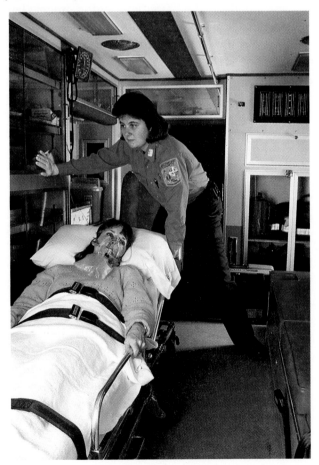

patient from the risk of additional injury. Second, if the patient is hostile, you need to protect yourself and the patient from potential injury.

In the case of nonhostile patients, follow these guidelines:

◆ Secure anyone sitting on the squad bench in a seat belt.

◆ Place a pregnant woman in the captain's chair, since the lateral force from sudden stops is much less pronounced there than on the squad bench. Make sure she is wearing a seat belt.

◆ If the patient is on a stretcher, use snug but comfortable straps across the lower chest (but not over the arms) and just above the knees. Then secure the stretcher to the bar. If local protocol allows it, use additional straps. If the patient's medical condition allows, elevate the head of the stretcher slightly to ease the patient's anxiety over sudden stops.

In the case of hostile patients, follow these guidelines:

◆ If a patient is so hostile as to need to be restrained, make sure law enforcement has been called or is on the scene. Remember, they are the experts in restraining hostile people, not firefighters.

◆ Never apply restraints maliciously. Laws vary from one state to another about how and when restraints can be used. The use of restraints should always be a last-resort measure to protect your safety. If you do use restraints, document why you felt they were necessary in your written report.

◆ Once you restrain a patient, never remove the restraints, even if the patient calms down. The patient should stay restrained until you arrive at the hospital.

◆ Whenever possible, use soft restraints. A quick and easy one is made by folding a gauze bandage in half, slipping it over your hand, and turning it over on itself. You can then loop these over the patient's wrists and ankles. Secure the ends in a bow-tie knot far enough from the patient's hands so they cannot be untied. Combine regular strapping over the chest and knees with soft restraints at the ankles and wrists.

◆ If an extremely violent patient can break gauze bandages or leather straps, place him or her face-down on a stretcher. Put a scoop stretcher over the patient. Then buckle the stretcher belts securely over the scoop. Make sure to tie the patient's arms down. This method allows for natural drainage and gives EMS personnel access for intravenous lines or blood pressure cuffs. Monitor the patient's ABCs with extreme care.

Securing Equipment

If an ambulance is involved in a crash, especially a rollover, any unsecured piece of equipment could become a projectile that can injure you and the patient. As part of the cleanup at the end of each ambulance run, secure all equipment. Stow equipment in appropriate storage areas, and secure doors shut with latches. Clamp masks, oxygen equipment, and other heavy items to appropriate brackets. Use straps to secure portable gear. Devise hooks to keep bench tops closed.

Performing CPR in a Moving Ambulance

While you are performing CPR in an ambulance, you are in a risky position. You cannot maintain three-point contact. You cannot brace yourself adequately, and you cannot secure yourself in a seat belt. To improve your safety while performing CPR in a moving ambulance, follow these guidelines:

◆ Position your feet at least at shoulder width for the best possible base.

◆ If you can, have someone sit on the squad bench, secured by a seat belt, and grasp the back of your belt. If the ambulance suddenly changes direction or accelerates, the person hanging onto you can keep you from catapulting.

◆ Try for as much bracing as you can. Bend your knees into the side of the stretcher, wedge your feet against the squad bench, or brace one knee against the stretcher and the other against the squad bench.

◆ Do not brace yourself with your head. Your neck will not withstand the pressure.

ON SCENE FOLLOW-UP

At the beginning of this chapter, you read that fire service First Responders were responding to the scene of a motor-vehicle collision. To see how chapter skills apply to this emergency, read the following. It describes how the call was completed.

SCENE SIZE-UP (CONTINUED)

We arrived on-scene just minutes after the state police. We positioned the rescue truck in a place that both protected us and gave traffic extra warning. When it was safe, we exited our vehicle, watching our backs for traffic. We sized up the scene for other hazards while we approached the crash vehicles. We saw a driver for each of the two vehicles present. Each of them had been wearing lap and shoulder belts. There were no passengers.

INITIAL ASSESSMENT

Neither of the drivers wanted to be transported to the hospital, but since the EMS transport was already responding and just minutes away, we did not cancel them. There didn't appear to be much damage to the vehicles. It was a minor sideswipe. The mechanism of injury certainly didn't seem severe. However, we knew that once the "rush" of being in a crash wore off and the drivers calmed down, they were more likely to realize that they were injured.

PHYSICAL EXAMINATION

We led the drivers over to a safe place on the side of the road. They allowed us to check their vital signs and examine them for possible injuries. Neither had any complaints. The exams turned up nothing.

PATIENT HISTORY

Both drivers denied any medical problems, and neither of them were taking any medications or had any allergies they were aware of.

ONGOING ASSESSMENT

We stayed with the drivers until the EMTs arrived on scene.

PATIENT HAND-OFF

When the ambulance arrived, we gave the crew the patients' vital signs and a rough description of the incident. Neither of the patients wanted to be transported. Both patients refused emergency care.

Kitty Kearney, the first state trooper on scene, complimented us on the positioning of our vehicle and use of safety cones around the collision scene. She told us it made her job much easier, and she was also able to keep traffic flowing instead of having to reroute it. We thanked her and smiled. We knew how dangerous that intersection could be, both to crash victims and rescuers.

Most basic training programs for rescuers teach how to react safely to a variety of dangers. However, the most common threat to safety is likely to be something as simple as oncoming traffic or the way an emergency vehicle is driven. Be prepared. Have the knowledge, equipment, and skills that allow you to meet the standards set by your EMS system for each phase of an emergency response.

Chapter Review

Remember to be part of the solution, not part of the problem. The time spent with a patient is perhaps the most notable and interesting part of being a fire service First Responder. However, if you are injured during the response, not only is there one more patient to care for, but patient care will suffer as well. Think about it. What would you focus on if your partner was injured on scene? The patient or your partner? It is critical, therefore, to remember to never ignore the basic tasks that make it possible for you to perform your job safely.

FIRE COMPANY REVIEW

Page references where answers may be found or supported are provided at the end of each question.

Section 1

1. What non-medical equipment should all First Responders have on hand? (p. 436)

2. What medical equipment should all First Responders have on hand? (p. 436)

Section 2

3. What are the six basic phases of an EMS response? (p. 437)

4. When should you check, restock, clean, and maintain your supplies? (p. 437)

Section 3

5. Why is it recommended that First Responders travel in pairs? (p. 439)

6. Why should emergency lights be used during a daytime response? (p. 440)

7. What are some reasons why it is a good idea to turn off your lights and siren when you approach an emergency scene? (p. 440)

8. What way to exit an emergency vehicle would give you the greatest protection? (pp. 441–442)

9. What are the basic guidelines for setting up flares at the scene of a motor-vehicle collision? (p. 442)

10. What are three goals of channeling traffic away from a scene? (pp. 442–443)

RESOURCES TO LEARN MORE

Peto, G., and W. Medve. *EMS Driving: The Safe Way*. Upper Saddle River, NJ: Brady/Prentice-Hall, 1992.

Wieder, M.A., ed. *Fire Department Pumping Apparatus Driver/Operator*, Eighth Edition. Stillwater, OK: International Fire Service Training Association, 1999.

PHASES OF AN EMERGENCY RESPONSE

PREPARATION
- Check equipment.
- Check vehicle.

DISPATCH
- Nature of call.
- Name, location, and call-back number of caller.
- Location of patient.
- Number of patients and severity of problem.
- Any pertinent special problems or considerations.

EN ROUTE TO THE SCENE
- Travel in a timely way.
- Travel in a safe way.

ARRIVAL ON SCENE
- Notify dispatch of arrival.
- Approach the scene cautiously.
- Park vehicle safely.
- Perform scene size-up.
- Request additional resources if necessary.
- Provide for patient assessment and care.
- Prepare patient for transport.

TRANSFER OF CARE
- Provide a concise and accurate patient report.
- Assist in packaging, or lifting and moving, patient.

POST-RUN ACTIVITIES
- Report to dispatch upon return to station.
- Clean and disinfect equipment.
- Replace disposable supplies.
- Change soiled clothing.
- Fuel emergency vehicle.
- Notify dispatch when ready for another call.

CHAPTER 24

Hazardous Materials

I **NTRODUCTION** *This chapter provides a brief overview of hazardous materials emergencies. Did you know that over 50 billion tons of hazardous materials are made in North America every year? To manage the risk to the public, our government has developed specific regulations that address nearly every aspect of the manufacturing, distribution, transportation, and use of such*

materials. Unfortunately, hazardous materials still may be spilled or released as a result of equipment failure, vehicle collisions, environmental conditions, and human error. The result can be the loss of life and property.

OBJECTIVES

Cognitive, affective, and psychomotor objectives are from the U.S. DOT's "First Responder: National Standard Curriculum." Enrichment objectives, if any, identify supplemental material.

Cognitive

7-1.6 Describe what the First Responder should do if there is reason to believe that there is a hazard at the scene. (p. 453)

7-1.7 State the role the First Responder should perform until appropriately trained personnel arrive at the scene of a hazardous materials situation. (pp. 453–458)

Affective

No objectives are identified.

Psychomotor

No objectives are identified.

Enrichment

◆ Explain what hazardous materials are. (p. 450)

◆ Identify the training required to respond to a hazardous materials emergency. (p. 453)

◆ Discuss how to recognize the presence of a hazardous material at the scene of an emergency. (pp. 450–453)

ON SCENE

DISPATCH

I am a production control clerk by trade. I work in aviation maintenance at a military installation. The hangar where I do my job has an emergency response team (ERT) trained to deal with fires, medical emergencies, and hazardous material spills that can occur at our hangar. Since I have always been a "fire buff," I decided to join the ERT. Although our training is kept basic in all areas, we are capable of handling just about any emergency that arises. When we are faced with an emergency beyond our capabilities, we call the fire department, which has more advanced training than our team.

I had just walked into my office one morning when my beeper went off, sounding the alarm for the ERT. "ERT, respond to the engine shop. Investigate a report of fumes and people with difficulty breathing. Time out 07:03."

SCENE SIZE-UP

I donned my bunker gear and proceeded to the area. As I approached the scene, I slowed my pace. I could see several people streaming out of the loading dock area. A few were bent over, trying to catch their breaths. Sam Hoffman, the manager of the engine shop, ran over to me. He told me that the cleaning people were working in the break room, adjacent to where the employees had been working. One of them mixed bleach, ammonia, and the "green stuff" together, which caused the problem.

> *Consider this situation as you read Chapter 24. What may be done to ensure scene safety? Are all those people injured? How can they be treated appropriately?*

IDENTIFYING HAZARDOUS MATERIALS

What Is a Hazardous Material?

Hazardous materials (dangerous goods), or **hazmats,** are those that in any quantity pose a threat or unreasonable risk to life, health, or property if not properly controlled. Hazmats include chemicals, wastes, and other dangerous products. The principal dangers they present are toxicity, flammability, and reactivity. Common hazmats are:

◆ Explosives.

◆ Compressed and poisonous gases.

◆ Flammable solids and liquids.

◆ Oxidizers.

◆ Corrosives.

◆ Radioactive materials.

◆ Poisons.

Hazardous materials often are transported directly to the user. While some travel by fixed pipeline (natural gas, for example), most go by rail or highway. That means hazmats can be the cause of an emergency anywhere.

Think about your community for example. Is there a farm or business that might have a hazardous material on the premises? Does a hospital in your area practice nuclear medicine? What chemicals does the local photo shop use to develop film? What about a lawn and garden company, discount department store, or home repair center? Do they stock fertilizers, insecticides, or pesticides? Do your grocery stores refrigerate produce with freezers cooled by ammonia? The local emergency planning committee should have an inventory of hazardous materials in your community.

Placards and Shipping Papers

Placards and labels are required on the outside of vehicles, packages, and containers that carry hazardous materials. (See Figures 24-1 and 24-2.) The driver of the vehicle also must have shipping papers, which identify the exact substance, quantity, origin, and destination.

A placard is usually a four-sided, diamond-shaped sign. Many are red or orange. A few are white or green. Whatever the color, the placard contains a four-digit number and a legend that identifies the material as flammable, radioactive, explosive, or poisonous.

Shipping papers, also known as "manifests" or "waybills," are another important means of identifying hazardous materials. If you can locate them, shipping papers have the name of the substance, the danger it presents, and a four-digit identification number.

> **COMPANY OFFICER'S NOTE**
>
> Never endanger yourself or others in order to retrieve shipping papers. The risk does not outweigh the benefit.

The National Fire Protection Association (NFPA) 704 system is used on fixed structures such as buildings. Its diamond-shaped symbol identifies danger with the use of colors and numbers. The color blue indicates a health hazard. Red indicates a fire hazard, and yellow means a reactivity hazard. White is used for information such as the need for protective equipment. The numbers used in this system are 0 to 4. For example, 1 in a blue diamond and 4 in a red diamond mean the material presents a low health hazard but is very flammable. (See Figure 24-3.)

Available Resources

The *North American Emergency Response Guidebook* is a special resource (Figure 24-4). It is distributed locally through emergency management agencies. It also is available from the Government Printing Office. Carry it in your vehicle at all times. It includes a table of commonly used DOT labels and placards, which is correlated with lists of chemicals. Each item in a list is keyed to specific emergency action instructions.

Material safety data sheets (MSDS) offer another resource. All employees working with hazardous materials have a right to know about the dangers of those materials. So, all manufacturers are required to provide MSDS on hazardous materials. These sheets generally name the substance, its physical properties, and fire, explosion, and health hazards. Emergency first aid also is listed.

The Chemical Transportation Emergency Center (CHEMTREC) is a toll-free 24-hour emergency

Hazardous Materials Warning Labels

DOMESTIC LABELING

General Guidelines on Use of Labels
(CFR, Title 49, Transportation, Parts 100-177)

- Labels illustrated above are normally for *domestic shipments*. However, some air carriers *may* require the use of International Civil Aviation Organization (ICAO) labels.
- Domestic Warning Labels *may* display UN Class Number, Division Number (and Compatibility Group for Explosives only) [Sec. 172.407(g)].
- Any person who offers a hazardous material for transportation MUST label the package, if required [Sec. 172.400(a)].
- The Hazardous Materials Tables, Sec. 172.101 and 172.102, identify the proper label(s) for the hazardous materials listed.

- Label(s), when required, must be printed on or affixed to the surface of the package near the proper shipping name [Sec. 172.406(a)].
- When two or more different labels are required, display them next to each other [Sec. 172.406(c)].
- Labels may be affixed to packages (even when not required by regulations) provided each label represents a hazard of the material in the package [Sec. 172.401].

**Check the Appropriate Regulations
Domestic or International Shipment**

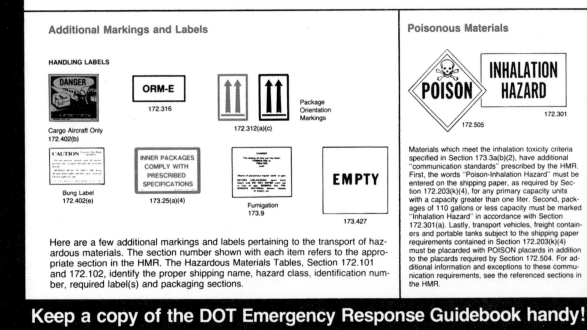

Additional Markings and Labels

HANDLING LABELS

Cargo Aircraft Only
172.402(b)

ORM-E
172.316

Package Orientation Markings
172.312(a)(c)

Bung Label
172.402(e)

INNER PACKAGES COMPLY WITH PRESCRIBED SPECIFICATIONS
173.25(a)(4)

Fumigation
173.9

EMPTY
173.427

Here are a few additional markings and labels pertaining to the transport of hazardous materials. The section number shown with each item refers to the appropriate section in the HMR. The Hazardous Materials Tables, Section 172.101 and 172.102, identify the proper shipping name, hazard class, identification number, required label(s) and packaging sections.

Poisonous Materials

POISON
172.505

INHALATION HAZARD
172.301

Materials which meet the inhalation toxicity criteria specified in Section 173.3a(b)(2), have additional "communication standards" prescribed by the HMR. First, the words "Poison-Inhalation Hazard" must be entered on the shipping paper, as required by Section 172.203(k)(4), for any primary capacity units with a capacity greater than one liter. Second, packages of 110 gallons or less capacity must be marked "Inhalation Hazard" in accordance with Section 172.301(a). Lastly, transport vehicles, freight containers and portable tanks subject to the shipping paper requirements contained in Section 172.203(k)(4) must be placarded with POISON placards in addition to the placards required by Section 172.504. For additional information and exceptions to these communication requirements, see the referenced sections in the HMR.

Keep a copy of the DOT Emergency Response Guidebook handy!

Figure 24-1 DOT requires packages and storage containers to be marked with specific hazard labels.

Hazardous Materials Warning Placards

DOMESTIC PLACARDING

Illustration numbers in each square refer to Tables 1 and 2 below.

1 EXPLOSIVES A	2 EXPLOSIVES B	3 BLASTING AGENTS	4 POISON GAS	5 FLAMMABLE GAS	6 NON-FLAMMABLE GAS	7 CHLORINE
8 OXYGEN	9 FLAMMABLE	10 COMBUSTIBLE	11 FLAMMABLE SOLID	12 FLAMMABLE SOLID	13 OXIDIZER	14 ORGANIC PEROXIDE
15 POISON	16 RADIOACTIVE	17 CORROSIVE	18 DANGEROUS			

WHITE SQUARE BACKGROUND FOR PLACARD
HIGHWAY
• Used for "HIGHWAY ROUTE CONTROLLED QUANTITY RADIOACTIVE MATERIALS." (Sec. 172.507)
RAIL
• Used for RAIL SHIPMENTS-"EXPLOSIVE A," "POISON GAS" and "POISON GAS-RESIDUE" placards. (Sec. 172.510(a))

Figure 24-2 DOT also requires display placards to be put on the outside of vehicles carrying hazardous materials.

phone service provided by chemical manufacturers. They advise rescuers on the nature of a product and the steps that need to be taken to manage an incident. They also may contact the shipper who will provide detailed information and field assistance. In the United States and Canada, the toll-free CHEMTREC number is 800-424-9300. For collect calls and calls from other points of origin, the number is 703-527-3887. CHEMTREC also can refer you to the proper authorities for incidents involving radioactive materials.

CHEMTEL, Inc., is an emergency response communications center. In the United States and Canada, its toll-free 24-hour phone number is 800-255-3924. For collect calls and calls from other points of origin, the number is 813-979-0626. CHEMTEL also can refer you to the proper authorities for incidents involving radioactive materials.

Another important resource is the regional poison control center and your own system's medical direction. They can provide guidance

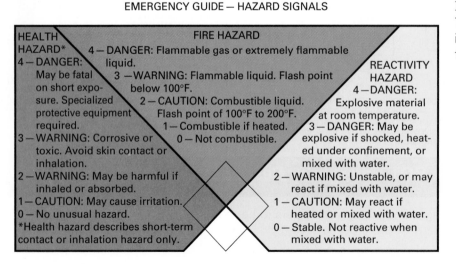

EMERGENCY GUIDE — HAZARD SIGNALS

HEALTH HAZARD*
4 — DANGER: May be fatal on short exposure. Specialized protective equipment required.
3 — WARNING: Corrosive or toxic. Avoid skin contact or inhalation.
2 — WARNING: May be harmful if inhaled or absorbed.
1 — CAUTION: May cause irritation.
0 — No unusual hazard.
*Health hazard describes short-term contact or inhalation hazard only.

FIRE HAZARD
4 — DANGER: Flammable gas or extremely flammable liquid.
3 — WARNING: Flammable liquid. Flash point below 100°F.
2 — CAUTION: Combustible liquid. Flash point of 100°F to 200°F.
1 — Combustible if heated.
0 — Not combustible.

REACTIVITY HAZARD
4 — DANGER: Explosive material at room temperature.
3 — DANGER: May be explosive if shocked, heated under confinement, or mixed with water.
2 — WARNING: Unstable, or may react if mixed with water.
1 — CAUTION: May react if heated or mixed with water.
0 — Stable. Not reactive when mixed with water.

Figure 24-3 The NFPA 704 System helps you to identify health, reactivity, and fire hazards.

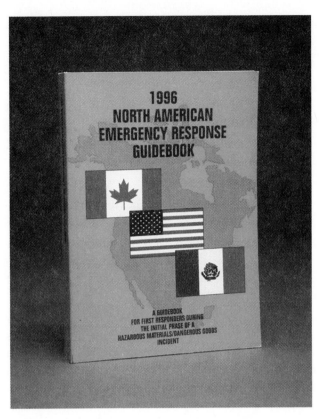

Figure 24-4 *North American Emergency Response Guidebook.*

for decontamination and treatment of patients affected by hazardous materials.

Training Required by Law

People can get hurt or lose their lives at hazmat emergencies. In fact, the Occupational Safety and Health Administration (OSHA) estimates that as many as 1.8 million workers may now be at risk for exposure to hazmats. Of these workers, approximately 84% are emergency responders.

The OSHA and the Environmental Protection Agency (EPA) have developed safety regulations. See the OSHA publication "29 CFR 1910-120—Hazardous Waste Operations and Emergency Response Standards (1989)." Four levels of training are identified:

◆ *First Responder Awareness.* This level of training is for those who are likely to witness or discover a hazmat emergency. They learn how to recognize a problem and how to call for the proper resources. No minimum training hours are required.

◆ *First Responder Operations.* This level of training is for those who initially respond to a haz-

mat emergency in order to protect people, property, and the environment. They learn how to keep at a safe distance and how to stop the emergency from spreading. A minimum of 8 hours of additional training are required.

◆ *Hazardous Materials Technician.* This level is for rescuers who actually plug, patch, or stop the release of a hazardous material. A minimum of 24 hours of training is required.

◆ *Hazardous Materials Specialist.* This level is for rescuers who want advanced knowledge and skills. They learn to provide command and support activities at the site of a hazmat emergency. A minimum of 24 hours of additional training beyond technician level is required.

The training levels outlined by OSHA have a fire service focus. To supplement them, the National Fire Protection Association has published two standards that deal with competencies for EMS personnel at hazmat incidents. They are NFPA 472 and NFPA 473.

SECTION **2**

GUIDELINES FOR A HAZARDOUS MATERIALS RESPONSE

Many hazmat incidents are dispatched as traffic accidents, poisonings, or unknown problem calls. Your initial actions build the crucial groundwork for the remainder of the incident. Your specific responsibilities include:

◆ Identifying the emergency as a hazmat incident.

◆ Identifying the hazardous materials, if you can do so safely.

◆ Establishing command and create control zones, which includes isolating the area and denying entry.

◆ Establishing a medical treatment sector.

As always, your first priority is your own safety. Never attempt a hazardous materials rescue unless you are properly trained and equipped. If you have no training, radio immediately for help. While waiting, protect yourself and bystanders by keeping away from the danger. Avoid contact with any unidentified material, regardless of the level of protection offered by your clothing and equipment.

COMPANY OFFICER'S NOTE

Remember, each and every fire that you respond to—whether it is a structure, wildland, or rubbish fire—is a hazardous materials incident.

Identifying the Hazmat Incident

Always consider the possibility of a hazmat incident. For example, if you are called to a car crash, ask yourself: Could hazardous materials be involved? Will gasoline or diesel fuel be leaking? What about radiator coolant or antifreeze? Is propane used to power this vehicle? If the call is for an unknown emergency, ask yourself: Is the call to the location of a previous hazmat incident? Is there an emergency plan for the location, which generally indicates that a risk exists?

Use all the pre-arrival information available to you to decide on the best course of action. Remember, most hazmat calls do not occur at a fixed location. They most often involve the transportation of a hazardous material.

As you approach the scene, use binoculars to begin your size up. Identify any placards on vehicles, buildings, or containers (Figure 24-5). Proceed as always with caution. Too many fire service First Responders discover a hazmat incident only after they are in the middle of it. Use all of your senses, coupled with a high index of sus-

Figure 24-5 Hazmat placard displayed on a truck.

picion. A number of visual clues can indicate a possible hazardous material:

◆ Smoking or self-igniting materials.

◆ Extraordinary fire conditions.

◆ Boiling or spattering of materials that have not been heated.

◆ Wavy or unusual vapors over a container of liquid material.

◆ Colored vapor clouds.

◆ Frost near a container leak (may indicate a liquid coolant).

◆ Unusual condition of containers (peeling or discoloration of finishes, unexpected deterioration, deformity, or unexpected operation of pressure-relief valves).

Remember you may not be able to see or smell a hazardous material. Some are odorless and colorless. Others have properties that can deaden your senses. Always assume that the area surrounding a spill or leak is dangerous.

Identifying the Hazardous Material

After you identify an emergency as a hazmat incident, station yourself uphill and upwind of the scene. This vantage point should prevent the vapors from overwhelming you. However, keep in mind that a safe area can become unsafe quickly. Once stationed, report your position and the situation to dispatch. Your report should include:

◆ Nature and exact location of the incident.

◆ Description of the incident, including any potential for fire or explosion.

◆ Number of patients involved.

◆ Request for additional help, such as fire, police, EMS, and hazmat support.

Also suggest the best way other emergency responders can approach the scene. Include instructions for a staging area (the safe area where all responders should check in and get orders). If possible, identify the hazardous materials and the severity of the situation.

◆ What is the material, its properties, and dangers? Look for placards, NFPA 704 numbers, or shipping papers. Then refer to your *Emergency Response Guidebook* for the name, properties, and dangers of the material.

◆ What are the sizes, shapes, kinds, and conditions of the containers?

◆ Is there imminent danger of the contamination spreading?

Although the hazmat team will be able to identify an unknown substance, you will be expected to make an initial identification.

Establishing Command

Your fire department should have a plan ready in case of a hazmat incident. Before such an emergency ever develops, all appropriate agencies need to know how forces will be mobilized to handle it. Generally, the plan addresses the worst possible scenario. That way, the community will be able to handle any emergency that arises. The following should be included:

◆ *One command officer,* who is responsible for all rescue decisions at every stage of the incident. All rescuers should be aware of who the command officer is. If the command officer hands over the decision-making power to someone else, all rescuers must be notified of the change.

◆ *Clear chain of command,* from each rescuer to the command officer.

◆ *Established system of communications,* which is used throughout the emergency. It should be one all rescuers are informed about, know how to use, and have access to.

◆ *Receiving facilities.* Choose the ones that are capable of handling large numbers of patients, have surgical capacity, and if possible have established decontamination procedures.

As the fire service First Responder on scene, activate your EMS system's plan and establish command. Stay in command until you are relieved by someone higher in the chain of command. The incoming incident commander will want to know the following:

◆ Nature of the problem.
◆ Identification of the hazardous materials.
◆ Kind and condition of the containers.
◆ Existing weather conditions.
◆ Whether or not there is the presence of fire.
◆ Time elapsed since the emergency occurred.
◆ What has been done by people on scene.
◆ Number of victims.
◆ Danger of victimizing more people.

Once transfer of command has occurred, be prepared to care for decontaminated patients or to support rescue personnel as directed.

Creating Control Zones

To prevent a hazmat incident from becoming worse, the danger area must be identified and isolated. That is usually done by designating three zones (Figure 24-6):

◆ *Hot (red) zone.* This is the most dangerous area. It can be entered only with the correct personal protective equipment. Initial or gross decontamination will be performed here.

◆ *Warm (yellow) zone.* This is the area immediately outside the hot zone. The proper protective equipment must be worn here (Figure 24-7). Once the immediate life-threats of the

Figure 24-6 Establish control zones.

Hot (Contamination) Zone
Contamination is actually present.
Personnel must wear appropriate protective gear.
Number of rescuers limited to those absolutely necessary.
Bystanders never allowed.

Warm (Control) Zone
Area surrounding the contamination zone.
Vital to preventing spread of contamination.
Personnel must wear appropriate protective gear.
Life-saving emergency care is performed.

Cold (Safe) Zone
Normal triage, stabilization, and treatment are performed.
Rescuers must shed contaminated gear before entering the cold zone.

Hazmat Team's Personal Protective Equipment

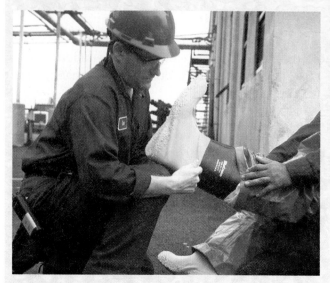

Figure 24-7a Putting on double boots.

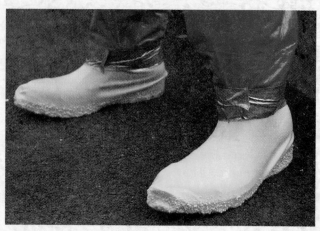

Figure 24-7b Boots taped and sealed.

Figure 24-7c Preparing the breathing tank.

Figure 24-7d Putting on a face mask.

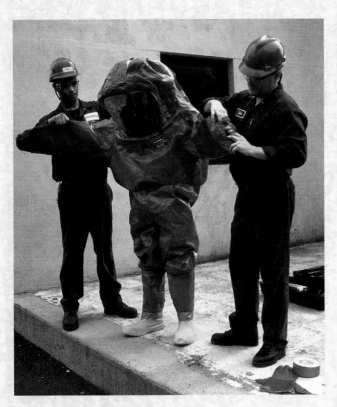

Figure 24-7e Attaching a hood.

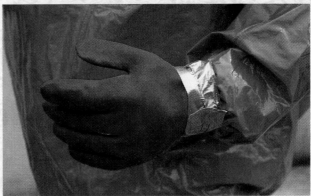

Figure 24-7f Taped and sealed gloves.

Figure 24-7g Fully suited hazmat team.

Figure 24-8 Rescuers in decontamination process.

patient are managed, complete decontamination of the patient is performed (Figure 24–8). This is also called the "decontamination corridor."

◆ *Cold (green) zone.* This is the outer perimeter. All contaminated clothing and equipment must be removed before entering.

Do not enter the warm (yellow) zone unless you are trained and equipped to do so. Instead, patients should be brought to you for emergency medical treatment in the cold (green) zone. Note that all people not necessary to the rescue should be kept away from the zoned areas.

Establishing a Medical Treatment Sector

All EMS personnel and equipment must be staged in the cold (green) zone. The establishment of a definable perimeter cannot be overstressed. In even relatively small incidents, victims may scatter and spread the contamination with them, resulting in an ever-increasing scene that quickly becomes unmanageable. Note that anyone exiting the hot (red) zone should be considered contaminated until proven otherwise.

If the hazardous materials can be identified, follow the treatment instructions given in the *North American Emergency Response Guidebook* or by the regional poison control center.

ON SCENE FOLLOW-UP

At the beginning of this chapter, you read that a member of a chemical plant's emergency response team was on the scene of a possible hazmat emergency. To see how chapter skills apply, read the following. It describes how the call was completed.

SCENE SIZE-UP (CONTINUED)

I radioed communications immediately: "ERT, we have a possible hazardous materials spill in the break room near the engine shop. Initiate the chemical incident plan." I went on to report, "I will be incident command. Please have emergency services respond to my location at the hangar door on the east side of the building. Advise them that we have chemical fumes from a mixture of bleach, ammonia, and a green-soap solution." Scanning the area quickly, I added, "We have approximately 12 people with trouble

breathing. Medical assistance is needed immediately."

At this point, I realized I had a large number of people who may leave the scene, so I asked the military police to do several things.

First, we had to get all of the injured people to go down the hall until help could arrive. Someone suggested that we ask them to wait outside, but I disagreed. I thought that if we let them outside, they would go to their cars and leave. Plus, that was where emergency services were staging. What with people leaving and EMS responding, well, I was afraid things would get out of hand.

Second, I asked the other military police to set up a perimeter at least 100 yards down the hall. I was amazed at the number of people who arrived, curious as to what was going on. Some even thought we were conducting a drill.

We were just getting things set up when the local fire department arrived. I quickly explained what happened and what had been done. The fire chief declared a hazardous materials incident and requested a hazmat team. After consulting with poison control, our ERT First Responders assisted the fire department paramedics in treating the patients—mainly with high-flow oxygen—and moving them to a triage point.

Several weeks after the incident I was commended by my employer for quick thinking and decisive action at the scene of an emergency. I guess he didn't realize how frightened I had been. I was glad I was able to put it all together.

> *A hazardous materials incident challenges the best in First Responders. Besides the usual patient-care issues, there is the additional safety problem. Be sure to follow all of your local protocols for such emergencies.*

Chapter Review

Understand that hazardous materials are everywhere. They can be found in more places than in tractor trailers with signs that warn of highly corrosive materials. Remember that even household cleaners can cause as deadly a reaction as any you can find on the highway.

The first way to stay safe from hazardous materials is to realize that you will encounter them. Sooner or later, it will happen. Be prepared.

FIRE COMPANY REVIEW

Page references where answers may be found or supported are provided at the end of each question.

Section 1

1. What is a hazardous material? (p. 450)
2. To what do the terms "placards" and "shipping papers" refer? (p. 450)
3. What is the NFPA 704 System? (p. 450)
4. What are some of the resources available to responders on the scene of a hazmat incident? (pp. 450, 452–453)
5. What are the four levels of hazmat training described by OSHA? (p. 453)

Section 2

6. What is a First Responder's first priority upon arrival at the scene of a hazmat incident? (p. 453)
7. What are some examples of visual clues that indicate a possible hazardous material? (p. 454)
8. After identifying an emergency as a hazardous materials incident, what information should your report to EMS dispatch include? (pp. 454–455)
9. What are "control zones"? Briefly describe them. (pp. 455, 458)

RESOURCES TO LEARN MORE

Hildebrand, M., G. Nool, and J. Yvorra. *Hazardous Materials: The Incident*. Stillwater, OK: Fire Protection Publications, 1995.

Wieder, M.A., ed. *Awareness Level Training for Hazardous Materials*. Stillwater, OK: Fire Protection Publications, 1995.

Wieder, M.A., ed. *Hazardous Materials for First Responders,* Second Edition. Stillwater, OK: International Fire Service Training Association, 1994.

FIRST RESPONDER RESPONSIBILITIES AT A HAZMAT INCIDENT

Identify the emergency as a hazmat incident.

Identify the hazardous materials, if you can do so safely.

Establish command.

Create control zones.

Establish a medical treatment sector.

Fireground Rehabilitation

*I*NTRODUCTION   *Firefighters die of stress- and heat-related illnesses far more often than they die of burns or fire-related injuries. Although fire fighting is a hazardous, physically demanding job, appropriate emergency scene rehabilitation, or "rehab," can reduce the risks.*

As a firefighter, you must be aware of your physical and mental limitations while on duty. Similarly, the physical and emotional well-being of your fellow firefighters is crucial to your performance and safety.

Cognitive, affective, and psychomotor objectives are from the U.S. DOT's "First Responder: National Standard Curriculum." Enrichment objectives, if any, identify supplemental material.

Cognitive

No objectives are identified.

Affective

No objectives are identified.

Psychomotor

No objectives are identified.

Enrichment

◆ Identify the types of heat-related emergencies relevant to firefighters. (pp. 464–465)

◆ Understand the principles of emergency incident rehabilitation. (pp. 467–471)

◆ Describe the role of the First Responder in the evaluation and care of firefighters in the rehabilitation area. (pp. 465–466)

◆ Describe the set up and components of a rehabilitation area. (pp. 466–467)

ON SCENE

DISPATCH

I was standing outside my house one Saturday morning, talking with my neighbor about how often we had been mowing our lawns with all the rain lately, when my fire department pager went off. "Box 24, structure assignment, 234 Tacoma Street. E2, E1 respond. This is a house on fire. Caller states that an adjoining structure also may be on fire." I quickly went inside my house, told my wife I had to go, and hurried to the station.

Putting on my turnout coat and pants at the station, I remarked to Malik, another firefighter, how we were really going to sweat today. With the temperature already in the upper 80s, and with humidity to match, I knew we were going to have to be really careful in order not to suffer some type of heat illness.

As we neared the location of the fire, we received orders to report to the command post.

SCENE SIZE-UP

As I approached the incident commander, one of his aides asked if we would set up the rehab sector.

We chose a location close enough to the incident to avoid causing the firefighters an undue amount of walking to reach us, yet upwind and a safe distance from the fire. We deployed our tarps and other supplies stored on our special rehab van. Then I observed the local Red Cross chapter's van arriving on scene. I waved them over to our location. They usually supplement our rehab efforts at large or prolonged incidents with plenty of bottled water and electrolyte replacement solutions.

Meanwhile, Malik was busy setting up our electrically powered exhaust fans to cool off firefighters as they entered rehab. He had already procured a neighbor's garden hose to provide cool showers over the heads of the personnel. At that moment, two EMT-Basic providers from a nearby department arrived and asked for an assignment. I assigned both of them to the evaluation and treatment process. Malik handled refreshments.

Satisfied that our sector was now operational, I contacted the command post and let them know we were ready. Five minutes later, a sweaty and red-faced firefighter entered our sector, and said he was feeling "dizzy."

Rehab of tired firefighters is likely to be an extremely important function, especially on a hot and humid day. Consider this emergency as you read Chapter 25. How would you proceed?

EMERGENCY INCIDENT REHABILITATION

On-scene **emergency incident rehabilitation (EIR)** is fairly new to fire service operations. The fact that stress- and heat-related emergencies are the main causes of on-duty firefighter deaths has made it an important part of emergency operations.

All fire departments should have formal standard operating procedures/guidelines (SOP/Gs) for EIR. They help to assure the well-being of personnel who are fighting a fire for a long stretch of time. At the scene of an emergency, the incident commander can put the SOP/Gs into action as soon as it becomes clear that they are needed.

When on duty at an emergency scene, be aware of your limitations, both physical and psychological. In fact, you also may be assigned to the rehabilitation sector/group to evaluate and monitor other firefighters. There you may be asked to recognize signs of stress and heat-related illnesses (Figure 25-1).

Remember that an exhausted firefighter risks not only his or her own life, but the lives of others as well. If you feel you are at your limit, or see others at theirs, your whole crew needs to report to the rehab sector/group for rest and evaluation.

Figure 25-1 Be alert for stress- and heat-related emergencies in yourself and other crew members. *(Source: Mark C. Ide)*

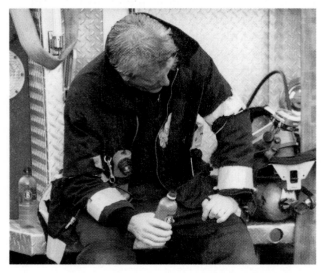

COMPANY OFFICER'S NOTE

Firefighters often are action-oriented people. They do not easily admit to any weaknesses, like being too tired or even injured. You may have to insist, even order, your personnel to the rehab sector to keep them safe and healthy.

Stress- and Heat-Related Emergencies

A heat-related emergency is a common danger for firefighters. The physical stress of fire fighting can result in cardiac emergencies and sudden death. Because the goal of EIR is to prevent or treat stress- and heat-related emergencies, a review of concepts important to remember at an emergency scene follows. (See Chapters 6 and 10 for cardiac emergencies. See Chapter 12 for a full discussion of heat-related emergencies.)

Mechanisms of Heat Loss

The body generates increased heat during periods of increased physical activity. Under normal conditions, the body rids itself of excess heat, mostly by way of convection and evaporation. However, both methods have their limitations. In convection, air moves heat away from the body. But as air temperature nears body temperature, cooling by convection is less effective. In evaporation, sweat helps to cool the body. But for this method to be effective, the air has to be low in humidity. As humidity increases, heat loss through evaporation decreases, too.

Heat-related emergencies are common in fire fighting for several reasons:

◆ *Fire.* Fire emits heat. In structural or wildland fires, heavy fire loads may produce extreme heat (Figure 25-2).

◆ *Activity.* The intense physical activity of fire fighting causes a firefighter's body to generate a great deal of heat.

◆ *Protective clothing.* Essential personal protective equipment impairs heat loss (Figure 25-3).

A combination of fire, activity, and protective clothing places firefighters at serious risk for heat-related emergencies.

The three most important heat-related emergencies that a firefighter risks are heat cramps,

Figure 25-2 Heat-related emergencies occur among crews fighting structural fires. *(Source: Mark C. Ide)*

heat exhaustion, and heat stroke. Of these conditions, heat stroke is the most serious. Failure to recognize and treat it may result in the firefighter's death.

Heat Cramps

Heat cramps usually develop during strenuous activity in a hot environment. Excessive sweating results in loss of electrolytes (especially sodium),

Figure 25-3 Protective clothing, though essential, can increase your risk of a heat-related emergency. *(Source: Howard M. Paul, Emergency! Stock)*

which contributes to muscle cramping. Heat cramps are usually not a serious problem. They respond well to rest, cooling, and the drinking of fluids. If a person with heat cramps is left untreated, heat exhaustion may develop.

Heat Exhaustion

Signs and symptoms of heat exhaustion vary with the amount of fluid loss. Early signs may include fatigue, light-headedness, nausea, vomiting, and headache. The firefighter's skin is usually pale and moist to the touch, with cool or normal temperature. If the condition is left untreated, signs and symptoms include increased heart and respiratory rate and, eventually, reduced blood pressure.

Without rehab, firefighters who engage in structural and wildland fire suppression often suffer from heat exhaustion. It is also common in hazmat operations where firefighters wear encapsulating suits.

Heat Stroke

Heat stroke occurs when the body can no longer regulate its temperature. It is a life-threat. Assume it in any rescuer who is working in a hot environment and has an altered mental status with hot skin that is dry or moist. Emergency care includes aggressive cooling.

SECTION 2

THE REHAB SECTOR/GROUP

Most EIR tasks are carried out in the rehab sector/group (Figure 25-4). Fire service First Responders usually play key roles in the staffing of this sector.

Role of the First Responder

Assigned to the rehab sector/group to perform a number of functions, First Responders:

◆ Provide a safe area in which fire and rescue crews can rest and receive **rehydration,** or replacement of water and electrolytes lost in sweating. During longer incidents, food is also to be supplied to crews.

◆ Identify firefighters and other rescuers entering rehab who are at risk for stress- and heat-related illness.

Figure 25-4 Rehab sector/groups should be established at all large, multi-jurisdictional or multiple-alarm fires.

♦ Medically monitor crews and determine whether they:

   ♦ Are fit to return to active fire/rescue duty.

   ♦ Require additional hydration and rest.

   ♦ Require transport to an emergency department for further evaluation and treatment.

♦ Assure accountability for firefighters and rescue personnel who enter and exit rehab.

♦ Update the status of rehab through timely reports to the safety officer or incident commander.

The materials and number of First Responders needed to perform these functions will be determined by the size of the incident, weather, and the number of personnel available.

Staffing

The rehab sector/group should have enough personnel to provide the appropriate medical care and to make sure there is sufficient and appropriate food and fluids.

Although First Responders may be assigned to the rehab sector/group, it is widely accepted that at a minimum, the area should be staffed by EMT-Basics. Advanced life support (ALS) units should be requested in large-scale incidents and when a trend toward serious stress- and heat-related illness is detected among firefighters. Often, outside EMS agencies provide the medical care. If so, it is essential that they have direct radio communication with incident command.

Personnel with no medical training can help provide food and fluids. In many volunteer fire departments, members of an "auxiliary" or "explorer post" fill this function. Volunteer relief agencies such as the American Red Cross also may offer food. When involved in EIR, they must follow the directions of incident command or the rehab sector/group supervisor.

Design and Location

The EIR area should be set up in a location that truly allows rest for the firefighters and other rescue personnel (Figure 25-5). A good rehab site:

♦ Is outside and upwind of the hot zone. This allows personnel to remove turnout gear and SCBAs safely.

♦ Permits prompt re-entry into emergency operations.

♦ Protects rescuers from environmental extremes. Locations such as a shady cool area in hot weather or a warm dry area during cold weather are preferred.

♦ Accommodates all those who may need rehab. Remember, when laying out the site that crews will need room to sit or lie down.

♦ Is free of vehicle exhaust.

♦ Is not immediately accessible to the media.

♦ Provides access to SCBA refill.

♦ Is close to an ambulance staging area in case transport is needed.

♦ Has a supply of running, drinkable water. This also allows rehab personnel to set up a cooling water spray in hot conditions.

♦ Is out of view of the work area, especially if the incident involves the recovery of fatalities.

The incident commander or the rehab sector/group supervisor usually selects a site. He or she must always take into account the possibility that the emergency scene may expand, such as with an uncontained fire. Making a good site selection is vital. Relocating during an emergency creates delays and confusion. Once the site is selected, its location should be made known to all incident personnel.

In large-scale incidents, more than one rehab sector/group may be needed. Rehab personnel at each site are to keep accurate logs of the entry and exit of crews.

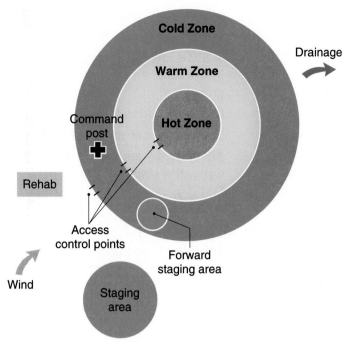

Figure 25-5 Location of the rehab sector/group relative to the hazard area.

Local fire department SOP/Gs describe a general physical layout for rehab sector/groups (Figure 25-6). The layout is to have a single entry/exit point, adjacent areas for logging crews in and out, and areas for performing medical evaluations. Most of the space within the rehab sector/group is devoted to two large areas: "rest and refreshment" and "medical evaluation/treatment." Large tarps of different colors may be used to distinguish the two.

Criteria for Entry

Local SOP/Gs detailing when crews must enter rehab may vary. However, the following guidelines are among the most commonly accepted.

First, whenever a firefighter or other rescuer feels that he has reached his limit, he should report to his immediate supervisor, who will notify incident command or his immediate supervisor. A fatigued firefighter places himself and other firefighters in that crew at risk. For this

Figure 25-6 Layout of a typical rehab sector/group.

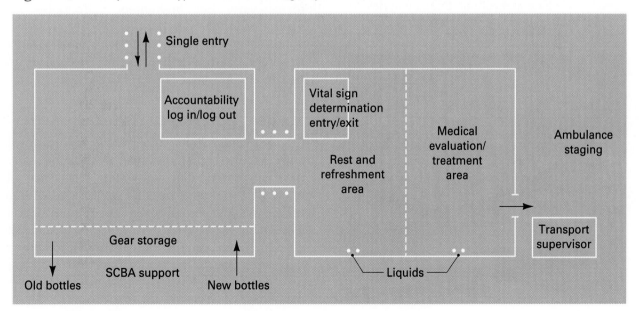

reason, the entire crew should report to rehab when one crew member becomes fatigued. This "whole crew to rehab" approach is a sound one. It assures good tracking of personnel on scene. It also recognizes that if one member of a crew is at his limits of endurance, then others in that crew may be too.

Another common indicator for mandatory rehab is the "two-cylinder rule." Under this standard, firefighters must report to the rehab sector/group after going through two 30-minute-capacity SCBA air cylinders.

Finally, some departments use a time limit. A common SOP/G says that after 45 minutes of active fire suppression duty, rehabilitation is mandatory.

Accountability

In this context, "accountability" refers to the process by which firefighters are tracked and accounted for at the scene of an incident. It can be life-saving. As you know, any delay in recognizing that a firefighter is missing could result in his or her death. Accurate accountability is especially important when there is a sudden change in fire conditions, such as a structural collapse.

Fire departments use various systems. For example, some use Velcro name tags (Figure 25-7), which are removed from turnout coats or helmet shields and placed on the command status board. Others might use magnetic name tags placed on pump panels.

Figure 25-7 Examples of name tags that help to assure accountability in the rehab sector/group. *(Courtesy Tactron (TM) Inc.)*

In the rehab sector/group, the following measures help to assure accountability:

◆ SOP/Gs are to direct that entire crews, not individuals, are assigned to the rehab sector/group.

◆ Crews should enter and exit rehab as a unit.

◆ The rehab staff must log crews in and out.

◆ All entry to and exit from rehab should be through a single access point.

The rehab sector/group supervisor must instantly be able to account for crews that have been in or are currently assigned to rehab. (See an example of a company check-in/check-out log in Figure 25-8.) If any crew members who have entered rehab need transport to the hospital, their names, units, and destinations must be recorded and made immediately available to incident command (Figure 25-9).

Triage

As noted, all crews enter rehab through a single entry point. There they are time-logged in by rehab staff. Upon entering, fire crews remove SCBAs, hoods, and turnout gear. Then each crew member is evaluated for injuries and stress- and heat-related illnesses. Vital signs are taken and recorded. Rehab staff also rapidly question crew members, being alert to symptoms of possible life-threats such as chest pain and shortness of breath.

When that is done, the rehab staff assign crew members to the rest and refreshment area or to the medical evaluation/treatment area. Local protocols vary, but firefighters with pulse rates greater than 120 per minute at entry, who have abnormal blood pressure, or who have sustained any injuries are assigned to the "medical evaluation/treatment" area (Table 25-1)

Medical Evaluation/Treatment Area

Crew members are triaged to the medical evaluation/treatment area because of possible stress- or heat-related illnesses. They also may be triaged to this area because of injuries. Patients triaged with abnormal vital signs require ongoing assessment while they rest, eat, and drink. After 20 minutes of cool-down with rest and rehydration, they are to be reassessed and vital signs recorded (Figure 25-10). Vital signs of most crew members will have returned to normal.

CREWS OPERATING ON THE SCENE: _____ *E8, E24, R1, L14, BAT6* _____

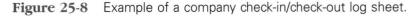

UNIT #	# PERSONS	TIME IN	TIME OUT		UNIT #	# PERSONS	TIME IN	TIME OUT
E8	4	1301	1335					

Figure 25-8 Example of a company check-in/check-out log sheet.

Some patients may still have elevated heart rates (greater than 100 beats per minute) after 20 minutes. These patients generally must remain in the rehab sector/group for additional rehydration. Some EIR SOP/Gs prohibit such personnel from returning to duty for the remainder of the incident or shift.

Ongoing Rehab Evaluations

The exact amount of time that crews spend in rehab depends on the level of exhaustion and

Figure 25-9 Information on crew members in the rehab sector/group must be instantly available to incident command. *(Courtesy Tactron (TM) Inc.)*

need for rest, food, and drink. At a minimum, crews should remain there for 20 minutes. Before returning to fire fighting or rescue duties, all crew members are to have a pulse and blood pressure check. If a crew member still has abnormal vital signs at this time, then the entire crew should remain in rehab for additional care and monitoring (Table 25-2).

All personnel who enter the rehab with injuries are to have those injuries promptly evaluated. Appropriate treatment should be given and transport to a medical facility provided, if needed. If a crew member is transported, other members of that crew potentially can go back on line. However, the crew and incident command are to be informed that a crew member has been transported.

Because a goal of EIR is to detect and prevent heat-related emergencies, many fire department SOP/Gs call for monitoring of a firefighter's oral temperature. Some rehabilitation SOP/Gs mandate an oral temperature on any rescuer who enters

Table 25-1

Entry Evaluation Findings Mandating Triage to the Medical Evaluation/Treatment Area

Heart Rate	>120
Blood Pressure	>200 systolic <90 systolic >110 diastolic
Injuries	Any

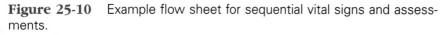

NAME / UNIT #	TIMES(S)	TIME /# Bottles	BP	PULSE	RESP	TEMP	SKIN	TAKEN BY	COMPLAINTS/CONDITION	TRANSPORT?
John Skoda	0940	NA	140/70	78	16	NA	NL	ETD	Swollen, deformed ankle	Yes/St. Mary's Hospital

Figure 25-10 Example flow sheet for sequential vital signs and assessments.

rehab with a heart rate greater than 110 to 120. If an oral temperature is greater than 100.6°F (38.1°C), then the rescuer is not permitted to put turnout gear or SCBA back on. For example, the Phoenix Fire Department Rehabilitation SOP/G states that any patient with a temperature higher than 101°F (38.3°C) must receive intravenous (IV) hydration and transport to the emergency department. Note that drinking cool fluids may affect the accuracy of an oral temperature, causing a false low reading.

In winter months, generalized hypothermia and local cold injuries must be considered during medical evaluation of fire and rescue crews. In these conditions, the rehab sector/group must provide crews with warmth and protection from ice and snow hazards. Remember, however, that stress- and heat-related illnesses also can occur in cool and cold weather.

During rehab and medical monitoring of crews, rehab sector/group personnel may detect

signs and symptoms of potential life-threats. Signs and symptoms include the following:

- Chest pain.
- Shortness of breath.
- Altered mental status.
- Skin that is hot and either dry or moist.
- Irregular pulse.
- Oral temperature greater than 101°F (38.3°C).
- Pulse greater than 150 at any time.
- Pulse greater than 140 after cool-down.
- Systolic blood pressure greater than 200 mm Hg after cool-down.
- Diastolic blood pressure greater than 130 mm Hg at any time.

If any one of these conditions is detected, the patient must be placed on high-flow oxygen, given the appropriate care, and transported immediately to the emergency department. Use of an advanced life support (ALS) unit is indicated for all of these conditions. However, transport from rehab should not be delayed if ALS is not available.

One other thing that rehab personnel must be on the lookout for when monitoring fire or rescue crews: a pattern of unusual complaints, illnesses, or injuries. Such patterns may indicate unexpected hazards involved in the incident. For example, if several patients at a fire scene complain of excessive salivation, runny noses, and diarrhea, it is a good indication that certain pesticides may be involved in the fire. Complaints of burning eyes could indicate the presence of metal gases. If

Table 25-2

Reevaluation Findings Mandating Continued Time in the Rehab Sector/Group

Heart Rate	>100
Blood Pressure	>160 systolic <100 systolic > 90 diastolic

unusual patterns are noted, immediately report the finding to the incident commander so that appropriate actions may be taken.

Hydration and Nourishment

Oral rehydration is essential during long periods of strenuous activity (Figure 25-11). Both water and electrolytes are lost from the body through sweating. Fluid replacement should, therefore, include both water and electrolytes. In addition, simple sugars (such as glucose or fructose) are important sources of energy.

Commercially available sports-activity beverages are ideal for fluid replacement in EIR. Some agency SOP/Gs recommend that they be diluted in a 50/50 mixture with water. Other agencies prefer to use the undiluted preparations. These fluids should be served either cool or at the ambient temperature, but not with ice and not ice cold. Beverages that are too cold can cause problems such as spasm of the esophagus and a sudden drop in heart rate.

No beverage containing caffeine or alcohol should be offered. They can alter cardiovascular performance and the body's heat-regulating mechanisms. Carbonated beverages also should be avoided, because they can cause digestive upset leading to nausea and vomiting.

Patients who are unable to tolerate oral rehydration due to nausea or vomiting should be

Figure 25-11 Liquids for rehydration must be available to rescuers.

assigned to the medical evaluation/treatment area. They will likely require intravenous (IV) hydration by ALS providers and transport to the emergency department.

During incidents that involve more than two to three hours of ongoing physical activity, fire and rescue crews should receive solid nourishment in rehab. Foods that are easily prepared, served, and digested are ideal. Soups or stews are often appropriate food choices. Fruits such as oranges and apples are good sources of fluids and sugars, too.

ON SCENE FOLLOW-UP

At the beginning of this chapter, you read that a rehab sector was established at the scene of a working house fire. A firefighter just entered the rehab sector, complaining of dizziness.

INITIAL ASSESSMENT

Our general impression of the firefighter was of a man in his 30s, who was sweaty and red-faced. His airway and breathing appeared to be okay. There were no obvious injuries or bleeding.

PHYSICAL EXAMINATION

We immediately took the firefighter to the evaluation and treatment sector and removed his heavy, sweat-soaked turnout coat and bunker pants. We asked him to lie down in the cool shady area, but he told us he felt more comfortable sitting. I asked if he was thirsty. He replied that he was, so I gave him a glass of cool water.

We checked his pulse, and noted that it was 98. His blood pressure was 138/94. Respirations were 24/minute. He was sweating, and although his face was red, overall his skin was pale. While drinking the glass of water, we turned on one of the exhaust fans and directed the stream of air directly at him for a cooling affect. He denied any pain or shortness of breath.

I also observed no injuries or burns to his body. I asked if he knew what day it was. He replied "Saturday." He also knew the correct date.

PATIENT HISTORY

He told me his name was John Salerno and that he was 35 years old. Because he had been a member of a neighboring department for only six months, we did not know each other. He told me he had been assigned to an exposure line protecting a house next to the burning structure, and estimated that he had been performing that task alone for approximately 45 minutes to an hour before being relieved. He was feeling dizzy for the last 10 minutes. He told me he was in excellent health, and took no medications and had no allergies.

ONGOING ASSESSMENT

After finishing his water, we encouraged John to drink a glass of the electrolyte solution, which he immediately did. He stated that these types of drinks never really quenched his thirst, but understood the need to replace certain electrolytes lost through sweating.

After drinking, his face did not appear to be as flushed as before, although his pulse remained right around 98 beats per minute. I asked if he was still dizzy. He said he was, but it did not seem as bad.

PATIENT HANDOFF

I asked Malik if he could contact a member of John's department and ask if John could report to the rehab sector.

A couple of minutes later, John's assistant chief arrived. I briefly reported what we had done. I also told John and the assistant chief that it would probably be a good idea for John to be evaluated in the hospital's emergency department. John was reluctant. Being a firefighter myself, I knew it can be hard to admit that you may be injured or ill. However, his assistant chief stated that he would definitely see that John was examined at the hospital. In fact, not wishing to take any chances, the assistant chief requested an EMS transport for John. John wasn't very happy about it, but agreed it was for his own good.

As the EMS transport arrived, I gave them this report:

"This is John Salerno, a 35-year-old firefighter with the Boonton Volunteer Fire Department. He was dressed in full turnout gear while protecting an exposure with a handline for approximately one hour when he began to feel dizzy. When he arrived at our sector, he was conscious, alert, and oriented. His skin was warm, pale, and he was sweating. His pulse was 98 and regular, BP was 134/94, and his respiratory rate was 24. He states he has no pertinent medical history, is not taking any medications, and is not allergic to anything. We removed his protective gear, placed him in the shade, cooled him with a fan, and rehydrated him with water and an electrolyte solution. Although he says his dizziness has subsided somewhat, he states he is still a little dizzy."

The transport team thanked me for my report and assisted John, despite his mild protests, to the back of their medic unit for the five-minute ride to the hospital. Turning my attention back to the fire scene, I noted that the fire was almost out, and that the rehab sector was becoming quite full of firefighters, most of whom were sitting and relaxing, drinking fluids, and some even joking around a bit.

The next day I happened to run into the assistant chief, who told me that John had been evaluated and released about an hour after arriving at the hospital. The doctors thought it was a mild form of heat exhaustion that had caused John's symptoms, and advised him to go home and rest inside with the air conditioning on.

> *Rehabilitating firefighters at any emergency incident is critical. You must be able to recognize when rehab is necessary and also how to perform the rehabilitation function properly. Remember, it is not burns and smoke inhalation that injure or kill most firefighters on the fireground. It's stress- and heat-related injuries.*

Chapter Review

One of the most difficult roles of the fire service First Responder is to treat a fellow firefighter at the scene of an incident. Even so, it is critical for you to have the basic knowledge and skills necessary for assessing and treating fellow firefighters when it is needed. It is also important for you to have a basic knowledge of the most common types of injuries and illnesses you may encounter.

Firefighters—and all emergency service personnel—encounter more stress and physically demanding challenges than almost all other occupations. The ability to take care of ourselves—and fellow firefighters—is critical if we are to maintain the ability to properly perform our job. We owe it to ourselves, our families, our department, and our community to do so.

FIRE COMPANY REVIEW

Page references where answers may be found or supported are provided at the end of each question.

Section 1

1. What is the primary function of emergency incident rehabilitation (EIR)? (p. 464)

2. What are the two most common mechanisms for heat loss regulation in the human body? Compare and contrast them. (p. 464)

3. What are the three most common heat-related emergencies the fire service First Responder is likely to encounter? (pp. 464–465)

4. What are the signs and symptoms common to heat-related emergencies? (pp. 464–465)

5. Which heat-related emergency poses the most serious threat to the fire service First Responder? (p. 465)

6. How should the fire service First Responder treat heat stroke? (p. 465)

Section 2

7. What are five functions that First Responders may perform in the rehab sector/group? (pp. 465–466)

8. When should firefighting crews enter the rehab sector/group? (pp. 467–468)

9. In the rehab sector/group, what does the term "accountability" mean? Also, explain why it is important. (p. 468)

10. What are the signs and symptoms of potentially life-threatening emergencies that may be detected during rehabilitation and medical monitoring of crews? (pp. 468–471)

RESOURCES TO LEARN MORE

Dickinson, E., and M.A. Wieder. *Emergency Incident Rehabilitation*. Upper Saddle River, NJ: Brady/Prentice-Hall, 2000.

Emergency Incident Rehabilitation. Federal Emergency Management Agency/U.S. Fire Administration, FA-114. Washington, D.C.: July 1992.

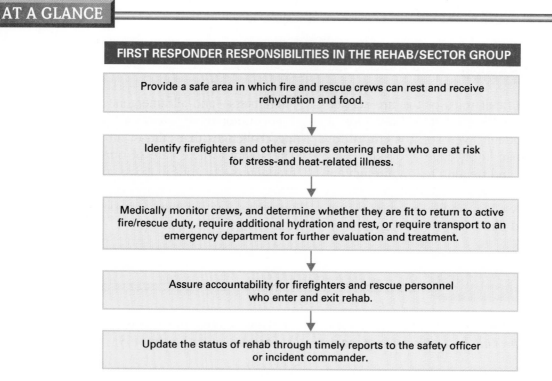

FIRST RESPONDER RESPONSIBILITIES IN THE REHAB/SECTOR GROUP

Provide a safe area in which fire and rescue crews can rest and receive rehydration and food.

↓

Identify firefighters and other rescuers entering rehab who are at risk for stress-and heat-related illness.

↓

Medically monitor crews, and determine whether they are fit to return to active fire/rescue duty, require additional hydration and rest, or require transport to an emergency department for further evaluation and treatment.

↓

Assure accountability for firefighters and rescue personnel who enter and exit rehab.

↓

Update the status of rehab through timely reports to the safety officer or incident commander.

EMS Rescue Operations

*I*NTRODUCTION   *Most of the time you will find your patients in safe, easily accessible locations where gaining access is no more than a knock on a door. However, there will be times when advanced rescue techniques must be used.*

When you arrive on the scene of a special rescue situation, remember: you did not cause the situation, you are there to remedy it. You and your team's safety must come first. This chapter provides an overview of how you can proceed safely and effectively. In practice, be sure to follow all local protocols.

Cognitive, affective, and psychomotor objectives are from the U.S. DOT's "First Responder: National Standard Curriculum." Enrichment objectives, if any, identify supplemental material.

Cognitive

7-1.3 Discuss the role of the First Responder in extrication. (p. 478)

7-1.4 List various methods of gaining access to the patient. (pp. 479–481)

7-1.5 Distinguish between simple and complex access. (pp. 479–480)

Affective

No objectives are identified.

Psychomotor

No objectives are identified.

Enrichment

◆ Describe the types of personal protective equipment recommended for rescue at the site of a vehicle collision. (p. 477)

◆ Discuss how to determine the number of patients at the scene of a vehicle collision. (p. 477)

◆ Describe basic goals of traffic control at the scene of a collision. (p. 478)

◆ State how a rescuer can recognize whether or not a vehicle is stable. (p. 478)

◆ Describe the basic steps of stabilizing an upright vehicle and an overturned vehicle. (p. 478)

◆ Describe key components of scene size-up in a water emergency. (pp. 482–483)

◆ Discuss some of the difficulties of assessing a patient who is in the water. (pp. 482–483)

◆ Describe emergency care of a near-drowning patient with no injuries to the spine. (p. 484)

◆ Describe emergency care of a near-drowning patient with suspected spine injuries. (pp. 484–488)

◆ List the hazards commonly associated with fast-moving water. (pp. 488–489)

◆ Discuss the differences between warm-water and cold-water rescues. (pp. 489–491)

ON SCENE

DISPATCH

We were relaxing on Thanksgiving evening at the fire station. Although we were on duty, our families had come down and provided us with a huge Thanksgiving dinner. It had been a quiet day when the alarm sounded. It was for a motor-vehicle collision with unknown injuries at the corner of Canyon Road and Sussex Turnpike. As we entered our rescue vehicle, I noted the time was 2200 hours.

SCENE SIZE-UP

Approaching the scene, we knew we would have to be extremely careful, as it was not only a holiday but a pitch black night as well. The highway patrol passed our vehicle just as we

were pulling up to the scene. We saw a small, red sports car sitting nose to nose with a large recreational vehicle. The front of the little car was collapsed like an accordion under the front axle of the RV. The front bumper of the RV was even with the windshield of the car. We saw a driver and a passenger in the RV, and both appeared to be responsive. We also saw two motionless young people in the front seat of the car. There was a considerable amount of blood coming from multiple face wounds.

Consider this emergency as you read Chapter 26. How would you proceed?

SECTION 1
MOTOR-VEHICLE COLLISIONS

Scene Safety

Personal Protective Equipment

All EMS responders working in or around a wrecked vehicle with an extrication in progress must wear the following:

◆ *Eye protection.* Goggles or safety glasses with side shields are best to prevent flying metal shards from penetrating the eyes. Safety glasses with side shields also can be used to protect you from blood-borne pathogens. The flip-down shield on a firefighter helmet is *not* adequate protection for the eyes.

◆ *Hand protection.* Firefighter gloves provide the best puncture protection, but they allow for poor manual dexterity. Although not offering as much protection, a pair of snugly fitting, leather work gloves is a good alternative.

◆ *Body protection.* A flame-retardant outer shell such as firefighter turnout gear, including coat and pants, provide some protection from fire and limited protection from sharp objects. All garments should have reflective trim to improve night recognition.

◆ *Foot protection.* Wear turnout gear with either short rubber or leather boots with lug soles to prevent slippage. Boots should be above ankle height to prevent glass from dropping in. They also should have steel toes.

Determine the Number of Patients

At the site of a vehicle collision, assess personal safety just as you would for every other type of emergency. Then identify the mechanism of injury or nature of illness. To determine the resources you need on scene, find out how many patients are involved. It may be difficult to locate all patients at first, but it is critical that you do. Use a systematic approach:

◆ If it is safe to enter the scene, ask a responsive patient to tell you how many others were involved in the crash.

◆ Question witnesses to see if a victim walked away from the scene.

◆ In case of a high-impact crash, search the surrounding area carefully. Look in ditches and tall weeds.

◆ Look for tracks in the earth or snow. A person who could get free from wreckage may be wandering aimlessly.

◆ Search the vehicle itself carefully. A patient may be wedged under the dashboard, for example.

◆ Look quickly for items that give clues to unaccounted for children, such as a lunch box, diaper bag, or extra jacket.

Combine this information with your scene size-up observations. Are there enough rescue personnel on scene? If not, send for help immediately. Continue to evaluate the situation for the most efficient, safest ways to help patients and protect rescue teams.

If a car is on fire, decide if you can remove the passengers quickly enough or if you should fight the fire. If the passengers are not trapped, move them first. If they cannot be extricated quickly, deal with the fire. That is, safely do what you can within your training and with the available equipment.

FIRE DRILL

On every extrication incident, you must make a decision. Ask yourself: Should I remove the victim from the vehicle, or should I remove the vehicle from the victim?

Control the Scene

The crash scene can involve environmental hazards as well as a great deal of confusion. If necessary, request law enforcement to help control the scene. For environmental hazards, a pumper with Class B foam capability may be necessary. While you wait for them to arrive, begin scene control. Quickly deal with bystanders by having them move out of the danger zone.

Spilled gasoline often is present at an auto crash. Allow no smoking on scene. Turn off all vehicle ignitions. If a pumper is on scene, pull the appropriate hose line for protection in the event of a fire. If a pumper is not yet on scene, prepare for its arrival by keeping everyone away from the danger. Set a fire extinguisher nearby.

Control Traffic

The basic goals of traffic control at the scene of a collision, are:

◆ To clear the scene so that emergency vehicles can get through quickly.

◆ To monitor regular traffic around the scene so no further crashes or injuries occur.

◆ To monitor traffic so that passing vehicles have a minimum of inconvenience.

Unless a distinct hazard justifies stopping all traffic, keep traffic moving. If the road is blocked, try to move traffic to an alternative route. Whatever you choose to do, make sure that motorists and pedestrians in the area know exactly what you want them to do. Keep rescue personnel well positioned along the roadway. Rescuers should always wear reflective clothing so they can be seen easily before and after dark. Use clear visual signals coupled with attention-getting devices such as flares or cones.

Flares should be set 10–15 feet (3 m to 5 m) apart and extend 100 feet (30 m) toward traffic. The pattern of fuses should lead traffic around the emergency. The danger zone includes at least a 50-foot (15 m) radius around the wrecked cars. When a crash occurs on a curve, consider the start of the curve as the edge of the danger zone. On a hill, one edge of the danger zone should be the crest of the hill. If the highway has two lanes, position flares in both directions. If heavy trucks travel the road, extend the flare string, because trucks take much longer than cars to stop. Note also:

◆ Look for and avoid spilled fuel, dry vegetation, and other combustibles before you ignite and position flares, especially on roadsides. If not positioned properly, flares will roll and may not be where you need them to be.

◆ Do not throw flares or cones out of moving vehicles.

◆ Never use a flare as a traffic wand. Molten phosphorus can splatter on you, causing severe burns.

Vehicle Stabilization

After all possible outside hazards are controlled, make the rescue setting as safe as possible. Remember, never enter an unstable vehicle. Always suspect that a vehicle is unstable until you have made it stable. Assume the vehicle is not stable if:

◆ It is on a tilted surface such as a hill.

◆ Part of it is stacked on top of another vehicle.

◆ It is on a slippery surface such as ice, snow, or spilled oil.

◆ It is overturned.

◆ It rests on its side.

Basic to stabilization is the use of cribbing (Figure 26-1). **Cribbing** is a system of wood or other supports used to prop up a vehicle. Wood is stacked in box-like squares and wedges to distribute pressure uniformly. To create a stable environment, the cribbing is arranged diagonally to the vehicle frame. Do not crib under wheels or tires, because the vehicle will tend to roll. Never stack cribbing higher than its length.

Any vehicle that can be moved easily during extrication or patient care needs to be stabilized. Place cribbing or step blocks under the frame. Excess movement of the vehicle could prove fatal for a patient with severe spine injuries and may injure the rescue team. To stabilize vehicles:

◆ *Upright vehicles.* For a vehicle that rests on all four wheels, place the gear selector in park or, if a standard shift, into reverse. Use blocks or wedges at wheels to prevent unexpected rolling. Chock wheels tightly against the curb when possible. To reduce the amount of movement even when you are using power tools, cut the tire valve stems so that the car rests on the rims.

◆ *Overturned vehicles.* To stabilize an overturned vehicle, place a solid object between the roof and the roadway. Use an object such as a wheel chock, spare tire, cribbing, or timber. If necessary, use a bumper jack to angle the vehicle against the solid object until it is stable. Hook a chain to the vehicle's axle. Then loop the chain around a tree or post.

Extrication of the Patient

The role of the fire service First Responder is to administer emergency medical care. You also must assure that the patient is removed in a way that minimizes further injury.

In certain circumstances, you may be required to take steps to gain access to the patient in a

Stabilizing with Cribbing

Figure 26-1a

Figure 26-1b

Figure 26-1c

Figure 26-1d

 COMPANY OFFICER'S NOTE

Every vehicle must be considered unstable until proven otherwise. Ask yourself: Is the ignition still on? Is the vehicle still in gear? If you suspect instability, do not enter the vehicle and do not attempt rescue. Adequate training will allow safe and effective operations on scene.

vehicle. Take only the steps you are trained to take. Call for additional assistance and rescue personnel if needed. In such cases, an incident management system should be established to ensure patient-care priorities.

Gaining Access

There are two basic ways rescuers can gain access to a patient. **Simple access** is access by which no tools are needed. **Complex access** is access that requires tools and specialized equipment.

Most emergencies do not present access problems. However, when you are confronted with one, quickly evaluate the situation and decide if simple or complex access is needed. For complex access, make sure only rescuers who have the proper training and equipment are used. Remember, rescuer safety comes first.

Crash scenes are charged with emotion. Although accessing the patient is critical, do not allow upset bystanders to cause you to enter a vehicle or area that is not safe. Enter the wreck only when the vehicle is stabilized and safe.

Doors

A door is always the access of choice. This is because it is the largest uncomplicated opening in a vehicle. Always start by testing the door handles.

First try to open the door nearest the patient. If the doors are locked, try to unlock them yourself. If that is not possible, and the patient is able, ask him or her to do so. Routinely unlock all other doors to allow access by other rescuers. If the doors cannot be opened, determine the best point of entry and proceed accordingly. For instructions on how to expose a lock mechanism, see Figure 26-2.

Windows

Car windows usually are made of tempered glass. Rear and side windows are designed to break into small granules. If you can, remove a window without breaking it. Fine particles of glass can stay unnoticed in a deep wound and cause damage after it closes. So cover the patient with a heavy safety blanket before breaking a window or as soon as possible.

If fixed windows are installed in U-shaped black plastic or rubber, remove the rim. Insert the point of a linoleum knife or similar tool into the molding at the midpoint of the glass. Keep the blade as flat against the glass as you can. Draw the knife across the top and down the side. Repeat on the other side. Soapy water will keep the blade moving easily. Work the end of a short pry bar behind the glass and pry it loose from the top. The window will pivot on its bottom edge.

If you must break a window, locate the window farthest from the patient. Give a quick hard thrust in the lower corner with a spring-loaded punch, Haligan tool, pick-headed axe, or other sharp object. (See Figure 26-3.) Use your gloved hand to carefully pull the glass outside the vehicle. Clear all glass away from the window opening. Before you crawl in, drape a heavy tarp or blanket over the door edge and the interior of the car just below the window.

Windshield

Most windshields are made of laminated safety glass, making them difficult to break safely. Sawing, chopping with an ax, or prying will remove the windshield regardless of make or design. Make sure the patient is covered with a blanket before attempting to remove a windshield.

Airbags

Most airbags are designed to inflate with a force of about four psi. If the airbag has released, there may be some residue. This is not harmful and can be washed off. If the airbag was not triggered by the crash, disconnect the negative side of the battery and the yellow airbag connector. Do not cut the connector or its wires. They will keep the shorting bar activated, preventing accidental triggering. Also, beware of undeployed airbags, which may pose a threat to First Responders. They may be located in the steering wheel, side panel, passenger-side dash, and doors.

Pinned Patients

If a patient is pinned beneath a vehicle, attempt to raise the vehicle just to the point where the victim may be removed. Once the vehicle is raised enough to access and remove the victim safely, stabilize the vehicle with cribbing or chains to ensure it does not move. Airbags work well to raise a vehicle. If they are not available, try a hydraulic tool or even a jack. When no other options exist, a large group of people may be used to lift the vehicle off the patient. However, this should be done as a last resort only.

If a patient has a body part through a window, pad the extruded part well with bandaging material. Then carefully use pliers or a knife to break or fold away the glass. Once the body part is free, care for it.

If a patient is jammed or pinned inside the vehicle, consider the following simple procedures:

Exposing a Lock Mechanism

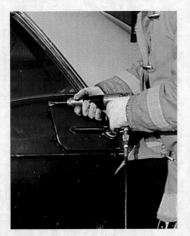

Figure 26-2a Cut around the handle and lock.

Figure 26-2b Pry open the cut area.

Figure 26-2c Operate rods and levers to pop open the lock.

◆ Move the front seat to give additional working space. It may be possible to lift the back seat entirely.

◆ Seat belts that will not open can be cut with shears or a knife. Support the dangling patient as you cut the belt.

◆ Displace the dashboard.

◆ Emergency Care

As in any emergency, your first priority is rescuer safety. Be sure the scene is safe, the vehicle is stable, and you are wearing the appropriate personal protective and BSI equipment before accessing the patient.

Patients in a motor-vehicle crash sometimes have blood pressure artificially maintained by vehicle parts that are pressing against them. Once the vehicle parts are removed, these patients become unstable quickly. If this occurs and the mechanism of injury is significant, call for advanced life support (ALS).

After gaining access safely, provide the same emergency care you would give to any trauma patient. Stabilize the head and neck. Monitor and protect the airway, breathing, and circulation,

Breaking a Car Window

Figure 26-3a Position a punch in the corner of the side window farthest away from the patient.

Figure 26-3b Push out the shattered tempered glass, and reach in to open the door or roll down the window.

as necessary. Be sure you have called for the necessary resources, such as advanced life support (ALS), extra ambulances, or medivac helicopter.

Remain with patients during a complex extrication. Besides continually monitoring their condition, talk to them. Explain what is occurring. Keep them as calm as possible. If their condition begins to deteriorate, advise the rescue crew. They may be able to change the approach to the incident and get the patients out more quickly. Also advise the EMS transport agency. During the process, be sure to protect yourself and the patient from the glass and flying debris. Use heavy blankets, a tarp, and a solid object like a backboard.

Immobilize the patient's spine during rescue. (Follow the precautions and procedures described in Chapter 18.) The only exception to this rule occurs when there is an immediate threat to life, such as fire, and an emergency move is required.

SECTION 2   WATER RESCUE

General Guidelines

Scene Size-up

In a water-related emergency, reaching the patient is your goal. Yet, you must do so with concern for your own safety as well as the safety of the other firefighters on scene. Water can conceal many hazards. Holes, sharp drop-offs, and underwater entanglements such as fallen trees or wire fences may not be visible from shore. The force of moving water also can be very deceptive. Never walk in fast-moving water over knee depth. Moving water in streams, rivers, even storm drains can push you over and hold you down.

Another concern regarding water emergencies is the possibility of coming in contact with hazardous materials. For example, a car in the water could leak oil or gas, which float on the surface. Such hazards pose a respiratory risk for both patients and rescuers. Floods can cause sewage to be released in normally safe waters. Risk of electrocution exists in flooded buildings or grounds. Severe bleeding of the patient also can pose the risk of infection to other patients and rescuers.

When determining how to respond to a water emergency, take into account the patient's condition, water conditions, and the resources on hand.

- Patient condition.
    - Is the patient responsive and able to assist in the rescue? If so, reaching out with a pole or throwing a rope may be the safest method of rescue.
    - Does the patient have any obvious injuries? If he is unstable, you may have to extricate him from the water before beginning care.
    - Is the patient on the surface, or is he submerged? The submerged patient may need basic life support immediately. He also may be difficult to locate.
- Water conditions.
    - *Visibility.* Can you see any potential hazards under the water? Can you see the patient and his injuries?
    - *Temperature.* For a cold-water drowning, you must continue resuscitation until the patient is rewarmed at the hospital. Note that even when the air is warm, the water may still be cold.
    - *Moving water.* Is it safe for you to enter the water? Will the location of the victim change?
    - *Depth of the water.* Can your feet touch the bottom so you can stand, or will additional equipment be needed?
    - *Other hazards.* Are there hazardous materials present, such as oil, gas, or sewage? Is there any risk of electrocution?
- Resources on hand.
    - How many rescuers are on scene? Are they trained in water rescue? Can they all swim? Does each have a personal flotation device?
    - Do you need any special rescue teams such as a dive team?

If a water emergency occurs in open, shallow water that has a stable, uniform bottom, attempt a rescue. However, never try a water rescue unless you meet all of the following criteria:

- You are a good swimmer.
- You are specially trained in water rescue.
- You are wearing a personal flotation device.
- You are accompanied by other rescuers.

If you meet all four criteria and your patient is responsive and close to shore, attempt a rescue. Use the "reach, throw, row, and go" strategy in the following order:

- *Reach.* Hold out an object for the patient to grab. Anything that will extend your reach will work. You can use a towel, shirt, backboard, or other strong object that will not break. Before holding out the object, make sure you have solid footing and will not slip in the water. Once the object is grabbed, pull the patient to shore.
- *Throw.* If the patient is too far to reach, then throw an object that floats (Figure 26-4). Use anything that will float and is heavy enough to throw. A thermos jug, a picnic cooler, or capped empty milk jug will do. This will give the patient support and give you more time to make the rescue. If possible, tie a rope to the object you throw. Toss the object to the patient, and pull on the rope to tow the patient in. Again, be sure of your own footing and stability.
- *Row.* If the patient is too far to reach or throw an object to from shore, then use a boat to get closer to the patient.
- *Go.* If reaching, throwing, and rowing are not possible, then swim to the patient. Be sure you are connected by rope to rescuers on shore. However, swim to the patient only if you are a good swimmer, specially trained in water rescue, wearing a personal flotation device, and accompanied by other rescuers.

Patient Assessment

Any injury occurring on land may also occur in water. However, it can be more difficult to detect and treat injuries in water. For example, bleeding is easily detected on dry ground. In the water, however, any bleeding that occurs may be immediately diluted and dispersed. So not only may it be difficult to judge how severe the bleeding is, it

Figure 26-4 Tie a sturdy rope to an object that floats. Throw it and pull the patient in.

also may not be possible to recognize bleeding in your patient. If the patient is wearing a wet suit, a large amount of blood can pool inside it before bleeding is recognized.

Possible fractures are difficult to identify, too. The water may be murky or dark, and its surface can distort visual images. Extremities, for example, may appear angulated or straight when they are not, just by the refraction. One way to deal with this problem is to assume fractures until proven otherwise in any unresponsive patient.

If the emergency is the result of a diving accident, or if the patient has been struck by a boat, water skier, surfboard, jet ski, or other object, or if the patient is unconscious, suspect a spine injury.

◆ *Emergency Care*

If the patient is responsive and you are sure there is no spine injury, follow these guidelines:

1. **Remove the patient from the water.** Do so as quickly as you can by any safe method possible.

2. **Complete the initial assessment.** Administer high-flow oxygen, if you are allowed to do so. Be prepared to suction.

3. **Conserve the patient's body heat.** Remove wet clothing. Place the patient on a blanket, and cover him with another blanket. If possible, move the patient to a warm environment. Do not allow the patient to walk.

4. **Perform a physical exam and gather a patient history,** while you are waiting for EMS transport.

If your patient is unresponsive and in shallow, warm water, maintain the airway but do not move the patient. If the patient is breathing, keep him or her in a face-up position. Support the patient's back (Figure 26-5). If there is a second rescuer present, also stabilize the patient's head and neck (Figure 26-6).

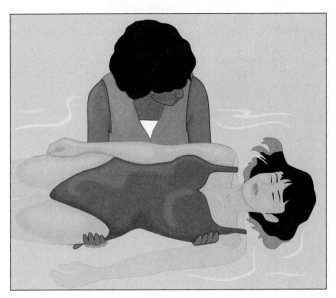

Figure 26-5 Hip and shoulder support.

If you find the unresponsive patient face-down in shallow water, you have to turn him. Perform the head-splint technique by following these steps (Figure 26-7):

1. **Get alongside the patient.**

2. **Position the patient's arms.** Extend them straight up alongside his head. Press the arms against the patient's head to create a splint. If necessary, move the patient forward to a horizontal position.

3. **Rotate the patient.** Bring the hand farthest away towards you and push away the hand that is closest. As you rotate the patient, lower yourself in the water until the water is at shoulder level.

4. **Maintain stabilization of the patient's head.** Use one hand to hold the patient's head between his arms. With your other hand, support the patient's lower back until help arrives.

HEAD-CHIN SUPPORT WITH TWO RESCUERS

Figure 26-6 In-line stabilization with two rescuers.

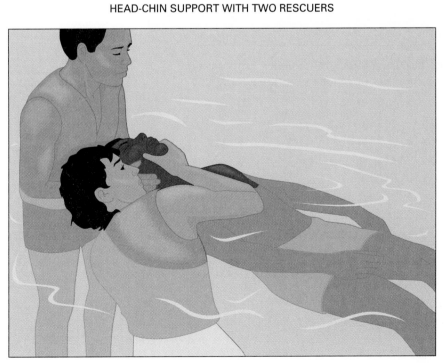

Head-Splint Technique

Figure 26-7a Position yourself alongside the patient.

Figure 26-7b Extend the patient's arms straight up alongside his or her head to create a splint.

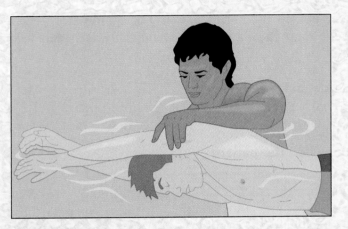

Figure 26-7c Begin to rotate the torso toward you.

Figure 26-7d As you rotate the patient, lower yourself in the water.

Figure 26-7e Maintain stabilization by holding the patient's head between his or her arms.

If the unresponsive patient is in water that is unsafe (deep, cold, or moving) or if the patient needs CPR, he or she should be positioned on a backboard. Once immobilized, the patient should be removed from the water. Note that ventilations can start in the water. Chest compressions cannot. Patients requiring CPR must be removed from the water first.

To turn a patient to a face-up position in deep water, perform the head-chin support technique. Follow these steps (Figure 26-8):

1. **Position yourself alongside the patient.**

2. **Position your arms.** Position one along the patient's spine, supporting the patient's head with your hand. Place your other arm along the patient's chest in line with the sternum, supporting the mandible with your hand. If necessary, move the patient forward to a horizontal position.

3. **Rotate the patient** by ducking under his body.

4. **Maintain manual stabilization** until a backboard is used to immobilize the spine.

To immobilize a patient, use a long backboard or other rigid support such as a water ski or surf board. Slide it under the patient. Let it float up until it is snugly against the patient's back. Apply a rigid cervical immobilization device. Then secure the patient to the backboard. Never try to support the patient's spine with anything that might bend or break, such as an air mattress or a Styrofoam float. As you are backboarding the patient, be sure to have enough rescuers helping. They need to make sure the patient's face does not become submerged.

After immobilization, lift the patient from the water, head first. If the patient is wearing a life jacket, leave it in place. Remember to pad under the patient's head to keep the spine in alignment.

NOTE: Always have a near-drowning patient taken to a hospital, even if you believe the danger has passed. Complications can develop and may be fatal as long as 72 hours after the incident.

Moving-Water Rescue

Many people are drawn to moving water, or "white water," for recreation. Those who are trained and experienced know how to read moving water. They understand the hazards and manage them. It is all part of their sport. Moving-water incidents usually occur when someone who is unaware of the dangers gets into the water. This is true of both victims and rescuers.

The force of moving water is measured by its depth, width, and velocity. For example, a river that is 200-feet wide, 4-feet deep, and moving at 20-feet per second will move about 16,000 cubic feet per second. That is roughly equal to 550 pounds of force.

Fast-moving water is dangerous. Certain river features make it even more so. They include the following:

◆ *Strainers*. These are obstructions that allow water to pass through but catch people and other objects. Some of the most common are trees and branches. If a strainer catches a swimmer, the force of the water can hold the swimmer there until hypothermia sets in and he or she tires and drowns.

◆ *Obstructions*. Another problem is any type of obstruction that a person can get pinned against, such as a bridge abutment. A person can easily become trapped against the object and be held there by the force of moving water. Again, hypothermia can set in and he or she can tire and drown.

◆ *Holes*. Not all water in a fast-moving river flows downstream. When water flows over a large object, a recirculating current or "hole" may form. When this happens, the current can keep recirculating a swimmer in its backwash until he or she tires and drowns. Holes are difficult to see from upstream. Large ones are very difficult to escape from.

◆ *Low-head dams* (Figure 26-9). These dams are only a few feet (one meter) high. They are built from concrete and have vertical abutments on each side. They are very difficult to see from upstream and often tend to be very wide. Water that flows over these dams can form a very large and uniform "hole" that extends across the river. If a boater gets too close to the "boil line," he or she can get caught in the recirculating current and capsize. The person and the boat then can get pushed to the bottom and back to the surface again and again.

◆ *Extremity entrapment* (Figure 26-10). Legs can get trapped between rocks or in other

Head-Chin Support Technique

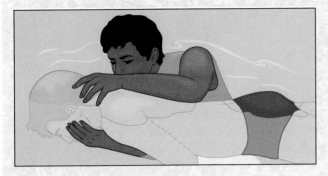

Figure 26-8a Support the patient's head with one hand and the mandible with the other.

Figure 26-8b Then rotate the patient by ducking under him or her.

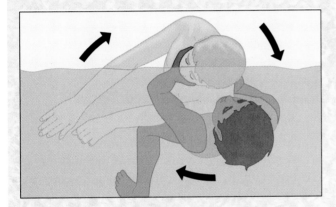

Figure 26-8c Continue to rotate until the patient is face up.

Figure 26-8d Maintain in-line stabilization until a backboard is used to immobilize the spine.

obstructions, especially in fast-moving water above a person's knees. Typically, this occurs to inexperienced folks who fall out of a boat and try to stand up. When this happens, the best thing to do is a back stroke with feet pointed downstream. Note that when an extremity gets caught, it must be extracted exactly in the same direction that it went in.

The basic, time-honored model for all water rescues is to reach, throw, row, and go. Remember, if you cannot easily effect a rescue using a simple shore-based technique (reach or throw), call for a team specializing in water rescue. Do not enter the water! Even special teams only try swimming or "live-bait" rescues as a last resort.

Ice Rescue

Judging the thickness of ice and overall safety is very tricky. Many factors can alter the thickness of the ice over large and small areas. For example, underground springs cause water turbulence from beneath and thinner ice above. So do decaying plant matter, schools of fish, and so on.

The "reach, throw, row, and go" method is used for all water rescues, including ice rescue. However, there are some differences you should note:

◆ *Reach and throw.* As a drowning patient becomes more hypothermic, he or she will be less and less able to hold onto a rope. An

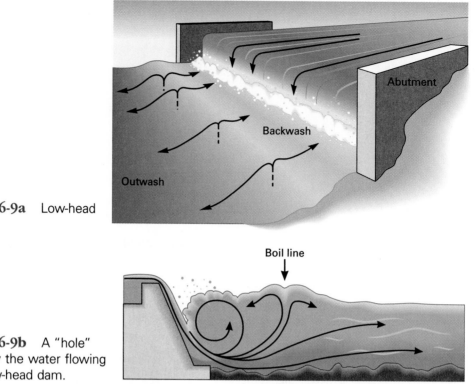

Figure 26-9a Low-head dam.

Figure 26-9b A "hole" formed by the water flowing over a low-head dam.

alternative technique is to throw an inflated fire hose. By using modified end caps and air from a SCBA tank, a fire hose can be inflated quickly and pushed out to the patient. Similarly, if you carry air bags for vehicle extrication, inflating and tossing these to the victim may also prove useful.

◆ *Row.* A conventional boat may not be able to break the ice as it moves, unless the ice is very thin. An option may be any flat-bottom boat or a small inflatable craft with ropes to

tether it and pull it from shore. Perhaps the best crafts for ice rescues are the air boat and hover craft. Either can be maneuvered over ice, water, or dry land.

◆ *Go.* When patients are too hypothermic to hang on, rescuers must go in to get them (Figure 26-11). This usually involves wearing a

Figure 26-11 An ice rescue. *(Source: Michael A. Gallitelli)*

Figure 26-10 Entrapment in fast moving water.

"dry" neoprene ice-rescue suit, which is tethered to shore. The rescuer then crawls, shuffles, or swims out to grab and pull the patient in.

When a person falls through ice, a fire service First Responder must immediately call for a special ice-rescue team, if available. Then, put on a personal flotation device and make reasonable attempts to reach or throw something to the patient from shore. If you are successful, the team can be canceled. If not, the team already on the way will have a chance to get to the scene in time to help the patient. Remember to prevent well intentioned bystanders from going onto the ice. More people to save may lead to more lives being lost.

ON SCENE FOLLOW-UP

At the beginning of this chapter, you read that fire service First Responders were on the scene of a collision with multiple patients. To see how chapter skills apply to this emergency, read the following. It describes how the call was completed.

SCENE SIZE-UP (CONTINUED)

This was a complex-access rescue situation. We immediately updated dispatch and requested our fire department's heavy rescue unit, as well as two ambulances. Then we positioned our vehicle about 100 feet (30 m) from the wreckage to give the rescue units space to pull in.

As soon as we determined it was safe to approach, we saw that both vehicles had all four wheels on solid ground and appeared to be stable. We made sure that both engines were turned off and brakes were on. We also looked for any smoke or leaks. We went ahead and stabilized both vehicles with cribbing.

INITIAL ASSESSMENT

We attempted to gain access to the patients and perform triage. We did as much of an initial assessment as we could from the open windows in the sports car. The occupants of the RV did not appear to be seriously injured. They were alert with minor complaints. We instructed them to remain in their seats until more help arrived. Most of the damage and injuries were to the sports car and its occupants.

We were taking spinal precautions, attempting to maintain airways, and control serious bleeding of the patients when the first ambulance arrived.

PATIENT HAND-OFF

We reported the number of patients and our initial observations to the EMS transport team, including which patients we believed were a first priority. While they proceeded to assess and care for the patients, we provided directions to the other responding units.

It took a while for the patients to be disentangled from the wreckage. Donning our bunker gear, we assisted the heavy rescue crew in performing the extrication.

After the patients were freed, we helped the EMS transport team to quickly reassess them and move them into the ambulances.

As in all emergency situations, remember that your first priority is your own safety. Do not enter the scene of an emergency until you have determined that it is safe. Do not attempt a complex rescue, unless you are trained and equipped to do so. Follow your local protocols.

Chapter Review

Whether on or off duty, you will come across rescue situations as a fire service First Responder that challenge your skills and ingenuity. Remember, it is not what you may use to accomplish a rescue, but the end result that is important. Paramount to that end result is you and your team's safety!

Always strive to be the rescuer, never a victim of your own rescue attempt. Maintain an attitude of safety. If you do, you will feel the satisfaction that comes from knowing you successfully helped a patient because of your training and ability to safely perform under pressure. That is what being in the fire service is all about.

FIRE COMPANY REVIEW

Page references where answers may be found or supported are provided at the end of each question.

Section 1

1. What are some types of personal protective equipment a rescuer should wear at the site of a vehicle collision? (p. 477)

3. How can a rescuer recognize whether or not a vehicle is stable? (p. 478)

3. What are some basic methods of stabilizing an upright vehicle? An overturned vehicle? (p. 478)

4. What is the difference between simple and complex access? (pp. 479–480)

5. What is always the access of choice? Explain your answer. (pp. 479–480)

6. What are some basic guidelines for the emergency medical care of patients who are trapped in wreckage? (pp. 481–482)

Section 2

7. What are some of the hazards a water emergency may pose for the rescuer? (pp. 482–483)

8. What criteria must you meet before you can attempt a water rescue? (p. 483)

9. What is the "reach, throw, row, and go" strategy for water rescue? (p. 483)

10. What are the general guidelines for emergency care of a near-drowning patient? (pp. 484–488)

RESOURCES TO LEARN MORE

Fire Service RESCUE, Sixth Edition. Stillwater, OK: Fire Protection Publications, 1996.

Grant, H.D., and J.B. Gargan. *Vehicle Rescue,* Second Edition. Upper Saddle River, NJ: Brady/Prentice-Hall, 1997.

Wieder, M.A., ed. *Principles of Extrication.* Stillwater, OK: International Fire Service Training Association, 1990.

FIRST RESPONDER'S ROLE IN VEHICLE RESCUE

SCENE SIZE-UP

- Don personal protective equipment.
- Determine number of patients.
- Control scene hazards, and prevent further injuries.
- Control traffic.
- Make rescue setting as safe as possible.

PATIENT EXTRICATION

- Gain access to patient, but only to the extent of your training.
- Administer emergency medical care.
- Assure that patient is removed in a way that minimizes further injury.

FIRST RESPONDER'S ROLE IN WATER AND ICE RESCUE

NEVER ATTEMPT A WATER RESCUE UNLESS YOU MEET ALL CRITERIA:

- You are a good swimmer.
- You are specially trained in water rescue.
- You are wearing a personal flotation device.
- You are accompanied by other rescuers.

If you meet all four criteria and your patient is responsive and close to shore, consider the "reach, throw, row, and go" strategy for rescue.

Glossary of Abbreviations

A

ABCs: airway, breathing, circulation.
ACLS: advanced cardiac life support.
ADA: Americans with Disabilities Act.
AED: automated external defibrillator.
AHA: American Heart Association.
AIDS: acquired immune deficiency syndrome.
ALS: advanced life support.
ATV: all-terrain vehicle.
AVPU: alert, verbal, painful, unresponsive.

B

BLS: basic life support.
BP: blood pressure.
BSA: body surface area.
BSI: body substance isolation.
BVM: bag-valve-mask device.

C

CDC: Centers for Disease Control.
CHEMTREC: Chemical Transportation Emergency Center.
CISD: critical incident stress debriefing.
CISM: critical incident stress management.
cm: centimeters.
COPD: chronic obstruction pulmonary disease.
CPR: cardiopulmonary resuscitation.
CVA: cerebral vascular accident.

D

DNR orders: do not resuscitate orders.
DOT: U.S. Department of Transportation.
DOTS: deformities, open injuries, tenderness, swelling.

E

EIR: emergency incident rehabilitation.
EKG: electrocardiogram.
EMD: emergency medical dispatcher.
EMS: emergency medical services.
EMT: emergency medical technician.

EMT-B: EMT-Basic.
EMT-I: EMT-Intermediate.
EMT-P: EMT-Paramedic.
EPA: Environmental Protection Agency.
ERT: emergency response team.
ETA: estimated time of arrival.

F

FBAO: foreign body airway obstruction.
FCC: Federal Communications Commission.
FGC: fire ground command.

H

hazmat: hazardous material.
HBIG: hepatitis B immunoglobulin.
HBV: hepatitis B virus.
HEPA respirator: high efficiency particulate air respirator.
Hg: mercury.
HIV: human immunodeficiency virus.

I

ICS: incident command system.
IFSTA: International Fire Service Training Association.
IMS: Incident Management System.
IV: intravenous.

L

L: liter.
LZ: landing zone.

M

m: meter.
MCI: multiple-casualty incident.
MCS: multiple-casualty situation.
ml: milliliters.
mm: millimeters.
MOI: mechanism of injury.
MSDS: material safety data sheets.

NFPA: National Fire Protection Association.
NHTSA: National Highway Traffic Safety Administration.
NIOSH: National Institute of Occupational Safety and Health.
NOI: nature of illness.
NREMT: National Registry of Emergency Medical Technicians.

— O —

O₂: oxygen.
OB kit: obstetrical kit.
OPQRRRST: onset, provocation, quality, region, radiation, relief, severity, time of pain.
OSHA: Occupational Safety and Health Administration.

— P —

PCR: prehospital care report.
PEA: pulseless electrical activity.

PMS-CR: pulse, movement, sensation, capillary refill.
PPE: personal protective equipment.
PSAP: public service answering point.
psi: pounds per square inch.

— S —

SAED: semi-automated external defibrillator.
SAMPLE: signs and symptoms, allergies, medications, pertinent medical history, last oral intake, events.
SCBA: self-contained breathing apparatus.
SIDS: sudden infant death syndrome.
SOP/G: standard operating procedure/guideline.
START system: simple triage and rapid treatment system.

— T —

TB: tuberculosis.
TIA: transient ischemic attack.

Glossary of Terms

A

abandonment: a legal term referring to discontinuing emergency medical care without making sure that another health-care professional with equal or better training has taken over.

abdominal cavity: the space below the diaphragm and continuous with the pelvic cavity.

abrasion: an open wound caused by scraping, rubbing, or shearing away of the epidermis.

abuse: improper or excessive action so as to injure or cause harm.

accessory muscles: additional muscles; in regard to breathing, these are the muscles of the neck and the muscles between the ribs.

activated charcoal: a finely ground charcoal that is very absorbent and is sometimes used as an antidote to some ingested poisons.

acute abdomen: a sharp, severe abdominal pain with rapid onset.

advance directive: a patient's instructions, written in advance, regarding the kind of resuscitation efforts that should be made in a life-threatening emergency.

afterbirth: the placenta after it separates from the uterine wall and delivers.

agonal respirations: reflex gasping with no regular pattern or depth; a sign of impending cardiac or respiratory arrest.

airway adjunct: an artificial airway.

alimentary tract: the food passageway that extends from the mouth to the anus.

altered mental status: a change in a patient's normal level of responsiveness.

alveoli: the air sacs of the lungs. *Singular* alveolus.

amniotic sac: a sac of fluid in which the fetus floats.

amputation: an injury that occurs when a body part is severed from the body.

anatomical position: a position in which the patient is standing erect with arms down at the sides, palms facing front.

aneurysm: an enlarged or burst artery.

antecubital space: the hollow, or front, of the elbow.

anterior: a term of direction or position meaning toward the front. *Opposite of* posterior.

aorta: major artery that starts at the left ventricle of the heart and carries oxygen-rich blood to the body.

apical pulse: an arterial pulse point located under the left breast.

arterial bleeding: recognized by bright red blood spurting from a wound.

arteries: blood vessels that take blood away from the heart.

arterioles: the smallest arteries.

artificial ventilation: a method of assisting breathing by forcing air into a patient's lungs.

asphyxia: suffocation.

aspirate: to inhale materials into the lungs.

assault: the threat of physical harm.

atria: the two upper chambers of the heart. *Singular* atrium.

auscultation: a method of examination that involves listening for signs of injury or illness.

autonomic nervous system: the part of the nervous system that handles involuntary activities.

avulsion: an open wound that is characterized by a torn flap of skin or soft tissue that is either still attached to the body or pulled off completely.

B

bag of waters: amniotic sac.

battery: unlawful physical contact.

behavior: the way a person acts or performs.

behavioral emergency: a situation in which a patient exhibits behavior that is unacceptable or intolerable to the patient, family, or community.

birth canal: a passage made of the cervix and vagina.

blanch: to lose color.

blood pressure: the amount of pressure the surging blood exerts against the arterial walls.

blood vessels: a closed system of tubes through which blood flows.

bloody show: the mucous plug that is discharged during labor.

blunt trauma: injuries caused by a sudden blow or force that has a crushing impact.

body armor: a garment made of a synthetic material that resists penetration by bullets.

body mechanics: the safest and most efficient methods of using the body to gain a mechanical advantage.

body substance isolation (BSI): a strict form of infection control based on the premise that all blood and body fluids are infectious.

brachial pulse point: an arterial pulse that can be felt on the inside of the arm between the elbow and the shoulder.

bracing: exerting an opposing force against two parts of a stable surface with your body; in EMS, usually refers to a safety precaution taken while riding in an ambulance patient compartment.

bronchi: the two main branches of the trachea, which lead to the lungs. *Singular* bronchus.

burn center: a medical facility devoted to treatment of burns, often including long-term care and rehabilitation.

burnout: a state of exhaustion and irritability caused by the chronic stress of work-related problems in an emotionally charged environment.

C
---

capillaries: the smallest blood vessels through which the exchange of fluid, oxygen, and carbon dioxide takes place between the blood and tissue cells.

capillary bleeding: recognized by dark red blood that oozes slowly from a wound.

capillary refill: the time it takes for capillaries that have been compressed to refill with blood.

cardiac arrest: the sudden cessation of circulation.

cardiac muscle: one of three types of muscles; makes up the walls of the heart.

carotid pulse point: an arterial pulse that can be felt on either side of the neck.

catheter: a hollow tube that is part of a suctioning system. *Also called* tonsil tip *or* tonsil sucker.

central nervous system: the brain and the spinal cord.

cerebrospinal fluid: a water cushion that helps to protect the brain and spinal cord from trauma.

cervical spine: the neck, formed by the first seven vertebrae.

cervix: the neck of the uterus.

chain of survival: term used by the American Heart Association for a series of interventions that provide the best chance of survival for a cardiac-arrest patient.

chief complaint: the reason that EMS was called stated in the patient's own words.

child: according to AHA standards, any patient who is age one to eight years old.

chronic: of long duration.

circulatory system: the system that transports blood to all parts of the body.

clamping injury: a soft-tissue injury usually caused by a body part being stuck in an area smaller than itself.

clavicle: the collarbone.

cleaning: the process of washing a soiled object with soap and water. *See* disinfecting *and* sterilizing.

closed wound: an injury to the soft tissues beneath unbroken skin.

coccyx: the tail bone, formed by four fused vertebrae. *Also called* coccygeal spine.

colicky pain: cramps that occur in waves.

command: person responsible for the management of a multiple-casualty incident.

competent: in EMS a competent adult is one who is lucid and able to make an informed decision about medical care.

complete foreign body airway obstruction: all air exchange has stopped because an object fully occludes the patient's airway.

complex access: the process of gaining access to a patient which requires the use of tools and specialized equipment.

consent: permission to provide emergency care. *See* expressed consent *and* implied consent.

constrict: get smaller.

contusion: a bruise; a type of closed soft-tissue injury.

cornea: the anterior part of a transparent coating that covers the iris and pupil.

cranium: the bones that form the top (including the forehead), back, and sides of the skull.

crepitus: the sound or feeling of bones grinding against each other.

cribbing: a system of wood or other materials used to support an object.

cricoid cartilage: shaped like a ring, this is the lowermost cartilage of the larynx.

critical incident: any situation that causes a rescuer to experience unusually strong emotions which interfere with the ability to function either during the incident or later.

critical incident stress debriefing (CISD): a session usually held within three days of a critical incident in which a team of peer counselors and mental health professionals help rescuers work through the emotions that normally follow a critical incident.

cross-finger technique: a method of opening a patient's clenched jaw.

crowing: a sound made during respiration similar to the cawing of a crow, which may mean the muscles around the larynx are in spasm.

crowning: the appearance of the baby's head or other body part at the opening of the birth canal.

crushing injury: an open or closed injury to soft tissues and underlying organs that is the result of a sudden blow or a blunt force that has a crushing impact.

cyanosis: bluish discoloration of the skin and mucous membranes; a sign that body tissues are not receiving enough oxygen.

––––––– **D** –––––––

debriefing: a technique used to help rescuers work through their emotions within 24 to 72 hours after a critical incident.

deep: a term of position, meaning remote or far from the surface. *Opposite of* superficial.

defibrillation: the process by which an electrical current is sent to the heart to correct fatal heart rhythms.

defusing: a short, informal type of debriefing held within hours of a critical incident.

dermis: second layer of skin. *See* epidermis *and* subcutaneous tissue.

diabetes: a disease in which the normal relationship between glucose (sugar) and insulin is altered.

diaphragm: a muscle, located between the thoracic and abdominal cavities, that moves up and down during respiration.

diastolic pressure: the result of the relaxation of the heart between contractions. *See* systolic pressure.

dilate: enlarge.

direct medical control: refers to an EMS medical director or another physician giving orders to

an EMS rescuer at the scene of an emergency via telephone, radio, or in person. *See* indirect medical control.

disinfecting: the process of cleaning plus using a disinfectant, such as alcohol or bleach, to kill microorganisms on an object. *See* cleaning *and* sterilizing.

distal: a term of direction or position, meaning distant or far away from the point of reference, which is usually the torso. *Opposite of* proximal.

Do Not Resuscitate (DNR) orders: documents that relate the wish of the chronically or terminally ill patient not to be resuscitated. *See* advance directive.

dorsalis pedis pulse: an arterial pulse point that can be felt at the top of the foot on the great toe side.

dressing: a covering for a wound.

drowning: death from suffocation due to immersion in water.

drug abuse: self-administration of one or more drugs in a way that is not in accord with approved medical or social practice.

duty to act: the legal obligation to care for a patient who requires it.

dyspnea: shortness of breath.

––––––– **E** –––––––

ecchymosis: black and blue discoloration.

emancipated minor: a minor who is married, pregnant, a parent, in the armed forces, or financially independent and living away from home with permission of the courts.

embolus: a mass of undissolved matter in the blood. *Plural* emboli.

emergency medical services (EMS) system: a network of resources that provides emergency medical care to victims of sudden illness or injury.

emergency move: a move made when there is immediate danger to the patient, usually performed by a single rescuer.

EMT-Basic (EMT-B): an emergency medical technician trained to the level above the EMS First Responder.

EMT-Intermediate (EMT-I): an emergency medical technician trained to a higher level than the EMT-Basic and First Responder.

EMT-Paramedic (EMT-P): the most highly trained emergency medical technician in EMS.

enhanced 9-1-1: with this type of 9-1-1 service, the EMS dispatcher is able to see the street address and phone number of a caller on a computer screen.

epidermis: outermost layer of skin. *See* dermis *and* subcutaneous tissue.

epiglottis: a leaf-shaped structure that prevents foreign objects from entering the trachea during swallowing.

epiglottitis: a bacterial infection of the epiglottis.

esophagus: a passageway at the lower end of the pharynx that leads to the stomach.

evisceration: the protrusion of organs from an open wound.

expiration: breathing out; exhaling.

expressed consent: permission that must be obtained from every responsive, competent adult patient before emergency medical care may be rendered.

external: a term of position, meaning outside. *Opposite of* internal.

extremities: the limbs of the body.

extrude: to push or force out.

eye orbits: eye sockets; the bones in the skull that hold the eyeballs.

F

fallopian tube: the tube or duct that extends up from the uterus to a position near an ovary.

femoral pulse point: an arterial pulse that can be felt in the area of the groin in the crease between the abdomen and the thigh.

femur: the bone in the thigh, or upper leg.

fibula: one of the bones of the lower leg.

finger sweep: a technique used to remove a foreign object from the mouth.

First Responder: the first person on the scene with emergency medical care skills, typically trained to the most basic EMS level.

flail chest: a closed chest injury resulting in the chest wall becoming unstable.

flail segment: an area of chest wall between broken ribs that becomes free-floating.

fontanel: a soft spot lying between the cranial bones of the skull of an infant.

freelancing: uncoordinated or undirected activity at the emergency scene.

frostbite: freezing or near freezing of a specific body part. *Also called* local cold injury.

full thickness burn: a burn that extends through all layers of skin and may involve muscles, organs, and bone.

G

gastric distention: inflation of the stomach.

genitalia: reproductive organs.

globe: eyeball.

glucose: a type of sugar.

grieving process: the process by which people cope with death.

guarding position: a position in which the patient is on his or her side with knees drawn up toward the abdomen.

H

hand-off report: a report of the patient's condition and the care that was given, made to the EMS personnel who take over patient care.

hazardous material: a substance that in any quantity poses threat or unreasonable risk to life, health, or property if not properly controlled.

hazmat: hazardous material.

head-tilt/chin-lift maneuver: a manual technique used to open the airway of an uninjured patient. *See* jaw-thrust maneuver.

Heimlich maneuver: a technique used to dislodge and expel a foreign body airway obstruction. *Also called* subdiaphragmatic abdominal thrusts *and* abdominal thrusts.

hematoma: a collection of blood beneath the skin.

hemodilution: an increase in volume of blood plasma resulting in reduced concentration of red blood cells.

hemothorax: collapse of the lungs caused by bleeding in the chest.

humane restraints: padded soft leather or cloth straps used to tie a patient down in order to keep the patient from hurting him- or herself and others.

humerus: the bone that extends from the shoulder to the elbow.

hydration: the addition of water.

hyperthermia: fever or raised body temperature.

hyperventilation: rapid breathing common to diseases such as asthma and pulmonary edema; the syndrome is common to anxiety-induced states.

hypoglycemia: low blood sugar.

hypoperfusion: *See* shock.

hypothermia: the overall reduction of body temperature. *Also called* generalized cold emergency.

hypoxemia: a condition caused by a deficiency of oxygen in the blood.

hypoxia: decreased levels of oxygen in the blood.

--------------- I ---------------

ilium: one of the bones that form the pelvis. *Plural* ilia.

immobilize: to make immovable.

impaled object: an object that is embedded in an open wound.

implied consent: the assumption that in an emergency a patient who cannot give permission for emergency medical care would give it if he or she could.

incontinent: unable to retain.

index of suspicion: an informal measure of anticipation that certain types of mechanisms produce specific types of injury.

indirect medical control: refers to EMS system design, protocols and standing orders, education for EMS personnel, and quality management. *See* direct medical control.

infant: according to AHA standards, a patient from birth to one year old.

infectious disease: a disease that can spread from one person to another.

inferior: a term of direction or position, meaning toward or closer to the feet. *Opposite of* superior.

inferior vena cava: the great vein that collects blood from the lower body and delivers it to the heart.

initial assessment: part of patient assessment, conducted directly after the scene size-up, in which the rescuer identifies and treats life-threatening conditions.

inspection: method of examination that involves looking for signs of injury or illness.

inspiration: breathing in; inhaling.

insulin: a hormone secreted by the pancreas, essential to the metabolism of blood sugar.

intercostal: between the ribs.

internal: a term of position, meaning inside. *Opposite of* external.

internal bleeding: bleeding that occurs inside the body.

involuntary muscle: *See* smooth muscle.

ischium: the lower portion of the pelvis or hip bone. *Plural* ischia.

--------------- J ---------------

jaw-thrust maneuver: a manual technique used to open the airway of an unresponsive patient who is injured or any patient who has suspected spine injury. *See* head-tilt/chin-lift maneuver.

--------------- K ---------------

kinematics of trauma: the science of analyzing mechanisms of injury.

--------------- L ---------------

labor: the term used to describe the process of childbirth.

laceration: an open wound of varying depth.

larynx: the voice box.

lateral: a term of direction or position, meaning to the left or right of the midline. *See* medial.

lateral recumbent position: the patient is lying on the left or right side.

level of responsiveness: mental status, usually characterized as alert, verbal, responsive to pain, or unresponsive.

ligaments: tissues that connect bone to bone.

litter: portable stretcher or cot.

local cold injury: freezing or near freezing of a specific body part. *Also called* frostbite.

log roll: a method of turning a patient without causing injury to his or her spine.

lumbar spine: the lower back, formed by five vertebrae.

--------------- M ---------------

manual traction: applying a pulling force to a body part in order to align it.

mechanism of injury (MOI): the force or forces that cause an injury.

meconium staining: a greenish or brownish color to the amniotic fluid, which means the unborn infant had a bowel movement.

medial: a term of direction or position, meaning toward the midline or center of the body. *See* lateral.

medical director: in EMS this person is the physician legally responsible for the clinical and patient-care aspects of an EMS system.

medical identification tag: medallion or bracelet that identifies a specific medical condition such as an allergy, epilepsy, or diabetes.

medical patient: a patient who is ill, not injured.

minor: any person under the legally defined age of an adult; usually under the age of 18 or 21.

mouth-to-barrier device ventilation: a technique of artificial ventilation that involves the use of a barrier device such as a face shield to blow air into the mouth of a patient.

mouth-to-mask ventilation: a technique of artificial ventilation that involves the use of a pocket mask with one-way valve to blow air into the mouth of a patient.

mouth-to-mouth ventilation: a technique of artificial ventilation that involves blowing air directly from the rescuer's mouth into the mouth of a patient.

multiple-casualty incident (MCI): any emergency where three or more patients are involved.

musculoskeletal system: a system made up of the skeleton and muscles, which help to give the body shape, protect the organs, and provide for movement.

myocardial infarction: heart attack.

N
---

nasal airway: *See* nasopharyngeal airway.

nasal cannula: an oxygen delivery device characterized by two soft plastic tips, which are inserted a short distance into the nostrils.

nasopharyngeal airway: an artificial airway positioned in the nose and extending down to the larynx. *Also called* nasal airway.

nasopharynx: the nasal part of the pharynx.

nature of illness (NOI): the type of medical condition or complaint a patient may be suffering.

neglect: refers to giving insufficient attention or respect to someone who has a claim to that attention and respect.

negligence: the act of deviating from the accepted standard of care through carelessness, inattention, disregard, inadvertence, or oversight that was accidental but avoidable.

nervous system: the body system that controls the voluntary and involuntary activity of the body; includes the brain, spinal cord, and nerves.

non-accidental trauma: injuries such as those caused by child abuse.

non-emergency move: a move made by several rescuers usually after a patient has been stabilized. *Also called* non-urgent move.

nonrebreather mask: an oxygen delivery device characterized by an oxygen reservoir bag and a one-way valve.

O
---

occlude: to block, close up, or obstruct.

occlusive dressing: a dressing that can form an air-tight and sometimes water-tight seal.

open injury: an injury to the soft tissues that is caused by a blow and results in breaking the skin.

oral airway: *See* oropharyngeal airway.

orbit: eye socket; the bones in the skull that hold the eyeball.

oropharyngeal airway: an artificial airway positioned in the mouth and extending down to the larynx. *Also called* oral airway.

oropharynx: the central part of the pharynx.

overdose: an emergency that involves poisoning by drugs or alcohol.

P
---

packaging: refers to getting the patient ready to be moved and includes procedures such as stabilizing impaled objects and immobilizing injured limbs.

palmar surface method: a method used to estimate the percent of body surface area involved in a burn injury.

palpation: method of examination that involves feeling for signs of injury or illness.

palpitations: a sensation of abnormal rapid throbbing or fluttering of the heart

paradoxical breathing: a segment of the chest moves in the opposite direction to the rest of the chest during respiration; typically seen with a flail segment.

paramedic: *See* EMT-Paramedic.

parietal pleura: the membrane that covers the internal chest wall.

partial foreign body airway obstruction: refers to an object that is caught in the throat but does not totally occlude the airway and prevent breathing.

partial thickness burn: a burn that involves both the epidermis and dermis.

patella: the knee cap.

patent airway: an airway that is open and clear of obstructions.

pathogens: microorganisms such as bacteria and viruses, which cause disease.

patient history: facts about the patient's medical history that are relevant to the patient's condition.

pediatric center: medical facility devoted to the treatment of infants and children.

pediatric patients: patients who are infants or children.

pelvic cavity: a space bound by the lower part of the spine, the hip bones, and the pubis.

pelvis: the hips.

penetration/puncture wound: an open wound that is the result of a sharp, pointed object being pushed or driven into soft tissues.

perfusion: refers to the circulation of blood throughout a body organ or structure.

perinatal center: medical facility devoted to the treatment of high-risk pregnant patients.

peripheral nervous system: the portion of the nervous system that is located outside the brain and spinal cord; the nerves.

personal protective equipment (PPE): equipment used by a rescuer to protect against injury and the spread of infectious disease.

pharynx: the throat.

placenta: a disk-shaped inner lining of the uterus that provides nourishment and oxygen to a fetus.

pleura: the membranes that enfold both lungs.

pleural cavity: the space between the visceral pleura and the parietal pleura.

pneumothorax: collapse of the lungs caused by air in the chest.

poison center: medical facility devoted to providing information for treatment of poisoning victims.

posterior: a term of direction or position, meaning toward the back. *Opposite of* anterior.

posterior tibial pulse: an arterial pulse point that can be felt behind the medial ankle bone.

potential crime scene: any scene that may require police support.

power grip: a technique used to get maximum force from hands while lifting and moving.

power lift: a technique used for lifting, especially helpful to rescuers with weak knees or thighs.

prehospital care: emergency medical treatment in the field before transport to a medical facility. *Also called* out-of-hospital care.

priapism: a constant erection of the penis.

prone: a position in which a patient is lying face down on his or her stomach. *Opposite of* supine.

protocols: written orders issued by the medical director that may be applied to patient care; a type of standing order.

proximal: a term of direction or position, meaning close or near the point of reference, which is usually the torso. *Opposite of* distal.

pubis: bone of the groin; the anterior portion of the pelvis.

public safety answering point (PSAP): location at which 9-1-1 calls are received.

pulmonary: concerning or involving the lungs.

pulmonary vein: vessel carrying oxygen-rich blood from the lungs to the left atrium of the heart.

pulse pressure: the difference between systolic blood pressure and diastolic blood pressure.

pulse: the wave of blood propelled through the arteries as a result of the pumping action of the heart.

pustules: raised areas of the skin that are filled with pus.

--- R ---

radial pulse point: an arterial pulse that can be felt on the palm side of the wrist.

radius: one of the bones of the forearm.

rape: sexual intercourse that is performed without consent and by compulsion through force, threat, or fraud.

rape trauma syndrome: a reaction to rape that involves four general stages: acute (impact) reaction, outward adjustment, depression, and acceptance and resolution.

rappelling: a special technique of getting down a cliff by means of a secured rope.

reasonable force: the minimum amount of force needed to keep a patient from injuring him- or herself and others.

recovery position: lateral recumbent position; used to allow fluids to drain from the patient's mouth instead of into the airway.

referred pain: pain felt in a part of the body that is different from its actual point of origin.

rehydration: replacement of water and electrolytes lost in sweating.

relative skin temperature: an assessment of skin temperature obtained by touching the patient's skin.

respiration: the passage of air into and out of the lungs.

respiratory arrest: the cessation of spontaneous breathing.

respiratory distress: shortness of breath or a feeling of air hunger with labored breathing.

respiratory system: organs involved in the interchange of gases between the body and the environment.

responsive: conscious; acting or moving in response to stimulus.

retraction: a pulling inward.

rule of nines: a method used to estimate the percent of body surface area involved in a burn injury.

S

sacrum: the lower part of the spine, formed by five fused vertebrae.

scapulae: the shoulder blades. *Singular* scapula.

scene size-up: an overall assessment of the emergency scene.

scope of care: actions and care legally allowed to be provided by a First Responder.

sector supervisor: at a multiple-casualty incident, the person in charge of overseeing a specific function such as transportation or rehab.

seizure: a sudden and temporary change in mental status caused by massive electrical discharge in the brain.

septum: a wall that divides two cavities.

sexual assault: any touch that the victim did not initiate or agree to and that is imposed by coercion, threat, deception, or threats of physical violence.

shock: a life-threatening, progressive condition that results from the inadequate delivery of oxygenated blood throughout the body.

shoulder girdle: consists of the clavicles and scapulae.

sign: any injury or medical condition that can be observed in a patient.

simple access: the process of gaining access to a patient without the use of tools.

skeletal muscle: one of three types of muscles; makes possible all deliberate acts such as walking and chewing. *Also called* voluntary muscle.

skull: a bony structure that houses and protects the brain.

smooth muscle: one of three types of muscles; found in the walls of tubelike organs, ducts, and blood vessels. *Also called* involuntary muscle.

sniffing position: position of a patient's head when the neck is flexed and the head is extended.

soft-tissue injuries: injuries to the skin, muscles, nerve, and blood vessels.

sphygmomanometer: instrument used to measure blood pressure. *Also called* blood pressure cuff.

spinal column: the column of bones, or vertebrae, that houses and protects the spinal cord.

spinal precautions: methods used to protect the spine from further injury; for First Responders, usually refers to the manual stabilization of the patient's head and neck until the patient is completely immobilized.

splint: a device used to immobilize a body part.

spontaneous abortion: miscarriage, or the loss of pregnancy before the twentieth week.

stabilize: to hold firmly and steadily.

standard of care: the care that would be expected to be provided to the same patient under the same circumstances by another First Responder who had received the same training.

standing orders: advance orders, rules, regulations, or step-by-step procedures to be taken under certain conditions; a type of indirect medical control.

status epilepticus: a seizure lasting longer than 10 minutes or seizures that occur consecutively without a period of responsiveness between them.

sterile: free of all microorganisms and spores.

sterilizing: process in which a chemical or other substance, such as superheated steam, is used to kill all microorganisms on an object. *See* cleaning *and* disinfecting.

sternum: breastbone.

stethoscope: instrument that aids in auscultating (listening) for sounds within the body.

stoma: a permanent surgically created opening that connects the trachea directly to the front of the neck.

stress: any change in the body's internal balance; occurs when external demands become greater than personal resources.

stridor: harsh, high-pitched sound made during inhalation, which may mean the larynx is swollen and blocking the upper airway.

stroke: loss of brain function caused by a blocked or ruptured blood vessel in the brain.

subcutaneous tissue: layer of fat beneath the skin.

sucking chest wound: open wound to the chest or back that bubbles or makes a sucking noise.

suctioning: using negative pressure created by a commercial device to keep the patient's airway clear.

superficial: term of position, meaning near the surface. *Opposite of* deep.

superficial burn: a burn that involves only the epidermis.

superior: a term of direction or position, meaning toward or closer to the head. *Opposite of* inferior.

supine: a position in which a patient is lying face up on his or her back. *Opposite of* prone.

symphysis pubis: the junction of the pubic bones on the midline in front; the bony eminence under the pubic hair.

symptom: any injury or medical condition that can only be described by the patient.

syrup of ipecac: a drug used to induce vomiting, usually in a patient who has ingested poison.

systolic pressure: the result of a contraction of the heart, which forces blood through the arteries. *See* diastolic pressure.

———————————— T ————————————

tendons: tissues that connect muscle to bone.

tension pneumothorax: a condition that is the result of an open chest wound, in which a severe build-up of air compresses the lungs and heart toward the uninjured side of the chest.

thoracic cavity: the space above the diaphragm and within the walls of the thorax. *Also called* chest cavity.

thoracic spine: the upper back, formed by 12 vertebrae.

thorax: the chest. *Also called* rib cage.

thrombus: a blood clot that obstructs a blood vessel.

tibia: one of the bones of the lower leg.

tongue-jaw lift: a technique used to draw the tongue away from the back of the throat and away from a foreign body that may be lodged there.

tourniquet: a constricting band used as a last resort on an extremity to apply pressure over an artery in order to control bleeding.

trachea: windpipe.

trauma center: a medical facility devoted to the treatment of injuries.

trauma patient: a patient who is injured.

triage: the process of sorting patients to determine the order in which they will receive care.

trimester: a three-month period.

tripod position: a position in which the patient is sitting upright, leaning forward, fighting to breathe.

———————————— U ————————————

ulna: one of the bones of the forearm.

umbilical cord: an extension of the placenta through which the fetus receives nourishment while in the uterus.

universal number: a phone number—usually 9-1-1— used in many areas to access emergency services including police, fire, rescue, and ambulance.

universal precautions: a form of infection control used against diseases spread by way of blood. *Also* BSI precautions.

unresponsive: unconscious; not acting or moving in response to stimulus.

uterus: the organ that contains the developing fetus.

———————————— V ————————————

veins: blood vessels that carry blood back to the heart from the rest of the body.

velocity: the speed at which an object moves.

venous bleeding: recognized by dark red blood that flows steadily from a wound.

ventilation: a method of assisting breathing by forcing air into a patient's lungs.

ventricles: the two lower chambers of the heart.

venules: the smallest kind of veins.

vertebrae: the 33 bone segments of the spinal column. *Singular* vertebra.

vesicles: small blisters or cysts that contain moisture.

visceral pleura: the membrane that covers the outer surface of the lungs.

vital signs: signs of life; assessments related to breathing, pulse, skin, pupils, and blood pressure.

voluntary muscle: *See* skeletal muscle.

W

wheals: itchy, raised, round marks on the skin that are red around the edges and white at the center.

withdrawal: a syndrome that occurs after a period of abstinence from the drugs or alcohol to which a person's body has become accustomed.

wound: a soft-tissue injury.

X

xiphoid process: the lowest portion of the sternum.

Index